Sustainability in Energy and Buildings: Future Perspectives and Challenges

Sustainability in Energy and Buildings: Future Perspectives and Challenges

Guest Editors

Sergio Montelpare
Camilla Lops

MDPI

Basel • Beijing • Wuhan • Barcelona • Belgrade • Novi Sad • Cluj • Manchester

Guest Editors

Sergio Montelpare
Geology and Engineering
Department
University G. d'Annunzio
of Chieti-Pescara
Pescara
Italy

Camilla Lops
Geology and Engineering
Department
University G. d'Annunzio
of Chieti-Pescara
Pescara
Italy

Editorial Office

MDPI AG
Grosspeteranlage 5
4052 Basel, Switzerland

This is a reprint of the Special Issue, published open access by the journal *Applied Sciences* (ISSN 2076-3417), freely accessible at: https://www.mdpi.com/journal/applsci/special_issues/sustainability_energy_buildings.

For citation purposes, cite each article independently as indicated on the article page online and as indicated below:

Lastname, A.A.; Lastname, B.B. Article Title. *Journal Name* **Year**, *Volume Number*, Page Range.

ISBN 978-3-7258-5767-8 (Hbk)
ISBN 978-3-7258-5768-5 (PDF)
https://doi.org/10.3390/books978-3-7258-5768-5

Contents

About the Editors

Sergio Montelpare

Sergio Montelpare is Full Professor of Applied Thermodynamics and Heat Transfer at the "G. d'Annunzio" University of Chieti-Pescara, Department of Engineering and Geology. His research focuses on energy efficiency in buildings, renewable energy systems, acoustics, HVAC systems, and thermo-fluid dynamics. Over the years, he has coordinated and participated in several national and international research projects, including PRIN, PON, and Horizon Europe programmes, addressing topics such as sustainable energy use, environmental quality, and digital innovation in building design. Professor Montelpare serves as Director of the Department of Engineering and Geology and is a member of the Academic Senate. He has authored more than 65 indexed journal papers with over 1190 citations and an H-index of 19. He is a member of the Editorial Board of the journal *Applied Sciences* and has acted as Guest Editor for several Special Issues on sustainability, wind energy, and climate change's impacts on buildings. His professional achievements reflect a strong commitment to advancing sustainable technologies and integrated approaches for the built environment.

Camilla Lops

Camilla Lops is a Research Fellow at the "G. d'Annunzio" University of Chieti-Pescara, Department of Engineering and Geology. She holds a Double Degree Ph.D. in Earth Systems and Built Environments from the same university and the Universitat Politècnica de Catalunya. Her research interests include building physics, energy modelling, computational fluid dynamics, renewable energy integration, and life cycle assessment for sustainable construction. Dr. Lops investigates multiscale thermophysical approaches for energy and comfort analysis in buildings, combining dynamic simulations, climate modelling, and artificial intelligence. She has participated in several national and European projects such as ENERGIDRICA, GENESIS, and necessARIA, focusing on energy efficiency and environmental quality in school and heritage buildings. Her work has been presented at international conferences and published in peer-reviewed journals, including *Applied Sciences, Energy and Buildings,* and *Advanced Engineering Materials.* Through her research, she promotes interdisciplinary innovation for the energy transition and resilience of the built environment.

Preface

This Reprint, "Sustainability in Buildings and Energy Systems: Future Perspectives and Challenges" gathers a selection of scientific contributions dedicated to one of the most pressing challenges of our time: the reduction in the energy consumption and environmental impact of the built environment. The aim of this Reprint is to provide an updated overview of innovative approaches, technologies, and methodologies that support the transition toward more sustainable buildings and energy systems.

Motivated by the growing global need to mitigate climate change and improve energy efficiency, this collection highlights interdisciplinary research spanning architecture, engineering, environmental science, and energy management. The studies presented address key aspects such as sustainable construction materials, advanced building envelopes, renewable energy integration, and life cycle assessment and cost analysis.

This Reprint is intended for researchers, professionals, and policymakers seeking a comprehensive understanding of current advances and future directions in sustainable building design and operation. By bringing together diverse perspectives and practical insights, it aims to foster collaboration and inspire continued innovation toward a more resilient and energy-efficient built environment.

Sergio Montelpare and Camilla Lops
Guest Editors

applied sciences

MDPI

Editorial

Sustainability in Energy and Buildings: Future Perspectives and Challenges

Camilla Lops * and Sergio Montelpare

Department of Engineering and Geology, University "G. d'Annunzio" of Chieti-Pescara, 65122 Pescara, Italy; sergio.montelpare@unich.it
* Correspondence: camilla.lops@unich.it; Tel.: +39-085-453-7258

How can we make the built environment more energy-efficient, sustainable, and resilient? This fundamental question has emerged as a critical priority for researchers, policymakers, and practitioners worldwide as the urgency of climate action intensifies. Buildings represent one of the most significant contributors to global energy consumption and greenhouse gas (GHG) emissions, accounting for approximately 32% of global energy demand and 34% of CO_2 emissions, with operational emissions reaching a record of 9.8 Gt and embodied carbon recorded at approximately 2.9 Gt in 2023 [1]. Specifically, in the European Union, buildings consume nearly 40% of total energy and contribute to approximately 36% of energy-related emissions [2,3]; similar patterns have emerged across developed economies globally. In rapidly urbanizing regions such as the Asia-Pacific and Sub-Saharan Africa, where urban populations are projected to double by 2050, the environmental impact from new construction and infrastructure development poses significant challenges for global emission reduction targets without coordinated sustainable building practices [4].

The scale and pervasive impact of buildings make them central to both challenges and opportunities arising from global sustainability transitions. Beyond their substantial environmental footprint, buildings fundamentally shape urban quality of life, influencing thermal comfort, indoor air quality, health outcomes, and community resilience to climate extremes [5,6]. This dual role positions the built environment as a unique platform for achieving co-benefits: reducing emissions while simultaneously enhancing human well-being and urban livability. International frameworks such as the Paris Agreement and the United Nations (UN) Sustainable Development Goals (SDGs) explicitly recognize buildings as critical points of intervention for climate mitigation and adaptation [7].

Contemporary buildings offer substantial opportunities for energy reduction through optimized envelope performance, renewable energy integration, and intelligent management systems, with leading examples demonstrating potential energy savings of 75–90% compared to their conventional counterparts [8,9]. However, despite technological advances and growing policy support, many existing building approaches continue to face significant limitations in achieving comprehensive sustainability. Current retrofit practices typically address only 30–40% of buildings' improvement potential, while new construction often falls short of net-zero targets due to cost constraints and regulatory gaps [10,11]. These limitations in scalability, affordability, and whole life cycle integration restrict broader adoption across diverse economic contexts, from developed markets to emerging economies, where rapid urbanization demands immediate solutions.

Overcoming these persistent barriers requires systemic approaches that extend beyond individual technologies to encompass holistic planning methodologies, supportive policy frameworks, innovative financing mechanisms, and unprecedented cross-sector collaboration. The transition toward a sustainable built environment demands integration across disciplines (from architectural design and engineering systems to urban planning, behavioral science, and circular economy principles) while addressing the diverse needs of global communities facing varying climatic, economic, and social conditions.

In response to these global demands, research in recent years has dramatically intensified its focus on advancing building technologies, design strategies, and management systems that can transform the built environment from an energy consumer into a driver of sustainability. The scale of this transformation is evident in market projections: the global smart building market was valued at USD 117.4 billion in 2024 and is projected to register a Compound Annual Growth Rate (CAGR) of approximately 21% until 2032 [12]. This rapid expansion reflects a fundamental shift in the construction industry toward sustainable practices across developed and emerging economies worldwide.

Contemporary research has increasingly converged on several transformative areas of innovation, with smart building materials leading the charge. These advanced materials are designed to promptly respond to environmental stimuli such as temperature, stress, and humidity, thereby enhancing both performance and resilience [13]. Notable examples include phase-change materials (PCMs) that regulate thermal conditions, self-healing concrete that extends structural durability, and shape-memory alloys that enable adaptive building components [14–17].

Advanced façade systems represent a dynamic frontier of innovation, combining responsive glazing technologies, building-integrated photovoltaics, and biomimetic designs that adapt to local climatic conditions while simultaneously maximizing daylight availability and ensuring thermal comfort [18–20]. In parallel, nature-based solutions (NbSs) are gaining unprecedented momentum in global building and urban practices, with applications ranging from green roofs and living walls to natural ventilation, rainwater harvesting, and urban biodiversity integration [21]. These approaches not only enhance building performance but also provide critical ecosystem services, such as improved air quality and mitigation of the urban heat island effect [22–24]. Complementing these strategies, circular economy (CE) principles are increasingly embedded in sustainable building design, encouraging material reuse, design for disassembly, and waste reduction across the construction life [25]. By prioritizing local sourcing and resource efficiency, these principles reduce transportation-related emissions while strengthening regional economies, creating a more resilient and regenerative built environment [26,27].

Building upon these approaches, life cycle assessment (LCA) methodologies have evolved into comprehensive frameworks (such as prospective life cycle assessment—pLCA) that evaluate impacts from material extraction through end-of-life disposal while also accounting for future scenarios, thus enabling evidence-based decision-making that optimizes both environmental performance and economic viability [28]. This holistic perspective extends well beyond traditional energy efficiency metrics to encompass embodied carbon, water consumption, material toxicity, and even social equity considerations. Consequently, buildings are increasingly envisioned not merely as minimized sources of environmental impact but as active contributors to regenerative urban systems, all while maintaining economic competitiveness across diverse global contexts, from established economies to rapidly urbanizing regions.

Finally, the application of artificial intelligence (AI) has emerged as a particularly promising frontier for advancing the sustainability of buildings and infrastructure. Its integration into energy optimization represents a critical pathway for reducing emissions. Recent studies suggest that, when combined with appropriate policy measures, AI could lower building energy use and associated emissions by as much as 40–90% by 2050, while more conservative estimates indicate reductions of 8–19% through automated design processes, predictive maintenance, and intelligent HVAC control systems [29–31]. Beyond direct energy efficiency, AI is also transforming the performance of renewable energy systems by enhancing forecasting accuracy and operational productivity [32]. AI-driven predictive analytics and monitoring tools have been shown to improve system uptime by 15–20% through early fault detection, translating into annual cost savings of approximately USD 10,000–30,000 per MW of installed capacity [33]. Furthermore, machine learning (ML) algorithms can optimize the interaction between smart building materials and environmental conditions; through this, they enable a more efficient response to thermal loads in phase-change materials and simultaneously coordinate with adaptive facade systems to minimize energy consumption while maximizing occupant comfort.

The studies presented in this Special Issue collectively reflect these innovative pathways, offering a broad range of approaches that demonstrate not only theoretical advances but also practical applications for the building industry. Their findings illustrate how emerging technologies can enhance performance while aligning with clean-label requirements and sustainability certification schemes. Importantly, they highlight the need for research that extends beyond isolated prototypes to consider large-scale deployment and long-term viability.

Future research should therefore prioritize the development of scalable implementation strategies for advanced sustainability technologies in both new construction and existing building retrofits, ensuring cost-effectiveness while maintaining performance standards. Attention must also be directed toward assessing the durability and long-term performance of integrated building systems, investigating user acceptance and market adoption, particularly for technologies with significant upfront costs or operational shifts, and rigorously evaluating the environmental benefits of sustainable building practices, including their potential contributions to global climate mitigation.

Regarding future outlook, the research findings highlighted in this volume clearly point toward several emerging directions. The path ahead will increasingly involve the integration of digital intelligence, advanced materials, and systemic sustainability frameworks, all of which are central to shaping resilient, low-carbon buildings and cities. We encourage researchers and practitioners to continue advancing these themes in the second volume, entitled *"Sustainability in Energy and Buildings: Future Perspectives and Challenges: 2nd Edition"*, where such topics will be at the forefront.

With continued progress in technology integration, circular economy strategies, and life cycle assessment methodologies, sustainable building practices hold immense potential to accelerate climate change mitigation, resource efficiency, and environmental protection in the built environment. Although the challenges are considerable, the opportunities are equally compelling. The contributions in this Special Issue reaffirm the vitality of the field and its essential role in driving the global transition toward sustainability.

Author Contributions: C.L.: writing—original draft preparation; S.M.: writing—review and editing. All authors have read and agreed to the published version of the manuscript.

Conflicts of Interest: The authors declare no conflicts of interest.

Abbreviations

The following abbreviations are used in this manuscript:

USD	United States dollar
GHG	Greenhouse gas emissions
UN	United Nations
SDG	Sustainable Development Goal

CAGR	Compound Annual Growth Rate
NbS	Nature-based solutions
CE	Circular economy
LCA	Life cycle assessment
pLCA	Prospective life cycle assessment
AI	Artificial intelligence
HVAC	Heating, ventilation, and air conditioning
ML	Machine learning

References

1. UNEP; GlobalABC. *Global Status Report for Buildings and Construction*; United Nations Environment Programme: Nairobi, Kenya; GlobalABC: Paris, France, 2024.
2. Council of the European Union. Fit for 55: Making Buildings in the EU Greener. 2025. Available online: https://www.consilium.europa.eu/en/infographics/fit-for-55-making-buildings-in-the-eu-greener (accessed on 9 September 2025).
3. European Environment Agency. *Buildings and Construction*; European Environment Agency: Copenhagen, Denmark, 2025.
4. U.S. Environmental Protection Agency. *Envisaging the Future of Cities. World City Report 2022*; U.S. Environmental Protection Agency: Washington, DC, USA, 2022; pp. 1–422.
5. Santamouris, M. Recent progress on urban overheating and heat island research. Integrated assessment of the energy, environmental, vulnerability and health impact. Synergies with the global climate change. *Energy Build.* **2020**, *207*, 109482. [CrossRef]
6. Schweiker, M.; Ampatzi, E.; Andargie, M.S.; Andersen, R.K.; Azar, E.; Barthelmes, V.M.; Berger, C.; Bourikas, L.; Carlucci, S.; Chinazzo, G.; et al. Review of multi-domain approaches to indoor environmental perception and behaviour. *Build. Environ.* **2020**, *176*, 106804. [CrossRef]
7. Intergovernmental Panel on Climate Change (IPCC). *Sixth Assessment Report (AR6) Climate Change 2022: Mitigation of Climate Change*; Cambridge University Press: Cambridge, UK, 2022; pp. 1–30.
8. IPHA. Active for More Comfort: Passive House. 2018, pp. 1–68. Available online: www.passivehouse-international.org (accessed on 7 September 2025).
9. Peper, S.; Feist, W. Energy efficiency of the Passive House Standard: Expectations Confirmed by Measurements in Practice. 2015. Available online: http://passiv.de/downloads/05_energy_efficiency_of_the_passive_house_standard.pdf (accessed on 7 September 2025).
10. International Energy Agency. *Net Zero by 2050: A Roadmap for the Global Energy Sector*; International Energy Agency: Paris, France, 2021; p. 70.
11. BPIE. Deep Renovation: Shifting from Exception to Standard Practice in EU Policy > BPIE—Buildings Performance Institute Europe. 2021. Available online: https://www.bpie.eu/publication/deep-renovation-shifting-from-exception-to-standard-practice-in-eu-policy/ (accessed on 7 September 2025).
12. Fortune Business Insights. Smart Building Market Size, Share & Growth Report [2032]. 2024. Available online: https://www.fortunebusinessinsights.com/industry-reports/smart-building-market-101198%0A (accessed on 8 September 2025).
13. Fraleoni-Morgera, A.; Afshani, M.; Lops, C.; Montelpare, S. Micro- and Nanocomposites for Thermal Insulation in Buildings. *Adv. Eng. Mater.* **2023**, *26*, 2301064. [CrossRef]
14. Suresh, C.; Kumar Hotta, T.; Saha, S.K. Phase change material incorporation techniques in building envelopes for enhancing the building thermal Comfort-A review. *Energy Build.* **2022**, *268*, 112225. [CrossRef]
15. Wang, X.; Li, W.; Luo, Z.; Wang, K.; Shah, S.P. A critical review on phase change materials (PCM) for sustainable and energy efficient building: Design, characteristic, performance and application. *Energy Build.* **2022**, *260*, 111923. [CrossRef]
16. Da Cunha, S.R.L.; de Aguiar, J.L.B. Phase change materials and energy efficiency of buildings: A review of knowledge. *J. Energy Storage* **2020**, *27*, 101083. [CrossRef]
17. Li, C.; Wen, X.; Cai, W.; Yu, H.; Liu, D. Phase change material for passive cooling in building envelopes: A comprehensive review. *J. Build. Eng.* **2023**, *65*, 105763. [CrossRef]
18. Lamberti, V.; Lehrer, D.; Betti, G.; Carlucci, F.; Fiorito, F. The Development of an Advanced Facade Map: An Evolving Resource for Documenting Case Studies. *Sustainability* **2024**, *16*, 10405. [CrossRef]
19. Zhou, Y.; Herr, C.M. A Review of Advanced Façade System Technologies to Support Net-Zero Carbon High-Rise Building Design in Subtropical China. *Sustainability* **2023**, *15*, 2913. [CrossRef]
20. Tahmasbi, F.; Khdair, A.I.; Aburumman, G.A.; Tahmasebi, M.; Thi, N.H.; Afrand, M. Energy-efficient building façades: A comprehensive review of innovative technologies and sustainable strategies. *J. Build. Eng.* **2025**, *99*, 111643. [CrossRef]

21. Bona, S.; Silva-Afonso, A.; Gomes, R.; Matos, R.; Rodrigues, F. Nature-Based Solutions in Urban Areas: A European Analysis. *Appl. Sci.* **2023**, *13*, 168. [CrossRef]
22. IUCN. *Global Standard for Nature-Based Solutions: A User-Friendly Framework for the Verification, Design and Scaling Up of NbS*, 1st ed.; IUCN: Gland, Switzerland, 2020. [CrossRef]
23. Programme U-HNHS. World Cities Report 2022: Envisaging the Future of Cities. 2022. Available online: https://unhabitat.org/wcr/ (accessed on 9 September 2025).
24. Tzortzi, J.N.; Guaita, L.; Kouzoupi, A. Sustainable Strategies for Urban and Landscape Regeneration Related to Agri-Cultural Heritage in the Urban-Periphery of South Milan. *Sustainability* **2022**, *14*, 6581. [CrossRef]
25. European Parliament. Circular Economy: Definition, Importance and Benefits | News | European Parliament. 2022. Available online: https://www.europarl.europa.eu/topics/en/article/20151201STO05603/circular-economy-definition-importance-and-benefits (accessed on 9 September 2025).
26. European Commission. COM 2020—Communication from the Commission to the European Parliament, the Council, the European Economic and Social Committee and the Committee of the Regions—A New Circular Economy Action Plan—For a cleaner and More Competitive Europe. 2020. Available online: https://eur-lex.europa.eu/legal-content/EN/TXT/?qid=1583933814386&uri=COM:2020:98:FIN (accessed on 9 September 2025).
27. Pineda-Martos, R.; Calheiros, C.S.C. Nature-Based Solutions in Cities—Contribution of the Portuguese National Association of Green Roofs to Urban Circularity. *Circ. Econ. Sustain.* **2021**, *1*, 1019–1035. [CrossRef]
28. Nurdiawati, A.; Mir, B.A.; Al-Ghamdi, S.G. Recent advancements in prospective life cycle assessment: Current practices, trends, and implications for future research. *Resour. Environ. Sustain.* **2025**, *20*, 100203. [CrossRef]
29. Ding, C.; Ke, J.; Levine, M.; Zhou, N. Potential of artificial intelligence in reducing energy and carbon emissions of commercial buildings at scale. *Nat. Commun.* **2024**, *15*, 5916. [CrossRef] [PubMed]
30. American Council for an Energy-Efficient Economy. Can Artificial Intelligence Get Us to Net-Zero Buildings? 2024. Available online: https://www.aceee.org/blog-post/2024/11/can-artificial-intelligence-get-us-net-zero-buildings (accessed on 9 September 2025).
31. Giovanardi, M.; Baietta, A.; Belletti, F.; Magnani, S.; Casadei, O.; Pracucci, A. Exploiting the Value of Active and Multifunctional Façade Technology through the IoT and AI. *Appl. Sci.* **2024**, *14*, 1145. [CrossRef]
32. Algburi, S.; Sabeeh Abed Al Kareem, S.; Sapaev, I.B.; Mukhitdinov, O.; Hassan, Q.; Khalaf, D.H.; Ibrahim Jabbar, F. The Role of Artificial Intelligence in Accelerating Renewable Energy Adoption for Global Energy Transformation. *Unconv. Resour.* **2025**, *8*, 100229. Available online: https://www.sciencedirect.com/science/article/pii/S2666519025000950 (accessed on 9 September 2025). [CrossRef]
33. Ukoba, K.; Olatunji, K.O.; Adeoye, E.; Jen, T.C.; Madyira, D.M. Optimizing renewable energy systems through artificial intelligence: Review and future prospects. *Energy Environ.* **2024**, *35*, 3833–3879. [CrossRef]

applied sciences

MDPI

Article

Research on the Implementation of a Heat Pump in a District Heating System Operating with Gas Boiler and CHP Unit

Damir Požgaj, Boris Delač *, Branimir Pavković and Vedran Medica-Viola

Faculty of Engineering, University of Rijeka, Vukovarska 58, 51000 Rijeka, Croatia;
damir.pozgaj67@gmail.com (D.P.); pavkovic.branimir@gmail.com (B.P.); vedran.medica@riteh.uniri.hr (V.M.-V.)
* Correspondence: boris.delac@riteh.uniri.hr; Tel.: +385-51651511

Featured Application

This study provides a practical framework for retrofitting existing third-generation district heating (DH) systems currently reliant on gas-fired boilers and CHP units with medium-scale heat pumps. The findings support utility companies, energy planners, and municipal DH operators in designing optimised hybrid systems that reduce carbon emissions, improve energy efficiency, and adapt to fluctuating energy prices. The results can guide investment decisions and policy development for sustainable DH system upgrades, particularly in urban areas seeking to meet decarbonisation targets.

Abstract

Given the widespread use of gas-fired boilers and combined heat and power (CHP) units in existing district heating (DH) systems, this study investigates the integration of medium-scale heat pumps (HPs) into such configurations. Fifteen DH system variants were analysed, differing in installed HP capacity, operational strategies, and the synchronisation of heat and electricity production with thermal demand. A dynamic simulation model incorporating real-world equipment performance was developed to assess energy efficiency, environmental impact, and economic viability under three distinct energy price scenarios. The results demonstrate that an HP sized to 17% of the total heating capacity of the DH system achieves a 54% decrease in primary energy consumption and a 68% decrease in emissions compared to the base system. Larger HP capacities enhance environmental performance and increase the share of renewable energy but also entail higher investment. An economic analysis reveals that electricity-to-gas price ratios strongly influence the cost-effectiveness of HP integration. Under favourable electricity pricing conditions, systems with HP operational priority achieve the lowest levelized cost of heating. The most economically viable configuration consists of 600 kW HP and achieves a payback period of 4.7 years. The findings highlight the potential for HPs to decarbonize DH systems while emphasising the importance of market conditions and system design in ensuring economic feasibility.

Keywords: district heating; combined heat and power; heat pump; simulation; primary energy; CO_2 emissions

1. Introduction

District heating (DH) systems play a crucial role in the decarbonisation of the energy sector. The improvement of DH systems through the use of heat pumps (HP) and renewable energy sources (RES) can lead to improved air quality and reduced dependence on fossil fuel imports. It is expected that the demand for accelerated decarbonisation and electrification of the energy sector will lead to an increased use of HPs.

A comprehensive definition and classification of DH systems based on the technology used and the period of implementation can be found in the literature [1]. Existing DH systems that rely on fossil fuels need to be upgraded to fourth-generation DH systems, which are characterised by lower operating temperatures in the pipe network, enabling lower heat losses and higher efficiency of HP. Alarnaout et al. [2] conducted theoretical analysis on complete replacement of gas boilers with small-scale high-temperature HPs which operate with synthetic refrigerant R1234-ze. The assessment of six distinct scenarios indicated that replacement ensures environmental and economic benefits for residential-use scenarios. In an analysis on the transition from the third to the fourth generation of DH system, Požgaj et al. [3] concluded that it is feasible to replace gas-fired boilers with commercially available medium-scale HPs utilising natural refrigerants. Volkova [4] theoretically evaluated three heating scenarios for a newly developed urban area in Estonia and concluded that low-temperature DH systems with large-scale HPs exhibited the lowest primary energy consumption and energy costs compared to high-temperature gas-boiler-based DH system.

The integration of combined heat and power (CHP) units into DH systems represents a transformative approach to energy generation, offering substantial efficiency, economic, and environmental advantages over conventional gas-boiler-based systems. CHP is a well-established technology that simultaneously produces electricity and thermal energy using a range of technologies and fuel sources, most commonly natural gas or coal. The viability of the utilisation of CHP technology within existing DH systems has been a subject of investigation by numerous authors [2,5–11]. These studies examined whether retaining existing CHP systems or replacing them with RES-based alternatives would be more feasible. Olympios et al. [12] found that a centralised DH system with CHP units is more advantageous than heating by decentralised gas boilers in London. In a recent study, Yu et al. [6] found that DH systems employing gas boilers and CHP units sized for base load heating can be economically competitive compared to individual heating systems utilising HPs. Reiter et al. [7] determined that it is feasible to replace the outdated CHP unit with solar thermal collectors and absorption HPs in a large-scale DH system in Graz. Popovski et al. [8] analysed the decarbonisation of a large-scale coal-fired CHP-based DH system in Germany, highlighting solar energy, biomass, waste incineration, and large-scale HPs as promising technologies. Despite their environmental benefits, large HPs were found to be economically uncompetitive without regulatory changes, prompting recommendations to lower electricity prices for HP operations.

The implementation of HP units within CHP-based DH systems plays a significant role in stabilising the electricity market, while it can also enhance energy efficiency and reduce environmental impact [5,13,14]. It can also improve system flexibility and reduce operational costs. In circumstances where electricity from RES is abundant and electricity prices are low, the utilisation of HPs becomes a priority. Conversely, when electricity from RES is insufficient and electricity prices are high, CHP units can operate profitably, generating both electricity and heat. Several studies have examined the integration of HPs into existing CHP-based DH systems. Ommen et al. [15] analysed various configurations for integrating HPs into DH systems incorporating gas-fired CHP units. They found that supplying HP-generated heat directly to consumers is more suitable for decentralised systems, whereas connecting the HP to the return pipe upstream of the CHP unit is better aligned with centralised system configurations. Capone et al. [16] investigated the placement of a large-scale HP within a DH system and concluded that incorrect placement has the potential to negate the positive effects by up to 63%. Levihn [14] provided practical insights into the DH network in Stockholm, where high-temperature HPs have been integrated with CHP plants since the 1970s. The study highlights that the combination of large-scale HPs and CHP enables flexible heat production and optimisation based on electricity prices. Aste et al. [9] examined the use of wood-biomass-powered CHP units and geothermal HP with photovoltaic (PV) power plants to achieve carbon dioxide (CO_2)-free heat and electricity production for a DH system near Milan, Italy. Barco-Burgos et al. [10] demonstrated that small-scale DH systems with CHP units offer greater potential for the integration of high-temperature HPs compared to medium- and large-scale systems. Their analysis of 12 different system configurations showed that the optimal integration method varies with system complexity, highlighting the need for configuration-specific research. Ommen et al. [11] performed an analysis on the implementation of booster HPs in ultra-low-temperature and low-temperature DH systems. The overall system performance improved when a central HP was utilised in the return flow of the DH system but also determined lower efficiency for the system when an HP was placed in parallel with a CHP plant. Helin et al. [17] indicated that large-scale HPs in CHP-based DH systems, utilised during periods of low electricity prices, are economically viable and contribute to the reduction of CO_2 emissions. The implementation of such HPs has the potential to reduce operating costs by 8–12% compared to DH systems based on CHP units.

While the literature has explored the integration of large-scale HPs either as replacements for, or in combination with, boiler or CHP-based DH systems, the potential of medium-scale HPs within existing third-generation DH networks that already utilize CHP and boiler technologies remains largely unaddressed. The modelling of HPs in these analyses also varies significantly, ranging from theoretical process models to data-driven approaches, both of which may not accurately represent commercially available products. Notably, research tends to focus on the operation of either CHP or heat pumps in isolation, with limited attention given to their synchronised operation within DH systems. Consequently, there is a clear research gap regarding the implementation and coordinated operation of medium-scale HPs alongside CHP and boiler units in existing third-generation DH systems.

This paper makes several key contributions to the field of DH system optimisation and decarbonisation. Specifically, it addresses the integration of medium-scale HPs into third-generation DH systems that are already operated with CHP and boiler units. The main research question of this study is as follows: Is it viable to implement medium-sized HPs to cover a fraction of the heating load in the DH system? This study is based on a comprehensive dynamic simulation model of the DH system encompassing consumers, distribution, and production subsystems. Unlike many previous works that rely on theoretical or simplified representations, this analysis incorporates the operational characteristics of commercially available equipment, thus providing more realistic insights into system performance. The research investigates various sizes of implemented HPs and explores multiple operational strategies, including prioritisation of technology and synchronised operation of CHP and HPs, as well as the management of produced electricity. Furthermore, the study evaluates the economic feasibility of these configurations under different gas and electricity price scenarios, reflecting the real-world volatility of energy markets.

2. Materials and Methods

2.1. Methods

This research contributes to the ongoing discussion surrounding the integration of medium-scale HPs into third-generation DH systems that are already operated with CHP and boiler units. The research does this by systematically evaluating all the DH subsystems, with a focus on the thermal energy production subsystem under controlled operation conditions. This study evaluates the energy performance, costs, and CO_2 emissions across a range of system configurations, operation regimes, and energy price scenarios. The primary flowchart of the model is presented in Figure 1.

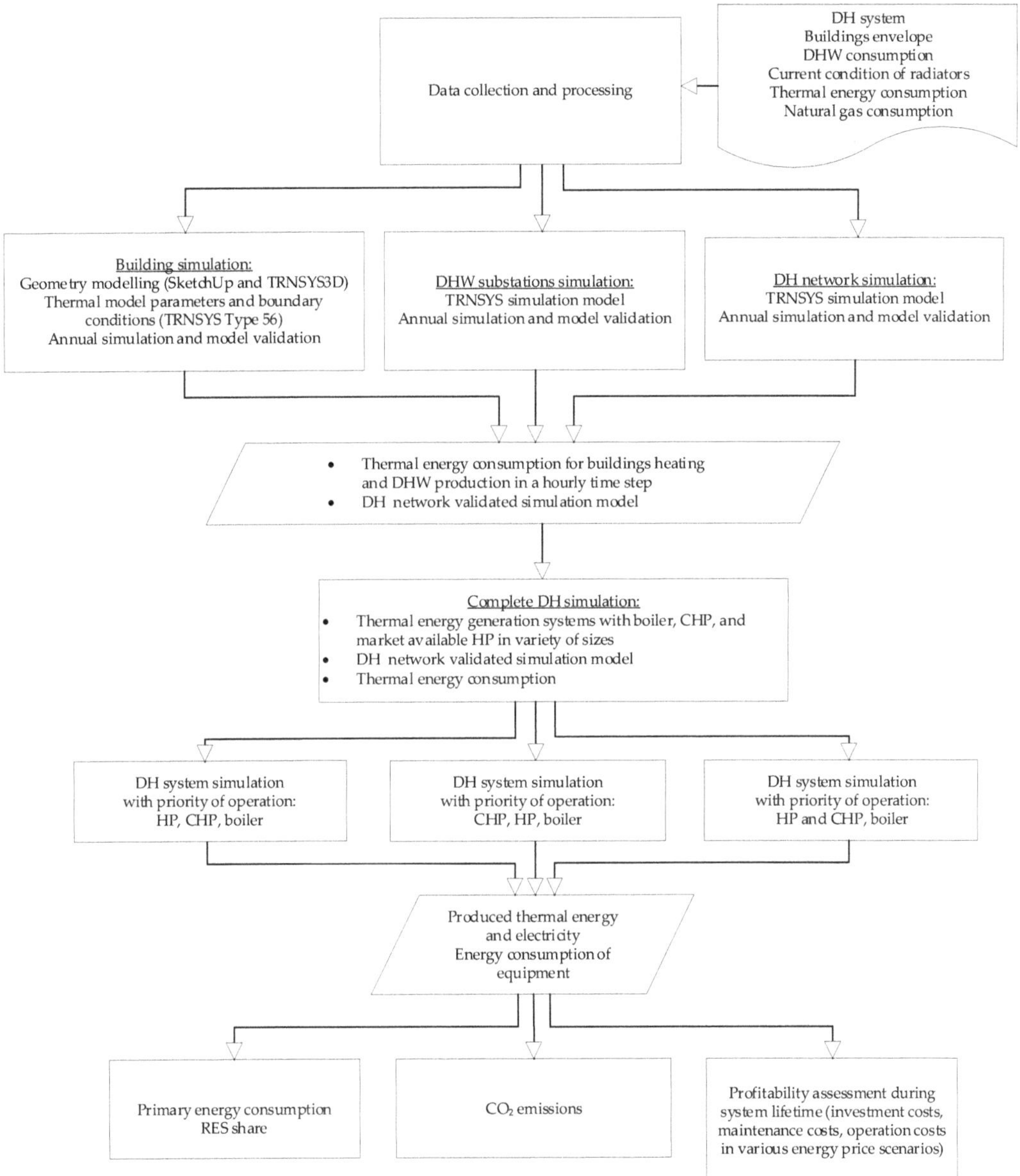

Figure 1. Flowchart for the modelling and simulation cycle.

To achieve a high degree of reliability, this study relies on the previous work of authors where a numerical model of a complete DH system was developed and validated [3]. The present study investigates the third-generation DH system in Croatia through a case study approach. The initial phase of the study entailed the aggregation and processing of all data pertinent to the DH system simulation. This encompassed the analysis of building design projects and the current state of the envelope and heating systems in buildings. Additionally, it involved the examination of DH data concerning equipment type and size, operational parameters, thermal energy utilisation in buildings, and fuel (natural gas) consumption.

Following the processing of the specified parameters, the TRNSYS 17 simulation environment was utilised to develop the dynamic simulation models. This decision was made due to the TRNSYS 17 environment's proven capability to model complex interactions within energy systems [18]. For the sake of clarity, details on this process are briefly elaborated here, with further elaboration available in previous works by the authors. The design of simulation models for systems employing a natural gas boiler, CHP unit, and HP was informed by the thermal energy consumption of the DH system and the requisite thermal energy capacity in design conditions. The boiler and CHP unit were regarded as existing equipment within the system, while the CHP units were considered in a range of sizes. The operation of the CHP unit was simulated based on data provided by the manufacturer, with particular reference to capacities and efficiency at partial load. The operation of HPs was simulated based on manufacturer data. This process incorporated the consideration of efficiency at partial load, as well as variable heat sink and source temperatures. The system control was analysed within the context of three distinct operational strategies. The first operational strategy prioritised the operation of HP over CHP. The second operational strategy prioritised the operation of CHP over HP. The third operational strategy involved the simultaneous operation of CHP and HP, with adjustments made to meet the required thermal energy demands. In all three of the considered operational strategies, the priority assigned to the boiler operation was the lowest. Simulations were performed for a whole year operation of the DH system to include a realistic scenario comprising peak load periods during the winter period and those with lower load during the transitional and summer period. Outputs from simulations provided data on the produced and consumed energy by each generator in the system, enabling the calculation of key performance indicators, including primary energy consumption, the proportion of renewable energy within the system, and CO_2 emissions. An economic analysis was conducted over the system's lifecycle, considering three energy price scenarios to evaluate the investment's profitability.

2.1.1. Control Logic for Technology Prioritisation in DH System Operation

DH systems often incorporate multiple heat generation units, such as CHP plants, HP, and boilers. In the simulations conducted, different operational strategies are analysed in which specific technologies are given priority, either operating individually (e.g., CHP or HP prioritised) or concurrently (e.g., CHP and HP operated in parallel).

The general control logic governing technology prioritisation is illustrated in Figure 2. This logic is applied at each simulation time step. At every time step, the required thermal capacity is calculated based on the system's supply and return water temperatures. This demand is then compared to the maximum available capacity of the prioritised heat generator. If the required capacity is less than the maximum output of the prioritised unit, that unit is activated and operates at partial load, while all other generators remain off. Conversely, if the required capacity exceeds the prioritised unit's maximum output, it is operated at full load, and the subsequent unit in the hierarchy is activated to supply the remaining demand, either at partial or full load, depending on the shortfall.

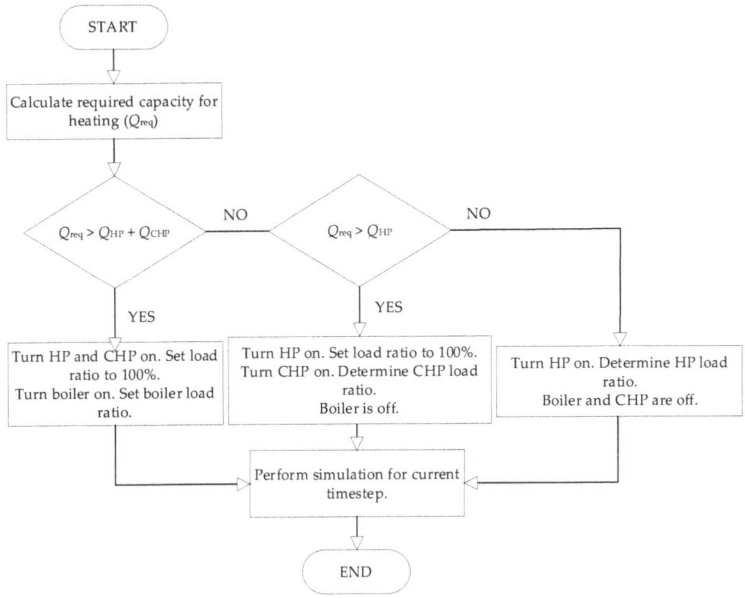

Figure 2. Control logic diagram for prioritised operation of HP, CHP, and boiler (highest to lowest priority).

2.1.2. Calculation of Energy Performance Indicators

The results from dynamic simulations were used to calculate the system's energy performance indicators. The scheme of the energy supply concept in the DH system is presented in Figure 3.

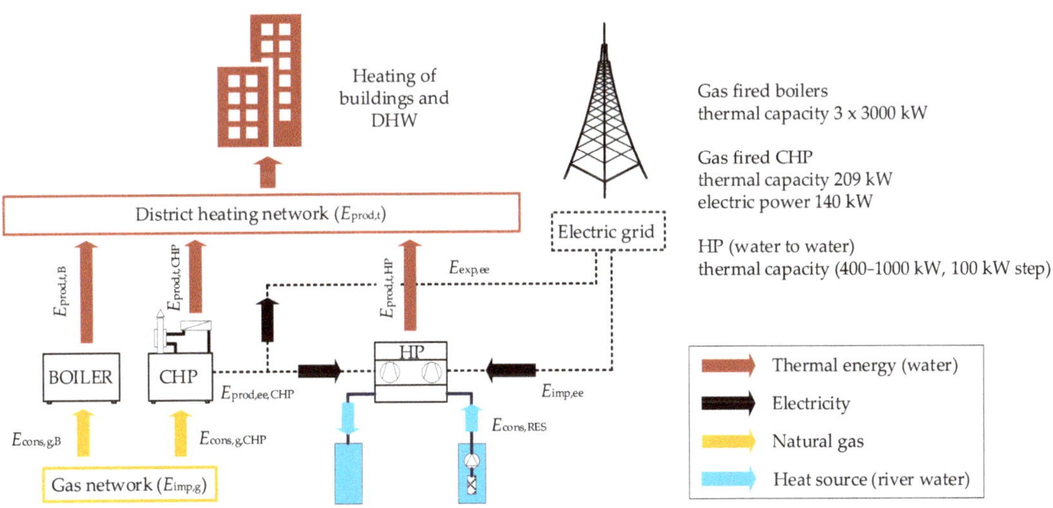

Figure 3. Scheme of the energy flows in DH system.

Energy flows are evaluated at each time step of the simulation. Energy consumption for the system operation is denoted as E_{cons}, where $E_{cons,g,B}$ represents natural gas consumption in the boiler, $E_{cons,g,CHP}$ represents natural gas consumption in the CHP unit, while the electricity consumption from the HP unit is presented as $E_{cons,ee,HP}$. The total produced thermal energy $E_{\mathrm{prod,t}}$, which is distributed across the DH network, is calculated as the sum of thermal energy produced by each thermal energy generator in the system:

$$E_{\mathrm{prod,t}} = \mathrm{SUM}(E_{\mathrm{prod,t,B}} + E_{\mathrm{prod,t,CHP}} + E_{\mathrm{prod,t,HP}}) \tag{1}$$

Natural gas is imported from the gas network; therefore, the imported energy to the system is calculated as follows:

$$E_{\mathrm{imp,g}} = \mathrm{SUM}(E_{\mathrm{cons,g,B}} + E_{\mathrm{cons,g,CHP}}) \tag{2}$$

The electricity produced by the CHP unit is denoted as $E_{\mathrm{prod,ee,CHP}}$, the electricity which is consumed by HP is denoted as $E_{\mathrm{cons,ee,HP}}$, the electricity which is imported from the grid is expressed as $E_{\mathrm{imp,ee}}$, while the electricity exported to the grid is $E_{\mathrm{exp,ee}}$.

Concerning strategies for management of produced electricity, for the case when the electricity produced by CHP is used for HP operation, the calculation is as follows:

$$\begin{gathered} \text{for } E_{\mathrm{cons,ee,HP}} > E_{\mathrm{prod,ee,CHP}} \\ E_{\mathrm{imp,ee}} = \mathrm{SUM}(E_{\mathrm{cons,ee,HP}} - E_{\mathrm{prod,ee,CHP}}) \end{gathered} \tag{3}$$

while the electricity which is exported to the grid is as follows:

$$\begin{gathered} \text{for } E_{\mathrm{cons,ee,HP}} < E_{\mathrm{prod,ee,CHP}} \\ E_{\mathrm{exp,ee}} = \mathrm{SUM}(E_{\mathrm{prod,ee,CHP}} - E_{\mathrm{cons,ee,HP}}) \end{gathered} \tag{4}$$

The exported and imported electricity for the case when the HP is operated by using electricity from the grid and the electricity produced by CHP unit is sold to the grid is calculated as follows:

$$E_{\mathrm{imp,ee}} = \mathrm{SUM}(E_{\mathrm{cons,ee,HP}}) \tag{5}$$

$$E_{\mathrm{exp,ee}} = \mathrm{SUM}(E_{\mathrm{prod,ee,CHP}}) \tag{6}$$

The primary energy consumption EP is calculated as the product imported and the primary energy factor $f_{\mathrm{imp,nren}}$ for each energy carrier. The calculation for electricity is performed in a way that imported energy is deducted by exported energy. Primary energy factors ($f_{\mathrm{ee,nren,i}}$) depend on the share of renewable, non-renewable, and imported energy components. The primary energy consumption for electricity is calculated with a factor of 1.614, while the primary energy for natural gas is calculated with a factor of 1.095. These values are determined at the national level [19].

$$EP = \mathrm{SUM}(E_{\mathrm{imp,ee}} - E_{\mathrm{exp,ee}}) \cdot f_{\mathrm{imp,nren,ee}} + \mathrm{SUM}(E_{\mathrm{imp,g}} \cdot f_{\mathrm{imp,nren,g}}) \tag{7}$$

The primary energy indicator EPI is defined as the ratio of total primary energy consumption to the total useful area of the buildings connected to DH system:

$$EPI = EP/A \tag{8}$$

The share of renewable energy RES is calculated using the following equation:

$$RES = \mathrm{SUM}\ E_{\mathrm{cons,res}} / E_{\mathrm{prod,t}} \tag{9}$$

In the Expression (9), $E_{\mathrm{cons,res}}$ represents the renewable energy produced in the system. For heat generators powered entirely by natural gas, such as boilers and CHP units,

$E_{s,cons,res}$ equals zero, and for systems using HP, it corresponds to the thermal energy derived from the source:

$$E_{cons,res} = E_{prod,t,HP} - E_{cons,ee,HP} \quad (10)$$

2.1.3. Calculation of the CO_2 Emissions

Emissions of CO_2 ($m_{CO2,i}$) for each energy carrier are calculated as follows:

$$m_{CO2,i} = E_{s,imp,i} \cdot f_{em,i} \quad (11)$$

The specific CO_2 emission for electricity is 0.2348 kg/kWh, whereas the specific emission for natural gas is 0.2202 kg/kWh. These values are defined at the national level [19].

2.1.4. Calculation of Costs

The total costs comprise investment costs for equipment, operating costs, and maintenance expenses. The present value of total costs is determined using the following equation:

$$C_t(\tau) = C_I + \text{SUM}\,[C_{a,i} \cdot Rd - V_{f,\tau}] \quad (12)$$

where τ represents the accounting period, C_I is the initial investment cost, C_a is the annual operating cost, R_d is the discount factor for each year, and $V_{f,\tau}$ is the residual value at the end of the evaluation period. The calculation is repeated annually over the accounting period.

The discount factor $Rd(p)$ is defined by the following:

$$Rd(p) = [1/(1 + r/100)]\,p \quad (13)$$

p represents the year, and r is the discount rate.

Levelized cost of heating C_{LH} is calculated for all scenarios and system variants considered in this study. It is defined as the ratio of the total system costs to the total thermal energy produced over the evaluation period, as expressed by the following equation:

$$C_{LH} = C_t(\tau)/E_{prod,t}(\tau) \quad (14)$$

2.2. Case Study

The research is conducted on the case study of one of the existing DH systems in Rijeka, Croatia. Constructed in the 1970s, the DH system services eight multi-story buildings and one primary school, covering a total heated area of 48,455 m^2. TRNSYS Type 56 multi-zone thermal model [18] is used to perform simulations of buildings and determine the required energy for heating. The methodology for developing and validating consumer, distribution, and production models has been detailed in a prior study by the authors [3].

The design load for heating of buildings was determined through dynamic simulation within the TRNSYS interface, consistent with the HRN EN ISO 12831-1:2017 standard [20]. The calculated design heating capacity was 2195 kW.

The total number of apartments in the DH system is 601, and the calculated capacity for DHW heating is 569 kW. The simultaneity factor of the heat load for heating buildings is 0.8 according to standard DIN 4708-2 [21]. Network heat losses were determined by dynamic simulation using TRNSYS Type 951. The heat pipelines' annual heat losses amounted to 167 kWh/m, with a peak hourly loss of 17.16 kW. Accordingly, the total heat capacity requirement in design conditions is 2342 kW. Characteristics of the DH system are outlined in Table 1.

Table 1. The main climate and location data and characteristics of the DH system considered in the study.

Climate data	City	Rijeka (Croatia)
	Longitude	E 14°26′31.834″
	Latitude	N 45°19′37.427″
	Design ambient temperature	−8 °C
	Annual average ambient temperature	15.8 °C
DH data	Trench total length	912 m
	Fuel type	Natural gas
	Annual fuel energy consumption (2019)	5,069,863 kWh
	Annual thermal energy supply into DH system (2019)	4,410,544 kWh
	Annual heating and DHW demand building (2019)	3,943,144 kWh
	Supply/return temperature	90/70 °C
DH consumers	Number of buildings	9
	Heated area	48,455 m^2
	Heated space volume	145,365 m^3
	Internal heat gains	6 W/m^2
	Infiltration/required ventilation rate	0.48 h^{-1}/1.32 h^{-1}
	Heating operation	Interrupted during the night
	Heating temperature set point	20 °C
	Number of apartments	601
	Number of persons	1455
	DHW set point	45 °C

The supply temperature in the DH network, under the assumption that all buildings undergo energetic renovation while retaining the current radiators, is 72 °C at an outdoor design temperature of −8 °C. During the heating season, the temperature of the supply pipe in the network varies in response to the external temperature, ranging from a maximum of 72 °C to a minimum of 65 °C. It does not fall below 65 °C due to demand for DHW in buildings.

In addition to the boiler and CHP-based systems, 15 variants of systems with implemented HP units were considered. These systems employ a gas-fired boiler, a CHP unit, and a water-to-water HP in different sizes. They are designed to meet the same heat load for space heating and DHW, maintaining a supply temperature within the DH network.

2.2.1. System with Gas-Fired Boiler

The gas-fired boiler system represents the original method of heat production established during the construction of the DH system. This system comprises three gas-fired boilers with a total heating capacity of 9 MW. The capacity is oversized, as it was based on the heating demand at the time of the system's installation, when the connected buildings were not yet renovated. Additionally, a reserve capacity was included for potential system expansion. The dynamic simulations of the gas-fired boiler operation were conducted using Trnsys Type 751 [22]. Figure 4 presents a schematic of a third-generation DH system based exclusively on oversized gas-fired boilers. This configuration represents the legacy system used for thermal energy production before any modernisation involving CHP units or HPs. It serves as one of the baseline scenarios against which all proposed modernisation variants are evaluated in terms of performance and cost-effectiveness.

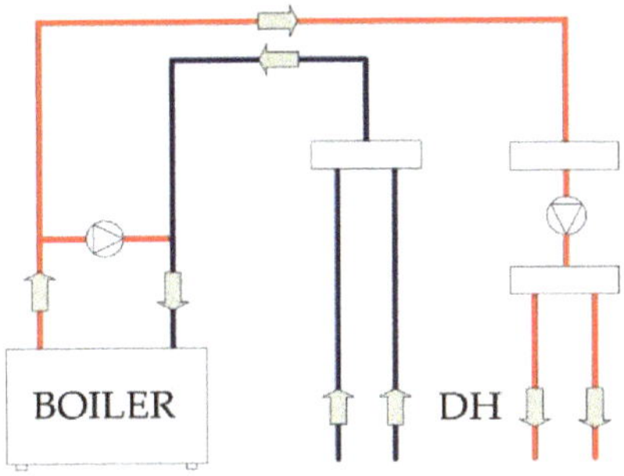

Figure 4. A schematic of a DH system with a gas-fired boiler.

2.2.2. System with Gas-Fired Boiler and CHP Unit

The heat production system consists of a gas-fired boiler and a CHP unit powered by a gas engine, as shown in Figure 5.

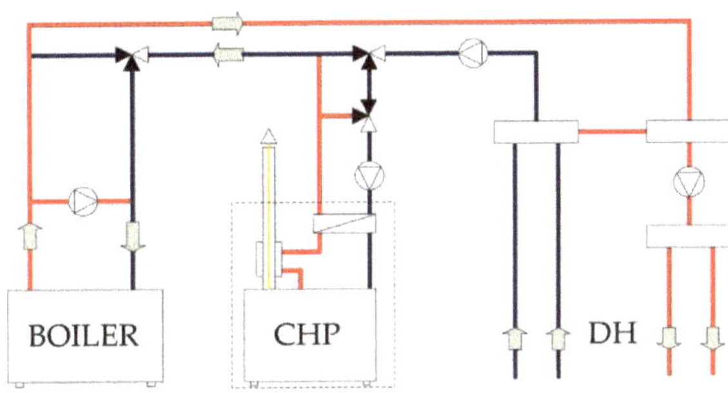

Figure 5. Schematic of a DH system with a gas-fired boiler and CHP unit.

The CHP unit is installed in the return line of the heating system, thereby ensuring priority in the production of thermal energy. The gas-fired boiler is activated when the CHP unit is unable to achieve the required water temperature. A comprehensive system simulation was performed to determine the heating energy required for space heating, DHW, and the heat losses in the distribution network during the year. The simulation results were integrated over an hourly time step to obtain the necessary thermal output. The data were then filtered based on the frequency of occurrence throughout the year and sorted from highest to lowest values.

The annual duration curve of the required heating capacity is illustrated in Figure 6. The area beneath this curve represents the total thermal energy required by the DH system. By the aforementioned figure and the established practice that the CHP device should achieve more than 5000 operating hours during the year, a CHP unit with a heating capacity of 209 kW was installed. This unit can operate at full load for 6685 h per year and even more if a partial load operation is introduced. The area of the corresponding rectangle illustrates the thermal energy produced by the CHP unit with a heating capacity of 209 kW.

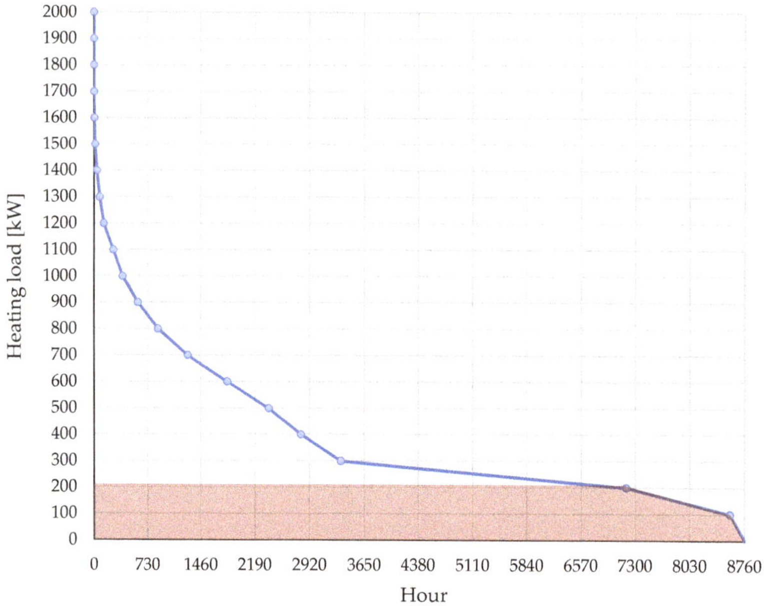

Figure 6. The annual duration of the required heating capacity and energy that can be covered by the operation of the CHP unit.

The technical specifications of the CHP unit with a gas engine are provided in Table 2 [23]. The CHP unit was modelled using the Trnsys Type 211 component developed by the authors [24]. This black-box model incorporates manufacturer catalogue data and allows partial load operation at specified increments.

Table 2. The technical specifications of the CHP unit with a gas engine.

CHP Load %	Electrical Output kW	Heating Output kW	Fuel Use kW
100	140	209	384
75	105	171	310
50	70	130	227

2.2.3. Systems with a Gas-Fired Boiler, CHP Unit, and HP

The implementation of a water-to-water HP has been identified as a potential enhancement to the aforementioned systems. The present study considers an HP that uses water from a nearby river as a heat source. The river's length is 19 km, and the abstraction of river water is located 18 km downstream. The river's relatively short length results in a consistent water temperature at the abstraction point, which remains almost constant throughout the year. In this case study, a constant water temperature of 8 °C was adopted. Temperature fluctuations within a few degrees Celsius would not significantly affect the efficiency of the HP during the year. The configuration of the HP is such that it is connected to the return line of the heating system immediately before the CHP unit, as illustrated in Figure 7. This configuration ensures a lower temperature of the inlet water to the HP compared to a configuration in which the HP is placed in the supply line after the CHP unit.

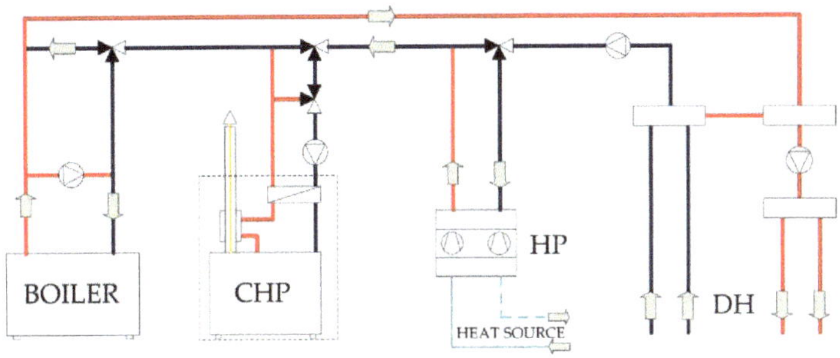

Figure 7. Schematic of a DH system with a gas-fired boiler, CHP and HP units.

Different operation strategies of equipment for heat and power generation were examined through variants of the system presented in Table 3, along with the thermal capacities of HP units considered in the study. Systems B and CHP-B in the table represent previously mentioned systems with a gas-fired boiler and a gas-fired boiler with a CHP unit.

Table 3. Overview of system variants considered in the study: configuration, operational priority, and energy utilisation.

System	Priority in Thermal Energy Production	HP Capacity	Electricity for HP Operation	Produced Electricity from CHP
B	Boiler	-	-	-
CHP/B	CHP	-	-	Sold to grid
CHP-HP/B	CHP-HP, boiler	400 kW	From CHP	Used for HP
HP/CHP/B	HP, CHP, boiler	400 kW–1000 kW (100 kW increment)	From grid	Sold to grid
CHP/HP/B	CHP, HP, boiler	400 kW–1000 kW (100 kW increment)	From grid	Sold to grid

The selection of HP is based on the required water temperature in the distribution network and the temperature of the heat source. When choosing an HP, the requirements of the current European legislation on the use of fluorinated greenhouse gases were taken into account, which will prohibit the installation of devices that work with a refrigerant that has a GWP of 150 or higher from the beginning of 2030 [25]. For this reason, the choice of HP is limited to devices that are available on the market and use natural refrigerants. Depending on the required heating capacity and temperature regime, units which are considered use R717 refrigerant. A range of HP units was selected using professional software GEA RTSelect [26]. All the HP units are selected under operating conditions of water at the condenser 70/60 °C and a heat source temperature of 8 °C. The smallest of the considered HPs has a heating capacity of 400 kW and requires 141.5 kW of electrical power. The criterion for selection of this unit was to match the required electrical power to the CHP electrical power at full load. The annual duration curve for this case is presented in Figure 8. It is visible that CHP with HP can cover the heating energy requirement for 7000 h in the year. Larger units are selected in increments of 100 kW heating capacity up to 1000 kW. Performance data and efficiency metrics for the HPs under variable temperature conditions and partial load operations, which are acquired using professional software, were used to simulate the operation of HPs using Trnsys Type 581 [22].

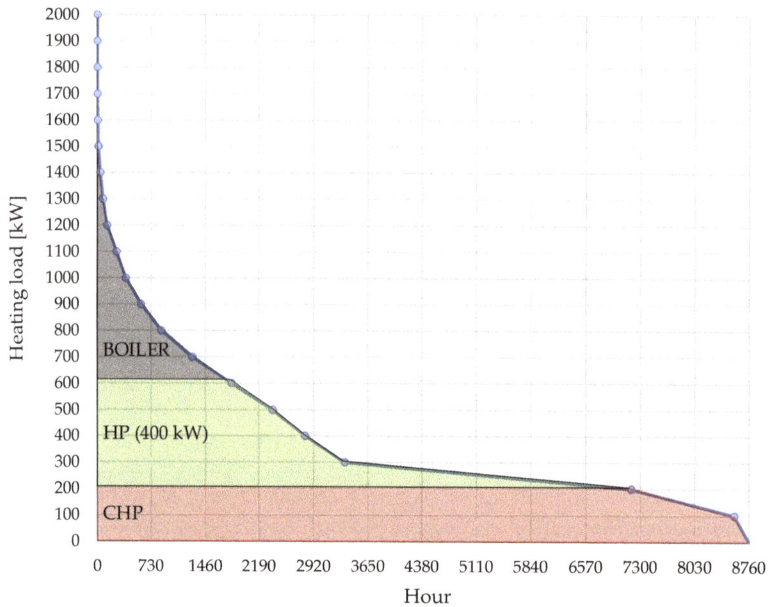

Figure 8. The annual duration of the required heating capacity and energy that can be covered by the operation of the CHP and HP unit with 400 kW heating capacity.

In the HP/CHP/B system, the HP has the priority for heat production. When the HP cannot achieve the required water temperature, the CHP unit is activated, followed by the boiler as needed. The HP draws electricity from the grid, while the electricity generated by the CHP unit is sold back to the grid. In the CHP/HP/B system, the CHP unit has heat production priority. When it cannot reach the necessary water temperature, the HP is switched on, followed by the boiler if required. Like the previous system, the HP uses grid electricity, and the electricity produced by the CHP unit is sold to the grid. In the CHP-HP/B system, the CHP unit and the HP operate simultaneously and adjust to maintain water temperature in the system. The HP's load is adjusted so that its required electrical power matches that of the CHP unit. The electricity produced by the CHP unit is used only for the HP operation.

2.2.4. Operational Cost Components: Energy, Equipment, and Maintenance

Electricity and natural gas are the energy sources considered in the analysis. The price of electricity varies depending on whether the energy is taken from the grid or delivered to the grid. In Croatia, the price of electricity for enterprise customers at medium voltage is set at 0.174 €/kWh for the high tariff and 0.105 €/kWh for the low tariff, as of 1 January 2025. The peak power charge is 3.865 €/kW, according to the National distributor for enterprise customers [27]. The price for electricity delivered to the grid is set at 0.125 €/kWh, as per the national legislation [28]. These electricity prices have been adopted for Scenario 1. A comparison of these prices with current prices reveals that at the beginning of 2024, the former was approximately 10% lower. For the second scenario, the electricity prices in effect at that time were adopted, with a high tariff of 0.161 €/kWh, a low tariff of 0.093 €/kWh, and a peak power charge of 3.451 €/kW, by the national distributor for enterprise customers [27]. The price for electricity delivered to the grid is set at 0.114 €/kWh by national legislation [28]. The electricity prices assumed for Scenario 3 are equivalent to those in Scenario 1. Heat energy producers may also have a different contractual model for electricity pricing, but this scenario is not included in this analysis because of business confidentiality.

The price of natural gas has undergone significant fluctuations in recent years. In this paper, the price of natural gas is analysed within three distinct scenarios. The gas price currently applied by the Republic of Croatia is utilised in Scenario 1. The gas price for heat producers is set at the same level as the purchase price, as defined by national guidelines [29]. For Scenario 1, a gas price of 0.0502 €/kWh was adopted, as reported by the Central European Gas HUB (CEGH) [30]. For Scenario 2, a gas price of 0.030 €/kWh was implemented, as observed at the beginning of 2024. The third scenario involved a gas price of 0.100 €/kWh, representing a substantial increase in gas costs. The highest average monthly gas price on the CEGH during 2022 was recorded in August 2022, reaching approximately 0.345 €/kWh. This surge was driven by supply disruptions and geopolitical tensions linked to the 2022 energy crisis.

In order to facilitate the comparison of scenarios, an indicator I_C, which represents the electricity-to-gas price ratio, has been introduced. It is calculated using the mean electricity price.

$$I_C = C_{ee(mean)}/C_g \tag{15}$$

The HP requires heat from a heat source for its operation. The analysis considered the use of water from a nearby river as a heat source. The fee for using river water is 0.013 €/m^3. Electricity, gas, and water use prices, as well as corresponding electricity-to-gas price indicators by scenarios, are shown in Table 4.

Table 4. Energy prices and electricity-to-gas price indicator.

	Scenario 1		Scenario 2		Scenario 3		Unit
	High	Low	High	Low	High	Low	
Electricity	0.174	0.105	0.161	0.093	0.174	0.105	€/kWh
	3.865		3.451		3.865		€/kW
Gas	0.050		0.030		0.100		€/kWh
Water	0.013		0.013		0.013		€/m^3
Ic	2.8		4.2		1.4		-

Systems with a boiler and CHP unit (B and CHP/B systems) have already been built, so the system's investment will not be considered. In contrast, investment costs for the proposed systems incorporating HP vary depending on the installed capacity. These costs were estimated based on data obtained from equipment manufacturers, including heat pumps, circulation pumps, and piping to the water abstraction point, as well as the associated labour costs for installation.

The connection of electrical equipment to the medium-voltage grid operated by the National distributor is classified as a complex interconnection; therefore, the grid connection cost has been incorporated [31]. Total investment costs as a function of the installed HP capacity are presented in Table 5.

Maintenance costs were considered for all evaluated systems. The maintenance costs for the gas-fired boilers are based on the actual maintenance costs for the boilers in system B, which amount to 100.000 € per year. The specific boiler maintenance cost, with a value of 0.031 €/kWh, is calculated by dividing the annual maintenance cost by the total thermal energy produced by the boilers (3226 MWh). The maintenance costs for gas-fired boilers in other systems are calculated according to the thermal energy produced by the boilers in each system. Maintenance costs for systems with gas-fired CHP units are calculated as

0.0138/kWh of electricity produced [32]. For systems integrating HPs, maintenance costs are determined as 4% of HP's investment value.

Table 5. Investment costs for systems.

System	HP Capacity	Investment Cost
CHP-HP/B HP/CHP/B CHP/HP/B	400 kW	540,000 €
	500 kW	610,000 €
	600 kW	690,000 €
	700 kW	760,000 €
	800 kW	830,000 €
	900 kW	900,000 €
	1000 kW	980,000 €

3. Results and Discussion

Dynamic simulations were used to determine the performance of each system. Among the many results that dynamic numerical simulation can provide, for this analysis, it was important to determine the energy flows of required and produced thermal energy by each generator, as well as the electricity taken and exported to the distribution network. The structure of the thermal energy produced in systems is shown in Figure 9. System B is based on the application of a natural gas boiler, which, as the results show, fully meets the system's thermal energy requirements. By introducing a CHP unit sized to cover base heating load while retaining the existing boiler (CHP/B system) and giving priority to the CHP unit, it is possible to cover half of the required heating needs with the CHP unit. The simultaneous operation of a 400 kW medium-scale HP and CHP unit, which has priority over boiler operation (CHP-HP/B) results in half of the thermal energy produced by the HP, 38% by the CHP unit, and 12% by the boiler. In this variant, the HP is operated according to the available electrical power from the CHP unit, resulting in a slightly larger share for the HP. When the priority in the production of heat energy is given to HP (HP/CHP/B systems), the greatest proportion of thermal energy is produced by the HP (72 to 99%), followed by the CHP unit, and the boiler. The share of energy produced by the HP depends on the installed capacity of the unit. The operation of a boiler or CHP unit is necessary for periods of operation at peak load. In variants where the system is configured to prioritize the operation of the CHP unit (CHP/HP/B systems), 50% of the thermal energy is produced by the CHP unit, while the HP, depending on the size of the unit, covers 38% to 50% of the energy requirement. Generally, in scenarios where a larger HP unit has been installed, only a small or negligible percentage of thermal energy is produced by the boiler. So, an HP sized to 43% of the heating load could replace a gas boiler in conditions of the reference year.

The share of renewable energy sources in the DH systems is determined based on simulation results, as illustrated in Figure 10. Systems incorporating only a boiler or a combination of boiler and CHP units do not utilize renewable energy sources. RES contributions are observed exclusively in configurations that include an HP. Systems where operational priority is assigned to the HP demonstrate a higher RES share compared to those where the CHP unit has priority. Moreover, the RES share increases with the installed HP capacity. The highest RES share, reaching 67%, is observed in the HP/CHP/B configuration, where a 1000 kW HP operates as the primary heat source. Even higher RES contributions could be achieved if the electricity consumed by the HP were sourced from on-site renewable generation.

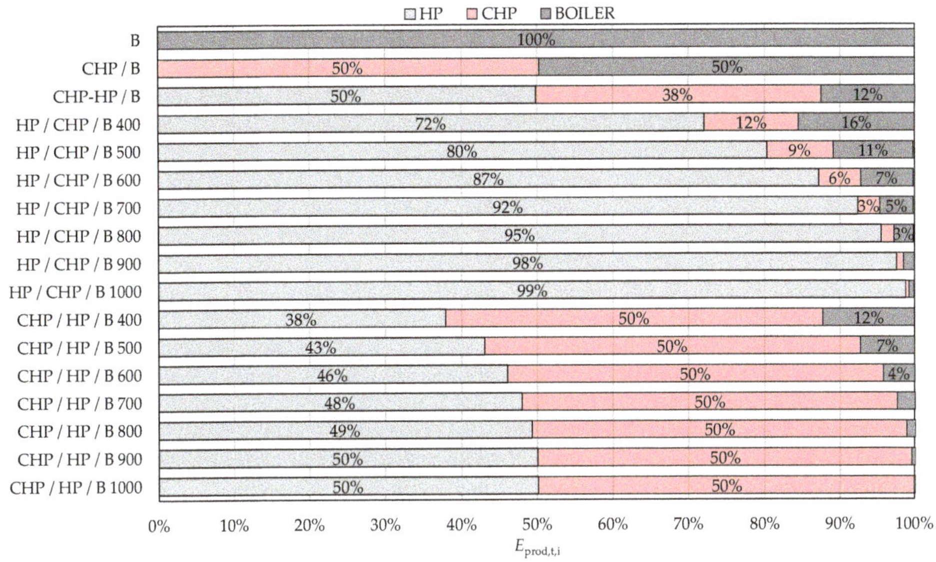

Figure 9. The structure of the thermal energy produced in systems.

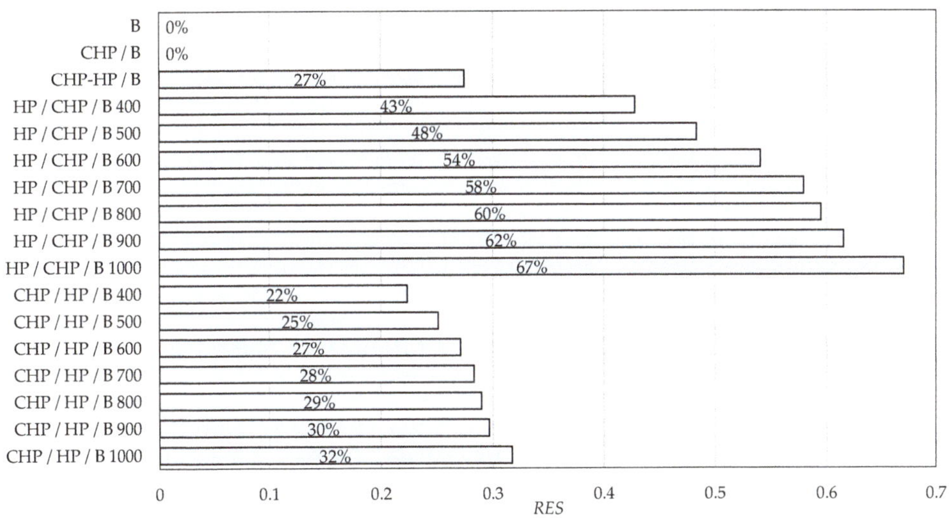

Figure 10. The share of RES by the systems.

The indicators of primary energy consumption (*EPI*) and CO_2 emissions (m_{CO2}) for analysed DH systems are shown in Figure 11. The introduction of a CHP unit to the DH system has been shown to have limited effectiveness in reducing primary energy consumption and CO_2 emissions, with no significant decrease observed in either variable. The highest energy performance index (*EPI*) recorded was 121 kWh/m^2 in system B, while system CHP/B demonstrated an *EPI* of 114 kWh/m^2. The integration of a medium-sized HP unit within the system has been observed to result in a substantial reduction in primary energy consumption. In the HP/CHP/B systems, where the HP has operational priority, the EPI is reduced up to 70% (1000 kW HP unit) compared to the base system with the boiler and CHP unit. Further reductions in *EPI* could be achieved through the deployment of large-scale HP units. In contrast, when the CHP unit holds operational priority, the

EPI reduction is less substantial, ranging from 41% with a 400 kW HP to 48% with a 1000 kW HP.

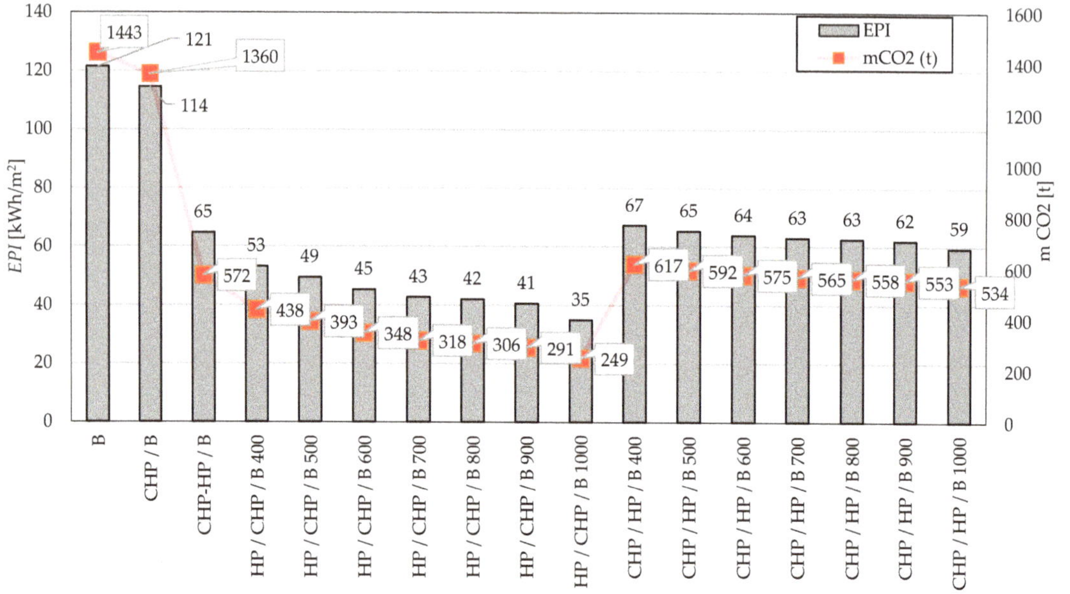

Figure 11. Indicators of primary energy consumption (*EPI*) and CO_2 emission.

The highest CO_2 emissions are produced in systems that utilize natural gas exclusively: 1443 tons of CO_2 per year in the gas-fired boiler system, and 1360 tons of CO_2 in the CHP/B system. In the systems where HP takes operational priority, annual emissions are, depending on the installed HP size, reduced in range from 71 to 82% compared to the CHP/B system. In the systems where CHP takes operational priority, annual emissions are reduced by a smaller amount compared to the systems with HP operational priority: reduction is in the range from 55 to 61% compared to the CHP/B system. Similar annual CO_2 emissions are achieved when the electricity generated by the CHP is utilised for the HP operation (CHP-HP/B system), which results in 572 tons of CO_2 emissions per year, representing a 58% reduction relative to the CHP-B system.

A full lifecycle assessment (LCA) was not conducted due to the unavailability of detailed data on materials, manufacturing, and supply chains. However, according to a recent LCA study on heat pumps [33], the use phase accounts for approximately 94.6% of total emissions, with other phases contributing marginally (~6%). Therefore, the operational CO_2 emissions presented in this study provide a reliable representation of the overall environmental impact.

The cost analysis is based on the assumption that the DH system operator invests its funds in the purchase of new equipment, i.e., does not make a loan for the investment. The initial costs for equipment, grid connection, maintenance, and purchased water for HPs are the same for all three price scenarios. However, the energy price differs due to different electricity and natural gas prices. The levelized cost of energy for heating C_{LH} is presented in Figure 12.

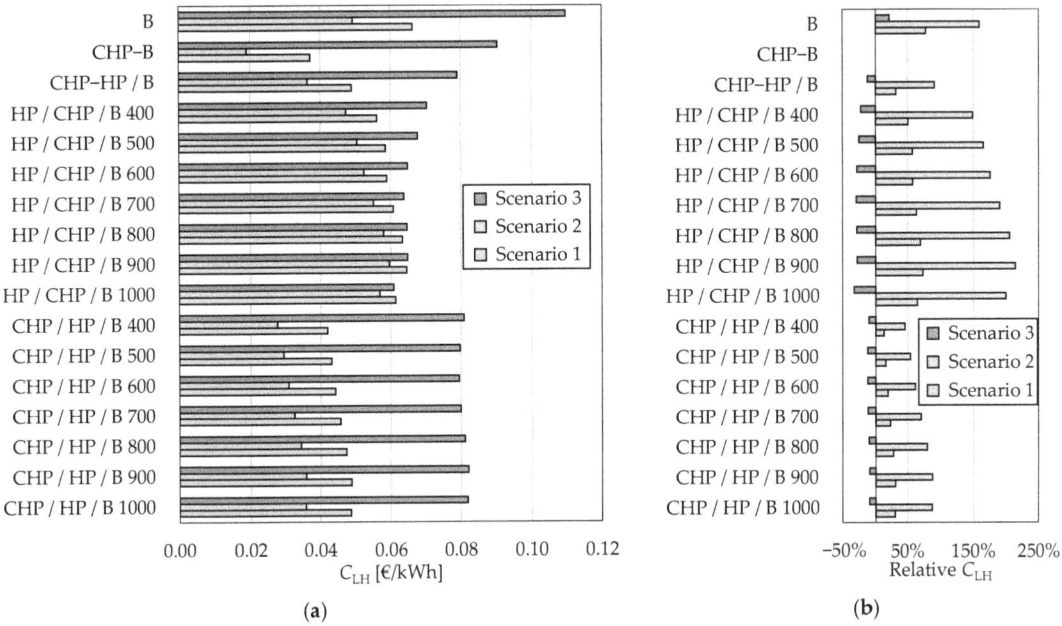

Figure 12. Comparison of system performance in different price variants in terms of (**a**) absolute values of levelized cost of energy for heating; (**b**) values of levelized cost of energy for heating relative to the base DH system.

In Scenarios 1 and 2, which reflect current and recent energy price conditions, the baseline DH system configuration CHP/B is found to be the most cost-effective. It achieves the lowest levelized cost of heating C_{LH} among all evaluated configurations, with values of 0.037 €/kWh and 0.019 €/kWh, respectively. Under these scenarios, high electricity prices favour the operation of the CHP unit, particularly due to the ability to export electricity to the grid. Consequently, systems incorporating HPs exhibit higher C_{LH} values compared to the baseline system.

The increase in costs is most pronounced in configurations where the HP has operational priority, with C_{LH} rising by up to 75% in Scenario 1 and by more than 200% in Scenario 2. Even in configurations where the CHP unit retains operational priority, the C_{LH} still increases by up to 31% in Scenario 1 and up to 88% in Scenario 2. The cost increase is more significant in systems with higher installed HP capacity. These results indicate that, under the current and recent electricity price conditions modelled in Scenarios 1 and 2, the installation of HPs, particularly large-scale units, is not economically viable. Additionally, the operation of a system relying solely on a gas-fired boiler under these pricing conditions proves economically unviable due to high fuel costs and the absence of electricity export revenues.

In contrast, Scenario 3 assumes a more favourable electricity price structure. Under this scenario, all system variants incorporating HPs achieve lower specific heating costs than the baseline DH system, which records a C_{LH} of 0.110 €/kWh. Cost reductions are more substantial in configurations where the HP has operational priority, with the greatest reduction of 33% observed in the system featuring the largest HP unit (1000 kW). As anticipated, systems prioritising CHP operation show more modest cost reductions, up to 12% compared to the baseline. These results suggest that under favourable electricity pricing conditions, the implementation of medium- to large-scale HPs may be economically feasible.

Accordingly, the long-term economic analysis focuses on Scenario 3, which reflects favourable electricity pricing conditions. To assess the feasibility of integration of HPs under these conditions, global costs over a 20-year period are calculated for each system configuration. The results for systems with prioritised operation of HP and CHP-HP are presented in Figure 13. Among these configurations, the CHP-HP/B system exhibits the highest levelized cost of heating C_{LH}, resulting in the longest payback period of 7 years. In contrast, systems with prioritised HP operation demonstrate improved economic performance, achieving payback periods ranging from 4 to 6 years. The shortest payback period—4.7 years—is observed in the system with a 600 kW HP, followed closely by the 500 kW and 700 kW configurations.

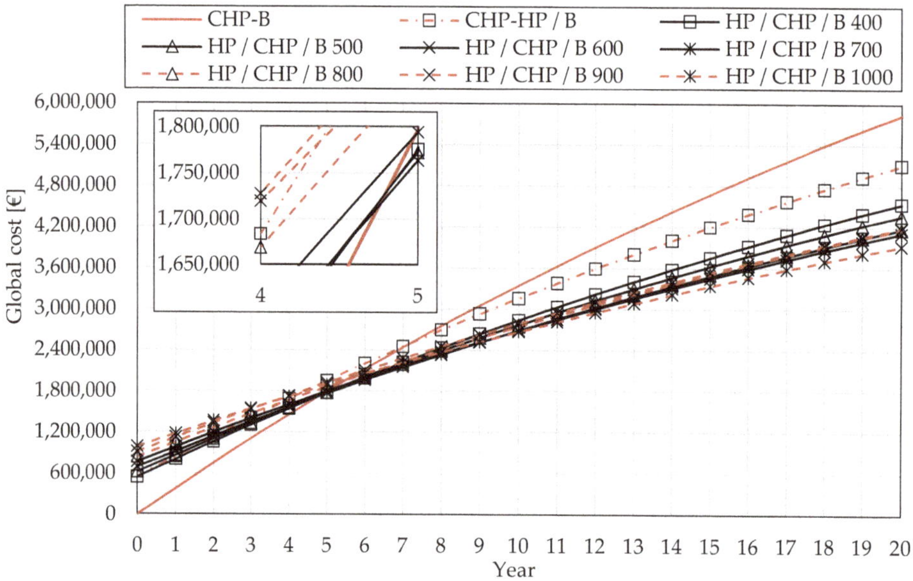

Figure 13. Global costs over the considered period for systems with prioritised CHP-HP and HP operation in price Scenario 3.

Figure 14 illustrates the results for systems with prioritised CHP operation under Scenario 3. Due to higher C_{LH} values, these configurations exhibit longer payback periods compared to those with prioritised HP operation. The shortest payback period in this group is 8 years, recorded for systems with 400 kW and 500 kW HP units. As the installed HP capacity increases, the payback period correspondingly extends, reaching up to 11.5 years for the system with a 1000 kW HP unit. Overall, the payback period increases with HP capacity. Furthermore, configurations in which the HP is integrated without operational priority are found to be economically unviable under the given conditions.

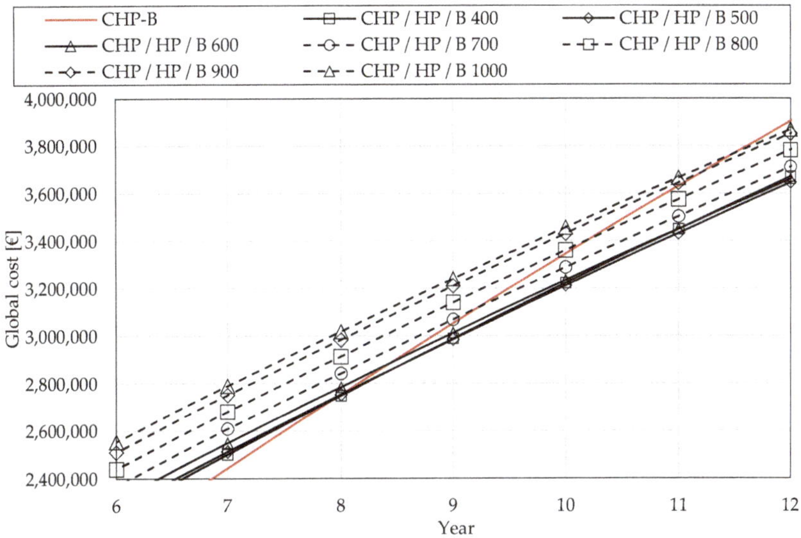

Figure 14. Global costs over the considered period for systems with prioritised CHP operation in price Scenario 3.

4. Conclusions

This study examined the viability of integrating medium-scale HPs into existing third-generation DH systems already comprising CHP units and gas boilers. A dynamic simulation model was developed to realistically represent the behaviour of the production, distribution, and consumption subsystems using commercially available technologies. The research assessed multiple system configurations and operational strategies under varying energy price scenarios to determine the energy, environmental, and economic impacts of HP integration.

The findings demonstrate that incorporating HPs into DH systems can significantly enhance environmental performance. When the HP is given operational priority, systems achieve substantial reductions in primary energy consumption (up to 70%) and CO_2 emissions (up to 82%) compared to the baseline configuration with CHP and boiler. Additionally, these configurations result in a higher share of RES, reaching up to 67%, especially with larger HP capacities. However, some reliance on the boiler or CHP remains necessary to cover peak thermal demands.

Despite these advantages, economic feasibility is highly dependent on the prevailing electricity and gas price ratios. Under current and recent pricing conditions in Croatia (Scenarios 1 and 2), where electricity prices remain high relative to gas, systems with HPs, particularly those with larger capacities, incur higher heating costs and longer payback periods. In these scenarios, configurations prioritising CHP operation remain the most cost-effective due to the opportunity to export electricity. The cost of heating in HP-priority systems can rise by up to 75% under such conditions, making them economically unviable in the short term.

Conversely, Scenario 3, characterised by more favourable electricity pricing, demonstrates that systems prioritising HP operation can achieve lower levelized costs of heating than the baseline CHP and boiler setup. In this scenario, HP systems exhibit payback periods of 4 to 6 years, with the shortest observed at 4.7 years for a 600 kW HP, confirming the potential for economic viability under supportive market conditions.

With regard to the generalisation of the results, it is important to note that the economic viability of HP integration depends significantly on local energy prices. Electricity-to-gas price ratios vary widely across Europe, and under conditions of high electricity prices, heat pumps become less economically attractive, while more favourable ratios enhance their competitiveness. In colder climates, the annual thermal energy demand of DH systems would be higher than in the Rijeka case and, conversely, lower in milder climates. However, since the analysed HP uses river water as a heat source, with relatively stable temperatures year-round, its coefficient of performance (COP) remains largely unaffected by ambient air temperature. Although specific national policies (e.g., carbon pricing or subsidies) may influence fuel price dynamics, they would not alter the technical performance of the system. Therefore, while the energy demand and cost-effectiveness may vary across regions, the technical findings remain broadly applicable in similar system configurations.

In conclusion, the integration of medium-scale HPs into DH systems can offer considerable environmental and energy efficiency benefits, especially when sized to cover a significant portion of the heating load. However, their economic justification is conditional on energy market structures. For widespread adoption, policy support or market reforms that reduce electricity prices relative to gas, alongside technical measures such as lowering system supply temperatures, will be essential. Further optimisation of DH system configurations and the adoption of low-temperature networks could enhance both the efficiency and cost-effectiveness of HP integration in future DH systems.

Future Work and Research Outlook

Several directions for future research have been identified to enhance the modelling framework and its decision-support capabilities:

1. Methodological enhancement through optimisation and hybrid approaches

To improve model adaptability and strengthen its decision-making value, future work will integrate optimisation algorithms and hybrid data-driven methods. These enhancements will enable automated identification of optimal system configurations and control strategies under varying boundary conditions, potentially improving both energy efficiency and economic outcomes.

2. Advanced financial modelling under dynamic market conditions

The current financial analysis assumes self-financing and static energy prices. Future studies will incorporate more realistic scenarios, including mixed financing models (e.g., loans, subsidies) and dynamic electricity and gas pricing. This will offer a more robust foundation for investment planning in an evolving energy market and policy landscape.

3. Progressive integration into fourth-generation DH paradigms

While this study supports the transition toward low-temperature DH through the inclusion of HPs, achieving full alignment with fourth-generation DH principles will require further development. These include lowering building-level supply temperatures and integrating on-site renewable electricity sources, such as PV, to power HPs. Future research will address these aspects to support broader decarbonisation and system flexibility.

Author Contributions: Conceptualisation, D.P., B.D., B.P., and V.M.-V.; methodology, D.P. and B.D.; software and simulation, D.P., B.D., and V.M.-V.; formal analysis, D.P.; investigation, D.P.; resources, D.P. and B.D.; writing—original draft preparation, D.P., B.D., and V.M.-V.; writing—review and editing, D.P., B.D., and V.M.-V.; visualisation, D.P., B.D., and V.M.-V.; supervision, B.D.; All authors have read and agreed to the published version of the manuscript.

Funding: This work has been supported by the University of Rijeka under the project Investigating the potential for the renewal of district heating systems using dynamic simulations (uniri-iskusni-tehnic-23-146).

Institutional Review Board Statement: Not applicable.

Informed Consent Statement: Not applicable.

Data Availability Statement: Dataset available on request from the authors.

Conflicts of Interest: The authors declare no conflicts of interest.

Abbreviations

The following abbreviations are used in this manuscript:

A	area (m^2)
C	cost (€)
COP	coefficient of performance (-)
E	energy (kWh)
EP	primary energy (kWh)
EPI	primary energy indicator (kWh/m^2)
f	energy factor (-)
SCOP	seasonal coefficient of performance (-)
RES	share of renewable energy (-)
m	mass (t)
CHP	combined heat and power
COP	coefficient of performance
DH	district heating
DHW	domestic hot water
GIS	geographic information system
GWP	global warming potential
HP	heat pump
PV	photovoltaic
RES	renewable energy sources
Subscripts	
a	annual
B	boiler
cons	consumption
d	discount
ee	electricity
em	emission
exp	exported
g	gas
i	energy carrier
I	initial investment
imp	imported
LH	levelized cost of heat
nren	non-renewable
p	year
prod	produced
r	discount rate
s	system
t	total
τ	accounting period

References

1. Lund, H.; Werner, S.; Wiltshire, R.; Svendsen, S.; Thorsen, J.E.; Hvelplund, F.; Mathiesen, B.V. 4th Generation District Heating (4GDH). *Energy* **2014**, *68*, 1–11. [CrossRef]
2. Alarnaot Alarnaout, G.; Navarro-Esbrí, J.; Mota-Babiloni, A. Operational, Economic, and Carbon Footprint Feasibility of a Moderately-High-Temperature Heat Pump as an Alternative to Conventional Boilers in Various Scenarios. *Energy Convers. Manag.* **2024**, *309*, 118424. [CrossRef]
3. Požgaj, D.; Pavković, B.; Delač, B.; Glažar, V. Retrofitting of the District Heating System Based on the Application of Heat Pumps Operating with Natural Refrigerants. *Energies* **2023**, *16*, 1928. [CrossRef]
4. Volkova, A.; Krupenski, I.; Pieper, H.; Ledvanov, A.; Latõšov, E.; Siirde, A. Small Low-Temperature District Heating Network Development Prospects. *Energy* **2019**, *178*, 714–722. [CrossRef]
5. Lund, H.; Østergaard, P.A.; Chang, M.; Werner, S.; Svendsen, S.; Sorknæs, P.; Thorsen, J.E.; Hvelplund, F.; Mortensen, B.O.G.; Mathiesen, B.V.; et al. The Status of 4th Generation District Heating: Research and Results. *Energy* **2018**, *164*, 147–159. [CrossRef]
6. Yu, H.; Bergaentzlé, C.; Petrović, S.; Ahlgren, E.O.; Johnsson, F. Combining Techno-Economic Modeling and Spatial Analysis for Heat Planning in Rural Regions: A Case Study of the Holbæk Municipality in Denmark. *Smart Energy* **2024**, *14*, 100144. [CrossRef]
7. Reiter, P.; Poier, H.; Holter, C. BIG Solar Graz: Solar District Heating in Graz—500,000 m^2 for 20% Solar Fraction. *Energy Procedia* **2016**, *91*, 578–584. [CrossRef]
8. Popovski, E.; Aydemir, A.; Fleiter, T.; Bellstädt, D.; Büchele, R.; Steinbach, J. The Role and Costs of Large-Scale Heat Pumps in Decarbonising Existing District Heating Networks—A Case Study for the City of Herten in Germany. *Energy* **2019**, *180*, 918–933. [CrossRef]
9. Aste, N.; Caputo, P.; Del Pero, C.; Ferla, G.; Huerto-Cardenas, H.E.; Leonforte, F.; Miglioli, A. A Renewable Energy Scenario for a New Low Carbon Settlement in Northern Italy: Biomass District Heating Coupled with Heat Pump and Solar Photovoltaic System. *Energy* **2020**, *206*, 118091. [CrossRef]
10. Barco-Burgos, J.; Bruno, J.C.; Eicker, U.; Saldaña-Robles, A.L.; Alcántar-Camarena, V. Review on the Integration of High-Temperature Heat Pumps in District Heating and Cooling Networks. *Energy* **2022**, *239*, 122378. [CrossRef]
11. Ommen, T.; Thorsen, J.E.; Markussen, W.B.; Elmegaard, B. Performance of Ultra Low Temperature District Heating Systems with Utility Plant and Booster Heat Pumps. *Energy* **2017**, *137*, 544–555. [CrossRef]
12. Olympios, A.V.; Pantaleo, A.M.; Sapin, P.; Markides, C.N. On the Value of Combined Heat and Power (CHP) Systems and Heat Pumps in Centralised and Distributed Heating Systems: Lessons from Multi-Fidelity Modelling Approaches. *Appl. Energy* **2020**, *274*, 115261. [CrossRef]
13. David, A.; Mathiesen, B.V.; Averfalk, H.; Werner, S.; Lund, H. Heat Roadmap Europe: Large-Scale Electric Heat Pumps in District Heating Systems. *Energies* **2017**, *10*, 578. [CrossRef]
14. Levihn, F. CHP and Heat Pumps to Balance Renewable Power Production: Lessons from the District Heating Network in Stockholm. *Energy* **2017**, *137*, 670–678. [CrossRef]
15. Ommen, T.; Markussen, W.B.; Elmegaard, B. Heat Pumps in Combined Heat and Power Systems. *Energy* **2014**, *76*, 989–1000. [CrossRef]
16. Capone, M.; Guelpa, E.; Verda, V. Optimal Installation of Heat Pumps in Large District Heating Networks. *Energies* **2023**, *16*, 1448. [CrossRef]
17. Helin, K.; Syri, S.; Zakeri, B. Improving District Heat Sustainability and Competitiveness with Heat Pumps in the Future Nordic Energy System. *Energy Procedia* **2018**, *149*, 455–464. [CrossRef]
18. TRNSYS Transient System Simulation Tool. 2013. Available online: https://www.trnsys.com/ (accessed on 15 May 2025).
19. Primary Energy Factors for Republic of Croatia. Available online: https://mpgi.gov.hr/UserDocsImages/dokumenti/EnergetskaUcinkovitost/meteoroloski_podaci/FAKTORI_primarne_energije.pdf (accessed on 3 February 2022).
20. *DIN EN 12831-1:2017-09*; Energy Performance of Buildings—Method for Calculation of the Design Heat Load. DIN Media GmbH: Berlin, Germany, 2017.
21. *DIN 4708-2:1994-04*; Zentrale Wassererwärmungsanlagen; Regeln Zur Ermittlung Des Wärmebedarfs Zur Erwärmung von Trinkwasser in Wohngebäuden. DIN Media GmbH: Berlin, Germany, 1994; p. 7.
22. *The TESS Component Libraries 17*; TESS: Madison, WI, USA, 2021.
23. *Technology Brochure-CHP Units for Heat and Power—Vitobloc*; Viessmann GmbH: Allendorf, Germany, 2019.
24. Delač, B.; Pavković, B.; Lenić, K.; Mađerić, D. Integrated Optimization of the Building Envelope and the HVAC System in NZEB Refurbishment. *Appl. Therm. Eng.* **2022**, *211*, 118442. [CrossRef]
25. Regulation (EU) 2024/573 of the European Parliament and of the Council of 7 February 2024 on Fluorinated Greenhouse Gases, Amending Directive (EU) 2019/1937 and Repealing Regulation (EU) No 517/2014; 2024. Available online: https://eur-lex.europa.eu/eli/reg/2024/573/oj/eng (accessed on 15 May 2025).
26. *GEA RTSelect Configuration and Engineering Software*, Version 13. 9. 2.; GEA Group: Düsseldorf, Germany, 2025.

27. HEP Elektra d.d. National Electricity Tariff Items for Supply the Enterprise Consumers. Available online: https://www.hep.hr/elektra/poduzetnistvo/tarifne-stavke-cijene-1578/1578 (accessed on 24 April 2025).
28. Official Gazette of the Republic of Croatia NN RH No. 83/2023—The Act on Amendments to the Law on RES and High-Efficiency Cogeneration. Available online: https://narodne-novine.nn.hr/clanci/sluzbeni/2023_07_83_1298.html (accessed on 24 April 2025).
29. Official Gazette of the Republic of Croatia NN RH No. 31/2023—The Regulation on Eliminating Market Disturbances. Available online: https://narodne-novine.nn.hr/clanci/sluzbeni/2023_03_31_531.html (accessed on 24 April 2025).
30. Central European Gas Hub. Available online: https://www.cegh.at/en/ (accessed on 24 April 2025).
31. National Electricity Operator HEP Elektra d.d. Complex Interconnection to the Medium-Voltage Grid. Available online: https://www.hep.hr/ods/pristup-mrezi/prikljucenje-na-mrezu-28/slozeno-prikljucenje/239 (accessed on 24 April 2025).
32. Schaumann, G.; Schmitz, K.W. *Kraft-Wärme-Kopplung*, 4th ed.; Springer: Berlin/Heidelberg, Germany, 2010.
33. Famiglietti, J.; Acconito, L.; Arpagaus, C.; Toppi, T. Environmental Life Cycle Assessment of Industrial High-Temperature to Residential Small-Size Heat Pumps: A Critical Review. *Energy Convers. Manag. X* **2025**, *26*, 100947. [CrossRef]

applied sciences

MDPI

Article

BEST—Building Energy-Saving Tool for Sustainable Residential Buildings

Marco Cecconi [1], Fabrizio Cumo [1], Elisa Pennacchia [1,*], Carlo Romeo [2] and Claudia Zylka [1]

1 Interdepartmental Research Centre Territory, Building, Restoration and Environment, Sapienza University of Rome, Via A. Gramsci, 53-00197 Rome, Italy; marco.cecconi@ingenergia.it (M.C.); fabrizio.cumo@uniroma1.it (F.C.); claudia.zylka@uniroma1.it (C.Z.)

2 ENEA—Italian National Agency for New Technologies, Energy and Sustainable Economic Development, Via Anguillarese, 301-00123 Rome, Italy; carlo.romeo@enea.it

* Correspondence: elisa.pennacchia@uniroma1.it

Abstract: The building and construction sector significantly impacts CO_2 emissions and atmospheric pollutants, contributing to climate change. Improving energy efficiency in buildings is essential to achieving carbon neutrality by 2050, as outlined in the European Green Deal. This study presents a decision-support tool for energy retrofit interventions in existing residential buildings. The methodological approach begins with the identification and classification of common roof and wall types in the national residential building stock, segmented by construction period, followed by defining optimized, pre-calculated standardized solutions. The performance evaluations of proposed solutions resulted in a matrix designed to guide practitioners in selecting pre-calculated, efficient, and sustainable prefabricated solutions based on energy performance criteria. The tool developed from this matrix enables preliminary energy assessment, offering an overview of potential retrofit interventions. It assists designers in identifying specific cases based on construction period, building type, and climate zone, allowing for the selection of standardized solutions, energy pre-analyses, energy and cost-saving simulations, and access to detailed performance sheets. Unlike other tools requiring extensive input on opaque envelope components and thermo-physical calculations, this tool streamlines the selection process of vertical and roof closures based on construction age and building type. Additionally, the tool estimates potential economic savings and the Net Present Value (NPV) of proposed insulation solutions, identifying available incentives for the intervention.

Keywords: energy retrofit solutions; building envelope technologies; building sustainability; innovative building analyses; residential case study

1. Introduction

It is widely recognized that the construction sector and the entire building industry significantly contribute to global CO_2 emissions and other atmospheric pollutants. Studies such as that of Alhazmi et al. highlight how the environmental performance of residential buildings, analyzed through the Life-Cycle Assessment method, confirms this concerning trend [1]. Globally, the Intergovernmental Panel on Climate Change (IPCC) attributes approximately 40% of global energy consumption and about 25% of total CO_2 emissions to this sector [2], which represent a critical factor in global warming and climate change [3,4].

In Europe, the situation appears particularly critical, with a largely obsolete and inefficient building stock (over 75%), as highlighted by the European Commission. Such inefficiency necessitates structural and energy retrofitting interventions to improve environmental performance and reduce the climate impact of buildings [5]. Studies have shown that residential buildings account for more than 25% of total energy consumption within the European Union [6].

The need for immediate action is further reinforced by the objectives set by the European Green Deal, which aims to achieve climate neutrality by 2050 [7], establishing an intermediate milestone of a 55% reduction in greenhouse gas emissions by 2030 [8]. COP27, held in Sharm el-Sheikh, reaffirmed the global commitment to the goals of the Paris Agreement, emphasizing the importance of mitigation measures to avoid catastrophic climate scenarios. This vision is consistent with the international orientation toward sustainable development, reflected in the adoption of green strategies and net-zero approaches. These criteria are crucial for revitalizing the construction sector and promoting the transition to a green economy [9].

In this context, addressing the built environment—particularly by targeting the existing building stock—emerges as a strategic priority for reducing the environmental footprint of human activities [10]. This direction has been formally embraced by the United Nations through the 2030 Agenda for Sustainable Development, adopted in 2015 by all member states. The Agenda sets out 17 Sustainable Development Goals (SDGs), among which are key objectives such as ensuring access to affordable, reliable, sustainable, and modern energy for all (Goal 7); making cities and human settlements inclusive, safe, resilient, and sustainable (Goal 11); and taking urgent action to combat climate change and its impacts (Goal 13). Against this backdrop, the European Directive 2018/844 represents a legislative milestone, replacing previous regulations and requiring member states to develop long-term national strategies for the energy retrofitting of buildings. The primary objective is to reduce emissions from the building sector in the European Union by 80–85% compared to 1990 levels, promoting the transformation of existing buildings into nearly zero-energy buildings (NZEB) [11]. This vision is further supported by the recent transition to positive energy buildings (PEB), which not only meet their own energy needs but also generate a surplus of energy, as demonstrated by emerging trends in the European building landscape [12]. The urgency of such interventions is not limited to the containment of emissions but also addresses climate resilience, energy efficiency, and the improvement of urban quality of life. In this context, eco-districts function not only as experimental platforms for sustainable urban development but also as strategic instruments for emissions reduction at the neighborhood scale. By integrating renewable energy systems, low-carbon mobility, energy-efficient buildings, and circular resource management, these districts contribute to the decarbonization of urban environments [13]. Their capacity to produce measurable reductions in greenhouse gas emissions while simultaneously enhancing urban livability positions them as key components of climate action strategies [14]. A systematic approach at European and global levels, combined with innovative technologies and sustainable management models, represents the key to effectively addressing the climate crisis through the transformation of the building sector [15].

Currently, numerous studies focus on evaluating the energy consumption of existing buildings. M. Sala-Lizarraga et al. [16] provide fundamental data that underscore the importance of optimizing building energy performance through energy analysis and thermoeconomics. Specifically, they state that the conventional energy efficiency of a building (based solely on energy balance) can exceed 90%, but actual energy efficiency is often below 30%. HVAC systems (heating, ventilation, and air conditioning) account for about 50% of a building's total energy consumption, while the building envelope (walls, roofs, windows, floors) is responsible for approximately 30–40% of total thermal losses. Poel et al. [17] analyze methods, criteria, and results related to the evaluation of energy performance in existing residential buildings, with particular attention to energy retrofitting interventions and their impact on overall energy efficiency and greenhouse gas emissions. The main shortcomings include poor thermal insulation quality and the presence of thermal bridges, which increase energy demands for heating and cooling.

Existing buildings also often use outdated HVAC systems, characterized by low efficiency and high consumption. Paraschiv Lizica et al. [18] explore the crucial role of thermal insulation in improving building energy efficiency, highlighting its benefits in terms of reducing energy consumption, CO_2 emissions, and operating costs.

Thermal insulation is recognized as one of the most effective strategies for reducing the energy demand of buildings by minimizing heat loss through the building envelope [19–21]. The widespread implementation of thermal insulation technologies can make a significant contribution to meeting global greenhouse gas reduction targets. According to Bazzocchi et al. [22], enhancing thermal insulation could lead to a reduction of approximately 20–30% in emissions from the building sector. A study conducted on a residential building in northern Iran demonstrated that implementing optimal thermal insulation led to a reduction in heat loss of 54.8% through the walls and 53.5% through the roof. The simultaneous application of appropriate insulation materials for both elements in regions characterized by moderate and humid climates resulted in the highest overall energy savings, reaching 47.2% across both summer and winter seasons. This approach not only significantly decreases energy losses related to heating and cooling but also improves the building's resilience to environmental conditions and contributes to lowering energy demand within the building sector [23].

In recent years, research has focused on developing innovative solutions that offer superior thermal performance, environmental sustainability, and a reduced ecological footprint. The research of Imane et al. [24] investigates the use of natural thermal insulation materials, such as hemp fiber and sheep wool, in desert regions. The results indicate that these materials can reduce building energy consumption by over 42%, highlighting their effectiveness in enhancing energy efficiency in challenging climates.

Various tools, such as SUSREF [25], IFORE [26], TABULA [27], EPIQR [28], Façade Refurbishment Toolbox [29], DanCTPlan [30] have developed methodologies for evaluating the energy performance of buildings. However, these tools do not adequately address the environmental sustainability of the proposed solutions to improve energy efficiency.

Over the years, several decision-support tools have been developed for sustainable retrofitting projects, including the Decision-Support Tool (DST) designed by Garufi et al. [31]. This tool assists in making informed decisions within the context of sustainable renovations in the social housing sector in the Netherlands, evaluating the effectiveness of interventions in terms of greenhouse gas emissions reduction, energy savings, and social benefits.

Similarly, the tool by Dall'O' et al. [32] focuses on the classification of energy performance of residential buildings at the urban level, proposing a methodology that integrates various parameters to analyze and improve energy efficiency of building stock. However, the tools designed by Garufi et al. and Dall'O' et al. present limitations related to their dependency on detailed and up-to-date data, which are challenging to obtain in some areas. Additionally, both tools require financial resources and specialized expertise, limiting their applicability in resource-scarce contexts. They are tailored to specific contexts, making it difficult to adapt them to different realities. Furthermore, they require constant updates to remain aligned with regulations but often do not fully consider the social costs and indirect impacts of renovations. Deng Z. et al. [33] propose the AutoBPS (Automatic Building Performance Simulation) tool, developed to support urban-scale energy modeling, with the aim of promoting energy efficiency at the city level.

The tool stands out for its automation of the modeling process, using publicly available data such as GIS and cadastral databases to generate detailed models without complex manual intervention. However, AutoBPS has some main limitations, such as the need for high-resolution data, which may not always be available or up to date in all cities. Moreover, the model requires significant computational power, making it less practical for large-scale city applications or real-time analyses. The tool designed by Godoy-Rangel et al. [34] supports the assessment and improvement of building energy performance, focusing on the climatic and socioeconomic specificities of northern Mexico. This tool mainly emphasizes energy efficiency and emissions but does not provide an in-depth analysis of socio-cultural and behavioral variables of occupants, which are crucial for comprehensively improving energy efficiency. Moreover, its limited applicability outside the northern Mexico region reduces its adaptability to other areas with different climatic and infrastructural conditions. TecSB [35] is a decision-support tool designed for both residential and non-residential buildings located exclusively in the city of Monterrey, Mexico. It consists of a free virtual platform that assesses energy efficiency measures for roofs and windows, as well as the potential for on-site energy generation.

The tool described by Arab et al. [36] employs the OTTV (Overall Thermal Transfer Value) index to evaluate building energy efficiency, a parameter that measures the amount of heat a building absorbs through its exterior surface. While this index is useful for determining the need for envelope improvement interventions, it has limitations, as it does not account for climate variability and occupant behavior patterns.

A comparison of the various tools previously described is presented in Table 1 below.

Table 1. A comparison of the various tools for energy efficiency.

Tool	Objective	Strengths	Weaknesses
SUSREF (Sustainable refurbishment of building façades and external walls)	Provides a systematic method for the sustainability refurbishment of building façades and external walls, combining environmental impact assessment, life-cycle cost analysis, and functional performance evaluation to guide the design and implementation of energy-efficient and environmentally responsible retrofit solutions.	Comprehensive environmental assessment; supports multi-criteria decision-making.	Focused mainly on external walls; limited scope beyond envelope elements; Requires specialized expertise.
IFORE (Innovation for Renewal)	Applies dynamic thermal modeling to evaluate retrofit scenarios in social housing.	Detailed thermal simulations; context-specific applications in UK and France.	Requires accurate input data; limited generalizability to other regions.
TABULA (Typology Approach for Building Stock Energy Assessment)	Standardized methodology for assessing the energy performance of the residential building stock in Europe.	Standardized approach; applicable at national and EU levels.	Simplified models may not capture all building-specific variables; does not provide the payback period.

Table 1. *Cont.*

Tool	Objective	Strengths	Weaknesses
EPIQR (Energy Performance Indoor environment Quality Retrofit)	These methods assist experts in conducting a comprehensive building diagnosis that encompasses building deterioration, energy performance, indoor environmental quality, and compliance with relevant standards and regulations.	User-friendly interface; includes economic and technical factors.	Requires regular data updates; limited integration with modern sustainability metrics.
Façade Refurbishment Toolbox	Supports design decisions in the energy upgrade of building façades.	Design-oriented; includes aesthetic and technical criteria.	Focused mainly on façade; lacks broader system-level assessment.
DanCTPlan energy planning tool	Decision-Making tool for sustainable energy planning and retrofitting in Danish communities and districts.	Scalable from building to district level; integrates planning tools.	Complex implementation; high data requirements.
Dutch Decision-Support Tool	Facilitates decision-making for sustainable renovation in social housing.	Considers environmental and social benefits; scenario analysis.	Context-specific; requires detailed and updated data.
Urban-scale Classification Tool	Classifies energy performance of residential stock at urban scale.	Supports urban energy planning; integrative data use.	Dependent on data availability; needs regulatory updates.
AutoBPS (Automated Building Performance Simulation)	Automates urban building energy modeling using public data.	Minimizes manual input; suitable for large-scale simulations.	High computational needs; sensitive to data resolution.
TecSB Sustainable Building Calculator	A free virtual platform that evaluates energy efficiency measures applicable to building roofs and windows, as well as the potential for on-site energy generation in Mexico.	Free accessibility and ease of use; integrated assessment of energy efficiency and solar energy generation; estimates of avoided greenhouse gas emissions and payback periods for investments.	Limited geographical applicability; requirement for detailed input data; limitations in efficiency strategies.
OTTV (Overall Thermal Transfer Value Tool)	Assesses energy efficiency of high-rise buildings using OTTV index.	Applicability in tropical climates.	Energy efficiency improvements limited to walls and windows; restricted geographical applicability; lack of economic analysis of the interventions.

The Proposed Tool: BEST

This research aims to develop BEST—Building Energy-Saving Tool for Sustainable Residential Buildings—an innovative decision-support tool for designing energy retrofitting interventions in the residential sector. This tool seeks to overcome the limitations of currently available tools by introducing a simplified and standardized approach to optimizing building energy performance. Compared to the tools analyzed earlier, BEST stands out for its ability to guide designers in identifying the most effective solutions without requiring complex inputs on the stratigraphy or thermo-physical parameters of building components. The key innovations introduced include:

- Automated solution selection: The tool enables the selection of vertical and horizontal closures based on predefined parameters such as the era of construction, building type, and climatic zone.
- Rapid pre-analysis of energy performance: Provides a preliminary assessment of energy-saving potential and compares proposed solutions in terms of energy performance and economic feasibility, in accordance with the current national regulatory framework.
- Advanced economic evaluation: Unlike other tools, BEST includes the calculation of NPV and the identification of applicable economic incentives.
- Focus on off-site construction: Proposes standardized solutions in line with the principles of industrialized and sustainable construction.

The methodology was developed by CITERA, the research center at Sapienza University of Rome, within the framework of the "Research of the Electricity System" program, in collaboration with ENEA and the Ministry of Economic Development. This initiative focused on the "Study of the state of the art of prefabricated solutions for roofing systems and the creation of a best practices catalog/Development of tools and support instruments for energy efficiency interventions in residential buildings undergoing refurbishment". The proposed methodological approach was validated through its application to a representative case study of the Italian building stock. The paper is structured as follows. First, the methodology adopted for the development of the tool is presented. Section 3 provides an overview of the case study used to validate the BEST tool. The results are presented in Section 4 and further analyzed in Section 5. Finally, the paper concludes with a discussion on key findings and potential future developments.

2. Materials and Methods

The overarching goal of this research is to enhance the energy efficiency of the national residential building stock, with the aim of developing high-performance buildings aligned with the energy transition objectives set by current regulations.

The study focuses on the development of a tool designed to assist designers in identifying the most suitable retrofit solutions based on the building's construction period, typology, structural characteristics, and climatic zone. The outcomes of previous research efforts have provided the foundation for the development of a decision-support tool for energy efficiency interventions in residential buildings undergoing renovation [37]. The tool is built on an Excel-based computational engine with macros in Visual Basic, integrating pre-calculated solutions through a dynamic matrix consisting of 718 configurations, in accordance with the parameters established by current regulations (Supplementary File S1).

Leveraging an optimized solution database, the system enables the selection of standardized, efficient, and sustainable technologies that meet the regulatory performance requirements for wintertime building performance, with reference to off-site construction practices. The tool offers a preliminary energy analysis, allowing users to assess the impact of retrofit interventions on the building envelope by providing insights into potential energy and cost savings. Through performance simulations, professionals can compare different optimized solutions, supported by a dynamic matrix that integrates criteria related to the existing building envelope and regulatory requirements. The calculations include heating and cooling energy demands, considering both the envelope and the most common HVAC system types. A key feature of the tool is its ability to select vertical and upper horizontal enclosures based on specific criteria, thereby simplifying the complexity associated with defining layer compositions. The proposed energy efficiency solutions, tailored to each case, allow for the calculation of economic savings, NPV, and available incentives. Additionally, users can download detailed reports containing stratigraphy information, simulation results, hygrothermal performance checks, and sustainability indicators. In the table below (Table 2) the functionalities of the BEST tool are presented.

Table 2. Functionalities of the tool and their descriptions.

Tool Functionality	Description
Development of the pre-retrofit energy model	Enables the creation of an energy model of the building in its current configuration
Assessment of current energy performance	Analyzes the building's energy performance using synthetic indicators
Selection of optimized retrofit solutions	Identifies the most effective retrofit measures based on consolidated data and technical criteria
Construction of the post-retrofit model and analysis of benefits	Simulates the behavior of the retrofitted building and evaluates both economic and technical benefits
Comparison of solutions and prioritization of interventions	Allows for the comparison of alternative scenarios to establish a hierarchy of improvement actions

The methodological approach underpinning this research is based on the definition of a flowchart that identifies the fundamental input data necessary to guide designers in selecting the most appropriate case scenario for the building to be retrofitted. This classification is based on construction period, building typology, structural characteristics, and climatic zone.

Figure 1 presents the methodological framework, outlining the key steps followed in the tool's development. The activities and findings from previous research, essential for generating a database of optimized pre-calculated solutions, are highlighted in grey.

In particular, these first three steps represent the preliminary phase of the development of BEST. This phase is focused on building the technical and knowledge base required to generate effective energy retrofit solutions. A detailed description is provided below:

- *Step 1—State of the art of the main types of Upper Horizontal Closures (UHC)*: The first step involves a comprehensive analysis of the state of the art and a typological classification of UHCs based on the buildings' construction periods. The objective of this step is to build an organized database that identifies the most recurrent construction solutions for each historical building era. The outcomes include a typology table of UHCs by construction age and a systematic classification matrix that will support the subsequent selection of appropriate retrofit interventions.
- *Step 2—Study of insulating materials through a sustainable approach*: The second step focuses on the study of insulating materials with a strong emphasis on sustainability. A wide range of materials are analyzed and selected based on environmental sustainability criteria and energy efficiency performance. This step prioritizes solutions that follow an off-site approach—meaning prefabricated and standardizable systems—so as to facilitate practical application while reducing both time and costs associated with the interventions.
- *Step 3—Simulations for the identification of solutions*: The third step consists of performing energy and performance simulations to evaluate the effectiveness of the selected insulating materials. The aim is to develop a validated dataset that will feed into the solution matrix outlined in Step 5. This matrix will serve as a key tool in

recommending retrofit strategies that are coherent with the specific characteristics of each building.

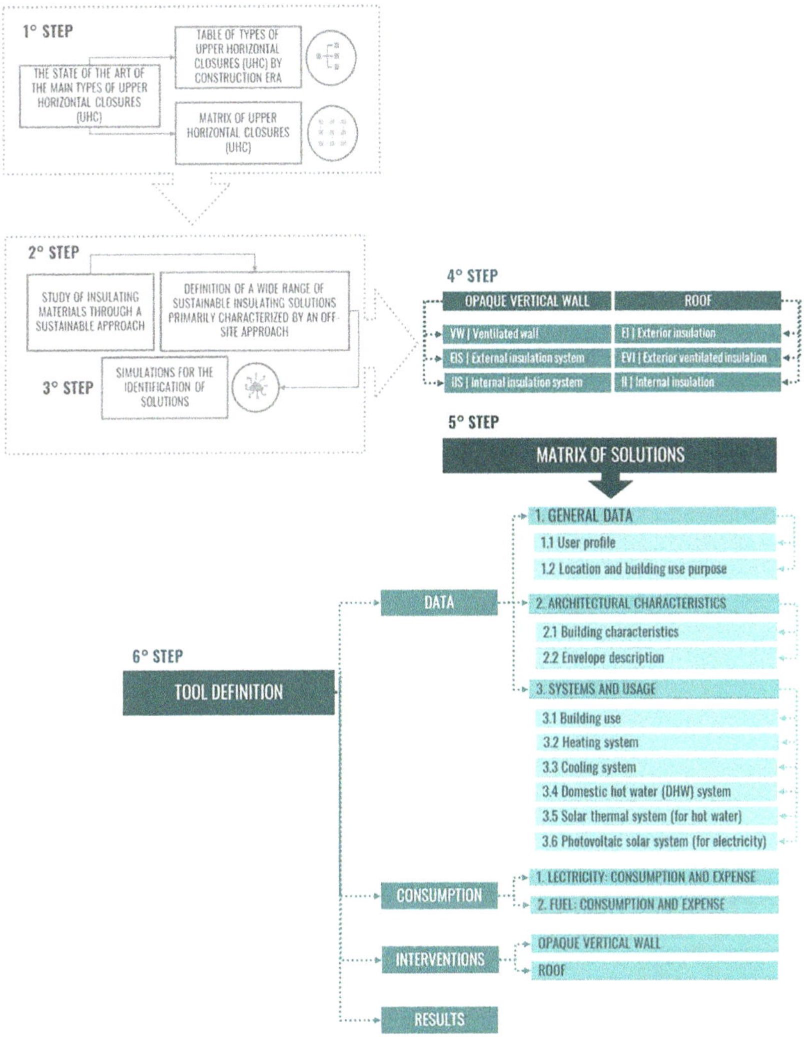

Figure 1. Methodology flowchart for tool development.

The first three steps of the previous research are followed by the subsequent stages of tool definition as outlined below:

- *Step 4—Opaque vertical wall and roof solutions*: In this step, the focus is on defining various insulation systems for the building envelope, specifically targeting opaque vertical walls and roofs.
- *Step 5—Matrix of solutions:* This step involves creating a matrix that organizes and presents the available solutions for building envelope insulation, based on the choices made in Step 4. The matrix serves as a decision-making tool, guiding the selection of the most appropriate solutions for each building's specific needs.
- *Step 6—Definition of the tool BEST:* In the final step, the tool is defined based on the data, materials, and solutions identified in the previous steps. This step consolidates

all the inputs and establishes the structure of the tool that will guide users in selecting appropriate retrofit solutions.

The tool BEST is structured as follows:

1. General data;
2. Architectural characteristics;
3. Systems and usage;
4. Energy consumption and costs;
5. Energy efficiency measures;
6. Results.

The first module of the BEST tool is designed for the systematic collection of data on the building under analysis. It is structured into three main sections: (1) general data, (2) architectural characteristics, and (3) systems and usage. The tool requires a set of fundamental input data to perform a reliable simulation. These minimum data inputs include location, building use (intended use), characteristics of heat-loss surfaces, type of thermal generators for heating/cooling/domestic hot water (DHW), use of renewable energy sources, and the installation period of the systems. These inputs are associated with simplifying assumptions based on standardized values derived from relevant UNI standards (e.g., UNI 10349, UNI TS 11300, UNI 10339).

All required inputs are mandatory for the simulation to proceed. The tool does not allow for missing or incomplete data among the essential inputs, thus ensuring consistency and reliability of the simulation results. The minimum required data and the related simplifying assumptions are as follows: (Table 3):

Table 3. Minimum required data and associated simplifying assumptions adopted in the first module of the BEST tool for preliminary energy analysis.

Data	Simplifying Assumptions
Location	Climatic data derived from UNI 10349 standard [38]
Building use type	Setpoint temperature data, system operation schedules, occupancy levels, air flow rates, domestic hot water demand, and internal loads derived from UNI TS 11300-1 and UNI 10339 [39] standards
Thermal envelope characteristics	Only one type of opaque or transparent surface per orientation is considered
Type of thermal generators for heating/cooling/DHW and renewable energy systems	Limited selection options with standardized efficiencies as defined by UNI TS 11300 standards
Installation period of building systems	Efficiency values for subsystems are selected based on assumed installation periods

The data collection process follows a sequential approach to ensure the completeness and consistency of the information, providing a solid foundation for subsequent simulation phases.

The tool BEST guides designers through a series of integrated steps:

1. Case study identification: Based on a structured database, the tool facilitates the identification of the building's current condition by taking into account its construction period, structural typology, and climatic zone. The tool includes 30 types of vertical perimeter walls and 30 roof configurations representative of the national residential building stock;

2. Preliminary energy analysis of opaque envelopes: It provides an overview of potential retrofitting opportunities in terms of energy savings and economic impact, addressing both heating and cooling energy demands;
3. Database of optimized solutions: Leveraging the results of prior Work Package activities, the tool suggests pre-calculated interventions for perimeter walls and for roof (Predominantly prefabricated insulation solutions (5 for the perimeter walls and 21 for the roofs) selected based on comparison parameters such as performance indicators, environmental impact, applicability, and cost, and validated through energy simulations conducted using certified BIM energy software to determine the required insulation material thicknesses);
4. Post-intervention simulation: BEST generates an energy model to evaluate the technical and economic performance of the proposed solutions. It also offers downloadable detailed performance sheets;
5. Prioritization of interventions: The tool ranks the available solutions to identify those with the best cost-benefit ratios.

2.1. BEST Tool Description

Users can access the tool via the internet using a browser as an interface; thus no specific hardware or software requirements are necessary. A simple internet connection and a browser are sufficient to fill in the data and run the calculation engine that generates the results.

The simplified interface allows for the description of building characteristics and their building systems in order to perform quick assessments of the potential interventions for the refurbishment of the opaque envelope of the residential building stock.

The BEST simulation tool was designed for application in the assessment of energy efficiency interventions for residential buildings, classified as follows:

- E.1 (1)—Residential buildings with continuous occupancy (e.g., dwellings, boarding schools, convents, penitentiaries, barracks);
- E.1 (2)—Residential buildings with intermittent occupancy (e.g., vacation homes and similar structures).

The computational tool aims to combine the simplicity of static methods with the accuracy of dynamic simulations, integrating additional functionalities to assess investment costs and their economic return, also considering the incentive schemes currently in force in Italy. For this reason, a simplified dynamic simulation approach was adopted, employing a single thermal zone to estimate the building's annual energy consumption in both electrical and thermal terms.

The "simplified dynamic simulation" implemented in the BEST tool is based on the Italian technical standards UNI/TS 11300 (Parts 1, 2, and 4). These standards define the underlying assumptions and boundary conditions, including standardized climatic data, internal conditions (such as setpoint temperatures and internal gains), operational schedules, building envelope characteristics, ventilation rates, and system efficiencies. The approach can be described as a quasi-steady-state model applied over discrete time steps, rather than a detailed transient simulation.

Regarding the climatic zone, one was selected arbitrarily for the purpose of the simulation. Given the objective of the study and the standardized nature of the input parameters, the specific choice of climatic zone does not substantially affect the methodological outcomes, or the validity of the comparative analysis carried out by the tool.

The BEST tool supports simulations for all Italian climatic zones, as defined by national regulations. The UNI/TS 11300 standards incorporate climate-specific parameters for each of the six climatic zones in Italy (from Zone A to Zone F), ensuring that energy assessments and retrofit recommendations are both accurate and contextually relevant to the building's geographical location. This climate sensitivity is essential to maintain the validity and applicability of the tool's outputs.

The BEST tool has been developed in alignment with key European and national energy policies, particularly the objectives outlined in the European Green Deal, the Energy Performance of Buildings Directive (EPBD), and national long-term renovation strategies. The reference values for steady-state thermal transmittance were derived from the Italian Ministerial Decree of 6 August 2020, *"Technical requirements for access to tax deductions for the energy retrofitting of buildings—so-called Ecobonus"* (Official Gazette No. 246, 5 October 2020), Annex E—Requirements for thermal insulation interventions, Table 1—Maximum allowable transmittance values for eligibility for tax incentives.

The periodic transmittance values, attenuation factor, time shift, and internal areal heat capacity obtained through the application of various thermal insulation solutions were compared with the limit values set by the Ministerial Decree on Minimum Requirements (26 June 2015) and the UNI EN ISO 13786:2008 standard [40].

2.1.1. Useful Energy Demand for Heating and Cooling

Heat Balance

The heat balance calculation algorithm implemented in the spreadsheet used in this study is more detailed than that prescribed by the UNI TS 11300 standards [41]. The development of a more refined method primarily arises from two limitations of the UNI TS 11300 standard:

- The inability to adequately assess intermittent use of the heating system;
- The assumption that the external temperature remains constant, equal to the 24 h average temperature.

The dynamic approach entails setting up an energy balance over a maximum time interval of one hour, which the procedure automatically reduces (multi-rate logic) based on the rate of temperature variation. This ensures a higher calculation density during periods of rapid change while optimizing computational efficiency during more stable periods.

The user can specify the maximum number of time steps per hour, which consequently defines the minimum thermal balance period. The energy balance accounts for the various thermal exchanges affecting the building, including heat transfer, ventilation, internal gains, and solar gains. These exchanges are classified as either "sensible" or "latent", as the method includes a detailed psychrometric analysis of air transformations.

For both winter and summer seasons, the heat balance equation takes the following form:

$$Q_{bil,sens} = Q_{tr+sol,o} + Q_{ve,sens} + Q_{\text{int},sens} + Q_{sol,t} + Q_{core} + Q_{restart} \quad (1)$$

$$Q_{bil,lat} = Q_{ve,lat} + Q_{\text{int},lat} \quad (2)$$

During the winter season, if the heat balance Equations (1) and (2) yield a positive result, it indicates that the incoming heat exceeds the outgoing heat, resulting in a null thermal load. Conversely, if the balance is negative, a thermal load must be met through the HVAC systems.

The procedure simulates each emission system as an air-based system, with appropriate values for recirculated and outdoor airflows. These values vary depending on the type of emission system, the temperature of the heat transfer fluid, the recirculation rate, the presence of heat recovery devices, the type of fans used, and other relevant parameters.

For each step, based on the installed emission system, the procedure determines the supply air state required to meet the thermal demands established by Equations (1) and (2). If the system cannot control relative humidity (e.g., radiators, fan coil units, radiant floor heating, etc.), the supply air state will compensate only for the sensible heat component. Otherwise, it will also account for humidity correction, either through humidification or dehumidification.

Once the supply air state is determined, the procedure constructs the actual psychrometric transformation performed by the HVAC systems, considering both ambient conditions and the specific characteristics of the installed equipment.

The thermal and cooling energy required for this transformation is then calculated, representing the net winter heating demand $Q_{H,nd}$ or summer cooling demand $Q_{C,nd}$. The psychrometric transformations performed follow a pattern similar to those illustrated in Figure 2.

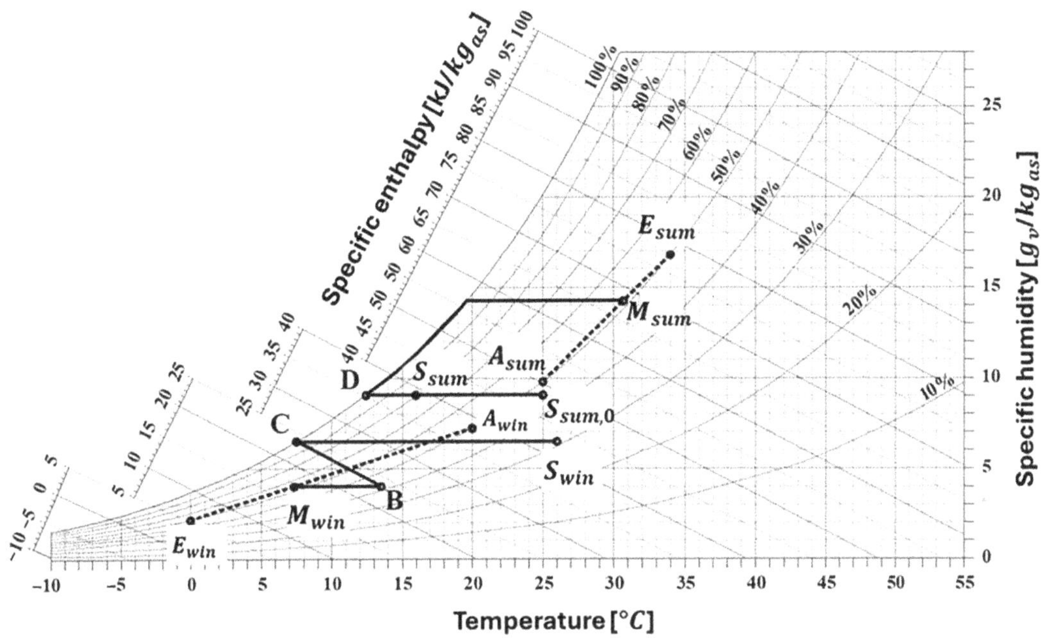

Figure 2. Example of Achievable Winter and Summer Psychrometric Transformations.

The psychrometric analysis integrated into the BEST tool is based on a detailed thermodynamic approach, grounded in the properties of moist air as defined in the ASHRAE technical handbooks (particularly the ASHRAE Handbook—Fundamentals). The psychrometric transformations are not resolved through approximate empirical correlations but through precise modeling that takes into account fundamental parameters such as dry bulb temperature, dew point temperature, relative humidity, absolute humidity, specific enthalpy, and specific volume.

In the thermal balance calculation, the procedure implements an hourly iterative resolution of the state point of the supply air, based on the sensible and latent loads to be compensated. It clearly distinguishes between systems capable of regulating humidity (Air Handling Units with air treatment) and those that operate solely on the sensible load.

For each time step (with a maximum interval of one hour and dynamic adjustments based on thermal variations), the software determines the required state point for the supply air, calculating the necessary psychrometric transformations based on the environmental conditions. These transformations include sensible cooling, dehumidification, heating, humidification, and their combinations.

The transformations are plotted in the psychrometric space, as shown in Figure 2 of the article, following the corresponding thermodynamic trajectories. The calculation of the thermal energy required for each transformation is performed by integrating the variations in specific enthalpy of moist air, also considering airflows and ventilation conditions (either natural or mechanical). This approach allows for accurate determination of the net thermal and cooling demands (*QH,nd* and *QC,nd*) necessary to meet the latent and sensible loads, explicitly distinguishing between internal contributions (occupants, equipment) and external contributions (ventilation, infiltration).

The approach described inherently and naturally accounts for hourly variations in environmental parameters, as well as the effects of the thermal inertia of building structures. Consequently, it does not require an approximation method for these effects, such as the utilization factor for heat gains.

Heat Exchange by Transmission

The heat exchange by transmission consists of the thermal transfer to the external air, the ground, and unconditioned spaces. It also includes the extra-flow caused by infrared radiation towards the sky dome, as well as the incoming solar thermal gains transmitted through opaque components.

The method is directly based on the numerical solution of the general heat exchange equation using finite difference schemes. The Fourier equation is presented as a second-order differential equation:

$$\frac{\partial^2\theta}{\partial x^2} = \frac{1}{\alpha}\frac{\partial\theta}{\partial t} \tag{3}$$

where:

$\alpha = \frac{\lambda}{\rho c_p}$ is the thermal diffusivity [m^2/s], which is the ratio between the material's thermal conductivity λ [W/mK] and the product of its density ρ [kg/m^3] and specific heat capacity c_p [J/kgK].

The analytical solution to the equation is complex and ill-suited for a computational procedure. For this reason, the dynamic heat exchange problem through a wall is typically solved using numerical methods. The finite difference method, sometimes referred to as the "Heat Balance" (HB) or "Conduction Finite Difference" (CondFD) method, is the most accurate approach among the existing solution methods. It does not present the issues of transfer function methods, such as the "Conduction Transfer Function Method" (CTF or TFM), nor the problems of impulse response methods like "Radiant Time Series" (RTS), which are prone to inaccuracies and instability when applied to components with high mass. The CondFD method is also the most flexible and can simulate special materials, such as phase change materials (PCMs), although these are not yet incorporated into the current version of the procedure. For the solution, an explicit forward finite difference scheme was chosen, which can be schematized as shown in Figure 3.

The spatial discretization of the wall was simplified by dividing it into three unequal parts. Two surface zones with very small thickness (approximately 3% each of the wall's thermal capacity), whose thermal capacity was aggregated with that of the internal and external spaces, and one internal zone containing the remainder of the thermal capacity. The assignment of 3% of the wall's thermal capacity to its surface zones represents a deliberate simplification aimed at reducing model complexity while preserving an acceptable level of accuracy. These surface zones correspond to the outermost layers of the wall material and possess a lower thermal mass compared to the deeper layers. As a result, although surface layers are more sensitive to temperature fluctuations and respond more rapidly to thermal variations, they store relatively little heat. In contrast, the inner regions of the wall have a greater capacity to absorb and release heat over time due to their higher thermal mass.

By allocating only a small fraction (3%) of the total thermal capacity to the surface zones, the model emphasizes the dominant role of the core wall layers in thermal dynamics, thus simplifying the computation without significantly compromising the physical realism of heat exchange processes.

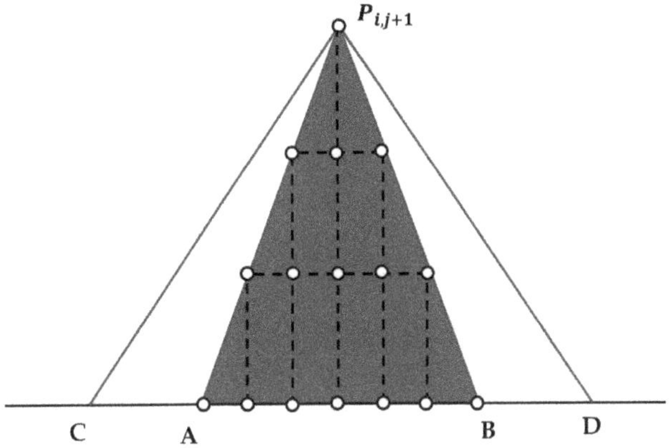

Figure 3. Explicit finite difference scheme used for solving the dynamic heat exchange of walls. On the x-axis, the thicknesses are represented, and on the y-axis, the times are plotted.

The surface zones were introduced to properly model the inertial effect of the wall's surface layers, which actively contribute to stabilizing the internal ambient temperature. Each wall has an internal temperature associated with the internal thermal capacity. This discretization can also be graphically represented as an equivalent electrical circuit (Figure 4).

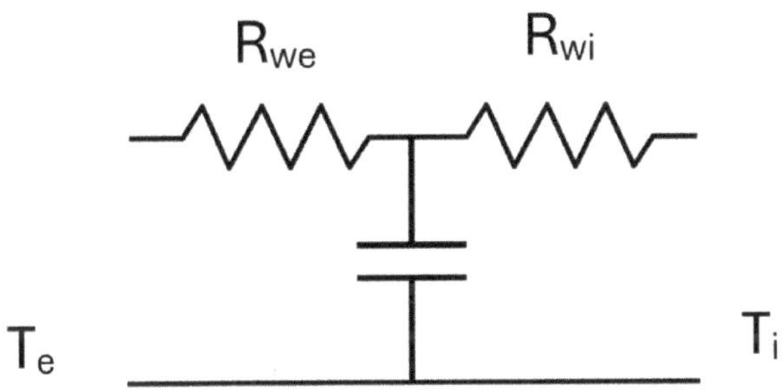

Figure 4. Equivalent electrical circuit for thermal modeling of the wall.

In this diagram: R_{we} and R_{wi} represent the thermal resistances [W/m^{2}K] of the outer and inner portions of the wall, respectively; Ct_w represents the thermal capacity of the inner part of the wall [J/kegs]; T_e and T_i represent the temperatures of the external and internal environments [°C], respectively.

This representation helps simplify the complex thermal behavior of the wall by modeling it with thermal resistances and capacities, similar to an electrical circuit with resistances and capacitors.

Heat Exchange by Ventilation

The heat exchange by ventilation is modeled on a monthly basis and consists of thermal exchange with the external air and unconditioned spaces. The calculation methodology involves precise computation of psychrometric transformations and the corresponding energy. For natural or mechanical ventilation without treatment, the heat exchange is calculated at each time step based on the actual temperatures and flow rates.

Internal Gains

The detailed approach involves the precise calculation of the individual sensible and latent heat gains from both occupants and all electrical and combustion appliances present. Internal gains are incorporated instant by instant into the building's dynamic energy balance as thermal flows directly injected into the internal air.

Solar Gains from Transparent Components

Solar gains are typically divided based on the component that the irradiation impacts. If the irradiation affects opaque components, such as walls or ceilings, part of it is absorbed by the wall and becomes part of the thermal balance of the wall, in the form of thermal flux injected at the surface of the wall.

Therefore, opaque irradiation enters the interior through conductive transfer across the opaque walls. If the irradiation affects transparent components, such as windows, most of it is transmitted inside and absorbed by the internal walls and furnishings. This second mode is clearly more direct, produces higher loads, and has very short propagation times.

The detailed approach calculates the solar fluxes $\Phi_{sol,op,mn,k}$ and $\Phi_{sol,w,k}$ in a manner similar to the regulatory calculation, with some differences (mainly for opaque components), and with the distinction that they are calculated on an hourly basis instead of monthly.

The solar gain from opaque components in the detailed method cannot be directly expressed with a formula, as it becomes part of the dynamic thermal balance of the opaque wall in the form of thermal flux injected into the wall's surface layer, between the external surface resistance and the resistance of the outer portion of the wall.

The solar gain through transparent components is calculated in a way similar to the standard method, as it is not subject to accumulation by the structures (internal accumulation is neglected).

Energy Demand for Domestic Hot Water Preparation

The energy demand related to domestic hot water consumption is calculated in full compliance with the UNI TS 11300-2 standard [42].

Solar Thermal and Photovoltaic Systems

The behavior of solar thermal or photovoltaic systems is simulated through a procedure largely derived from the UNI TS 11300-4 technical standard [43].

2.1.2. Results Generated by the Tool

For each intervention (or combination of multiple interventions), the following information can be displayed:

- The percentage reduction in total primary energy consumption compared to the baseline scenario (ex-ante);
- The initial investment required for the implementation of the interventions;
- The NPV of the investment over a 20-year period, compared to the current state. This represents the economic value generated by the investment over 20 years, net of the initial expenditure and taking into account the effects of inflation;
- The Payback Time (PBT) of the investment, compared to the current state;
- The amount of CO_2 emissions avoided annually (kg/m^2 year) with respect to the baseline scenario, as a result of the proposed solutions.

3. Case Study

To evaluate the operational capabilities of the proposed tool, a simulation was conducted on a selected case study, the details of which are provided in Supplementary File S2.

The validation of the BEST tool was performed using a single-storey residential building with a net usable floor area of 61.23 m^2, located in Trevignano Romano, in the province of Rome. Constructed in 1950, the building falls within Climate Zone D (Figure 5).

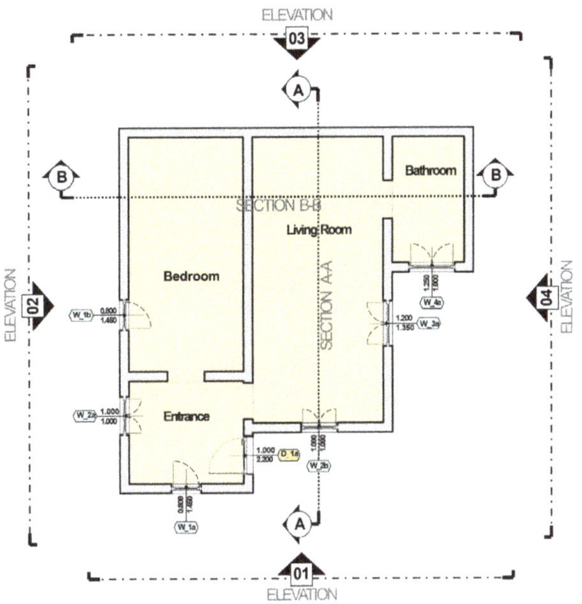

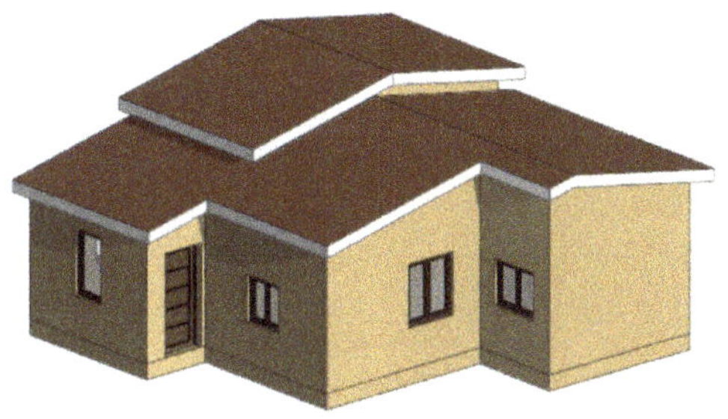

Detailed building data	
Gross built volume m^3	314.23
Gross dispersing surface area m^2	303.33
Surface-to-volume ratio (S/V ratio) 1/m	0.97
Net volume m^3	226.28
Net usable floor area m^2	61.23

Figure 5. Floor plan of the case study, three-dimensional model, and detailed data.

The vertical external wall (EW) of the dwelling consists of a functional layer composed of 21 cm of solid brick masonry, flanked on both sides by 2.5 cm of plaster, resulting in a total thickness of 25 cm. The lower horizontal closure (LHC) is characterized by a functional layer composed of a 25 cm reinforced concrete slab, a 6 cm concrete screed, a 1 cm cement mortar bedding layer, and a 2 cm ceramic tile floor finish, for a total thickness of 34 cm. The upper horizontal closure consists of a functional stratigraphy composed of a 16 cm hollow-block and concrete slab, a 4 cm layer of reinforced concrete, a 1 cm cement mortar bedding layer, and a final layer of terracotta tiles, with a total thickness of 25.5 cm. The window frames are made of wood and equipped with single glazing. In the first section of the tool, the province and municipality in which the building is located can be selected via a dropdown menu. This step is essential for the automatic retrieval of degree days and the corresponding climate zone, which are crucial for the subsequent identification of potential energy efficiency solutions. After identifying the construction period of the building, the tool suggests the most likely technical elements defining the building envelope, based on the national residential building stock matrix.

For the case study under analysis, the technical components most representative of the existing envelope (Figure 6) are as follows:

- EW05: uninsulated solid brick masonry, typical of the period 1900–1950, with a steady-state thermal transmittance (U-value) of 1.95 W/m^{2}K;
- LHC07: concrete ground floor slab, typical until 1975, with a steady-state thermal transmittance of 1.96 W/m^{2}K;

- UHC05: pitched roof in hollow-block concrete, typical from 1930 to 1975, with a steady-state thermal transmittance of 1.82 W/m^{2}K.

NATIONAL RESIDENTIAL BUILDING STOCK MATRIX

ID	Construction Era Class								Building Type: Isolated Buildings		Building Type: Multi-family Buildings					
	1. Up to 1900	2. From 1901 to 1920	3. From 1921 to 1945	4. From 1946 to 1960	5. From 1961 to 1975	6. From 1976 to 1990	7. From 1991 to 2005	After 2005	Single-family	Two-family	In line	Tower	Row	With balcony	Gallery	Block
111			x						x		x		x	x	x	x

ID	Construction Type: Load-bearing Masonry				Construction Type: Frame		Building Envelope Characteristics		
	Stone	Stone and Brick	Brick	Contrete	Wood	Reinforced Contrete	Upper Horizontal Closures	Vertical Closures	Lower Horizontal Closures
111			x				UHC.05 - Pitched roof in clay tiles without insulation	EW.05 - Solid brick masonry (25 cm)	LHC.07 - Concrete slab on ground

ID	Characteristics of Ante Operam Upper Horizontal Closures: Upper Horizontal Closures -Ante Operam		Morphological		Geometric		Accessibility					Insulation Solutions
	Thickness (cm)	Thermal Transmittance U = W/m²K	Continuous	Discontinuous	Horizontal plane 1-5%	Inclined plane >5%	Class A	Class B	Class C	Class D	Class E	SOLUTION SHEET LINK
111	25.5	1.82		x		x	x	x				SS_05

Figure 6. Extract from the National Residential Building Stock Matrix, highlighting the most representative technical components of the existing envelope for the case study.

The simulation of the building's current condition performed using the tool produced results related to thermal and energy performance indicators, annual CO_2 emissions, and energy-related operating costs (Figure 7).

As is condition

Code	0
Name	Baseline Scenario
Description	Pre-intervention condition of the building
Global energy performance index [kWh/m²·year]	485.12
Useful thermal performance index for heating [kWh/m²·year]	328.40
Useful thermal performance index for cooling [kWh/m²·year]	5.75
Energy performance index for heating [kWh/m²·year]	461.88
Energy performance index for cooling [kWh/m²·year]	0.00
Energy performance index for domestic hot water (DHW) [kWh/m²·year]	23.23
Energy performance index for ventilation [kWh/m²·year]	0.00
Energy performance index for lighting [kWh/m²·year]	0.00
Energy performance index for transport systems [kWh/m²·year]	0.00
CO emissions [kg/m²·year]	90.75
Annual gas expenditure [€/year]	3053.00
Annual electricity expenditure [€/year]	0.00

Figure 7. Extract from the BEST tool with the as-is condition of the case study building.

4. Results

The matrix provided by the tool identifies nine insulation solutions for the roof: three non-ventilated external insulation systems (applied to the upper surface), four ventilated roof solutions, and two internal insulation options (applied to the underside).

For the vertical envelope, six solutions are proposed: two external thermal insulation composite systems, two ventilated façade types, and two internal insulation options. Given that the building is not subject to any heritage or urban planning constraints, all insulation solutions can be applied externally.

The solutions selected for the case study correspond to those with the simplest installation processes that also result in the lowest steady-state thermal transmittance values for both the roof and vertical envelope.

The tool allows users to download technical datasheets (Supplementary File S3), which include:

- The code of the roof and vertical wall typologies, the historical period in which they were most prevalent, and the stratigraphy to which the proposed solution is applicable;
- The code of the selected insulation solution;
- The type, thickness, and thermal conductivity of the proposed insulating material;
- A drawing of the insulation stratigraphy, along with a description of its individual layers;
- Thermo-physical properties of the existing building components (walls and roof) and the proposed insulation layers;
- Calculated performance values under both winter and summer conditions;
- Surface condensation risk assessment (isotherm analysis);
- Interstitial condensation risk assessment (Glaser diagram—critical month);
- A radar chart comparing key performance parameters, including steady-state transmittance (U-value), time lag (Φ), internal surface periodic thermal capacity (Cip), dynamic transmittance (Yie), and insulation thickness;
- Verification of compliance with regulatory limits for both heating and cooling seasons.

For the case study, three different retrofit interventions aimed at improving the thermal performance of the opaque envelope were selected: INT.1, INT.2, and INT.3, the latter being a combination of the first two.

- INT.1: The first energy retrofit intervention involves external insulation of the roof using solution IE.01.b, which consists of 10 cm thick rigid polyurethane foam panels with a thermal conductivity of 0.023 W/mK. This measure reduces the steady-state U-value to 0.20 W/m^{2}K, complying with the maximum transmittance limits set out in Annex E—Thermal Insulation Requirements, which refers to the UNI EN ISO 6946 standard [44] for calculating the thermal transmittance of opaque structures (excluding thermal bridges).
- As shown by the results, this intervention requires an investment of €14,148.40, achieves a CO_2 emissions reduction of 21.18 kg/m^2·year, a yearly energy cost saving of €713.00, and yields a positive 20-year NPV of €3546.00, corresponding to a Payback Time (PBT) of 19 years and 10 months. The PBT does not account for interest, inflation, or variations in gas prices.
- INT.2: The second intervention targets thermal insulation of the vertical envelope using an ETICS system identified as SC.01, consisting of 12 cm thick rigid hydrophobic glass wool panels with low vapor permeability and a thermal conductivity of 0.032 W/mK. This reduces the steady-state U-value to 0.233 W/m^{2}K, also in compliance with Annex E.
- According to the results, this intervention requires an investment of €16,395.32, achieves a CO_2 emissions reduction of 35.55 kg/m^2·year, a yearly saving of €1193.00,

and yields a positive 20-year NPV of €1343.54, corresponding to a PBT of 13 years and 9 months.

- INT.3: The third intervention combines both previous strategies (INT.1 + INT.2). This combined solution, with a total investment of €30,543.72, achieves a CO_2 emissions reduction of 56.04 kg/m^2·year, a yearly cost saving of €1883.00, but results in a negative 20-year NPV of −€2525.73, indicating a PBT exceeding 20 years. This means that under a 3% discount rate, the investment cannot be fully recovered within the 20-year analysis period.

Figure 8 shows an excerpt of the simulation results generated by the BEST tool.

Evaluated interventions			
Code	INT.01	INT.02	INT.03
Name	Roof Thermal Insulation	Wall Thermal Insulation	Combined Wall and Roof Thermal Insulation
Description	External roof insulation with rigid polyurethane foam panels (IE.01.b)	External thermal insulation system with glass wool panels (EI.01)	Combination of wall insulation with glass wool panels (EI.01) and roof insulation with rigid polyurethane foam panels (IE.01.b)
Global energy performance index [kWh/m^2·year]	371.87	295.09	185.53
Useful thermal performance index for heating [kWh/m^2·year]	244.57	185.60	103.34
Useful thermal performance index for cooling [kWh/m^2·year]	7.44	11.71	17.11
Energy performance index for heating [kWh/m^2·year]	348.64	271.64	162.08
Energy performance index for cooling [kWh/m^2·year]	0.00	0.00	0.00
Energy performance index for domestic hot water (DHW) [kWh/m^2·year]	23.23	23.45	23.45
Energy performance index for ventilation [kWh/m^2·year]	0.00	0.00	0.00
Energy performance index for lighting [kWh/m^2·year]	0.00	0.00	0.00
Energy performance index for transport systems [kWh/m^2·year]	0.00	0.00	0.00
CO emissions [kg/m^2·year]	69.57	55.20	34.71
Annual gas expenditure [€/year]	2,340.00	1,860.00	1,170.00
Annual electricity expenditure [€/year]	0.00	0.00	0.00
Investment [€]	14,148.40	16,395.32	30,543.72
NPV (20 years) [€]	3,546.00	1,343.54	-2,525.73
PTB [years]	19.83 years	13.75 years	> 20 years
CO emissions reduction [kg/m^2·year]	21.18	35.55	56.04
Consumption reduction [%]	23.35	39.17	61.76

Figure 8. Extract from the BEST tool with summarized data for the three different intervention solutions related to the thermal efficiency of the opaque envelope.

5. Discussion

5.1. Strengths and Practical Applications

The result described enables the achievement of significant benefits for the national electrical system and its users, such as:

- Reduction in energy consumption: The support provided for optimizing energy efficiency interventions in national residential buildings leads to a decrease in energy demand at the national level, contributing to the overall improvement of the electrical system's efficiency.
- Improvement of system stability: The widespread implementation of energy efficiency interventions positively impacts the stability of the national electrical system by mitigating energy demand peaks and promoting a more rational and sustainable use of available resources.
- Economic benefits for users: The cost-benefit analysis provided by the tool ensures the identification of targeted and optimized solutions, capable of maximizing the cost-benefit ratio. This leads to a significant reduction in energy costs for citizens and families, improving the economic sustainability of the interventions.
- Promotion of environmental sustainability: The large-scale implementation of energy efficiency interventions contributes to reducing CO_2 emissions, supporting the achievement of sustainability goals set at both European and national levels. This approach produces positive effects on both the environment and the quality of life within communities.
- Support for the spread of innovative technologies: The tool promotes prefabricated and optimized technological solutions, encouraging the growth of the advanced and sustainable technology market. This dynamic stimulates innovation in the construction sector and strengthens the competitiveness of the businesses involved.
- Strategic planning tool for institutions: The tool enables the creation of priority rankings for interventions, providing a valuable resource for energy and urban planning at the territorial level. This allows institutions to plan targeted actions, ensuring efficient resource allocation and a positive impact on the management of building stock.

To ensure broad accessibility and ease of use, a structured web-based application was developed.

The interface is organized into clearly defined sections that, in a sequential manner, describe the building–plant system based on its key characteristics. Particular emphasis was placed on minimizing the need for technical input from users. Where input is required, data entry is facilitated through guided, multiple-choice menus designed to streamline the process and support non-expert users.

5.2. Limitations and Challenges

At present, the developed tool is exclusively applicable to the residential sector. It was designed to streamline user input by accepting only data that meet predefined specifications, with constraints in place to prevent the entry of values outside the acceptable ranges established by the system.

In many instances, data input is limited to selections from predefined options available through dropdown menus. Consequently, expert users and technical professionals may experience limitations due to these predefined choices.

Regarding building systems, the tool's structure is inherently rigid, as it does not support the integration of systems beyond those currently embedded in its internal database.

The estimation of intervention costs is based on targeted evaluations derived from regional price lists or analyses of market trends, leading to the identification of characteristic parametric costs depending on the type of intervention (€/kW, €/m^2, €/m^3). These costs are inclusive of VAT and cover materials, components, and installation. These costs require annual updating.

Currently, the tool does not allow users to simulate interventions involving the replacement of the existing heating and cooling systems; instead, it focuses exclusively on measures related to the building's opaque envelope.

6. Conclusions and Future Developments

The current condition of the European residential building stock is marked by widespread obsolescence, inadequate thermal insulation, and high energy consumption, making it one of the primary contributors to greenhouse gas emissions in the built environment [5]. Energy retrofitting buildings within the European Union would result in substantial energy savings and a 26% reduction in energy consumption [45].

In Italy, the initial regulation aimed at reducing energy consumption in buildings was introduced in 1976 (Law 373/1976) [46]; however, it was characterized by limited restrictions and was frequently neglected due to insufficient enforcement mechanisms. It was not until 1991 (Law 10/1991) that the first comprehensive code targeting thermal performance standards was established, by which time more than 80% of the existing residential building stock was already constructed [47]. As a result, a substantial portion of Italy's residential properties continues to exhibit high energy consumption levels [20]. Therefore, a significant portion of the residential building stock is in urgent need of deep renovation to meet current energy performance standards and contribute to the EU's decarbonization goals. Renovation efforts must prioritize thermal insulation, modernization of HVAC systems, and the integration of renewable energy sources in order to reduce both energy demand and greenhouse gas emissions [48]. These buildings are often characterized by outdated construction techniques, inefficient HVAC systems, and poorly performing building envelopes. This context highlights the pressing need for scalable, accessible, and standardized retrofit strategies capable of supporting the transition toward nearly zero-energy buildings (nZEBs) and, in the longer term, positive energy buildings.

In response to this challenge, the BEST was developed as a decision-support instrument designed to simplify the preliminary energy analysis and the selection of retrofit solutions for residential buildings.

Through the classification of typical building envelope components and the integration of optimized, pre-calculated interventions, the tool offers practical guidance to designers and stakeholders, significantly reducing the complexity usually associated with energy performance assessments. Its application to a representative case study from the Italian residential context demonstrated the tool's effectiveness in identifying sustainable and cost-efficient interventions, with measurable benefits in terms of energy savings, CO_2 emissions reduction, and economic return. Future developments aim to extend the application of the tool to other building typologies, thereby broadening its scope and enhancing its contribution to the sustainability of the built environment as a whole.

The potential of the tool extends beyond individual building analysis. Upcoming developments of BEST will include the integration of geo-referencing functionalities, enabling the spatial mapping of energy performance levels across the national building stock. The integration with GIS platforms will allow the identification of territorial clusters with poor energy performance, supporting data-driven decision-making and the strategic allocation of public resources. This approach may serve as an effective instrument for urban and energy planning at the municipal or regional scale.

Moreover, the scope of the tool will be expanded to simulate interventions on building systems, particularly the replacement or upgrading of existing HVAC systems. Given that these systems account for a substantial portion of a building's overall energy consumption, their inclusion will enable a more holistic assessment of retrofit strategies. This evolution will enhance the tool's capacity to estimate comprehensive energy savings and provide a more robust basis for investment decisions.

To ensure long-term applicability, future versions of the tool will incorporate automatic update mechanisms for energy prices, material costs, and incentive schemes, thereby ensuring the reliability of economic assessments and alignment with evolving regulations and market conditions.

The BEST tool represents a significant step toward operationalizing the objectives of the European Green Deal in the residential sector. Its adaptability, user-friendliness, and analytical capacity make it a valuable resource for both professionals and policymakers, facilitating a broad transition toward energy-efficient and low-carbon buildings.

The integration of geo-referenced data and the inclusion of building system interventions will further enhance its impact, enabling comprehensive assessments at both the building and territorial levels.

Supplementary Materials: The following supporting information can be downloaded at: https://www.mdpi.com/article/10.3390/app15126817/s1, Supplement S1: National residential building stock matrix (Excel file); Supplement S2: Operational manual of the BEST Tool (PDF); Supplement S3: Technical sheet of Thermal Insulation Solutions (PDF).

Author Contributions: Conceptualization, M.C., F.C., E.P., C.R. and C.Z.; methodology, E.P. and C.Z.; software, M.C.; validation, E.P., C.R. and C.Z.; formal analysis, E.P. and C.Z.; investigation, E.P. and C.Z.; resources, E.P. and C.Z.; data curation, E.P., C.R. and C.Z.; writing—original draft preparation, E.P. and C.Z.; writing—review and editing, E.P., C.R. and C.Z.; visualization, F.C., E.P., C.R. and C.Z.; supervision, C.R.; project administration, F.C. and E.P.; funding acquisition, F.C. and E.P. All authors have read and agreed to the published version of the manuscript.

Funding: This research was funded by the Ministry of Economic Development, in the program for "Research of the Electricity System" in cooperation with ENEA on the project "High-Efficiency Buildings for the Energy Transition", 2022–2024, Director's decree of 15 September 2022.

Institutional Review Board Statement: Not applicable.

Informed Consent Statement: Not applicable.

Data Availability Statement: The data presented in this study are openly available in https://www.ricercasistemaelettrico.enea.it/archivio-documenti.html?task=download.send&id=5226:studio-stato-dellarte-di-soluzioni-prefabbricate-per-sistemi-di-copertura-e-creazione-catalogo-best-practices&catid=597 (accessed on 6 February 2025).

Conflicts of Interest: The authors declare no conflicts of interest.

References

1. Alhazmi, H.; Alduwais, A.K.; Tabbakh, T.; Aljamlani, S.; Alkahlan, B.; Kurdi, A. Environmental Performance of Residential Buildings: A Life Cycle Assessment Study in Saudi Arabia. *Sustainability* **2021**, *13*, 3542. [CrossRef]
2. Hong, J.; Shen, G.Q.; Guo, S.; Xue, F.; Zheng, W. Energy use embodied in China's construction industry: A multi-regionalinput-output analysis. *Renew. Sustain. Energy Rev.* **2016**, *53*, 1303–1312. [CrossRef]
3. Wang, T.; Seo, S.; Liao, P.C.; Fang, D. GHG emission reduction performance of state of-the-art green buildings: Review of two case studies. *Renew. Sustain. Energy Rev.* **2016**, *56*, 484–493. [CrossRef]
4. Shad, R.; Khorrami, M.; Ghaemi, M. Developing an Iranian green building assessment tool using decision making methods and geographical information system: Case study in Mashhad city. *Renew. Sustain. Energy Rev.* **2017**, *67*, 324–340. [CrossRef]
5. Economidou, M.; Todeschi, V.; Bertoldi, P.; D'Agostino, D.; Zangheri, P.; Castellazzi, L. Review of 50 years of EU energy efficiency policies for buildings. *Energy Build.* **2020**, *225*, 110322. [CrossRef]
6. European Commission. EU Energy in Figures. Statistical Pocketbook 2018. Available online: https://euneighbourseast.eu/wp-content/uploads/2021/07/mjab18001enn.en_.pdf (accessed on 28 May 2025).

7. European Commission. European Green Deal. Available online: https://commission.europa.eu/strategy-and-policy/priorities-2019-2024/european-green-deal_it (accessed on 6 February 2025).
8. European Parlament. Ridurre le Emissioni di Anidride Carbonica: Obiettivi e Politiche dell'UE. Available online: https://www.europarl.europa.eu/pdfs/news/expert/2018/3/story/20180305STO99003/20180305STO99003_it.pdf (accessed on 6 February 2025).
9. Tucci, F.; Amadei, F.; Pani, M.M.; Romano, G. *Verso la Neutralità Climatica di Architetture e Città Green: Sperimentazioni e Casi di Studio nel Nord e Mittel Europa*, 1st ed.; FancoAngeli: Milano, Italy, 2023; pp. 9–34.
10. Fiore, P.; Donnarumma, G.; Falce, C. A Tool for the Evaluation of Energy-Environmental Retrofit Interventions on Opaque Walls Using ETICS. *Sustainability* **2021**, *13*, 4063. [CrossRef]
11. Senate of the Republic. Attuazione Della Direttiva (UE) 2018/844 Concernente la Prestazione Energetica Nell'edilizia e L'efficienza Energetica Atto del Governo 158. Available online: https://www.senato.it/service/PDF/PDFServer/BGT/01143499.pdf (accessed on 6 February 2025).
12. Magrini, A.; Lentini, G.; Cuman, S.; Bodrato, A.; Marenco, L. From nearly zero energy buildings (NZEB) to positive energy buildings (PEB): The next challenge—The most recent European trends with some notes on the energy analysis of a forerunner PEB example. *Dev. Built Environ.* **2020**, *3*, 100019. [CrossRef]
13. Poli, I.; Ravagnan, C.; Ricci, L. A Planning Framework for Urban Resilience toward Climate Adaptation and Mitigation: Potentials and Limits of "Eco-Districts". *Urban Sci.* **2022**, *6*, 49. [CrossRef]
14. Nematchoua, M.K.; Sadeghi, M.; Reiter, S. Strategies and scenarios to reduce energy consumption and CO_2 emission in the urban, rural and sustainable neighborhoods. *Sustain. Cities Soc.* **2021**, *72*, 103053. [CrossRef]
15. Hafez, F.S.; Sa'di, B.; Safa-Gamal, M.; Taufiq-Yap, Y.H.; Alrifaey, M.; Seyedmahmoudian, M.; Stojcevski, A.; Horan, B.; Mekhilef, S. Energy Efficiency in Sustainable Buildings: A Systematic Review with Taxonomy, Challenges, Motivations, Methodological Aspects, Recommendations, and Pathways for Future Research. *Energy Strategy Rev.* **2023**, *45*, 101013. [CrossRef]
16. Sala Lizarraga, J.M.P.; Picallo-Perez, A. 12—Design and optimization of the envelope and thermal installations of buildings. In *Exergy Analysis and Thermoeconomics of Buildings. Design and Analysis for Sustainable Energy Systems*; Butterworth-Heinemann: Oxford, UK, 2020; pp. 911–1005.
17. Poel, B.; Cruchten, G.; Balaras, C.A. Energy performance assessment of existing dwellings. *Energy Build.* **2007**, *39*, 393–403. [CrossRef]
18. Paraschiv Lizica, S.; Paraschiv, I.S.; Ion, V.I. Increasing the energy efficiency of buildings by thermal insulation. *Energy Procedia* **2017**, *128*, 393–399.
19. Santos, P.; Abrantes, D.; Lopes, P.; Mateus, D. Experimental and Numerical Performance Evaluation of Bio-Based and Recycled Thermal Break Strips in LSF Partition Walls. *Buildings* **2022**, *12*, 1237. [CrossRef]
20. Ali, A.; Issa, A.; Elshaer, A. A Comprehensive Review and Recent Trends in Thermal Insulation Materials for Energy Conservation in Buildings. *Sustainability* **2024**, *16*, 8782. [CrossRef]
21. Jezierski, W.; Sadowska, B.; Pawłowski, K. Impact of Changes in the Required Thermal Insulation of Building Envelope on Energy Demand, Heating Costs, Emissions, and Temperature in Buildings. *Energies* **2021**, *14*, 56. [CrossRef]
22. Bazzocchi, F.; Ciacci, C.; Di Naso, V. Evaluation of Environmental and Economic Sustainability for the Building Envelope of Low-Carbon Schools. *Sustainability* **2021**, *13*, 1702. [CrossRef]
23. Amani, N. Energy efficiency of residential buildings using thermal insulation of external walls and roof based on simulation analysis. *Energy Storage Sav.* **2025**, *4*, 48–55. [CrossRef]
24. Imane, S.Z.; Benoudjafer, I.; Benoudjafer, I. Towards the use of natural thermal insulation on the exterior of vertical building walls in desert regions: A case study in the city of Bechar, Algeria. *J. Therm. Eng.* **2024**, *10*, 308–320.
25. Häkkinen, T. Systematic method for the sustainability analysis of refurbishment concepts of exterior walls. *Constr. Build. Mater.* **2012**, *37*, 783–790. [CrossRef]
26. Sdei, A.; Tittelein, P.; Lassue, S.; McEvoy, M.E. Dynamic Thermal Modeling of Retrofitted Social Housing in England and France. In Proceedings of the CLIMA 2013: 11th REHVA World Congress and the 8th International Conference on Indoor Air Quality, Ventilation and Energy Conservation in Buildings, Prague, Czech Republic, 16–19 June 2013; Karel Kabele, M.U., Suchý, K., Lain, M., Eds.; Society of Environmental Engineering (STP): Prague, Czech Republic, 2013; p. 6882.
27. Ballarini, I.; Corgnati, S.P.; Corrado, V. Use of reference buildings to assess the energy saving potentials of the residential building stock: The experience of TABULA project. *Energy Policy* **2014**, *68*, 273–284. [CrossRef]
28. Flourentzou, F.; Genre, J.-L.; Roulet, C.-A. Epiqr-tobus: A new generation of refurbishment decision aid methods. In *Towards Sustainable Building*; Maiellaro, N., Ed.; Springer Science + Business Media: Dordrecht, The Netherlands, 2001; Volume 61, pp. 161–169.
29. Konstantinou, T. *Facade Refurbishment Toolbox: Supporting the Design of Residential Energy Upgrades*, 1st ed.; A + BE | Architecture and the Built Environment: Delft, The Netherlands, 2014; pp. 1–420.

30. Jradi, M. A Decision-Making Tool for Sustainable Energy Planning and Retrofitting in Danish Communities and Districts. *Energies* **2025**, *18*, 692. [CrossRef]
31. Garufi, D.; de Vries, B. Decision support tool for sustainable renovation projects in the dutch housing corporations. In Proceedings of the CISBAT 2015, International Conference, Lausanne, Switzerland, 9–11 September 2015; Future Buildings & Districts—Sustainability from Nano to Urban Scale; Centre de Recherches en Physique des Plasmas: Lausanne, Switzerland, 2015; pp. 987–992.
32. Dall'O', G.; Galante, A.; Torri, M. A methodology for the energy performance classification of residential building stock on an urban scale. *Energy Build.* **2012**, *48*, 211–219. [CrossRef]
33. Deng, Z.; Chen, Y.; Yang, J.; Causone, F. AutoBPS: A tool for urban building energy modeling to support energy efficiency improvement at city-scale. *Energy Build.* **2023**, *282*, 112794. [CrossRef]
34. Godoy-Rangel, C.; Gijon-Rivera, M.; Belmont, Z.V.C.; Mousavi, N.; Rivera-Solorio, C.I.; Mousavi, S. An energy efficiency tool to evaluate energy consumption and emissions in buildings in northern Mexico. In Proceedings of the XXVIII Congreso Internacional Anual de la SOMIM, Bogotà, Colombia, 21–23 September 2022.
35. Godoy-Rangel, C.; Rivera-Solorio, C.I.; Gijón-Rivera, M. TecSB: An Open Web Tool of Energy Efficiency and Solar Energy Integration in Mexican Buildings. *Sustainability* **2023**, *15*, 3630. [CrossRef]
36. Arab, Y.; Hassan, A.S.; Al-Absi, Z.A.; Achour, H.; Witchayangkoon, B.; Qanaa, B. Retrofitting of a High-Rise Residential Building for Energy Efficiency with OTTV as an Assessment Tool. *J. Adv. Res. Fluid Mech. Therm. Sci.* **2023**, *102*, 110–119. [CrossRef]
37. Pennacchia, E.; Romeo, C.; Zylka, C. Towards High-Efficiency Buildings for Sustainable Energy Transition: Standardized Prefabricated Solutions for Roof Retrofitting. *Sustainability* **2024**, *16*, 3850. [CrossRef]
38. *UNI 10349-1:2016*; Heating and Cooling of Buildings—Climatic Data—Part 1: Monthly Means for Evaluation of Energy Need for Space Heating and Cooling and Methods for Splitting Global Solar Irradiance into the Direct and Diffuse Parts and for Calculate the Solar Irradiance on Tilted Planes. UNI: Milan, Italy, 2016.
39. *UNI 10339:1995*; Impianti Aeraulici al Fini di Benessere. Generalità, Classificazione e Requisiti. Regole per la Richiesta D'offerta, L'offerta, L'ordine e la Fornitura. UNI: Milan, Italy, 1995.
40. *UNI EN ISO 13786:2008*; Prestazione Termica dei Componenti per Edilizia—Caratteristiche Termiche Dinamiche—Metodi di Calcolo. UNI: Milan, Italy, 2008.
41. *UNI/TS 11300-1*; Energy Performance of Buildings—Part 1: Evaluation of Energy Need for Space Heating and Cooling. Ente Nazionale Italiano di Unificazione: Milan, Italy, 2014.
42. *UNI/TS 11300-2*; Energy Performance of Buildings—Part 2: Determination of Primary Energy Demand and Efficiencies for Space Heating, Domestic Hot Water Production, Ventilation, and Lighting. Ente Nazionale Italiano di Unificazione: Milan, Italy, 2019.
43. *UNI/TS 11300-4*; Energy Performance of Buildings—Part 4: Use of Renewable Energy Sources and Other Generation Methods for Space Heating and Domestic Hot Water Production. Ente Nazionale Italiano di Unificazione: Milan, Italy, 2016.
44. *ISO 6946:2017*; Building Components and Building Elements—Thermal Resistance and Thermal Transmittance—Calculation Methods. ISO: Geneva, Switzerland, 2017.
45. Biere-Arenas, R.; Spairani-Berrio, S.; Spairani-Berrio, Y.; Marmolejo-Duarte, C. One-Stop-Shops for Energy Renovation of Dwellings in Europe—Approach to the Factors That Determine Success and Future Lines of Action. *Sustainability* **2021**, *13*, 12729. [CrossRef]
46. Ministry for Ecological Transition. Strategy for Energy Retrofitting of National Building Stock. Available online: https://energy.ec.europa.eu/system/files/2021-12/2020_ltrs_italy_-_en.pdf (accessed on 28 May 2025).
47. Georgescu, E.-S.; Georgescu, M.S.; Macri, Z.; Marino, E.M.; Margani, G.; Meita, V.; Pana, R.; Cascone, S.M.; Petran, H.; Rossi, P.P.; et al. Seismic and Energy Renovation: A Review of the Code Requirements and Solutions in Italy and Romania. *Sustainability* **2018**, *10*, 1561. [CrossRef]
48. Babiarz, B.; Krawczyk, D.A.; Siuta-Olcha, A.; Manuel, C.D.; Jaworski, A.; Barnat, E.; Cholewa, T.; Sadowska, B.; Bocian, M.; Gnieciak, M.; et al. Energy Efficiency in Buildings: Toward Climate Neutrality. *Energies* **2024**, *17*, 4680. [CrossRef]

applied sciences

MDPI

Article

Characterization and Energy Performance of Rice Husk Fiber Insulation Applied by the Blowing Technique in an Industrialized Modular Housing System

Karin Rodríguez Neira [1,2,*], Carlos Javier Rojas-Herrera [3,4], Juan Pablo Cárdenas-Ramírez [3], Joaquín Torres Ramo [1] and Ana Sánchez-Ostiz [1]

[1] School of Architecture, University of Navarra, 31009 Pamplona, Spain; jtorram@unav.es (J.T.R.); aostiz@unav.es (A.S.-O.)

[2] Departamento de Arquitectura—Facultad de Arquitectura y Urbanismo, Universidad Nacional de San Agustín de Arequipa, Arequipa 04001, Peru

[3] Facultad de Arquitectura, Construcción y Medio Ambiente, Universidad Autónoma de Chile, Temuco 4810101, Chile; carlos.rojas@uautonoma.cl (C.J.R.-H.); juanpablo.cardenas@uautonoma.cl (J.P.C.-R.)

[4] Departamento de Ingeniería de Obras Civiles, Universidad de La Frontera, Temuco 4811230, Chile

* Correspondence: krodriguez.2@alumni.unav.es; Tel.: +51-958910255

Abstract: The construction sector plays a key role in climate change due to its high energy consumption and greenhouse gas emissions. Developing environmentally friendly building materials with low environmental impact is essential to improving energy efficiency. Insulation derived from agricultural waste is particularly promising due to its low ecological footprint, responsible resources use, and potential for integration into various construction systems. This study evaluates the potential of rice husk fiber as a thermal insulating material applied through the blowing technique in the Skylark 250 modular system. Rice husk fiber was morphologically and thermally characterized using scanning electron microscopy (SEM), while its thermal behavior was analyzed by thermogravimetric analysis (TGA) alongside a fire behavior assessment. Additionally, energy simulations were conducted to compare the thermal performance of rice husk fiber with other insulating materials when integrated into a building's thermal envelope. The results showed an average thermal conductivity of 0.040 W/mK, a U-value of 0.17 W/m^2K, and a heating demand of 9.56 kWh/m^2-year when applied to the modular system. The material also exhibited good fire resistance, with a smoldering velocity of 3.40 mm/min. These findings highlight rice husk fiber's potential as a sustainable insulation material for modular construction, contributing to energy efficiency and climate change mitigation.

Keywords: natural fiber insulation; blowing insulation technique; rice husk; modular construction; energy-efficient buildings; sustainable building materials

1. Introduction

The construction sector has a significant environmental impact, primarily due to its extensive consumption of non-renewable resources. This consumption leads to land degradation, the generation of substantial solid waste, and increased levels of dust, gas emissions, and noise pollution. Notably, the sector is responsible for approximately 50% of global carbon dioxide emissions, 50% of solid waste production, and nearly 50% of total energy and natural resource consumption [1]. To promote sustainable development in the built environment, the use of locally sourced, renewable, and recycled materials has become increasingly important [2]. In particular, the adoption of natural materials has gained traction in recent years, driven by a growing emphasis on energy conservation, sustainable architecture, and the challenges associated with the disposal of conventional insulation materials such expanded polystyrene (EPS) [3]. Among the alternative resources, agricultural residues have emerged as a promising, renewable, and cost-effective option for sustainable construction practices [4]. These materials offer thermal insulation potential while supporting circular economy principles. Derived from recycled industrial and agricultural waste, natural fibers not only enhance thermal comfort but also help to reduce environmental impact, aligning with the fundamental principles of sustainable development and circular economy practices [5]. The use of bio-based materials contributes to material recycling and contributes positively to their environmental impact [6]. The growing shift toward these materials is driven by their potential to reduce the overall environmental footprint of construction processes.

Thermal insulation remains one of the most effective strategies for achieving energy efficiency in residential, commercial, and industrial buildings. Advanced insulation material reduces the demand for heating and cooling, improves the thermal resistance of building envelopes, and lowers both energy consumption and greenhouse gas emissions [7]. Eco-friendly insulation materials are particularly advantageous, as they are typically non-toxic, do not emit harmful chemicals or volatile organic compounds (VOCs), and tend to have a lower carbon footprint compared to traditional materials. Moreover, their production is often less energy-intensive, they generate fewer emissions, and many of them are recyclable or biodegradable, which enhances their reuse potential and reduces waste [5,8].

Rice is among the most widely cultivated crops globally, feeding nearly half of the world's population [9]. Rice husk represents approximately 20% of global rice production, generating nearly 200 million tons of this agricultural waste annually. Additionally, the slow biodegradability of this residue poses significant environmental concerns, emphasizing the importance of valorizing this natural fiber [10]. Given these environmental challenges, recent research efforts have focused on developing innovative insulation materials derived from rice husk, with examples including insulation panels produced via the pulping method, exhibiting a thermal conductivity of 0.037 W/mK and 137 Kg/m^3 [11], as well as composites combined with various binders to create panels exhibiting favorable thermal and properties in the range of 107–378 kg/m^3 and 0.0045–0.108 W/mK [12–18]. Significant research has also been conducted on utilizing rice husk ash to improve certain concrete properties, where it has been used as a substitute for Portland cement. A hollow insulated block, where rice husk ash and rice husk insulation mixed with aluminous cement served as the binder, has been examined [19]. Rice husk is notably characterized by its high silica content compared to other agricultural residues [10]. The promising results observed in other studies advocate for further exploration of rice husk's properties and its potential applications in energy-efficient construction [17]. Considering the potential of this agricultural residue, it is essential to investigate its integration into broader contexts, such as industrialized construction systems.

Industrialized construction represents a transformative approach that integrates technological solutions to automate processes, enhance supply chain efficiency, and promote sustainability [20]. Despite challenges related to initial invest costs, the limited integration of digital tools, and the need for specialized expertise, this construction approach has demonstrated strong potential in improving efficiency, product standardization, modularization and mass production. Within this broader framework, there is a growing need for further research into environmentally friendly and economically viable insulation materials [21]. Among these systems, modular construction stands out for its ability to reduce material waste, energy consumption, and environmental impact while also decreasing construction time and cost. Consequently, it is increasingly recognized as a promising strategy for sustainable development in the construction industry [22]. Moreover, modular systems facilitate mass customization by allowing products to be tailored to specific consumer requirements without incurring substantial additional costs or delays [23]. The Skylar 250 system, an open-source construction model and the primary system used in WikiHouse, is a modular, industrialized solution designed for rapid assembly, accessibility, and low energy consumption. Using subtractive manufacturing, this system enables efficient and precise prefabrication, which significantly reduces both construction time and material waste [24]. A notable advantage of this system is its lower environmental impact compared to conventional construction methods. It generates significantly fewer greenhouse gas emissions due to its reduced air conditioning demand, estimated at approximately 2000 kWh/year, compared to about 4200 kWh/year which is required by traditional systems [25].

Previous studies involving modular construction systems have demonstrated that the incorporation of sustainable insulation materials, such as wheat straw, can effectively meet thermal performance standards, achieving a thermal transmittance of 0.45 W/m^2K for walls. In addition, energy simulations have shown that optimized modular designs using sustainable insulation can significantly reduce total energy demands, reaching values as low as 55 kWh/m^2 year [25]. Other research conducted in Temuco, Chile reported heating energy demands ranging from 40 to 82 kWh/m^2 year in buildings where thermal insulation was improved, regardless of the specific material used. These results represent substantial improvements when compared to buildings without enhanced thermal insulation [26].

These findings emphasize the importance of passive design strategies in reducing energy consumption. Furthermore, the integration of appropriate thermal insulation materials, particularly those based on natural fibers, can enhance energy efficiency and indoor thermal comfort, ultimately benefiting homeowners and contributing to the reduction in environmental impacts associated with construction [25,27].

In this context, rice husk fiber, a renewable agricultural by-product, offers a promising alternative for building insulation. Its natural, biodegradable properties make it suitable for modern construction systems, particularly for use in building components such as walls and roofs. When applied through the blowing technique, rice husk fiber enables uniform distribution, preserves internal porosity, and improves thermal performance. This technique allows for insulation to be injected without removing the internal or external cladding of a building component. It creates internal air pockets whose characteristics vary depending on the type of fiber used, enhancing the system's overall thermal efficiency [28–30]. Moreover, the blowing technique has proven to be a rapid and efficient application method, significantly reducing installation time. This is especially true in modular construction, where minimizing on-site labor and assembly time is essential [25]. In addition to lowering labor costs, this technique often employs materials that are less expensive than traditional insulation options, making it an economically viable choice for sustainable housing projects [31].

The objective of this study is to evaluate the potential of rice husk fiber as a thermal insulation material in industrialized modular construction systems. To achieve this, our research investigates its technical and environmental viability as a sustainable alternative to conventional insulation materials, applying the blowing technique within modular building components. The methodology combines experimental laboratory testing to characterize the key physical and thermal properties of rice husk fiber, such as density and thermal conductivity. Additionally, the fire behavior of the material is assessed through combustion analysis. Finally, energy simulations were conducted using Desing Builder software version 5.5.2.007 to evaluate the thermal and environmental performance of modular buildings insulated with rice husk fiber, with a specific focus on the energy efficiency benefits achieved by the incorporation of this sustainable material.

2. Materials and Methods

This section presents the methods used to evaluate the physical properties of the rice husk agro-waste, including thermal conductivity, density, moisture content, and fire behavior. It details the experimental setup and parameters for determining optimal blowing conditions, such as nozzle aperture, airflow speed, and power. Additionally, it outlines the procedures for measuring energy consumption and performing simulations in the Design Builder software. This comprehensive approach ensures a thorough understanding of the performance of rice husk fiber within modular construction systems.

2.1. *Sample Preparation and Material Application*

The rice husk was obtained through "Arrocera Mantul", located in a commune in the Maule region of Chile. This product was obtained after the milling phase in the process to obtain rice; tests were conducted and in the loose material, it was found that there was preliminary conductivity 0.041 W/mK, densities of 130–170 kg/m^3, and moisture of around 10.68%.

For the blowing process, The X-FLOC equipment, model M99-DS, was used, which can be seen in Figure 1. This equipment has a nominal power of 3.6 kW, nominal blowing power of 440 m^3/h (adjustable 2 × 1.8 (kW)), and a variable opening range from 1 to 10. "For the application of the material by means of the blowing technique, 2 × 4" pine wood specimens were used, whose dimensions were 60 × 40 × 9 cm in length, width, and thickness, respectively.

Figure 1. The X-FLOC equipment.

2.2. *Physical Characterization*

2.2.1. Thermal Stability

Thermogravimetric analysis (TGA) was performed using a TGA/DSC STA 6000, Perkin Elmer, Waltham, MA, USA, to evaluate the thermal stability of natural rice husk fibers, following a methodology like that of previous studies on lignocellulosic fibers [32–34]. The purge gas was nitrogen (N_2) set at a flow rate of 40 mL/min. The mass of the sample was 20.233 mg. The temperature program involved heating from 25 to 120 °C at a rate of 50 °C/min, followed by a 3 min hold at 120 °C. Subsequently, the temperature was increased from 120 to 950 °C at a rate of 100 °C/min, with cooling from 950 to 450 °C at 100 °C/min. At 450 °C, a gas switch from nitrogen to oxygen (O_2) was performed with a flow rate of 40 mL/min. The heating then continued from 450 to 800 °C at 100 °C/min, concluding with a 3 min isothermal hold at 800 °C [32].

2.2.2. Thermal Conductivity

The thermal conductivity of the prototypes was measured using the KD2 Pro instrument manufactured by Decagon Devices Inc., Pullman, WA, USA, which has been utilized in other studies analyzing fiber-based thermal insulation materials for construction [11,30,35]. The device uses an interchangeable sensor that is inserted into the material, providing a thermal conductivity value. The measurement process is based on the transient line heat source method, enabling the quick analysis of homogeneous materials, with measurement intervals of 1 s during a 90 s heating and cooling cycle. The KD2 Pro complies with IEEE 442-1981 [36] specifications and ASTM D5334-00 standards [37], with an accuracy of ±0.001 W/mK. To ensure the stabilization of the sensor, a 10 min gap was maintained between analyses. Variations in the results may have occurred depending on fiber properties such as length, diameter, density, moisture content, and arrangement.

2.2.3. Density and Moisture

The density was calculated manually by correlating the volume of the test specimens, as previously described, with the mass of material insufflated in each of the four repetitions carried out. The mass was measured in kilograms and the volume was recorded in cubic meters, resulting in density values expressed in Kg/m^3.

The moisture content was measured using the Boeco moisture analyzer, Hamburg, Germany, model BMA H50, as shown in Figure 2. This device operates with halogen light and functions within a temperature range of 10–40 °C, providing results with an accuracy of 0.001%. The measurement process using this analyzer is illustrated in [28].

Figure 2. Process of measuring moisture.

2.2.4. Surface Analysis

The morphological analysis of the rice husk and the prototypes was conducted using scanning electron microscopy (SEM) with a Hitachi VPSEM SU 3500 microscope, Tokyo, Japan. The analysis was carried out under the following conditions: magnifications of 40–100–200–500× for the surface view and 30–100–200–500× for the cross-sectional view. The BSE detector was used, with settings of 10 KeV, WD12 mm, and 30 Pa, as specified by the laboratory equipment guidelines [25,33].

2.3. *Fire Behavior*

To analyze the fire behavior of the samples of rice husk, a fire reaction test and a flameless test to assess smoldering combustion were performed.

2.3.1. Fire Reaction Test

The rice husk fiber was placed in a cylindrical metal container and the fiber was subjected to a portable butane gas torch capable of reaching temperatures up to 1200 °C, with the flame applied directly to the center of the fiber for a duration of three seconds and at a constant distance. After the flame was withdrawn, the ignition time—the interval between flame removal and the appearance of ignition—was measured, along with the subsequent extinguishment time. This process was repeated continuously for two minutes, with all ignition and extinguishment events recorded on video for later analysis. This study aimed to observe the fiber's behavior before heating, during ignition, and after flame extinction. By analyzing the footage, the average ignition time and total number of ignitions were determined, providing insights into the flammability and combustion characteristics of rice husk fiber under repeat heat exposure. Following this fire behavior evaluation, a drip analysis based on the UNE 23-725-90 standard [38] was performed.

2.3.2. Smoldering Combustion Analysis Test

The experimental setup utilized in this study was based on prior research that examined the upward smoldering process using other combustible materials under a comparable experimental configuration [39].

The experimental setup in Figure 3 consists of an electric panel, a hot plate, a cylindrical reactor, seven thermocouples, and a data logger. The reactor was a perforated cylindrical steel structure with a diameter of 80 mm (D) and a height of 120 mm (L). During testing, temperatures along the central axis of the reactor were monitored using 5 thermocouples positioned at distances of 0.5, 3, 6, 9, and 12 cm from the hot plate, and an additional thermocouple was placed directly on the hot plate. After recording the temperature data, the smoldering velocities and the maximum temperature reached during the combustion process were determined.

Figure 3. Smoldering setup and samples. (**A**) Rice husk, (**B**) recycled textiles, (**C**) cellulose.

Smoldering is a phenomenon of combustion without flame, which is especially crucial in combustible porous materials. During the experiments, the hot plate was heated for one hour until it reached a predetermined temperature of 360 °C, after which it was left to cool down to room temperature for another hour. The process was monitored using an infrared camera, capturing images every 5 min [39,40]. The smoldering velocity was calculated by analyzing the temperature data recorded by the thermocouples, specifically by identifying the time it took for each thermocouple position to reach a temperature of 270 °C. In order to perform the experiment, rice husk, recycled textiles, and cellulose were selected to compare their fire behavior. As shown in Figure 3, these materials exhibit similar thermal insulation properties: cellulose ranges between 0.035 and 0.040 W/mK [41], and recycled textiles are in the range of 0.33 to 0.39 [42], making them suitable reference materials for evaluating rice husk fiber's insulation performance. Additionally, due to their physical characteristics, these materials are also suitable for application via the blowing technique within modular construction systems.

2.4. Thermal Performance of the Insulating Envelope and Energy Simulation

2.4.1. Thermal Inertia

To evaluate the impact of the rice husk fiber material on thermal inertia, a component of the Skaylark 250 system, specifically the block M version 0.1, was selected (Figure 4). The thermal inertia of the material was analyzed using the Ubakus online tool, which adheres to the procedures outlined in the DIN 4108 [43] and DIN EN ISO 6946 [44] standards. These standards define methods for assessing thermal resistance and energy efficiency in building systems.

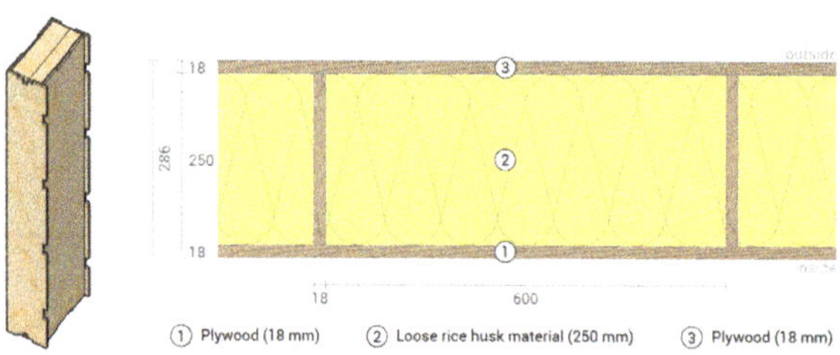

Figure 4. Details of block M wall element.

Ubakus employs finite element methods to calculate thermal resistance and transmittance, offering a reliable approach for evaluating and optimizing the thermal performance of the construction materials. For thermal capacity calculations, the tool determines the temperature profile within the material for two internal temperatures differing by 1 °C under static conditions ($t \rightarrow \infty$). From these profiles, the average temperature difference (dT) is calculated for each layer, and the stored heat (dQ) is computed using the formula:

$$dQ = dT \cdot \text{density} \cdot \text{specific heat capacity} \cdot \text{thickness} \quad (1)$$

The key thermal parameters considered included time lag (the delay between external and internal temperature peaks), decrement factor (the ratio of temperature amplitudes), the thermal capacity of the inner layers, and the heat storage capacity. A 24 h simulation with sinusoidal external temperature variations ranging from 15 °C to 35 °C provided temporal temperature profiles, enabling a comprehensive analysis of the material's thermal damping properties [25,29]. For the thermal inertia assessment, rice husk fiber was analyzed and compared with recycled PUR, as both materials are suitable for application through the blowing technique.

2.4.2. Energy Simulation

DesignBuilder version 5.5.2.007 is a comprehensive building simulation software that utilizes the open-source EnergyPlus version 8.6 engine to simulate indoor microclimates and energy consumption. The software allows users to create detailed building models with ease, offering a user-friendly interface. It enables users to define every aspect of the building's systems, including schedules, construction materials, lighting, HVAC systems, and window-to-wall ratios, providing a thorough and customizable modeling experience [45].

Scenario Setup

The simulation sequence is as follows: A modular housing unit, which features an area of 14.75 m^2 (Figure 5) and is located in Temuco, Chile within thermal zone F according to the Chilean building standards [46] and has a Köppen climate classification of Cfb, is built using Skylark 250. The modular housing unit will be modeled to evaluate energy consumption, indoor thermal comfort, and CO_2 emissions under three different scenarios. The first scenario uses rice husk fiber thermal insulation, the second employs a recycled PUR insulation material, and the third considers the absence of any insulation. The modular unit will have a standardized design to ensure comparability between scenarios. The parameters for energy simulation in DesignBuilder are presented in Table 1. Simulations will be carried out for the city of Temuco, using weather data derived from the EnergyPlus climate file database provided by the U.S. Department of Energy. Parameters such as outdoor Dry-Bulb Temperature, wind speed, and relative humidity will be enforced via DesignBuilder's pre-installed weather file. Additionally, the simulations will allow for the assessment of the thermal envelope's efficiency under each scenario, providing a comparative analysis of energy performance and carbon footprint.

Figure 5. Modular housing unit.

Table 1. Parameters for energy simulation in DesignBuilder.

Parameters for Energy Simulation in DesignBuilder		
Parameters	**Tipe**	**Characteristics**
Activity	Residential	0.02 (people/m^2)—residential, residential occ schedule
Construction	Project template	Walls, roofs, and floors of Skylar 250 system have model infiltration of 0.7 (ac/h)
Opening	Project template	30% wall to window ratio—double glazing, clear, 6 mm/6 mm
Lighting energy	Project template	5.00 W/m^2—100 lux
HVAC		ON, heating COP 0.5 and cooling COP 4.5, fuel electricity from grid

3. Results and Discussion

3.1. Material Application Results

The rice husk was injected into the test panel by means of the blowing technique (Figure 6A) using an X-Floc M99-DS blowing machine X-Floc Dämmtechnik-Maschinen GmbH, Renningen, Germany. The correct application was achieved using the following configuration: 80% blowing power, corresponding to 352 m^3/h, with an opening at the second level and the hopper operating at 80% capacity. It was necessary to maintain a consistent feed of the rice husk fiber to prevent blockages in the machine. Experimental testing confirmed that, under these conditions, the material could be efficiently blown without causing blockages in the hose or nozzle. This configuration allowed for uniform fiber distribution and density to be reliably achieved in each required application, contributing to consistently low thermal conductivity. Additionally, this method ensured that the thermal insulation material filled the entire empty space, as shown in Figure 6B. As demonstrated in the studies [28,30], this technique ensures a uniform distribution of fibers without the need for chemical additives, thereby reducing the material's environmental impact. Additionally, the blowing method offers flexibility in application, making it suitable for both new construction and renovation work [30].

(A) (B)

Figure 6. (**A**) Rice husk blown into the specimen, (**B**) Rice husk applied by the blowing technique.

3.2. Physical Characterization

3.2.1. Thermal Stability Analysis

The proximal analysis carried out on rice husk in its natural state is detailed in Figure 7. According to the TGA, the fiber began to lose weight at 120 °C. This was observed, corresponding to 7.88% being mainly related to moisture, and subsequently, a more significant mass loss of 55.13% was observed between 120 °C and 759 °C, associated with the volatile compounds hemicellulose and cellulose. Once the environmental change to oxygen took place, carbon losses of 17.07% were observed between 758 and 798°, corresponding to lignin. Finally, the percentage of ashes was 19.92%, and this was displayed at 800 °C. The rice husk fiber presented stability until 120 °C, indicating that the temperature used for applying or processing should have been lower than this value to avoid thermal degradation. Similar results are shown for other natural fibers [30,32].

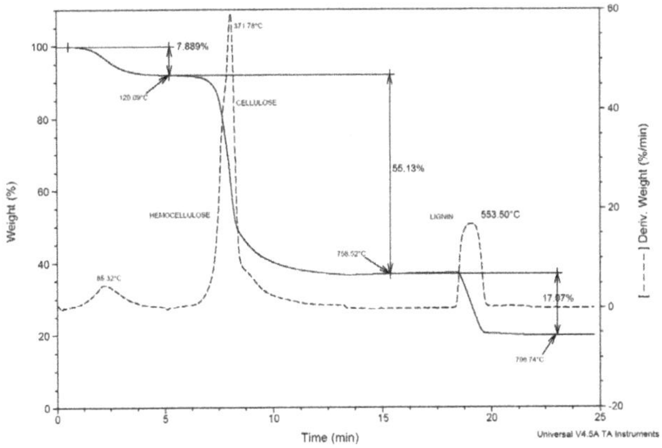

Figure 7. Thermal stability of rice husk (TGA).

The differential curve TGA(DTGA) shows in its shoulders the three stages mentioned. The first one corresponds to moisture with a peak at 85.32 °C, the second one with a peak at 371 °C corresponds to cellulose, and the last one with a peak at 553 °C corresponds to lignin. As can be seen in Figure 7, the rice husk stays thermally stable up to 120 °C; above this temperature, thermal degradation starts, but the rice husk shows a temperature peak for maximum degradation at 371, which is similar to that reported in other studies of rice husk TGA [12,13]. In addition, based on this analysis, the rice husk fiber was found to remain stable up to 120 °C, indicating that its processing temperature should remain below this threshold to avoid thermal degradation. In this regard, it was noted that the blowing application process did not cause a rise in the material's temperature, ensuring that no degradation occurred during its application.

3.2.2. Thermal Conductivity Results

The thermal conductivity is measured [M], as shown in Table 2, with an overall average value of 0.040 ± 0.001 W/mK. Figure 8 shows the thermal conductivity measurements of the insufflated samples; these results indicate that the prototypes exhibit favorable thermal performance, suggesting that the material is suitable for building insulation and supporting the potential for developing new insulation materials based on agricultural waste. This result is comparable with other rice husk materials, with values of 0.37 to 0.042 W/mK [11]; and with other lignocellulosic materials, like wheat straw, hemp, eucalyptus bark fiber, algae, and mushroom insulation material, which are in the range of 0.32 to 0.40 W/mK [35,47–49]. The thermal conductivity values obtained for rice husk are comparable to those conventional insulation materials currently available on the market, like mineral wool (0.033–0.038 W/m·K) and expanded polystyrene (0.035–0.040 W/mK) [50,51], underscoring its potential as a competitive alternative for thermal insulation.

Table 2. Results of thermal conductivity measurements.

	Thermal Conductivity [W/mK]				AVG	SD
Measurement (M)	M1	M2	M3	M4		
Sample 1	0.040	0.035	0.037	0.041	0.038	0.0028
Sample 2	0.041	0.041	0.042	0.040	0.041	0.0008
Sample 3	0.040	0.041	0.040	0.040	0.040	0.0005
Sample 4	0.041	0.040	0.041	0.038	0.040	0.0014
					0.040	0.0014

Figure 8. Measurement of thermal conductivity.

3.2.3. Density and Moisture Content Results

The density values averaged 130 kg/m^3, with the lowest measurements of 121 kg/m^3 corresponding to sample 1, with 121 kg/m^3 shown in Figure 9. The moisture content of rice husk was found to have an average of 10.23 $\pm$ 0.1%, which is consistent with the behavior observed in other natural fiber materials. This moisture level is considered desirable, as it supports the preservation of the thermal properties of insulation materials. Excessive moisture can negatively impact thermal conductivity, increase condensation, and reduce biological resistance. Conversely, maintaining low-to-moderate moisture levels enhances the material's biological resistance, making it well suited for use in building applications.

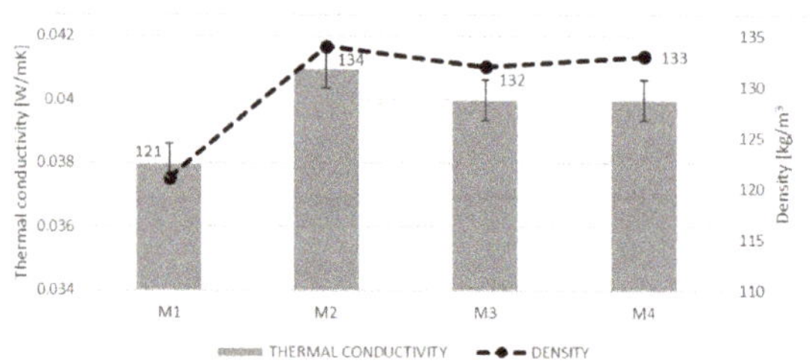

Figure 9. Relationship between thermal conductivity and density of samples.

Materials with open porosity, such as rice husk, facilitate moisture movement driven by vapor pressure gradients, gravity, and capillary action. However, in well-designed systems, vapor diffusion is the dominant mechanism, preventing excessive moisture levels and contributing to the material's effectiveness as a thermal insulator in energy-efficient buildings [52].

Furthermore, the results suggest that rice husk insulation would not promote the proliferation of pathogens if installed in homes in regions such as Chile. Previous studies, such as those conducted by Soto M. et al. (2023), have shown that similar natural fibers, like cellulose combined with wheat straw, do not support fungal growth or region-specific pathogens within comparable moisture ranges [28].

Figure 9 presents the measurement results, showing that the samples exhibit a thermal conductivity below 0.041, demonstrating their potential as a thermal insulation material. This value is comparable to conventional insulating materials, reinforcing their suitability for energy-efficient construction applications [53]. Unlike traditional insulation materials, where a higher density reduces thermal conductivity, lignocellulosic fibers show the opposite trend, with conductivity increasing as density rises [28]. This highlights the importance of optimizing density to maintain insulation efficiency properties.

3.2.4. Surface Morphology Analysis

Figure 10 presents three microscopic images of the rice husk, accentuating its suitability as a thermal insulation material. Image (A) shows the surface of the rice husk fiber, revealing a well-defined porous structure. This structure is further detailed in image (B), where the voids on the fiber surface are clearly visible, contributing to its ability to enhance thermal insulation performance [28,54]. Finally, the longitudinal view in image (C) illustrates the entire surface of the fiber, characterized by well-distributed pores, which play a key role in achieving optimal thermal insulation levels. These microscopic observations confirm that rice husk possesses optimal properties for use in insulation material applications.

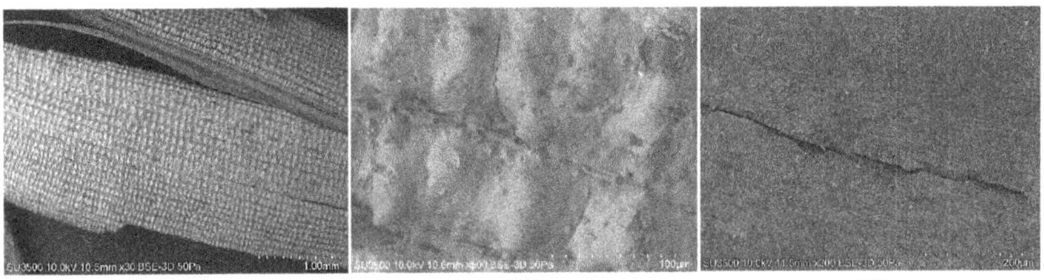

Figure 10. SEM. (**A**) and (**B**) show the internal longitudinal external; (**C**) shows the internal longitudinal internal.

3.3. Fire Behavior

3.3.1. Fire Reaction Test Results

Figure 11 shows the flame extinction time, for a total of 23 applications, reaching a total of 90.34 s., which nearly doubles the number of tests reported in previous research on wood dust [29]. The average of all the extinguishing times is 3.93, with the first ignition time being 3.93 s. A significant observation was made at the 9 s mark, when a carbonized protective layer (Figure 12) began to form on the surface of the sample. This layer acted as a thermal barrier, preventing further combustion and allowing the material to endure 23 flame applications without complete degradation. This behavior aligned with previous TGA findings, where the initial moisture release was followed by the decomposition of cellulose and hemicellulose, leading to the formation of a char layer that enhanced fire resistance.

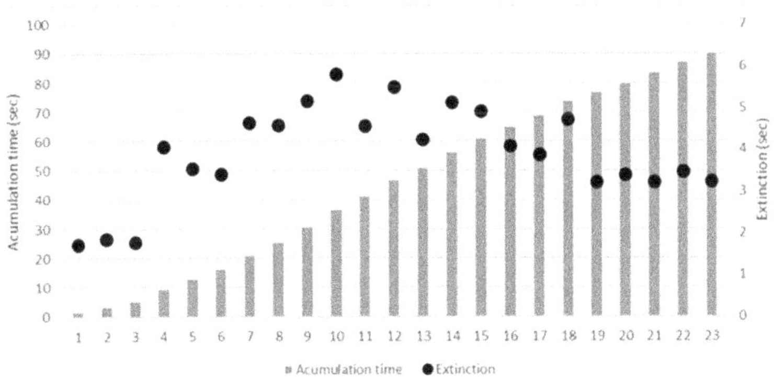

Figure 11. Extinction times and accumulation times.

Figure 12. Carbonization of the rice husk.

Furthermore, a previous study [11] indicates that this self-protective carbonization effect in rice husk is primarily due to its high silica content. Silica contributes to the material's resistance to combustion by forming a stable insulating layer that limits heat penetration and delays thermal degradation.

These results demonstrate the effectiveness of the material in delaying flame propagation compared to traditional passive fire protection strategies due to its self-protective charring mechanism. This natural flame-retardant behavior suggests its potential application in fire-resistant materials. The ability to withstand nearly twice as many flame applications as wood dust [29] indicates its suitability for use in thermal insulation and fire-resistant construction material.

3.3.2. Smoldering Combustion Performance

For this smoldering test, the reactor was filled with samples of (A) rice husk fiber, (B) recycled textile, and (C) cellulose, which were evaluated. During the experiments, the evolution of smoldering was visualized with an infrared camera. Figure 13 shows the results: The images are presented in a sequence of seven pictures every 10 min from 50 min to 110 min. The red color corresponds to temperatures around 240–261 °C. In samples (A) and (B), the evident smoldering begins at 65 min, while with cellulose, it starts at 55min.

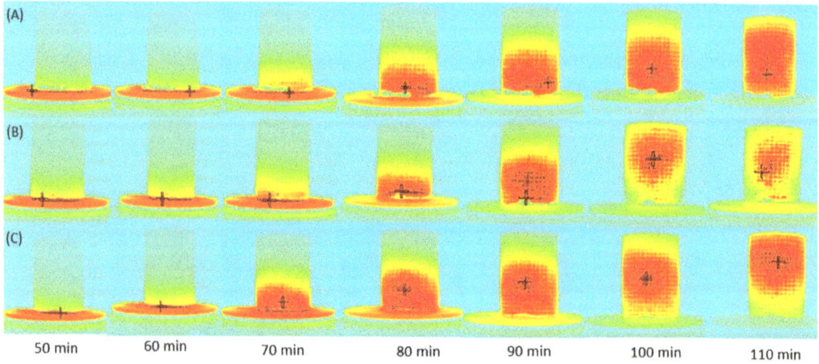

Figure 13. Smoldering process thermographies. (**A**) Rice husk, (**B**) recycled textile, and (**C**) cellulose.

Figure 13 shows that after conducting the smoldering test on three samples, it was observed that rice husk fiber resisted smoldering for 60 min, reaching a temperature of 343.2 °C at the first thermocouple at 0.5 cm before the process began. Similarly, the recycled textile sample also withstood smoldering for 60 min, but at a lower temperature of 270.3 °C. In contrast, the cellulose sample initiated smoldering earlier, at 55 min, with a recorded temperature of 308.9 °C. These results suggest that the variations in the thermal behavior of the samples can be attributed to their composition, density, and structural properties, and the performance of rice husk fiber highlights its potential as a more fire-resistant insulation material.

Figure 14 shows the temperatures recorded by the thermocouples on samples tested to a pre-set temperature of 360 °C; the dashed lines correspond to the temperatures measured at the hotplate surface. From these data, the smoldering velocity can be evaluated by determining the times at which each thermocouple position reaches a specific temperature, which here is chosen as 270 °C. (A) shows better behavior in the smoldering test than the (B) both in the temperature and velocity of smoldering, reaching maximum temperatures of 617 °C, while the recycled textile reaches the maximum at 652 °C, and the cellulose at 708 °C.

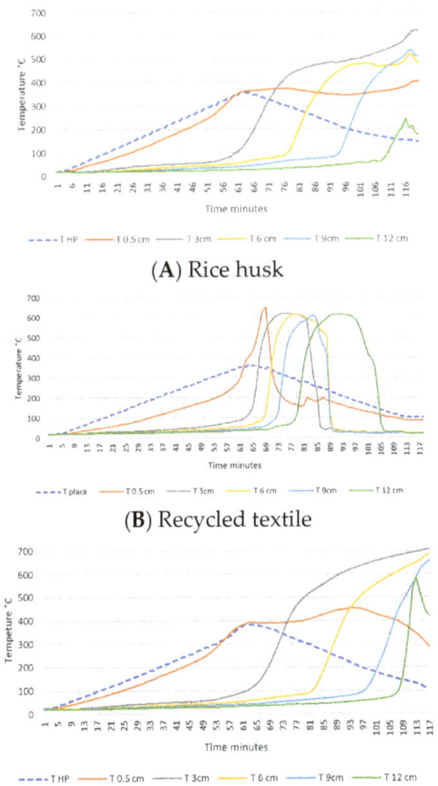

Figure 14. Temperature evolution for thermocouples located at each 3 cm along material. Pre-set temperature of the hot plate is 360 °C. (**A**) Rice husk, (**B**) recycled textile, (**C**) cellulose.

Figure 15 shows that the smoldering velocity is highest in the recycled textile at 6.30 mm/min, while the rice husk records a lower velocity of 3.40 mm/min, a result consistent with a previous study on rice panels produced using the pulping method [11]. The recycled textile exhibits a similar behavior to the corn pith [40], while cellulose demonstrates the slowest smoldering rate at 1.91 mm/min. This slower combustion is attributed to the presence of boron salts as a flame retardant, which are commonly added to commercially available insulation materials.

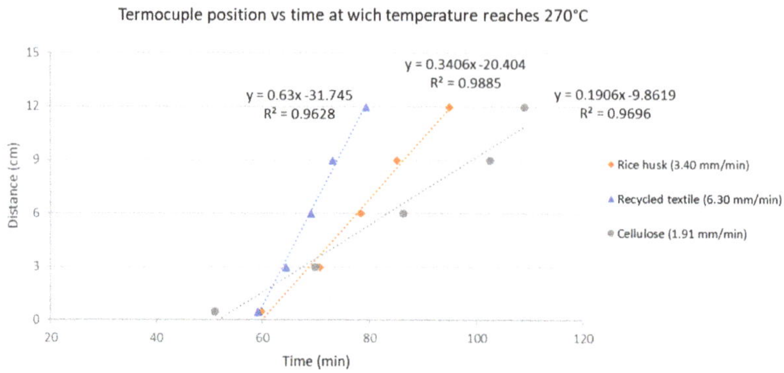

Figure 15. Smoldering propagation velocities are calculated as the slope of the obtained linear regression.

3.4. Thermal Performance of the Insulating Envelope and Energy Simulation

3.4.1. Thermal Inertia Results

Table 3 shows the findings from the thermal inertia analysis, indicating that the rice husk fiber used in the test samples possesses an energy storage of 98 (KJ/m^{2}K). This means that, when incorporated into a building's component, this building will exhibit over 30% of the energy storage capacity compared to the use of recycled PUR, and similar behavior was observed in a previous study with sawdust material [29].

Table 3. Results of thermal performance.

Property	Unit	Wall M Rice Husk	Wall M Recycled PUR
U-value	W/m^{2}K	0.17	0.201
Time lag	h	14.50	8.50
Decrement factor	_	40.30	11
Heat storage capacity (whole component)	KJ/m^{2}K	98	69
Thermal capacity of inner layers	KJ/m^{2}K	49	34

The thermal transmittance results indicate that rice husk fiber with a transmittance value of 0.17 W/m^{2}K in the wall component outperforms recycled PUR, which has a thermal transmittance of 0.201 W/m^{2}K. Both materials exhibit transmittance values well below the maximum limit of 0.45 W/m^{2}K established by the atmospheric decontamination plan (PDA) for walls in Temuco [55]. The superior performance of rice husk fiber suggests a greater ability to reduce heat transfer, enhancing the thermal efficiency of the building envelope. This lower transmittance contributes to a more stable and comfortable environment, minimizing temperature fluctuations regardless of the external weather conditions.

Figure 16 shows the surface temperature behavior when the insulation material is changed from recycled PUR to rice husk fiber. Notably, replacing the insulation significantly increases the temperature wave delay from 8.5 h to 14.5 h, enhancing thermal storage capacity, reducing energy consumption, and improving indoor thermal comfort. This demonstrates the potential to enhance the energy efficiency of buildings [56].

Figure 17 shows the temperature behavior when recycled PUR insulation materials are replaced with rice husk fiber. A notable reduction in temperature fluctuation amplitude occurs, demonstrating its effectiveness in mitigating the impact of external temperature peaks. Such thermal performance improvements are typically not observed with conventional insulation materials like recycled polyurethane, which tend to closely follow external temperature fluctuations [29].

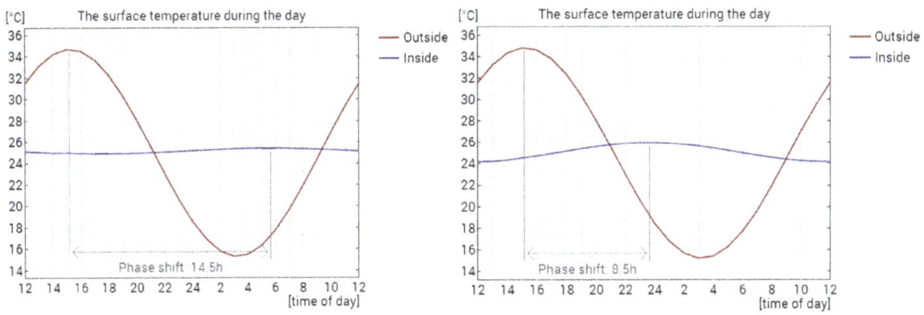

(A) (B)

Figure 16. Surface temperature of (**A**) rice husk and (**B**) recycled PUR during the day.

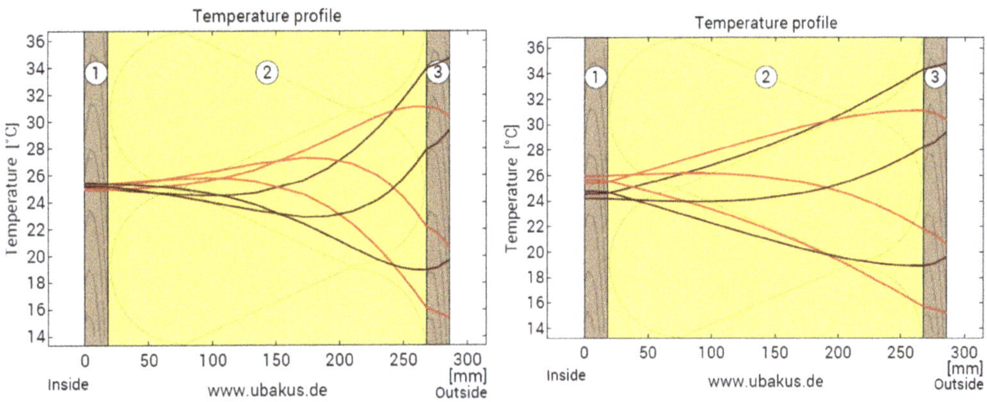

(A) (B)

① Plywood (18mm) ② Thermal insulation (250 mm) ③ Plywood (18 mm)

Temperature at 3pm, 11am and 7am

Temperature at 7pm, 11pm and 3am

Figure 17. Temperature profiles of (**A**) rice husk and (**B**) recycled PUR during the day.

3.4.2. Energy Simulation Results

To evaluate the energy efficiency of different insulation materials, a simulation was conducted considering three different scenarios. The first scenario incorporated rice husk fiber insulation (A), the second scenario utilized recycled polyurethane insulation, and the third scenario represented the non-insulated case, serving as a baseline to assess the impact of insulation on energy consumption (Figure 18).

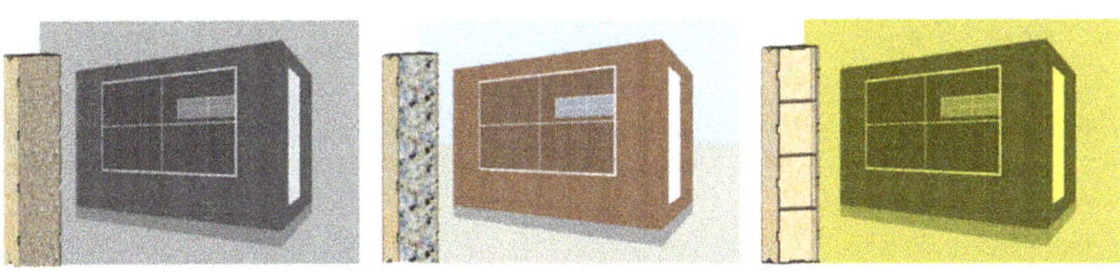

(A) (B) (C)

Figure 18. Different scenarios simulated. (**A**) Rice husk fiber, (**B**) recycled PUR, (**C**) non-insulated.

Temperature Homogeneity

The results presented in Figure 19 indicate that rice husk fiber maintains indoor temperatures within the comfort range throughout the year, demonstrating its effectiveness in stabilizing indoor thermal conditions and its provision of superior thermal regulation compared to the other scenarios. Recycled PUR insulation performs better than the non-insulated scenario by reducing temperature fluctuations; however, it is still below the lower comfort zone [55] during winter, indicating that while it offers some insulation, it is not as effective as a rice husk fiber. The non-insulated scenario exhibits the highest temperature variations, closely following outdoor temperatures, which indicates the importance of using insulation materials to enhance indoor thermal stability and energy efficiency. Additionally, rice husk fiber insulation shows the best performance in extreme temperature conditions, preventing excessive heat loss in winter and minimizing overheating in summer, making it a viable and sustainable option for energy-efficient construction climates like Temuco. The rice husk within the modular system achieves thermal comfort levels comparable to dwellings that comply with the technical buildings code standard [26]. Overall, these findings suggest that the rice husk fiber is a highly effective natural insulation material, confirming its potential application in sustainable building systems for enhanced thermal efficiency and occupant comfort.

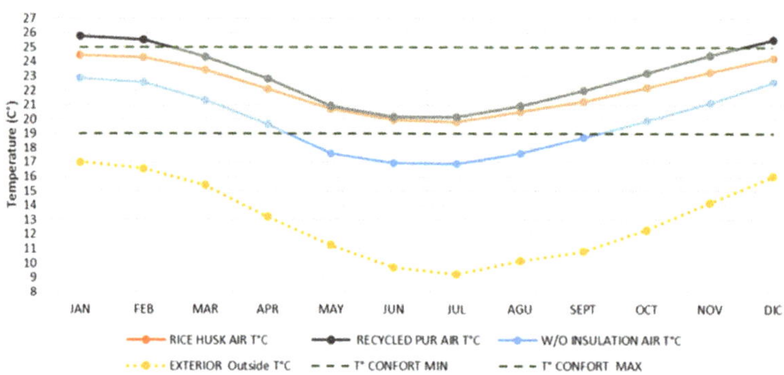

Figure 19. Annual indoor temperature profiles for rice husk, recycled PUR, and no insulation scenarios.

Energy Consumption and CO_2 Emissions

The results presented in Figure 20 illustrate the energy consumption in kWh required for heating and cooling the modular unit under different insulation scenarios. The data clearly indicate that the rice husk fiber exhibits the lowest energy consumption, outperforming both recycled PUR and the non-insulated case. The non-insulated scenario shows significantly higher energy demand, particularly during the winter months, where heating consumption peaks at well above 643 kWh. This highlights the inefficiency of an uninsulated module at maintaining indoor thermal stability, requiring excessive energy to compensate for heat loss. The rice husk fiber demonstrates the highest energy efficiency. In Table 4, the annual consumption and the rice husk scenario that maintains lower heating and cooling demand compared to the other two scenarios is shown, with 9.56 kWh/m^2-year of heating demand, which is close to the 11.70 kWh/m^2-year reported by [25] in a similar construction using wheat straw insulation. The heating energy demand observed using rice husk fibers is significantly lower compared to other thermally improved dwellings in Temuco, with a range between 40 and 80 kWh/m^2-year [26]. This represents an approximate reduction of 81%, underscoring rice husk fiber's substantial potential for enhancing energy efficiency in buildings.

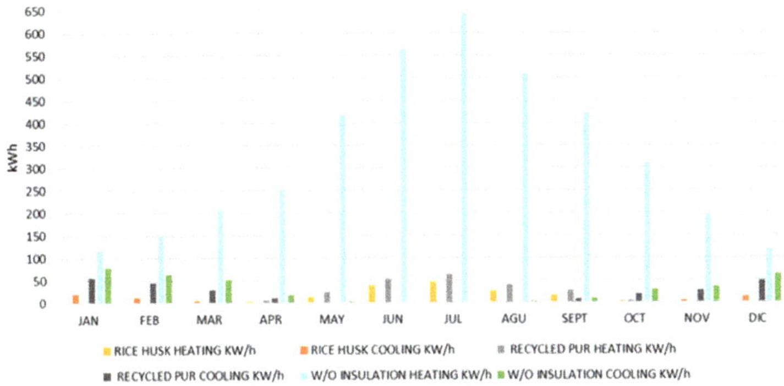

Figure 20. Energy consumption in kW/h required for heating and cooling the modular unit under different insulation scenarios.

Table 4. Annual consumption results for heating, cooling, and associated CO_2 emissions.

Scenario	Heating Demand (kWh-year)	Cooling Demand (kWh-year)	Heating Demand (kWh/m^2-year)	Cooling Demand (kWh/m^2-year)	Total Demand (kWh/m^2-year)	CO_2 Emissions (kg eq-year)
Rice husk insulation	140.96	55.35	9.56	3.75	13.31	1956.81
Recycled PUR insulation	217.00	241.00	14.71	16.34	31.05	2116.04
Without Insulation	3909.14	350.06	265.03	23.73	288.76	4418.92

This confirms its superior thermal performance, reducing the need for mechanical heating in winter and cooling in summer. The results suggest that rice husk fiber not only improves indoor comfort but also enhances energy efficiency, making it a more sustainable and effective alternative for insulation in modular construction systems.

Additionally, it achieves the lowest CO_2 emissions, associated with heating and cooling consumption, of 1956.81 kg per year, representing a decrease of about 60% compared to the scenario without insulation, as shown in Table 4, significantly reducing the building's carbon footprint.

4. Conclusions

This study demonstrates that rice husk fiber is a viable and sustainable alternative to conventional thermal insulation materials, particularly when integrated into modular construction systems. Applied through the blowing technique, rice husk not only enhances the energy efficiency of buildings but also supports environmental sustainability by promoting the reuse of agricultural waste.

From a thermal performance perspective, the results demonstrate that rice husk fiber possesses favorable insulating properties. The measured thermal conductivity is 0.040 W/mK, a value comparable to conventional insulation material. Thermal simulations further indicate that its integration into modular construction, specifically through the blowing technique, can significantly improve building energy efficiency. In this context, the use of rice husk fiber achieves a U-Value of 0.17 W/m^{2}K and reduces heating and cooling energy demand by more than 50% down to 9.56 kWh/m^2-year compared to an uninsulated module. These findings highlight its potential for contributing to energy-efficient construction. Future studies should address the long-term durability and potential degradation of the material under variable and moisture conditions.

In terms of fire safety, the rice husk fiber insulation exhibits superior performance compared to the other evaluated materials. It demonstrates a smoldering velocity of

3.40 mm/min, lower than that of the other tested samples. This performance is attributed to its relatively high silica content, which promotes the formation of a protective char layer that enhances fire resistance. However, to broaden its applicability in more demanding conditions, further research on eco-friendly fire-retardant treatments is recommended.

From an environmental perspective, rice husk fiber aligns well with the principles of sustainable construction. Its use promotes the reuse of agricultural by-products and avoids the need for chemical additives. When applied through the blowing technique, it allows for efficient and uniform installation, supporting ease of implementation in sustainable building projects. In this study, the use of rice husk fiber results in a 55.7% reduction in CO_2 emissions compared to the uninsulated baseline module. Nonetheless, future studies should evaluate the potential environmental and occupational impacts associated with particulate emissions during the blowing process.

To maximize the potential of rice husk fiber insulation, it is necessary to optimize the blowing application method to ensure long-term stability. In addition, evaluating its performance under diverse climatic conditions and exploring potential hybrid formulations with other bio-based materials could further improve its thermal and fire-resistance properties.

In summary, rice husk fiber insulation stands out a promising bio-based solution for promoting energy efficiency and sustainability in the construction sector.

Author Contributions: K.R.N.: Conceptualization, Methodology, Formal analysis, Investigation, Writing—original draft preparation, Validation, Writing—review and editing, Project administration. C.J.R.-H.: Writing—original draft preparation, Methodology, Visualization. J.P.C.-R.: Conceptualization, Methodology, Writing—review and editing, Resources, Supervision. J.T.R.: Writing—review and editing, Validation, Resources, Supervision. A.S.-O.: Writing—review and editing, Validation, Supervision. All authors have read and agreed to the published version of the manuscript.

Funding: This research received no external funding.

Institutional Review Board Statement: Not applicable.

Informed Consent Statement: Not applicable.

Data Availability Statement: The original contributions presented in this study are included in the article. Further inquiries can be directed to the corresponding author.

Acknowledgments: The authors would like to thank the Sustainable Buildings 4.0 group of the Civil Engineering Department of Frontera University for allowing the use of its facilities and equipment for the measurements carried out in this research, and thank the Polytechnic University of Catalonia for allowing fire behavior tests to be conducted in their laboratories.

Conflicts of Interest: The authors declare no conflicts of interest.

References

1. Vasilca, I.S.; Nen, M.; Chivu, O.; Radu, V.; Simion, C.P.; Marinescu, N. The management of environmental resources in the construction sector: An empirical model. *Energies* **2021**, *14*, 2489. [CrossRef]
2. Dikmen, N.; Ozkan, S.T.E. *Unconventional Insulation Materials*; IntechOpen: London, UK, 2016. [CrossRef]
3. Korjenic, A.; Petránek, V.; Zach, J.; Hroudová, J. Development and performance evaluation of natural thermal-insulation materials composed of renewable resources. *Energy Build.* **2011**, *43*, 2518–2523. [CrossRef]
4. Sangmesh, B.; Patil, N.; Jaiswal, K.K.; Gowrishankar, T.P.; Selvakumar, K.K.; Jyothi, M.S.; Jyothilakshmi, R.; Kumar, S. Development of sustainable alternative materials for the construction of green buildings using agricultural residues: A review. *Constr. Build. Mater.* **2023**, *368*, 130457. [CrossRef]
5. Ajabli, H.; Zoubir, A.; Elotmani, R.; Louzazni, M.; Kandoussi, K.; Daya, A. Review on Eco-friendly insulation material used for indoor comfort in building. *Renew. Sustain. Energy Rev.* **2023**, *185*, 113609. [CrossRef]
6. Chen, L.; Zhang, Y.; Chen, Z.; Dong, Y.; Jiang, Y.; Hua, J.; Liu, Y.; Osman, A.I.; Farghali, M.; Huang, L.; et al. Biomaterials technology and policies in the building sector: A review. *Environ. Chem. Lett.* **2024**, *22*, 715–750. [CrossRef]

7. Aditya, L.; Mahlia, T.I.; Rismanchi, B.; Ng, H.M.; Hasan, M.H.; Metselaar, H.S.C.; Muraza, O.; Aditiya, H.B. A review on insulation materials for energy conservation in buildings. *Renew. Sustain. Energy Rev.* **2017**, *73*, 1352–1365. [CrossRef]
8. Islam, S.; Bhat, G. Environmentally-friendly thermal and acoustic insulation materials from recycled textiles. *J. Environ. Manag.* **2019**, *251*, 109536. [CrossRef]
9. Bin Rahman, A.N.M.R.; Zhang, J. Trends in rice research: 2030 and beyond. *Food Energy Secur.* **2023**, *12*, e390. [CrossRef]
10. Soltani, N.; Bahrami, A.; Pech-Canul, M.I.; González, L.A. Review on the physicochemical treatments of rice husk for production of advanced materials. *Chem. Eng. J.* **2015**, *264*, 899–935. [CrossRef]
11. Rodríguez Neira, K.; Cárdenas-Ramírez, J.P.; Rojas-Herrera, C.J.; Haurie, L.; Lacasta, A.M.; Torres Ramo, J.; Sánchez-Ostiz, A. Assessment of Elaboration and Performance of Rice Husk-Based Thermal Insulation Material for Building Applications. *Buildings* **2024**, *14*, 1720. [CrossRef]
12. Wang, S.; Li, H.; Zou, S.; Zhang, G. Experimental research on a feasible rice husk/geopolymer foam building insulation material. *Energy Build.* **2020**, *226*, 110358. [CrossRef]
13. Buratti, C.; Belloni, E.; Lascaro, E.; Merli, F.; Ricciardi, P. Rice husk panels for building applications: Thermal, acoustic and environmental characterization and comparison with other innovative recycled waste materials. *Constr. Build. Mater.* **2018**, *171*, 338–349. [CrossRef]
14. Marques, B.; Tadeu, A.; António, J.; Almeida, J.; de Brito, J. Mechanical, thermal and acoustic behaviour of polymer-based composite materials produced with rice husk and expanded cork by-products. *Constr. Build. Mater.* **2020**, *239*, 117851. [CrossRef]
15. Antunes, A.; Faria, P.; Silva, V.; Brás, A. Rice husk-earth based composites: A novel bio-based panel for buildings refurbishment. *Constr. Build. Mater.* **2019**, *221*, 99–108. [CrossRef]
16. Muthuraj, R.; Lacoste, C.; Lacroix, P.; Bergeret, A. Sustainable thermal insulation biocomposites from rice husk, wheat husk, wood fibers and textile waste fibers: Elaboration and performances evaluation. *Ind. Crops Prod.* **2019**, *135*, 238–245. [CrossRef]
17. Pascual, A.M.; Romero, M.L.; Serra, E.; Guerrero, J.C.; Perez, R. Sustainable insulation panel for buildings made of rice husks and posidonia. *Constr. Build. Mater.* **2024**, *445*, 137983. [CrossRef]
18. Palumbo, M.; Navarro, A.; Avellaneda, J.; Lacasta, A.M. Characterization of thermal insulation materials developed with crop wastes and natural binders. *World Barcelona* **2014**, *28*, 1.
19. Ma, W.; Kolb, T.; Rüther, N.; Meinlschmidt, P.; Chen, H.; Yan, L. Physical, Mechanical, thermal and fire behaviour of recycled aggregate concrete block wall system with rice husk insulation. *Energy Build.* **2024**, *320*, 114560. [CrossRef]
20. Sánchez-Garrido, A.J.; Navarro, I.J.; García, J.; Yepes, V. A systematic literature review on modern methods of construction in building: An integrated approach using machine learning. *J. Build. Eng.* **2023**, *73*, 106725. [CrossRef]
21. Rojas-Herrera, C.; Martínez-Soto, A.; Avendaño-Vera, C.; Carrasco, R.C.; Barbato, N.R. Industrialized Construction: A Systematic Review of Its Benefits and Guidelines for the Development of New Constructive Solutions Applied in Sustainable Projects. *Appl. Sci.* **2025**, *15*, 2308. [CrossRef]
22. Nguyen, T.D.H.N.; Moon, H.; Ahn, Y. Critical Review of Trends in Modular Integrated Construction Research with a Focus on Sustainability. *Sustainability* **2022**, *14*, 12282. [CrossRef]
23. Bonvoisin, J.; Halstenberg, F.; Buchert, T.; Stark, R. A systematic literature review on modular product design. *J. Eng. Des.* **2016**, *27*, 488–514. [CrossRef]
24. Open Systems Lab. Available online: https://www.wikihouse.cc/blocks/skylark-250 (accessed on 14 October 2024).
25. Quidel, G.R.; Acuña, M.J.S.; Herrera, C.J.R.; Neira, K.R.; Cárdenas-Ramírez, J.P. Assessment of Modular Construction System Made with Low Environmental Impact Construction Materials for Achieving Sustainable Housing Projects. *Sustainability* **2023**, *15*, 8386. [CrossRef]
26. Martinez-Soto, A.; Jimenez-Gallardo, C.; Villarroel-Lopez, A.; Reyes-Riveros, A.; Höhl, J. Toward Sustainable Indoor Environments: Assessing the Impact of Thermal Insulation Measures on Air Quality in Buildings—A Case Study in Temuco, Chile. *Sustainability* **2024**, *16*, 547. [CrossRef]
27. Altan, H.; Ozarisoy, B. An Analysis of the Development of Modular Building Design Elements to Improve Thermal Performance of a Representative High Rise Residential Estate in the Coastline City of Famagusta, Cyprus. *Sustainability* **2022**, *14*, 4065. [CrossRef]
28. Soto, M.; Rojas, C.; Cárdenas-Ramírez, J.P. Characterization of a Thermal Insulating Material Based on a Wheat Straw and Recycled Paper Cellulose to Be Applied in Buildings by Blowing Method. *Sustainability* **2023**, *15*, 58. [CrossRef]
29. Rojas-Herrera, C.; Martínez-Soto, A.; Avendaño-Vera, C.; Cárdenas-R, J.P. Characterization and utilization of sawdust waste generated from advanced manufacture for its application as a thermal insulation in sustainable buildings using the blowing technique. *J. Build. Eng.* **2024**, *88*, 109217. [CrossRef]
30. Javier, R.H.C.; Karin, R.N.; Pablo, C.-R.J. Valorization of Wheat Crop Waste in Araucanía, Chile: Development of Prototype of Thermal Insulation Material for Blowing Technique and Geographical Analysis. *Buildings* **2023**, *13*, 1152. [CrossRef]
31. Salonen, T.; Fischer, H.; Korjenic, A. Chopped Straw as an Insulation Material: The Influence of Different Blow-In Technologies and Flame Retardants on Hygrothermal Properties. *Buildings* **2023**, *13*, 2555. [CrossRef]

32. Soto, M.; Vera, M.; Parra, K.; Rojas, C.; Cárdenas, J.P. Potential of the Residual Fibers of Pisum Sativum (PS), for use in a Development of a Thermal Insulator Material. In *IOP Conference Series: Earth and Environmental Science*; Institute of Physics Publishing: Bristol, UK, 2020. [CrossRef]
33. Rojas, C.; Cea, M.; Iriarte, A.; Valdés, G.; Navia, R.; Cárdenas-R, J.P. Thermal insulation materials based on agricultural residual wheat straw and corn husk biomass, for application in sustainable buildings. *Sustain. Mater. Technol.* **2019**, *20*, e00102. [CrossRef]
34. Herrera, C.R.; Lemus, M.C.; Diaz, F.E.R.; Ramirez, J.P.C. Physical, chemical and mechanical characterization of a prototype insulating material based on eucalyptus bark fiber. *IEEE Lat. Am. Trans.* **2018**, *16*, 2441–2445. [CrossRef]
35. Aravena, M.; Almonacid-Muñoz, L.; Rojas-Herrera, C.; Herrera, H.; Cárdenas-Ramírez, J.P.; Veliz Reyes, A.; Sagredo-Saez, C. Evaluation of the Thermal Insulation Potential of Post-Harvest Blocks Using the Native Strain of the Edible Mushroom Pleurotus ostreatus. *Buildings* **2024**, *14*, 3908. [CrossRef]
36. *IEEE Std 442-1981*; IEEE Guide for Soil Thermal Resistivity Measurements. IEEE: Piscataway, NJ, USA, 1980; pp. 1–16.
37. *D5334-00*; Standard Test Method for Determination of Thermal Conductivity of Soil and Soft Rock by Thermal Needle Probe Procedure. ASTM International: West Conshohocken, PA, USA, 2017. [CrossRef]
38. *UNE 23-725-90-2-3*; Principle of the Method [in Spanish]. Spanish Association for Standardization (AENOR): Madrid, Spain, 1990.
39. Rosa, A.C.; Teixeira, I.; Lacasta, A.M.; Haurie, L.; Soares, C.A.; Tam, V.W.; Haddad, A. Experimental Design for the Propagation of Smoldering Fires in Corn Powder and Cornflour. *Eng* **2022**, *4*, 15–30. [CrossRef]
40. Palumbo, M.; Lacasta, A.M.; Navarro, A.; Giraldo, M.P.; Lesar, B. Improvement of fire reaction and mould growth resistance of a new bio-based thermal insulation material. *Constr. Build. Mater.* **2017**, *139*, 531–539. [CrossRef]
41. Hurtado, P.L.; Rouilly, A.; Vandenbossche, V.; Raynaud, C. A review on the properties of cellulose fibre insulation. *Build. Environ.* **2016**, *96*, 170–177. [CrossRef]
42. Hadded, A.; Benltoufa, S.; Fayala, F.; Jemni, A. Thermo physical characterisation of recycled textile materials used for building insulating. *J. Build. Eng.* **2016**, *5*, 34–40. [CrossRef]
43. *DIN 4108-10*; Thermal Insulation and Energy Economy in Buildings–Part 10: Application-Related Requirements for Thermal Insulation Materials. Deutsches Institut für Normung (DIN): Berlin, Germany, 2015.
44. *DIN EN ISO 6946:2017*; Building Components and Building Elements–Thermal Resistance and Thermal Transmittance–Calculation Methods. Deutsches Institut für Normung (DIN): Berlin, Germany, 2017.
45. DesignBuilder LATAM; Official Website. Available online: https://designbuilder-lat.com (accessed on 14 October 2024).
46. *NTM 011/2:2014*; Thermal Insulation–Requirements for the Application of Insulation Materials [in Spanish, Draft Standard]. Instituto Nacional de Normalización (INN): Santiago, Chile, 2014.
47. Cintura, E.; Nunes, L.; Esteves, B.; Faria, P. Agro-industrial wastes as building insulation materials: A review and challenges for Euro-Mediterranean countries. *Ind. Crops Prod.* **2021**, *171*, 113833. [CrossRef]
48. Khalife, E.; Sabouri, M.; Kaveh, M.; Szymanek, M. Recent Advances in the Application of Agricultural Waste in Construction. *Appl. Sci.* **2024**, *14*, 2355. [CrossRef]
49. Rojas Herrera, C.J.; Rodríguez Neira, K.; Cárdenas-Ramírez, J.P. Evaluation of Two Chilean Native Macroalgae: 'Pelillo' (Gracilaria chilensis) and 'Lamilla' (Ulva sp.) for Thermal Insulation Application. *Buildings* **2023**, *13*, 2622. [CrossRef]
50. Fedorik, F.; Zach, J.; Lehto, M.; Kymäläinen, H.R.; Kuisma, R.; Jallinoja, M.; Illikainen, K.; Alitalo, S. Hygrothermal properties of advanced bio-based insulation materials. *Energy Build.* **2021**, *253*, 111528. [CrossRef]
51. Zhou, Y.; Trabelsi, A.; El Mankibi, M. A review on the properties of straw insulation for buildings. *Constr. Build. Mater.* **2022**, *330*, 127215. [CrossRef]
52. Mehrez, I.; Hachem, H.; Jemni, A. Thermal insulation potential of wood-cereal straws/plaster composite. *Case Stud. Constr. Mater.* **2022**, *17*, e01353. [CrossRef]
53. Schiavoni, S.; D'Alessandro, F.; Bianchi, F.; Asdrubali, F. Insulation materials for the building sector: A review and comparative analysis. *Renew. Sustain. Energy Rev.* **2016**, *62*, 988–1011. [CrossRef]
54. Charai, M.; Sghiouri, H.; Mezrhab, A.; Karkri, M. Thermal insulation potential of non-industrial hemp (*Moroccan cannabis sativa* L.) fibers for green plaster-based building materials. *J. Clean Prod.* **2021**, *292*, 126064. [CrossRef]
55. Padre, T.Y.; Casas, L. *Construcción + PDA: Temuco y Padre Las Casas*; Ministerio de Vivienda y Urbanismo (MINVU): Santiago, Chile, 2018; ISBN 978-956-398-525-2.
56. Kośny, J.; Thakkar, J.; Kamidollayev, T.; Sobkowicz, M.J.; Trelles, J.P.; Schmid, C.; Phan, S.; Annavajjala, S.; Horwath, P. Dynamic Thermal Performance Analysis of PCM Products Used for Energy Efficiency and Internal Climate Control in Buildings. *Buildings* **2023**, *13*, 1516. [CrossRef]

applied sciences

MDPI

Article

Comparative Lifecycle Assessment of Renewable Energy Investments in Public Buildings: A Case Study of an Austrian Kindergarten Under Atypical Operational Conditions

Georgia Kousovista, Giannis Iakovides, Stefanos Petridis, Nikolaos-Charalampos Chairopoulos, Angelos Skembris, Maria Fotopoulou *, Despina Antipa, Nikolaos Nikolopoulos and Dimitrios Rakopoulos

Centre for Research and Technology Hellas, Chemical Process and Energy Resources Institute, 52 Egialias Str., 15125 Athens, Greece; s.petridis@certh.gr (S.P.); n.chairopoulos@certh.gr (N.-C.C.); a.skembris@certh.gr (A.S.); rakopoulos@certh.gr (D.R.)

* Correspondence: fotopoulou@certh.gr

Abstract: This paper investigates the environmental and economic impacts of energy-efficient renovations, specifically focusing on the integration of photovoltaic (PV) systems in a public kindergarten. Leveraging the VERIFY platform, this study employs Life Cycle Assessment (LCA) and Life Cycle Costing (LCC) methodologies to evaluate building performance over a 25-year analysis period under three distinct scenarios: a low-usage period during the COVID-19 pandemic (2021), a normal-usage period under post-pandemic conditions (2024) with the realized investment, and a hypothetical scenario with a PV installation size that allows for appropriate reductions alongside favorable financial outcomes. The pandemic-induced occupancy reduction led to atypical energy demand patterns, with lower self-consumption and increased electricity exports to the grid, affecting the financial viability of PV investments. By incorporating post-pandemic operational data, a meaningful comparison of energy efficiency measures under constrained and stable operating conditions is conducted, addressing the impact of fluctuating demand on long-term energy investment sustainability. The results highlight that system sizing and energy reconciliation policies (net metering, net billing) significantly influence financial outcomes. The PV system achieved a Levelized Cost of Electricity (LCOE) of EUR 0.0811–0.0948/kWh, with payback periods ranging from 6.01 to 14.66 years, depending on operational intensity. The findings demonstrate that while PV systems contribute to emission reductions and cost savings, their economic feasibility depends on occupancy stability and policy frameworks. This study provides insights for optimizing renewable energy investments in public buildings, demonstrating the importance of considering dynamic operational conditions in lifecycle assessments.

Keywords: LCA; LCC; economic analysis; energy upgrade; net metering; net billing; reconciliation strategies; roof-slanted PV system

1. Introduction

The pressing global challenges of climate change, rapid urbanization, and resource depletion have led to a growing demand for sustainable, energy-efficient building practices. Life Cycle Assessment (LCA) and Life Cycle Costing (LCC) have emerged as key methodologies for evaluating environmental and economic impacts across the entire building lifecycle, from initial design through operation and end-of-life stages [1–4]. These frameworks are particularly valuable for assessing energy consumption, greenhouse gas

(GHG) emissions, and the financial feasibility of renewable energy systems and efficient materials [5]. Despite their benefits, traditional LCA and LCC models often lack the adaptability needed to account for dynamically changing conditions, such as those introduced by external disruptions like the COVID-19 pandemic [6,7].

The COVID-19 pandemic significantly disrupted building operations, revealing challenges in maintaining energy efficiency and meeting sustainability targets. With shifts in occupancy patterns, energy demand fluctuated dramatically, affecting the performance of building systems and renewable energy installations [8,9]. These disruptions underscored the need for adaptive analysis frameworks capable of addressing short-term operational changes and long-term sustainability goals [10–12]. Accurate assessments of building renovations during this period require methodologies that incorporate scenario-based modeling and simulation tools to reflect dynamically changing conditions [7,13,14].

Advancements in digital technologies, such as digital twins, are transforming the early design stages of urban and building projects by enhancing transparency and data accuracy. Saadé et al. (2022) [8] emphasized the importance of integrating circular economy principles with LCA to achieve comprehensive sustainability insights early in the design process. Similarly, Fnais et al. (2022) [9] highlight challenges in data collection and the need for robust methodologies in building LCA. Digital twins provide a virtual representation of systems, enabling real-time monitoring and iterative sustainability evaluations. By addressing gaps in data quality and collection, these technologies enhance tracking capabilities and support more accurate, transparent assessments. Adaptive energy systems that integrate photovoltaic (PV) panels [12,15–17], battery energy storage systems, and heating, ventilation, and air-conditioning (HVAC) systems have been shown to significantly enhance energy efficiency and cost savings in residential and office buildings. Zhao et al. (2023) [13] demonstrated that combining PV systems with battery storage reduces energy costs by optimizing real-time energy use in office buildings, resulting in up to a 30% improvement in energy utilization efficiency. Similarly, Linssen, Stenzel, and Fleer (2015) [14] highlighted how variations in consumer load profiles influence the economic and energy performance of PV-battery systems, underscoring their potential to align energy generation and demand effectively. These integrations not only improve operational efficiency but also reduce reliance on grid electricity, leading to lower overall costs and improved energy performance. Despite these advancements, many LCA and LCC tools still predominantly rely on static estimations of the use-stage performance of the devices analyzed, limiting their capacity to evaluate dynamically changing operational conditions [11,18].

To address these challenges, this study utilizes the VERIFY platform [19–21], a web-based tool integrating LCA and LCC methodologies for building renovation assessments. VERIFY combines static and dynamic data with simulation-based evaluations to provide scenario analyses of environmental and economic impacts. The platform aligns with international standards such as ISO 14040, ISO 14044, ISO 15686-5, and EN 15978 [22–28] and incorporates frameworks like the EU's Level(s) [29–31] for energy performance and global warming potential (GWP) indicators. These features make it particularly relevant for analyzing long-term building performance under varying conditions, including disruptions like the COVID-19 pandemic [32–35].

The case study presented in this paper is part of the AURORAL project (https://www.auroral.eu/ (accessed on 1 August 2024), Grant Agreement No. 101016854), a European initiative running for 48 months, designed to advance smart communities by leveraging digital services and fostering interoperability. This program aims to enhance connectivity and establish a digital ecosystem comprising smart objects and interoperable service platforms, promoting innovation, applications, and services tailored to rural settings. By bridging the digital divide between rural and urban areas, AURORAL contributes to economic development, job creation, and addressing critical societal challenges in underrepresented regions. However, the COVID-19 pandemic introduced significant challenges during the project's implementation, such as delays in pilot deployments, restricted field activities, and disruptions in stakeholder engagement across the seven European regions. These limitations impacted the project's ability to collect real-world data and validate cost-efficient cross-domain solutions under normal conditions, which may have influenced the project's impact, as evaluated through its stated key performance indicators (KPI) and target values. In the presence of major unforeseen disruptions, the evaluation of such activities' efficacy requires systematic frameworks that can combine both data- and model-driven approaches and scenario-based analytic capabilities to estimate the potential outcomes of investments in the energy upgrade of the European building stock even in the presence of limited field data.

This study examines the renovation of a public kindergarten in Austria conducted during the AURORAL project utilizing the VERIFY platform. The analysis spans a hypothetical 25-year period and evaluates two distinct operational scenarios: a low-usage case, based on field data collected during 2021 during the COVID-19 pandemic when building occupancy was significantly reduced, and a high-usage case that reflects stable conditions with full occupancy, using the economic conditions prevalent in 2024. By integrating LCA and LCC methodologies, this research aims to demonstrate the benefits of scenario-based analysis in situations with atypical social and economic conditions, which would otherwise prevent a reliable estimation of an energy upgrade investment's impact. It also highlights the need for such analyses to adapt to dynamic operational conditions and assess long-term energy and environmental performance. The findings offer insights into the impact of reduced building usage on energy performance and emissions, validate scenario-based approaches for evaluating sustainability, and provide recommendations for integrating renewable energy systems to ensure long-term cost-efficiency and environmental resilience in public buildings.

2. Comparative Review of Life Cycle Analysis Tools for Building Assessment

2.1. Existing LCA/LCC Tools and Methods

The field of building assessment has produced various LCA tools that evaluate the environmental and economic impacts of construction and renovation projects. While these tools have significantly advanced sustainability practices, they exhibit certain limitations, particularly when applied to dynamic scenarios such as building renovations under atypical conditions. This section provides a comparative overview of widely used LCA tools, identifying their strengths and shortcomings to set the stage for how VERIFY addresses these gaps in the next subsection.

Several established tools are frequently used for LCA and LCC in building assessments, each focusing on specific aspects of sustainability and performance. SimaPro, for instance, is a versatile tool applied across various industries, including construction, offering robust data modeling and environmental impact assessment capabilities [36]. However, its broad applicability often requires extensive customization for construction-specific applications and lacks modules tailored to operational energy performance or renovation scenarios. Similarly, the Athena Impact Estimator for Buildings, designed for the North American construction sector, facilitates detailed assessments of embodied energy and emissions through its comprehensive material database [37]. Despite its strengths in material impact analysis, Athena's limited capability for operational energy modeling and lifecycle economic evaluations restricts its scope for comprehensive building performance analyses. RETScreen, on the other hand, is widely used for renewable energy feasibility studies and performance analysis, offering an integrated platform to evaluate energy efficiency measures, cost savings, and emission reductions. While RETScreen excels in energy performance and financial analysis, it is less suited for detailed lifecycle environmental assessments [38].

In Europe, One Click LCA software is widely used for its integration with Building Information Models (BIM), streamlining LCA processes for architects and construction managers [39]. Its alignment with European standards and extensive material database makes it particularly effective for material impact evaluations. However, as of v1.0.4.0, its primary focus on material impacts limits its flexibility in addressing scenario-based analyses or dynamic simulations, which are essential for assessing renovation strategies. Tally, another tool integrated with BIM, excels in evaluating embodied carbon and resource use during early design stages [40]. While valuable for preliminary assessments, Tally's limited applicability to post-construction evaluations and operational energy modeling reduces its suitability for comprehensive lifecycle analyses of building renovations.

Despite their contributions, existing tools often face limitations when applied to dynamic building renovations. Most are designed for static evaluations, which restricts their ability to simulate changing operational conditions, such as occupancy shifts or evolving energy demands. For example, many tools emphasize material-focused analyses or short-term performance, making them less effective for evaluating long-term impacts or responding to disruptions like the COVID-19 pandemic. Additionally, limited scenario flexibility and a lack of advanced simulation capabilities hinder the comprehensive assessment of lifecycle costs, payback periods, and renovation strategies. Tools optimized for specific regional markets further reduce their adaptability to global or multi-regional contexts.

2.2. VERIFY as a Dynamic LCA/LCC Tool

The abovementioned challenges underline the necessity of tools that integrate environmental and economic assessments with dynamic, scenario-based modeling to better support decision-making in building renovations. VERIFY addresses these gaps by combining robust data integration, advanced simulation capabilities, and scenario flexibility. VERIFY enables stakeholders to assess the impacts of renovations across a range of conditions, such as atypical disruptions like the COVID-19 pandemic, by dynamically integrating data on energy usage, energy tariffs, and renewable energy contributions. Unlike static tools, VERIFY's interoperability with energy modeling tools supports long-term energy performance projections and thermal simulations over extended periods, providing accurate insights into potential renovation impacts. Additionally, VERIFY's alignment with European and international standards ensures compliance with regulatory requirements and facilitates the generation of standardized reports suitable for certification processes.

VERIFY is a web-based platform developed to address the challenges of integrating environmental and economic assessments for building renovations, particularly under dynamic and atypical conditions. Its hybrid model data-driven LCA and LCC approach offers a holistic framework for the evaluation of environmental and financial implications of renovation strategies. By utilizing both static and dynamic data, VERIFY enhances the accuracy and relevance of its assessments, addressing limitations observed in traditional tools. The modular design within VERIFY ensures that components such as building systems, renewable technologies, and energy storage are analyzed both independently (in terms of their lifecycle impact and costs) and jointly (in terms of their contribution to the building's energy consumption and production patterns) while considering their contributions to the overall lifecycle performance. This modular functionality allows for both targeted system-level optimizations and holistic assessments, enabling stakeholders to evaluate short-term operational impacts and long-term sustainability goals with precision.

In the context of project AURORAL, VERIFY was integrated into the broader AURORAL platform to support the LCA and LCC analyses for the project's building renovation projects. Figure 1 presents the architecture of the overall platform. It should be noted that, while in this case, the AURORAL platform was used as a data source, any source provisioning either real-time or historical data of hourly granularity can be used for VERIFY's analyses.

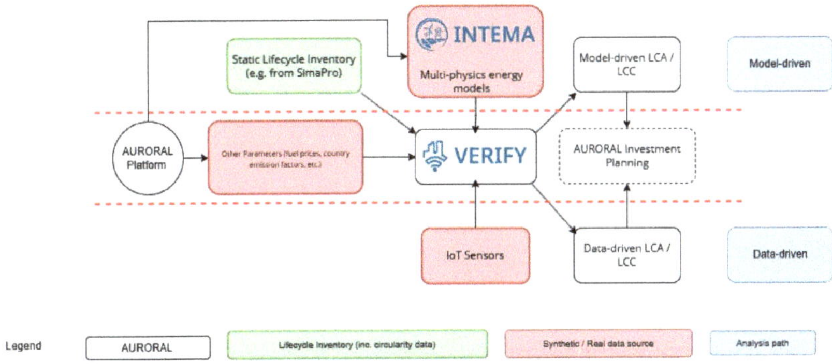

Figure 1. Integration of INTEMA and VERIFY within AURORAL Framework for Comprehensive LCA and LCC Analysis.

For cases where no historical data are available, VERIFY relies on physics-based energy simulation software to acquire representative time series for the building's operational profile. This is achieved through VERIFY's integration with INTEMA, a building energy modeling tool developed by the Centre for Research and Technology Hellas and validated through its application in multiple modeling scenarios [21,41–46]. INTEMA, developed in Modelica [47], conducts accurate building energy performance simulations. The underlying Modelica library consists of two main packages: the first deals with the passive elements of the building, accounting for various heat transfer phenomena such as conduction, convection, short-wave, and long-wave radiation. The second package includes dynamic models for the building's active systems, such as HVAC, electrical generation, and energy storage. The variables considered for estimating the energy load include, among others, (i) solar radiation; (ii) the presence of people, their activities, and the use of the space; (iii) psychrometric charts and HVAC system operations; and (iv) power generation from electrical systems, if applicable. The flexibility of the Modelica language and the ability to develop custom components ensure precise simulations of energy demand and efficiency under various operating conditions, providing an accurate foundation for environmental impact assessments.

In cases where energy billing data are available for building systems that include heat pumps to meet their heating and cooling demands, a custom method for time-series disaggregation has been implemented, whereby synthetic hourly-based time series of the building's energy usage are adjusted using historical actual energy consumption data from the bills. This approach allows for an accurate estimation of the heat pump's consumption as a portion of the total demand. This estimation is crucial for the LCA of the heat pump, as it provides a solid basis for evaluating its operation by combining both simulated data and actual meter readings.

In addition to simulation data from INTEMA, VERIFY also utilizes additional system parameters, such as energy profiles and system efficiencies, and integrates these values with static lifecycle inventory data and additional external dynamic parameters like fuel prices and emission factors. VERIFY synthesizes these inputs to perform comprehensive LCA and LCC assessments, producing insights into the environmental and economic impacts of renovation scenarios.

3. Methodological Framework

3.1. Scenario Definition

3.1.1. Building Configuration and Renovation Action

As part of the project "AURORAL: a European initiative aimed at enhancing smart communities through the integration of digital tools and interoperability" (https://www.auroral.eu (accessed on 1 August 2024)/, Grant Agreement No. 101016854), a case study was conducted on a public kindergarten located in Southern Burgenland, Austria. This single-story building, constructed primarily with bricks, spans a total floor area of 168.85 m^2 and operates year-round. The kindergarten's energy demands, particularly for heating during winter and cooling in summer, are covered using conventional energy systems, namely, a heat pump and basic air conditioning.

The energy upgrade action implemented under the AURORAL project focused on the installation of a photovoltaic (PV) system as a key step toward sustainability. The roof-slanted PV system was designed to reduce reliance on grid electricity by generating renewable energy during daylight hours. The building operates under a net billing buyback scheme, enabling it to export surplus electricity to the grid while remaining connected for additional energy needs. This targeted sustainable energy measure represents a practical application that constitutes a basic use case for demonstrating how scenario-based analysis can be used to analyze and optimize the cost-efficiency of sustainability investments in real-world contexts.

Table 1 outlines the basic structural and operational characteristics of the SBA1 Kindergarten, providing an overview of the building.

Table 1. Basic Key Attributes of SBA1 Kindergarten.

Parameter	Details
Project Name	SBA1 Kindergarten
Location	Southern Burgenland, Austria (Latitude: 47.0145, Longitude: 16.4579)
Building Type	Public Kindergarten
Building Structure	Single-story, primarily brick construction
Total Floor Area	168.85 m^2
Operation Schedule	Year-round
External Wall Material	Brick
Number of Floors	1
Operational Parameters	Net billing buyback scheme
Primary Energy Source	Electricity

Table 2 provides detailed information about the primary building components and systems, including the heating and cooling systems, insulation, glazing, and the domestic hot water (DHW) system. These details reflect the existing infrastructure and its associated costs and performance metrics.

Table 2. Building Components and Systems.

Component	Details
Heating System	Heat Pump Air-to-Water
Thermal Rating (kW)	9.6
Heating Efficiency	2.5
Lifetime (years)	20
Heating Purchase Cost (EUR)	6720.00
Heating Maintenance Cost (EUR/year)	490.56
Cooling System	Basic Air Conditioning
Thermal Rating (kW)	7.5
Cooling COP	2.5
Lifetime (years)	20
Cooling System Purchase Cost (EUR)	1298.08
Cooling System Maintenance Cost (EUR/year)	51.92
Insulation Material	Extruded Polystyrene
Thickness (mm)	100
Surface Area (m^2)	110.3
Coverage	All Walls
Lifetime (years)	100
Insulation Purchase Cost (EUR)	3131.90
Glazing Type	Double Glazing (Double/Float 4–16–4 (Air))
Glazing Opening Surface (m^2)	160.61
Frame Material	Aluminium Frame (20% Frame Coverage)
Glazing Lifetime (years)	40
Glazing Purchase Cost (EUR)	51,652.18
Glazing Maintenance Cost (EUR/year)	25.83
Tank for Domestic Hot Water	Copper Tank
Tank Capacity (L)	600
Electricity Rating (kW)	5
Lifetime (years)	20
DHW Purchase Cost (EUR)	1681.79
DHW Maintenance Cost (EUR/year)	16.82

Table 3 summarizes the energy upgrades implemented in the SBA1 Kindergarten, focusing on the photovoltaic (PV) system and its associated environmental metrics.

Table 3. Energy Upgrade Details (PV system).

PV System Parameter	Details
Mounting	Roof-slanted PV System
PV Panel Material	Polycrystalline-Si
Reference Capacity (kWp)	20.14
PV Purchase Cost (EUR)	19,334.40
Lifetime of PV System (years)	30

The values in Table 4 for Primary Energy (PE) demand and Global Warming Potential (GWP) (Stage A quantities) were obtained using SimaPro v9.5.0.2 software with the EcoInvent database v3.9.1 [48], and the same impact assessment method was applied to all components. The Life Cycle Inventory (LCI) data are consistent across components, ensuring comparability.

Table 4. Primary Energy Demand and GHG Emissions (Stage A) for Key Building Systems.

Component	Stage A—PE Demand (GJ)	Stage A—GWP ($kgCO_2$-eq)
Heat Pump Air-to-Water	6.32	1526.4
Basic Air Conditioning	0.0525	3525
Insulation	74.44	3187.31
Glazing	62.48	5460.74
Domestic Hot Water	10.1	868
PV System	1770	11,882.6

Table 5 presents the CO_2 equivalent (CO_2-eq) emission factors for Austria, as officially reported for each EU Member State, from the dataset provided by the Joint Research Centre (JRC) of the European Commission for the years 1990 to 2021 [49]. Since VERIFY dynamically adjusts the CO_2-eq emission factor in its analysis every year, a Logistic Regression model was developed to estimate the corresponding emission factors beyond 2021 based on historical values. This model fits the historical data and generates future projections (extrapolations) of emission factors based on the characteristic trend observed. The model is based on the logistic function, as shown in Equation (1) [50]:

$$f(x) = \frac{L}{a - \exp(-k\ ) \cdot (x - median(x)))} \quad (1)$$

where the parameters L, k, and α are estimated through curve-fitting on the historical data. The curve fitting was performed using the nonlinear least squares method, minimizing the sum of squared residuals between the observed and predicted values. The algorithm was implemented using Python 3.8.

Table 5. Country-Specific Electricity and Emissions Factors for Austria.

Parameter	Value	Units
Primary Energy Factor	1.91	KWh_P/kWh
CO_2-eq Emission Factors	2021: 0.281 2022: 0.225 2023: 0.192 2024: 0.312	kg/kWh

The analysis period for the kindergarten spans 25 years, during which the CO_2-eq emission factor for years beyond 2024 is assumed to remain consistent with the 2024 value.

3.1.2. Analysis Period

The performance of the kindergarten following the installation of the photovoltaic (PV) system was evaluated over a 25-year period through two distinct scenarios, each representing different building usage patterns and operational conditions.

The first scenario focuses on a low-usage case, reflecting the conditions during the COVID-19 pandemic. For this scenario, real building energy consumption data were collected during 2021, a period marked by reduced building occupancy and irregular usage patterns caused by lockdowns and restrictions. These atypical operational conditions resulted in fluctuating energy demands, and the analysis examines how the PV system mitigated energy costs and enhanced resilience under limited usage. The second scenario represents a high-usage case, corresponding to normal, stable operational conditions with full building occupancy. This analysis begins in 2024 and models performance under consistent energy demand over the remaining timeline. By comparing the two cases, this study evaluates the long-term sustainability and financial benefits of the PV system under normal usage patterns without real field data, including operational cost savings and reductions in greenhouse gas emissions.

These analyses, conducted within the AURORAL project, align with broader efforts to foster digital transformation and implement smart energy solutions in private and public infrastructure. The use of VERIFY ensures a robust, data-driven evaluation of energy and economic performance under both dynamic and stable conditions, supporting sustainable decision-making and long-term planning.

3.2. Data Availability

3.2.1. Economic Context: Electricity and Inflation Rates

To accurately assess the economic performance of the kindergarten renovation project, electricity prices and inflation rates during the analyzed periods were incorporated into the LCC analysis. These metrics, sourced from Eurostat, reflect the real-world conditions impacting the project's financial sustainability. Table 6 presents key electricity prices and inflation rates for Austria for 2021 and 2024, including a 2024 (without subsidy) column to estimate prices under unsubsidized conditions [51–53].

Table 6. Electricity Prices and Inflation Rates of Austria (source Eurostat (https://ec.europa.eu/eurostat/databrowser/view/nrg_pc_204/default/table?lang=en (accessed on 16 October 2024))).

Prices	Average of 2021	2024	2024 (Without Subsidy)
Import Price (EUR/kWh)	0.2251	0.2731	0.3878
Export Price (EUR/kWh)	0.1421	0.2946	0.2946
Buyback Price (EUR/kWh)	0.0635	0.0771	0.0836
Inflation Rate (%)	2.76	3.48	

Austria's electricity export prices are determined by wholesale market prices, particularly the day-ahead market, which drives cross-border electricity trade and influences arbitrage opportunities between countries. In the absence of precise export price data, we assumed that the export price would equal Austria's wholesale price. The values used include the 2021 average price (from both semesters) and the 2024 first semester price. For the import price, all components, such as taxes, levies, and network charges, were included to reflect the full cost borne by consumers. The 2021 and 2024 values incorporate Austria's Eco-Social Tax Reform, which introduced carbon pricing, further increasing energy costs [54,55]. The buyback price for 2021 was provided by AURORAL project partners. To estimate the base price for 2024, we applied a 21.34% increase derived from observed price trends. This adjustment yields a 2024 base buyback price of 0.0771 EUR/kWh. The "2024 (without subsidy)" column estimates the impact of removing Austria's energy subsidies and reductions. This scenario highlights potential cost increases for end-users under an unsubsidized framework [56].

These values highlight significant fluctuations influenced by external shocks, including the COVID-19 pandemic and the war in Ukraine. During the pandemic, global supply chain disruptions, shifts in energy demand, and government interventions created volatility in energy markets, initially leading to lower prices but subsequently driving increases as economies reopened. The war in Ukraine further exacerbated energy price inflation in Europe due to disruptions in natural gas supplies, impacting electricity prices and Austria's inflation rates. These factors were integrated into VERIFY's analysis to ensure that financial assessments reflect the turbulent economic conditions and provide a realistic understanding of energy performance under both historical and projected market scenarios.

Additionally, this analysis considers multiple scenarios for low usage and high usage conditions with varying PV system sizes and corresponding CAPEX, as summarized in Table 7. As the exact cost of the PV system installation was not covered within the AURORAL project, the cost was estimated based on national benchmarks for PV system prices in Austria. These benchmarks were sourced from the International Energy Agency Photovoltaic Power Systems Programme (IEA-PVPS) for 2021 and 2023 [57–59].

Table 7. PV Characteristics Across Usage Scenarios.

Case	Electrical Demand (kWh/year)	PV Production (kWh/year)	CAPEX (EUR)
Low Usage	7739	24,066.99	27,158.79
High Usage (Large PV)	10,608.35	24,066.99	29,897.83
High Usage (Small PV)		13,431.66	16,687.37

3.2.2. Data for Energy Demand and Performance

This analysis examined the kindergarten's energy demand and the performance of its PV system under three scenarios. During the low-usage case, energy demand variations were influenced by changes in occupancy and usage patterns due to pandemic-related restrictions. Austria, including the Southern Burgenland region, experienced multiple lockdowns between March 2020 and December 2021, with schools and public buildings facing closures and limited operations during this period [60]. Detailed time series data, consisting of 8760 hourly final energy consumption measurements for the year 2021, were sourced from the AURORAL platform. These data covered key energy-related components, including the air conditioning (AC) system, heat pump, domestic hot water (DHW) system, PV system, and electrical consumption for lighting and other uses. The analysis specifically relied on the 2021 dataset, as it provided the most complete and representative time series available. This annual dataset served as the foundation for evaluating the building's energy demand and the performance of its systems under hypothetical reduced usage conditions for a period of 25 years.

The high-usage period modeled stable operational conditions for 25 years (2024–2049), focusing on the sustainability and economic performance of the PV system. In the absence of operational data for this extended period, the annual electricity demand for the kindergarten was estimated at 10,608.38 kWh, based on regional benchmarks for European schools [61] reporting an average energy use of 120 kWh/m^2/year. Using the kindergarten's total floor area of 168.85 m^2, this value reflects the projected energy demand under normal operational conditions. To further simulate system interactions, hourly time series data were generated using INTEMA.building, providing a detailed representation of the building's energy demand and interactions among its systems.

In addition, as the energy consumption patterns indicated that the PV-based system was overprovisioned considering the actual needs of the building, a third hypothetical scenario was considered, with a PV system that was more rationally sized based on the expected annual electricity demand of the building. This scenario is referred to as "high-usage (low PV)" in the rest of the text.

To better understand the energy performance under both low- and high-usage conditions, the analysis considered two distinct energy policies: net metering buyback and net billing buyback. The net metering buyback scheme allows the electricity generated and exported to the grid to directly offset the electricity consumed on a 1:1 energy basis. At the end of the billing period, any surplus energy can either roll over as a credit for future use (rolling credit) or be compensated at a predefined rate. This approach generally maximizes economic benefits when energy generation and consumption are balanced over the billing period. The net billing buyback scheme, on the other hand, values the exported energy at a lower rate than the imported energy price. Instead of offsetting energy directly, the policy balances the monetary difference between exported and imported electricity. Both policies were examined under various reconciliation periods (yearly, quarterly, monthly, and daily) to analyze their economic impacts. In this context, "reconciliation" refers to the timeframe over which the balance between exported and imported electricity is calculated. For example, yearly reconciliation aggregates surplus energy credits over the year to offset annual consumption costs. Conversely, monthly or daily reconciliation calculates balances over shorter periods, which can reduce the economic benefit for users when energy generation and consumption fluctuate significantly.

The results highlight how these policies influence the building's energy costs under both low-usage conditions (using 2021 data during the COVID-19 period) and high-usage conditions (using 2024 data under stable operation). These policies align with established methodologies and are supported by the literature, such as the comparative assessment of net metering and net billing systems by Dufo-López and Bernal-Agustín (2015) [56]. By considering different reconciliation strategies and energy export schemes, the analysis provides a comprehensive evaluation of the economic performance of the PV system across varying building usage scenarios, different socioeconomic conditions affecting energy prices, as well as different billing schemes. The large number of parameters and potential combinations to be considered highlights the advantages of scenario-based approaches over the traditional solutions outlined in Section 2.1.

3.3. *Key Performance Indicators*

The evaluation of the kindergarten energy upgrade project is grounded on a robust framework of Key Performance Indicators (KPIs) designed to capture both the environmental and economic impacts of the energy upgrade. These KPIs align with the European Level(s) framework, ensuring compliance with standardized sustainability benchmarks. By focusing on critical aspects such as energy efficiency, carbon emissions, and financial performance, the KPIs provide a comprehensive methodology for assessing the renovation's outcomes over the building's lifecycle.

An overview of the equations, variables, and methodologies for these KPIs is provided in Table 8, while a detailed explanation of the equations, variables, and methodologies is available in the Supplementary Materials.

Table 8. Key Performance Indicators.

KPI Name	Equation	Units	Description
		Environmental KPIs	
Global Warming Potential (GWP)	$L_{GWP} = \frac{L_{GHG}}{Useful\ area}$ with $L_{GHG} = I_{GHG} + \sum_{i=1}^{N} \left(O_{GHG}^{[i]}\right)$	kg CO_2-eq/m^2	L_{GHG}: Total greenhouse gas emissions over the lifecycle *Useful Area*: Building area with heating/cooling access $O_{GHG}^{[i]}$: Operational GHG emissions of the building's components in year i; I_{GHG}: infrastructure (embodied) GHG emissions.
Primary Energy Demand (PED)	$L_{PE} = I_{PE} + \sum_{i=1}^{N} \left(O_{PE}^{[i]}\right)$	kWh	I_{PE}: Energy demand during infrastructure/construction phases $O_{PE}^{[i]}$: Operational/maintenance energy demand during building's use stage
		Economic KPIs	
Lifecycle Costs (LCC)	$LCC = I_C + \sum_{i=1}^{N} \left(O_{NC}^{[i]}\right) - V_R$	EUR	I_C: Infrastructure costs (CAPEX) $O_{NC}^{[i]}$: operational net costs, including energy consumption and maintenance V_R: Residual value of components at end of lifecycle
Whole Life Cost (WLC)	$WLC = I_C + \sum_{i=1}^{N} \left(O_{C}^{[i]}\right) - V_R$	EUR	I_C: Infrastructure costs (CAPEX) $O_{C}^{[i]}$: operational costs (energy consumption/maintenance and operational revenues due to electricity exports from RES V_R: Residual value of components at end of lifecycle
Renewable Energy System (RES) Payback Period	$T_P = T_L + t_r$	Years	T_L: Last year before cumulative savings meet or exceed RES investment t_r: Remaining fraction of the year required for full payback of RES investment
Levelized Cost of Electricity (LCOE)	$LCOE = \frac{\sum_{i=1}^{N} \frac{I_{C,GEN}^{[i]} + O_{C,GEN,MN}^{[i]} + O_{C,GEN.NEI}^{[i]}}{(1+r)^i}}{\sum_{i=1}^{N} \frac{SC^{[i]} + EX^{[i]}}{(1+r)^i}}$	EUR/kWh	$I_{C,GEN}^{[i]}$: generator infrastructure costs $O_{C,GEN,MN}^{[i]}$: annual generator maintenance costs (which include replacement costs if the analysis period exceeds the generators' lifetime) $O_{C,GEN.NEI}^{[i]}$: costs of fuel used for electricity generation (applicable only in the case of electricity generators using other fuel) $SC^{[i]}$ and $EX^{[i]}$: total energy self-consumed and exported by the building r :project discount rate
Net Present Value (NPV)	$NPV = \sum_{i=0}^{N} \frac{R_i - C_i}{(1+r)^i}$	EUR	R_i: revenue or savings in period I C_i: cost in period i, when $i = 0$ the C_0 is the initial investment r: project discount rate i: period index
Return on Investment (ROI)	$ROI = \frac{NPV}{Initial\ Investment}$	ratio	NPV: Net Present Value *Initial Investment*: capital expenditure at the beginning of the project

4. Results and Discussion

4.1. Results

4.1.1. Comparison of Energy Performance: Low and High Usage of the Kindergarten

To compare the results between the low-usage and high-usage scenarios of the building, three 25-year analyses were conducted using the VERIFY platform. The analysis examined key metrics, including electrical demand, self-consumption levels, and energy exported to the grid. Two sets of results were generated to facilitate comparison: the energy performance during the COVID-19 period (2021) and the estimated performance under stable, post-pandemic conditions (2024). The objective was to highlight differences in energy usage patterns and evaluate the impact of the pandemic on the kindergarten's energy management.

Figure 2 illustrates the energy performance of the kindergarten in 2021, reflecting the low-usage conditions resulting from the COVID-19 pandemic mitigation measures. The blue line illustrates the daily export of electricity to the grid, while the green line depicts the self-consumption levels within the building. During the summer months, when solar irradiance was higher, the PV system generated significantly more electricity, leading to notable peaks in energy exported to the grid. However, self-consumption levels remained consistently low throughout the year due to the building's limited operations under COVID-19 conditions. This low internal energy usage highlights the effect of restrictions, as the kindergarten was not fully utilized during this period. The consistent gap between production (export) and consumption underscores the underutilization of generated energy for internal needs, emphasizing the impact of external disruptions on energy management and, potentially, the grid.

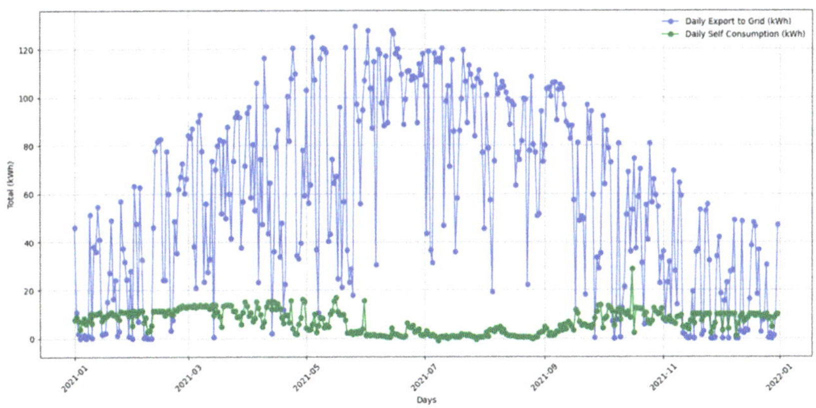

Figure 2. Daily Export and Self-Consumption Patterns for Low Usage (2021).

Figure 3 depicts the estimated daily energy performance of the kindergarten throughout 2024, representing high-usage conditions under stable and normal operational circumstances. The green line represents daily self-consumption levels, while the blue line indicates daily electricity export to the grid. During the winter months, when the kindergarten operates at full capacity, energy demand is higher due to heating and regular building use. The PV production during this time is primarily used for self-consumption, as shown by consistently higher green line values, while export levels to the grid remain minimal. In contrast, during the summer months, when the building is closed for school holidays, energy demand drops significantly. As a result, the PV system's output is largely exported to the grid, reflected by higher blue line peaks.

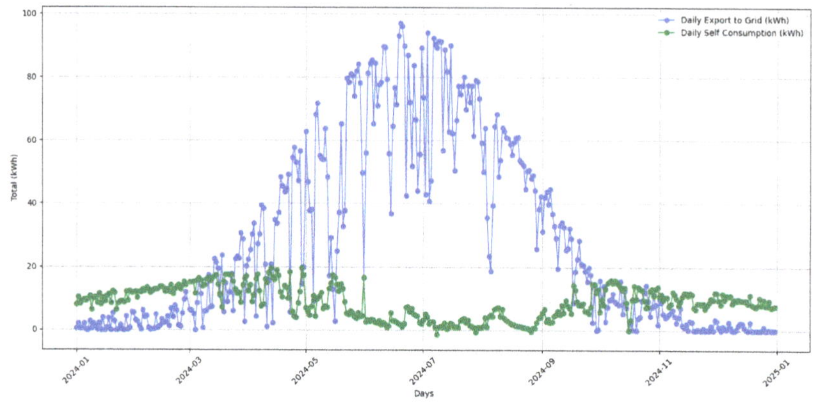

Figure 3. Daily Export and Self-Consumption Patterns for High Usage (2024).

Figure 4 presents the aggregated monthly energy consumption and production patterns for the kindergarten during 2021, characterized by low usage. The blue bars represent electricity exported to the grid; the green bars indicate self-consumed electricity generated by the PV system, and the orange bars show the building's total electrical demand. Throughout this period, monthly electrical demand remained reduced and relatively stable, reflecting limited operations and lower occupancy caused by pandemic restrictions. Electricity exports fluctuated significantly, peaking during the summer months (June to August) when increased solar irradiance boosted PV production. Conversely, exports were lower in the winter months due to reduced solar production. Self-consumption levels remained consistent throughout the year, corresponding to the building's baseline energy demand, such as energy needed for maintenance systems, which persisted regardless of occupancy. The aggregated data also highlight that electricity imports occurred when PV production could not fully meet demand, even though substantial amounts of energy were exported during periods of surplus generation.

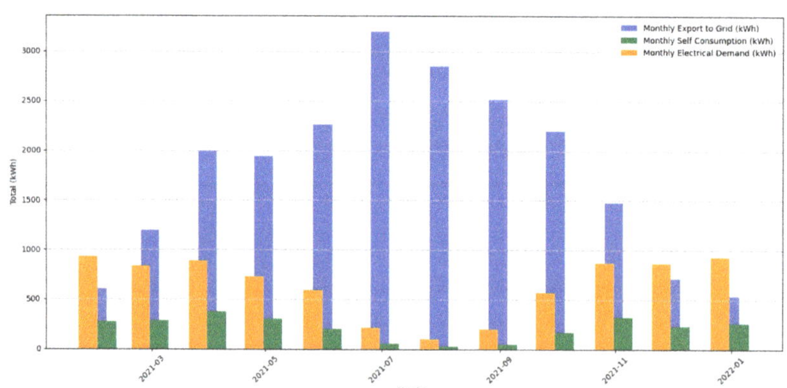

Figure 4. Monthly Export to Grid, Self-Consumption, and Electrical Demand during Low Usage (2021).

Figure 5 depicts the estimated monthly energy performance of the kindergarten under normal operational conditions, simulating a fully functional year (2024) without pandemic disruptions. The blue bars represent electricity exported to the grid; the green bars indicate self-consumed electricity, and the orange bars reflect the total electrical demand. Under high usage, the monthly electrical demand was significantly higher than during the low usage scenario, reflecting the kindergarten's full operational schedule with regular occupancy and activity levels. Self-consumption was also notably higher, particularly during the winter months, as the PV system supported the increased energy needs of the building, including heating and lighting. During the summer months (June to August), the kindergarten's energy demand dropped due to school holidays, leading to reduced self-consumption. However, the PV system continued to generate electricity at high levels due to optimal solar conditions. This resulted in a marked increase in electricity exported to the grid (blue bars), as surplus energy was fed into the grid when internal demand was minimal.

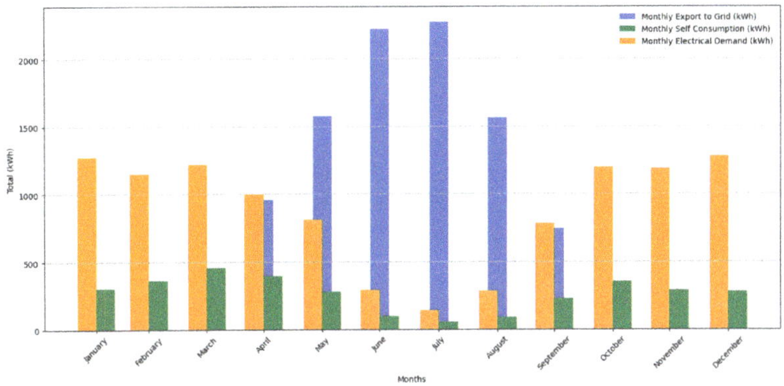

Figure 5. Monthly Export to Grid, Self-Consumption, and Electrical Demand Under High Usage (2024).

To further assess the impact of the kindergarten's energy upgrade across the three usage scenarios, the key performance indicators (KPIs) were summarized to provide a comprehensive evaluation of the environmental and economic outcomes. These KPIs, derived from the VERIFY platform, offer a detailed analysis of the building's performance, highlighting the effects of the implemented energy sustainability measures under differing operational conditions.

The comparison presented in the tables showcases the variations and improvements achieved during the low-usage period and the high-usage period, as well as the building's baseline performance without any upgrades. Table 9 focuses on the environmental KPIs, comparing results across these scenarios. The findings reveal differences in PED and GWP, offering valuable insights into the environmental benefits of the energy upgrade and its effectiveness under varying usage conditions.

Table 9. Environmental Outcomes of Low and High Usage of the kindergarten for 25 years of analysis.

Environmental KPIs	Without PV		Low Usage		High Usage	
	Total	Annual Avg.	Total	Annual Avg.	Total	Annual Avg.
Lifetime Primary Energy Demand (kWh)	547,941	21,917.63	385,040	15,401.61	510,024	20,400.98
Lifetime Primary Energy Demand Per m^2 (kWh/m^2)	3245.13	129.81	2280.37	91.21	3020.57	120.82
Lifetime Global Warming Potential ($kgCO_2$-eq)	61,868.07	2474.72	45,975.52	1839.02	56,796.02	2271.84
Lifetime Global Warming Potential per m^2 ($kgCO_2$-eq/m^2)	366.41	14.66	272.29	10.89	336.37	13.45

The results reveal significant differences in both Primary Energy Demand (PED) and Global Warming Potential (GWP) across the analyzed scenarios, reflecting the impact of operational intensity and energy upgrades. For Primary Energy Demand (PED), the baseline case (without PV installation) records the highest lifetime PED, reflecting the building's full reliance on grid electricity. In the low-usage scenario, the lifetime PED was 29.7% lower than the baseline, driven by reduced occupancy and decreased energy demand during periods of limited kindergarten usage. Conversely, in the high-usage scenario, lifetime PED shows a 6.9% reduction compared to the baseline but increases relative to the low-usage case. This rise in PED under high usage is attributed to greater operational intensity and consistent energy demands under normal conditions. Under low-usage conditions, the GWP decreases by 25.7% compared to the baseline, primarily due to lower energy consumption during periods of limited occupancy. In the high-usage scenario, lifetime GWP is 8.2% lower than the baseline but 23.5% higher than the low-usage case, reflecting the increased energy consumption associated with higher operational intensity under normal usage conditions.

The observed values also account for the impact of gradual component degradation on energy performance and environmental outcomes over time. The KPIs highlight the interplay between operational energy demands, the efficiency of the renewable energy system, and lifecycle events such as maintenance and component replacements. While the results confirm the kindergarten's ability to achieve emission reductions through PV system upgrades, they also emphasize the importance of ongoing system optimization to sustain these benefits over the long term.

The trends observed in the KPIs are further contextualized and analyzed in the subsequent discussion of the PED and GHG emissions plots, providing deeper insights into the system's performance under varying operational conditions.

Figures 6 and 7 illustrate the annual trends in PED and GWP over the 25-year analysis period, highlighting key lifecycle events and differences between low-usage and high-usage scenarios:

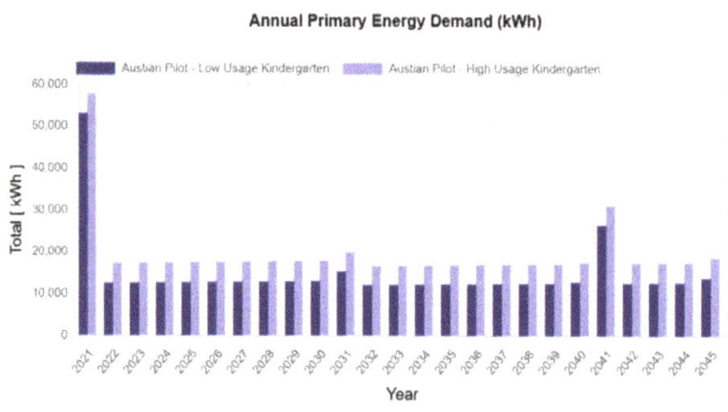

Figure 6. Annual PE Demand for Low and High Usage over 25 years of analysis.

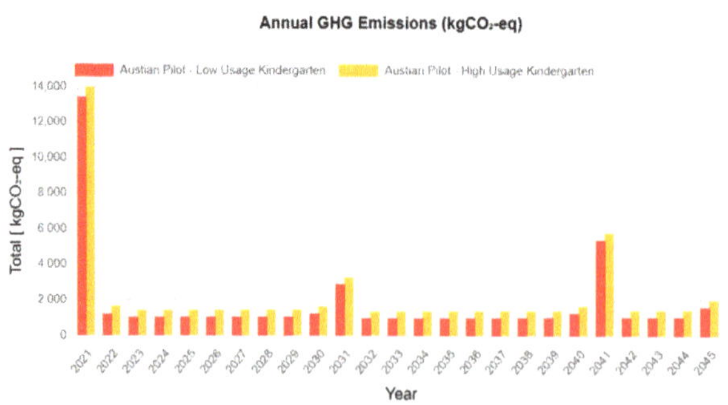

Figure 7. Annual GHG Emissions for Low and High Usage over 25 years of analysis.

- Initial Peak (2021): The sharp peak observed in 2021 reflects the embodied energy associated with the installation of the PV system. This includes energy requirements for the manufacturing, transportation, and installation processes, which significantly increase PED and GWP in the first year;
- Steady Trend with Gradual Increase: Following the initial installation, PED and GWP stabilize but exhibit a gradual increase over time. This trend is driven by the 0.5% annual degradation assumed in the PV system's performance. As the system's efficiency declines, its renewable energy output decreases, causing a greater reliance on grid electricity to meet the building's energy demand;
- Spikes (2031 and 2041): Distinct spikes are expected to occur in 2031 and 2041, corresponding to the replacement of the air conditioning (assumed to occur in 2031) and the heat pump (in 2041) systems. These lifecycle events temporarily increase PED due to the embodied energy required for the manufacturing and installation of the replacement components.

4.1.2. Comparison of Economic Performance: Low and High Usage of Kindergarten

The economic evaluation employs LCC methodologies to provide detailed insights into the long-term financial impacts of the renovation measures. By capturing data on the building's economic performance, the assessment ensures a robust and holistic understanding of its energy systems. The findings validate the renovation strategy, demonstrating that the kindergarten remains energy-efficient and economically viable throughout its lifecycle. To further analyze cost-effectiveness, the evaluation incorporates two distinct schemes:

- Net metering—where exported energy offsets imported energy on a 1:1 basis;
- Net billing—where exported energy is valued at a lower rate than imported energy.

Both schemes were examined under various reconciliation periods: yearly, quarterly, monthly, and daily, allowing for a detailed comparison of their economic performance under differing operational conditions. Table 10 summarizes the economic outcomes of the kindergarten's energy upgrade for both the low-usage and high-usage scenarios, including the case with a smaller PV system under high-usage conditions. This combined analysis provides a comprehensive perspective on the financial performance and cost-effectiveness of the implemented energy measures, particularly focusing on the PV system installation and its impact across varying system sizes and operational conditions.

The results from Table 10 highlight significant differences in economic performance between the net metering buyback and net billing buyback schemes across all usage scenarios, with variations becoming more pronounced as reconciliation frequency shortens.

Under net metering, surplus energy offsets imported energy costs on a 1:1 basis, resulting in better economic outcomes, especially with yearly reconciliation. In high-usage scenarios, shorter reconciliation intervals (e.g., daily) increase LCC and WLC by 17.6% and 22.6%, respectively, compared to yearly intervals. In contrast, net billing, which values exported energy at a lower rate, shows smaller cost variations with shorter intervals, achieving slightly lower LCC (0.4%) and a 3.5% shorter payback period under daily reconciliation compared to net metering. These results highlight net metering's advantage in long intervals and net billing's stability with shorter intervals.

In the low-usage scenario, net metering outperforms net billing due to its reliance on surplus energy exports, which directly offset costs. Yearly reconciliation under net metering results in shorter payback periods and lower WLC compared to daily reconciliation, while net billing sees higher LCC and extended payback periods as surplus energy is undervalued. These findings emphasize net metering's advantage in low-demand settings, where it achieves greater savings than net billing.

Table 10. Economic Outcomes of Low and High Usage of the kindergarten for 25 years of analysis.

Net Metering Buyback	Life Cycle Costs (LCC, EUR)	Annual Avg. LCC (EUR/year)	Whole Life Cycle Costs (WLC, EUR)	Annual Avg. WLC (EUR/Year)	PV Payback Period (Years)	Net Billing Buyback	Life Cycle Costs (LCC, EUR)	Annual Avg. LCC (EUR/Year)	Whole Life Cycle Costs (WLC, EUR)	Annual Avg. WLC (EUR/Year)	PV Payback Period (Years)
High Usage											
Yearly	60,882.41	2435.30	34,148.97	1365.96	9.04	Yearly	60,882.41	2435.30	33,450.65	1338.03	8.9
Yearly without subsidy	60,882.41	2435.30	32,908.38	1248.39	6.34	Yearly without subsidy	60,882.41	2435.30	36,190.72	1447.63	6.63
Quarterly	62,971.64	2518.87	35,648.38	1425.94	9.28	Quarterly	62,176.11	2487.04	34,405.78	1376.23	9.05
Monthly	67,190.23	2678.61	38,676.01	1547.04	9.9	Monthly	66,668.47	2666.74	37,722.44	1508.90	9.7
Daily	71,589.03	2863.56	41,832.96	1673.96	10.65	Daily	71,320.82	2852.83	41,157.22	1646.29	10.28
Low Usage											
Yearly	58,143.37	2325.74	31,085.44	1243.42	12.94	Yearly	58,143.37	2325.74	34,363.04	1374.52	14.46
Quarterly	58,143.37	2325.74	31,085.44	1243.42	12.94	Quarterly	59,471.80	2378.87	35,097.94	1403.91	14,82
Monthly	59,498.15	2379.93	32,058.04	1282.32	13,31	Monthly	61,765.93	2470.63	36,366.78	1454.67	15.56
Daily	63,275.30	2531.01	34,769.67	1390.79	14.66	Daily	64,080.84	2563.23	37,647.24	1505.89	16.38
High Usage Small PV											
Yearly	47,671.95	1906.88	34,446.04	1377.84	6.01	Yearly	47,671.95	1906.88	33,709.74	1348.39	5.91
Quarterly	69,072.04	2762.88	49,804.58	1992.18	9.3	Quarterly	68,796.94	2751.88	49,306.10	1972.24	9.13
Monthly	69,072.04	2762.88	49,804.58	1992.18	9.3	Monthly	68,796.94	2751.88	49,306.10	1972.24	9.13
Daily	70,683.86	2827.35	50,961.36	2038.45	9.7	Daily	70,311.96	2812.48	50,424.62	2016.99	9.5

In the high-usage scenario with a smaller PV system, both net metering and net billing result in significantly higher costs and longer payback periods due to reduced energy generation. For daily reconciliation, net metering shows a 48.2% increase in LCC and a 61.4% longer payback period compared to yearly reconciliation, while net billing exhibits similar trends with slightly shorter payback periods (1.7% lower) across all intervals. However, WLC remains consistently higher for shorter reconciliation periods, underscoring the inefficiencies of balancing energy production and demand over smaller intervals.

However, in certain cases where the values remain identical across different reconciliation periods or schemes, this outcome occurs because the PV system generates sufficient surplus energy to cover the building's demand. At the end of the reconciliation period, any remaining credits ensure that grid electricity is not required, resulting in the same LCC, WLC, and payback periods across those conditions.

The LCOE is a key indicator for assessing the economic viability of the PV systems [62] implemented in the kindergarten's energy upgrade. The results in Table 11 represent the average cost of electricity generated over the lifetime of the PV system, considering the installation, operational, and maintenance expenses.

Table 11. Levelized Cost of Electricity Results.

Scenarios	LCOE (EUR/kWh)
High Usage	0.0885
Low Usage	0.0811
High Usage (Small PV)	0.0948

The LCOE values for all scenarios demonstrate the economic feasibility and competitiveness of the PV system installation. For comparison, the Fraunhofer Institute for Solar Energy Systems) reported that, as of 2021, the LCOE for PV systems in Germany ranged from EUR 0.0312 to EUR 0.1101 per kWh, depending on system size and location [63]. The kindergarten's LCOE values fall well within this range, confirming the financial viability of the energy upgrade. The low-usage scenario achieves the lowest LCOE due to reduced energy demand, which allows the PV system to meet a larger share of the building's needs while exporting surplus energy to the grid. In the high-usage scenario with a larger PV system, the LCOE remains financially viable, showing a 9.1% increase compared to the low-usage case, driven by higher energy demand and greater reliance on grid electricity. However, in the high-usage scenario with a smaller PV system, the LCOE rises further, reflecting a 16.9% increase compared to the low-usage scenario and a 7.1% increase relative to the high-usage case with the larger PV system. This increase highlights the impact of reduced PV capacity, which limits energy production and spreads fixed costs over a smaller output.

Figure 8 from the ROI analysis reveals that net metering consistently outperforms net billing across all usage cases, particularly under favorable conditions like yearly reconciliation, where surplus energy offsets cost on a 1:1 basis. In contrast, net billing undervalues exported energy, leading to comparatively lower returns. Yearly reconciliation achieves the highest ROI for both policies by aggregating surplus energy over an extended period, significantly reducing costs. However, as reconciliation intervals shorten to quarterly, monthly, or daily, ROI declines due to surplus energy being penalized or unused, which diminishes its cost-saving potential. For instance, in the High Usage scenario under net metering, ROI decreases from 2.29 (yearly) to 1.89 (daily), marking a 17.4% drop. When it comes to maximizing the financial benefits of the investment, the analysis also highlights the importance of appropriately selecting the PV system size under the constraint of covering the building's energy needs. As observed in the High Usage Low PV case, the payback period is 6.01 years (yearly), and ROI is significantly higher compared to the full High Usage scenario, underscoring the importance of appropriately sized PV systems to optimize cost-effectiveness and ensure balanced economic outcomes.

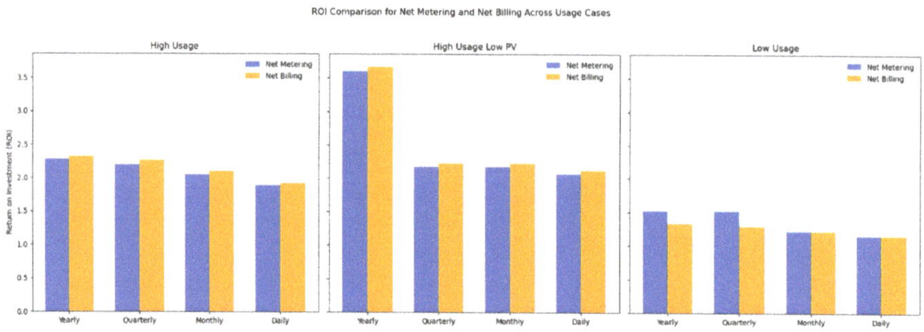

Figure 8. ROI in Different Usage Cases and Schemes.

4.2. Discussion

The results of this study underscore the significant environmental and economic advantages of the kindergarten's energy upgrade, providing critical insights into its performance under both constrained and stable operational conditions. The analysis highlights the impact of pandemic-induced operational disruptions on energy performance, emphasizing that energy efficiency measures must be evaluated across different usage patterns to fully assess their long-term viability. For the environmental KPIs, it is essential to recognize that RES alone is insufficient to achieve meaningful sustainability improvements, especially if a country's energy mix already incorporates a high share of renewables. The real change must come from enhancing the building's energy efficiency, which reduces overall energy demand and ensures that RES integration maximizes environmental benefits. Without such improvements in a building's energy performance, even extensive RES installations may fail to deliver optimal sustainability gains, as the excess energy consumption will offset potential environmental benefits. This reinforces the need for a holistic approach, combining renewable energy solutions with efficiency upgrades to achieve substantial reductions in PED and GWP [5,57,64].

In Austria, where energy policy favors net billing over net metering [65], the economic outcomes of PV systems are influenced by reconciliation periods, system size, and usage conditions. Yearly reconciliation delivers the most favorable results, minimizing LCC and payback periods by effectively utilizing surplus energy credits. Conversely, shorter reconciliation periods (e.g., monthly or daily) increase costs and extend payback periods due to reduced energy balancing efficiency. PV system size also plays a critical role; smaller systems achieve higher Return on Investment (ROI) and shorter payback periods by optimizing energy production and minimizing grid reliance. Larger systems, however, lead to up to 49% higher WLC, emphasizing the need for appropriately sizing PV installations to align production with demand. These findings highlight the variability in economic benefits across consumption patterns and scenario parameters, such as electricity prices and subsidies. The complexity increases when comparing energy upgrades across countries, illustrating the challenge of establishing consistent EU-wide policy measures.

By integrating LCA and LCC with scenario-based evaluations, this study provides a robust framework for assessing the long-term impacts of renewable energy systems and retrofitting measures. Future research could build on this by exploring additional scenarios (e.g., incorporating additional operational conditions, such as seasonal demand fluctuations or hybrid usage models), incorporating real-time grid emissions and electricity price data to improve the accuracy of economic and environmental assessments under dynamic market conditions, and applying findings to larger building portfolios with different energy efficiency solutions to generalize findings across diverse contexts.

5. Conclusions

This study assessed the environmental and economic impacts of renovating a public kindergarten in Austria as part of the AURORAL project, with a focus on integrating PV systems and evaluating the outcomes of such investments under different socioeconomic conditions affecting the building's energy profile. Using LCA and LCC methodologies, this analysis highlighted the potential benefits of renewable energy systems in reducing the building's GWP and examined operational costs under varying usage conditions and installation sizes. By analyzing both a low-usage scenario (2021, during COVID-19) and a high-usage scenario (2024, post-pandemic), this study provides insights into how pandemic-related disruptions have the potential to affect the utilization and financial outcomes of energy sustainability measures, highlighting the need for comprehensive scenario-based analysis methodologies for the maximization of sustainability investment outcomes. During the low-usage period, the PV system met energy needs despite reduced operational demand, resulting in the lowest LCOE. However, the economic benefits were limited, as the lower energy demand reduced the revenue generated from surplus energy sold back to the grid. Under high-usage conditions, the renovation demonstrated more significant energy savings, emission reductions, and improved payback periods, emphasizing its potential for long-term economic and environmental benefits. The findings also highlighted the importance of appropriately sizing PV systems; depending on the pricing scheme and export rates, larger systems might not yield optimal economic returns over their lifetime. Furthermore, the results underscore the influence of favorable energy policies in enhancing the financial viability of PV installations and minimizing lifecycle costs, providing critical incentives for investments in RES systems.

This research contributes to the broader field of sustainable building practices by validating the importance of adaptive analysis frameworks that account for varying operational conditions and evolving energy policies. These findings are particularly relevant for municipalities and policymakers, as public buildings, such as educational facilities, represent significant energy expenditures [66–68]. Expanding the use of tools like VERIFY across similar projects in the public sector could accelerate the transition toward energy-efficient infrastructure, delivering cost savings and supporting climate action goals. Future research will build upon these findings by incorporating real-time energy data to enhance the accuracy of energy demand forecasts and PV system performance evaluations. Additionally, exploring the integration of energy storage solutions could help optimize self-consumption and reduce grid dependence, further improving economic feasibility. Analyzing the role of emerging energy technologies, such as battery storage and smart grid systems, could provide valuable insights into long-term energy sustainability investments. Furthermore, conducting an economic sensitivity analysis, assessing the impact of fluctuating energy prices, inflation rates, and subsidy changes, could enhance the understanding of financial risks and investment resilience in renewable energy projects. Such evaluations would help policymakers and stakeholders develop robust financial strategies for large-scale energy-efficient building renovations under dynamic economic conditions. The conclusions drawn from this study provide a robust foundation for advancing sustainability initiatives, ensuring that public buildings align with climate goals and financial efficiency benchmarks while contributing to broader sustainable city strategies and the development of sustainable, resilient infrastructure.

Supplementary Materials: The following supporting information can be downloaded at: https://www.mdpi.com/article/10.3390/app15052330/s1.

Author Contributions: Writing—original draft, G.K., G.I.; writing—review and editing, G.K., G.I., and A.S.; software, S.P. and N.-C.C.; conceptualization, A.S. and M.F.; project administration, D.A.; supervision, N.N. and D.R. All authors have read and agreed to the published version of the manuscript.

Funding: This research was funded by project AURORAL, Horizon 2020: Grant Agreement No. 101016854.

Institutional Review Board Statement: Not applicable.

Informed Consent Statement: Not applicable.

Data Availability Statement: The data presented in this study are available on request from the corresponding author. The data are not publicly available due to IP and confidentiality issues.

Conflicts of Interest: The authors declare no conflict of interest.

References

1. Curran, M.A.; Young, S. Report from the EPA conference on streamlining LCA. *Int. J. Life Cycle Assess.* **1996**, *1*, 57–60. [CrossRef]
2. Arulnathan, V.; Heidari, M.D.; Doyon, M.; Li, E.P.H.; Pelletier, N. Economic Indicators for Life Cycle Sustainability Assessment: Going Beyond Life Cycle Costing. *Sustainability* **2023**, *15*, 13. [CrossRef]
3. Nicita, A.; Squadrito, G.; Maggio, G. Life-Cycle Cost (LCC) Applied to Hydrogen Technologies: A Review. *Int. J. Life Cycle Assess.* **2023**, *29*, 46–79. [CrossRef]
4. World Economic Forum. Retrofitted Buildings: Accelerating the Energy Transition Through Sustainability. 2024. Available online: https://www.weforum.org/stories/2024/01/retrofitted-buildings-energy-transition/ (accessed on 1 September 2024).
5. Quaschning, V. *Understanding Renewable Energy Systems*; Earthscan: London, UK, 2005. [CrossRef]
6. Zulfiqar, F.; Muhammad-Sukki, F.; Abu-Bakar, S.H.; Bani, N.A.; Mas'Ud, A.A. Effect of COVID-19 on Photovoltaic Demand and Installation. *IET Conf. Proc.* **2023**, *2023*, 174–179. [CrossRef]
7. Gui, X.; Gou, Z.; Zhang, F.; Yu, R. The Impact of COVID-19 on Higher Education Building Energy Use and Implications for Future Education Building Energy Studies. *Energy Build.* **2021**, *251*, 111346. [CrossRef]
8. Saadé, M.; Erradhouani, B.; Pawlak, S.; Appendino, F.; Peuportier, B.; Roux, C. Combining Circular and LCA Indicators for the Early Design of Urban Projects. *Int. J. Life Cycle Assess.* **2022**, *27*, 1–19. [CrossRef]
9. Fnais, A.; Rezgui, Y.; Petri, I.; Beach, T.; Yeung, J.; Ghoroghi, A.; Kubicki, S. The Application of Life Cycle Assessment in Buildings: Challenges, and Directions for Future Research. *Int. J. Life Cycle Assess.* **2022**, *27*, 627–654. [CrossRef]
10. Braulio-Gonzalo, M.; Bovea, M.D. Environmental and Cost Performance of Building's Envelope Insulation Materials to Reduce Energy Demand: Thickness Optimisation. *Energy Build.* **2017**, *150*, 527–545. [CrossRef]
11. Bierer, A.; Götze, U.; Meynerts, L.; Sygulla, R. Integrating Life Cycle Costing and Life Cycle Assessment Using Extended Material Flow Cost Accounting. *J. Clean. Prod.* **2015**, *108*, 1289–1301. [CrossRef]
12. Jiménez-Gómez, L.M.; Velásquez-Henao, J.D. A Comprehensive Analysis of Real Options in Solar Photovoltaic Projects: A Cluster-Based Approach. *Heliyon* **2024**, *10*, e35984. [CrossRef]
13. Zhao, G.; Clarke, J.; Searle, J.; Lewis, R.; Baker, J. Economic Analysis of Integrating Photovoltaics and Battery Energy Storage System in an Office Building. *Energy Build.* **2023**, *284*, 112885. [CrossRef]
14. Linssen, J.; Stenzel, P.; Fleer, J. Techno-Economic Analysis of Photovoltaic Battery Systems and the Influence of Different Consumer Load Profiles. *Appl. Energy* **2015**, *153*, 333–346. [CrossRef]
15. Mao, D.; Yang, S.; Ma, L.; Ma, W.; Yu, Z.; Xi, F.; Yu, J. Overview of Life Cycle Assessment of Recycling End-of-Life Photovoltaic Panels: A Case Study of Crystalline Silicon Photovoltaic Panels. *J. Clean. Prod.* **2023**, *434*, 140320. [CrossRef]
16. Khairi, N.H.M.; Akimoto, Y.; Okajima, K. Suitability of Rooftop Solar Photovoltaic at Educational Building Towards Energy Sustainability in Malaysia. *Sustain. Horiz.* **2022**, *4*, 100032. [CrossRef]
17. Rotas, R.; Fotopoulou, M.; Drosatos, P.; Rakopoulos, D.; Nikolopoulos, N. Adaptive Dynamic Building Envelopes with Solar Power Components: Annual Performance Assessment for Two Pilot Sites. *Energies* **2023**, *16*, 2148. [CrossRef]
18. Spudys, P.; Osadcha, I.; Morkunaite, L.; Manhanga, F.C.; Georgali, P.Z.; Klumbyte, E.; Jurelionis, A.; Papadopoulos, A.; Fokaides, P. A comparative life cycle assessment of building sustainability across typical European building geometries. *Energy* **2024**, *302*, 131693. [CrossRef]

19. Seitaridis, A.; Mamounakis, I.; Tagkoulis, N.; Iliadis, P.; Bellos, E.; Papalexis, C.; Sougakis, V.; Nikolopoulos, N. An Innovative software platform for efficient energy, environmental and cost planning in buildings retrofitting. In *IFIP International Conference on Artificial Intelligence Applications and Innovations*; Maglogiannis, I., Iliadis, L., Macintyre, J., Cortez, P., Eds.; Springer International Publishing: Cham, Switzerland, 2023; pp. 217–228. [CrossRef]
20. Skopetou, N.; Zestanakis, P.A.; Rotas, R.; Iliadis, P.; Papadopoulos, C.; Nikolopoulos, N.; Sfakianakis, A.; Koroneos, C. Energy Analysis and Environmental Impact Assessment for Self-Sufficient Non-Interconnected Islands: The Case of Nisyros Island. *J. Clean. Prod.* **2024**, *447*, 141647. [CrossRef]
21. Bellos, E.; Iliadis, P.; Papalexis, C.; Rotas, R.; Mamounakis, I.; Sougkakis, V.; Nikolopoulos, N.; Kosmatopoulos, E. Holistic Renovation of a Multi-Family Building in Greece Based on Dynamic Simulation Analysis. *J. Clean. Prod.* **2022**, *381*, 135202. [CrossRef]
22. *BS EN 15978:2011*; Sustainability of Construction Works—Assessment of Environmental Performance of Buildings—Calculation Method. BSI Standards Publication: London, UK, 2011. Available online: https://knowledge.bsigroup.com/products/sustainability-of-construction-works-assessment-of-environmental-performance-of-buildings-calculation-method?version=standard (accessed on 1 September 2024).
23. *ISO 14040:2006*; Environmental Management—Life Cycle Assessment—Principles and Framework (2nd ed.). ISO: Geneva, Switzerland, 2006. Available online: https://www.iso.org/standard/37456.html (accessed on 1 September 2024).
24. *ISO 14044:2006*; Environmental Management—Life Cycle Assessment—Requirements and Guidelines. ISO: Geneva, Switzerland, 2006. Available online: https://www.iso.org/standard/38498.html (accessed on 1 September 2024).
25. *BS EN 15804:2012+A2:2019*; Sustainability of Construction Works—Environmental Product Declarations—Core Rules for the Product Category of Construction Products. British Standards Institution: London, UK, 2013. Available online: https://shop.bsigroup.com (accessed on 1 September 2024).
26. *BS ISO 15686-5:2017*; Buildings and Constructed Assets—Service Life Planning—Part 5: Life-Cycle Costing. British Standards Institution: London, UK, 2017. Available online: https://knowledge.bsigroup.com/products/organization-and-digitization-of-information-about-buildings-and-civil-engineering-works-including-building-information-modelling-bim-information-management-using-building-information-modelling-health-and-safety-information-1?version=standard&creative=730824407410&keyword=building%20information%20modelling%20iso&matchtype=b&network=g&device=c&gad_source=1&gclid=Cj0KCQiAwtu9BhC8ARIsAI9JHalpeL9SZHKyzAEQzD6md7Y_vzMfZfV6bWQwQkLk2NEjqP3yy7DXqQ0aAjz0EALw_wcB&gclsrc=aw.ds (accessed on 1 September 2024).
27. *prEN 15459:2006*; Energy Efficiency for Buildings—Economic Evaluation Procedure for Energy Systems. European Committee for Standardization: Brussels, Belgium, 2006. Available online: https://www.cen.eu (accessed on 1 September 2024).
28. *BS EN 16627:2015*; Sustainability of Construction Works—Assessment of Economic Performance of Buildings—Calculation Methods. British Standards Institution: London, UK, 2015. Available online: https://www.en-standard.eu/bs-en-16627-2015-sustainability-of-construction-works-assessment-of-economic-performance-of-buildings-calculation-methods/?srsltid=AfmBOorchUHeLxVubXRAYpAatxgwcLrsbgIkSsmu78Hap37nwpsLO7Sc (accessed on 1 September 2024).
29. Dodd, N.; Donatello, S.; Cordella, M. Level(s) Indicator 1.1: Use Stage Energy Performance User Manual: Introductory Briefing, Instructions and Guidance (Publication version 1.2). European Commission, Joint Research Centre. 2021. Available online: https://susproc.jrc.ec.europa.eu/product-bureau//sites/default/files/2021-01/UM3_Indicator_1.1_v1.1_25pp.pdf (accessed on 1 September 2024).
30. Dodd, N.; Donatello, S.; Cordella, M. Level(s) Indicator 1.2: Life Cycle Global Warming Potential (GWP) User Manual: Introductory Briefing, Instructions, and Guidance (Publication version 1.1). European Commission, Joint Research Centre. 2021. Available online: https://susproc.jrc.ec.europa.eu/product-bureau/sites/default/files/2021-01/UM3_Indicator_1.2_v1.1_37pp.pdf (accessed on 1 September 2024).
31. Dodd, N.; Donatello, S.; Cordella, M. Level(s) Indicator 6.1: Life Cycle Costs User Manual: Introductory Briefing, Instructions, and Guidance (Publication version 1.1). European Commission, Joint Research Centre. 2021. Available online: https://susproc.jrc.ec.europa.eu/product-bureau/sites/default/files/2021-01/UM3_Indicator_6.1_v1.1_21pp.pdf (accessed on 1 September 2024).
32. Huang, B.; Wang, Y.; Huang, Y.; Xu, X.; Chen, X.; Duan, L.; Yu, G.; Li, Z.; Liu, H.; Kua, H.W.; et al. Life Cycle Cost Analysis of Solar Energy via Environmental Externality Monetization. *Sci. Total. Environ.* **2022**, *856*, 158910. [CrossRef]
33. De Wolf, C.; Cordella, M.; Dodd, N.; Byers, B.; Donatello, S. Whole Life Cycle Environmental Impact Assessment of Buildings: Developing Software Tool and Database Support for the EU Framework Level(s). *Resour. Conserv. Recycl.* **2022**, *188*, 106642. [CrossRef]
34. Amoruso, F.M.; Schuetze, T. Carbon Life Cycle Assessment and Costing of Building Integrated Photovoltaic Systems for Deep Low-Carbon Renovation. *Sustainability* **2023**, *15*, 9460. [CrossRef]
35. Herrando, M.; Elduque, D.; Javierre, C.; Fueyo, N. Life Cycle Assessment of Solar Energy Systems for the Provision of Heating, Cooling and Electricity in Buildings: A Comparative Analysis. *Energy Convers. Manag.* **2022**, *257*, 115402. [CrossRef]

36. PRé Sustainability. SimaPro Software for Life Cycle Assessment. 2023. Available online: https://simapro.com (accessed on 1 September 2024).
37. Athena Sustainable Materials Institute. Athena Impact Estimator for Buildings. 2023. Available online: https://athenasmi.org/our-software-data/impact-estimator/ (accessed on 1 September 2024).
38. Wikipedia contributors. RETScreen. Wikipedia, The Free Encyclopedia. 2024. Available online: https://en.wikipedia.org/wiki/RETScreen (accessed on 1 September 2024).
39. One Click LCA Ltd. One Click LCA: Life Cycle Assessment Software. 2023. Available online: https://www.oneclicklca.com (accessed on 1 September 2024).
40. KieranTimberlake Research Group. Tally: Life Cycle Assessment in Building Information Modeling. 2023. Available online: https://choosetally.com (accessed on 1 October 2024).
41. Kitsopoulou, A.; Ziozas, N.; Iliadis, P.; Bellos, E.; Tzivanidis, C.; Nikolopoulos, N. Energy performance analysis of alternative building retrofit interventions for the four climatic zones of Greece. *J. Build. Eng.* **2024**, *87*, 109015. [CrossRef]
42. Ziozas, N.; Kitsopoulou, A.; Bellos, E.; Iliadis, P.; Gonidaki, D.; Angelakoglou, K.; Nikolopoulos, N.; Ricciuti, S.; Viesi, D. Energy Performance Analysis of the Renovation Process in an Italian Cultural Heritage Building. *Sustainability* **2024**, *16*, 2784. [CrossRef]
43. Kitsopoulou, A.; Zacharis, A.; Ziozas, N.; Bellos, E.; Iliadis, P.; Lampropoulos, I.; Chatzigeorgiou, E.; Angelakoglou, K.; Nikolopoulos, N. Dynamic Energy Analysis of Different Heat Pump Heating Systems Exploiting Renewable Energy Sources. *Sustainability* **2023**, *15*, 11054. [CrossRef]
44. Apostolopoulos, V.; Mamounakis, I.; Seitaridis, A.; Tagkoulis, N.; Kourkoumpas, D.-S.; Iliadis, P.; Angelakoglou, K.; Nikolopoulos, N. An integrated life cycle assessment and life cycle costing approach towards sustainable building renovation via a dynamic online tool. *Appl. Energy* **2023**, *334*, 120710. [CrossRef]
45. Bellos, E.; Iliadis, P.; Papalexis, C.; Rotas, R.; Nikolopoulos, N.; Kosmatopoulos, E.; Halmdienst, C. Dynamic investigation of centralized and decentralized storage systems for a district heating network. *J. Energy Storage* **2022**, *56*, 106072. [CrossRef]
46. Violidakis, I.; Atsonios, K.; Iliadis, P.; Nikolopoulos, N. Dynamic modeling and energy analysis of renewable heating and electricity systems at residential buildings using phase change material based heat storage technologies. *J. Energy Storage* **2020**, *32*, 101942. [CrossRef]
47. Modelica Association. Modelica: A Unified Object-Oriented Language for Physical Systems Modeling. Modelica Association. 2024. Available online: https://modelica.org/ (accessed on 1 September 2024).
48. ecoinvent Association. Ecoinvent Database v3. Ecoinvent. 2024. Available online: https://ecoinvent.org/ (accessed on 1 September 2024).
49. Joint Research Centre (JRC). Emission factors for CO_2 equivalent (CO_2-eq) emissions in EU Member States (1990–2021). European Commission. 2021. Available online: https://data.jrc.ec.europa.eu/dataset/919df040-0252-4e4e-ad82-c054896e1641 (accessed on 1 September 2024).
50. Seber, G.A.F.; Wild, C.J. *Nonlinear Regression*; John Wiley & Sons: Hoboken, NJ, USA, 2003. [CrossRef]
51. Eurostat. Electricity Prices for Household Consumers (NRG_PC_204). European Commission. 2023. Available online: https://ec.europa.eu/eurostat/databrowser/view/nrg_pc_204/default/table?lang=en (accessed on 1 September 2024).
52. Inflation.eu. Historic Inflation Austria—CPI Inflation Austria 2021. 2023. Available online: https://www.inflation.eu/en/inflation-rates/austria/historic-inflation/cpi-inflation-austria-2021.aspx (accessed on 1 September 2024).
53. European Central Bank. The Impact of the Pandemic and the War: Insights from the ECB's Consumer Expectations Survey. Economic Bulleti. 2022. Available online: https://www.ecb.europa.eu/press/economic-bulletin/focus/2022/html/ecb.ebbox202204_01~68ef3c3dc6.en.html (accessed on 1 September 2024).
54. International Energy Agency (IEA). *Eco-Social Tax Reform Act (Austria ETS)*; IEA: Paris, France, 2024. Available online: https://www.iea.org/policies/13998-eco-social-tax-reform-act-austria-ets (accessed on 1 September 2024).
55. Taxually. Austria VAT Rate and Compliance 2024. Taxually. 2024. Available online: https://www.taxually.com/blog/austria-vat-rate-and-compliance-2024 (accessed on 1 September 2024).
56. Dufo-López, R.; Bernal-Agustín, J.L. A comparative assessment of net metering and net billing policies. Study cases for Spain. *Energy* **2015**, *84*, 684–694. [CrossRef]
57. International Energy Agency (IEA). *Cost of Capital by Project Type in Selected Countries*; IEA: Paris, France, 2022. Available online: https://www.iea.org (accessed on 1 September 2024).
58. Amann, C.; Torres, P.; Boldizsár, G.; Hofer, G.; Stumpf, W.; Leutgöb, K.; van Nuffel, L. Support to Primary Energy Factors Review (PEF): Final Report. European Commission Directorate-General for Energy. 2023. Available online: https://ec.europa.eu (accessed on 1 September 2024).
59. International Energy Agency Photovoltaic Power Systems Programme (IEA-PVPS). National Survey Report of PV Power Applications in Austria 2023. 2023. Available online: https://iea-pvps.org/wp-content/uploads/2024/10/National-Survey-Report-of-PV-Power-Applications-in-Austria-2023.pdf (accessed on 1 September 2024).

60. Wikipedia Contributors. COVID-19 Pandemic in Austria. Wikipedia, The Free Encyclopedia. 2024. Available online: https://en.wikipedia.org/wiki/COVID-19_pandemic_in_Austria (accessed on 1 September 2024).
61. Brychkov, D.; Goggins, G.; Doherty, E.; Romero, N.; Roudil, N.; Di Trani, A.; Singh, A.; Smit, S.; McLoughlin, E.; Lima, R.d.C.R.; et al. A systemic framework of energy efficiency in schools: Experiences from six European countries. *Energy Effic.* **2023**, *16*, 21. [CrossRef]
62. Kabeyi, M.J.B.; Olanrewaju, O.A. The levelized cost of energy and modifications for use in electricity generation planning. *Energy Rep.* **2023**, *9*, 495–534. [CrossRef]
63. Fraunhofer Institute for Solar Energy Systems ISE. "Levelized Cost of Electricity—Renewables Clearly Superior to Conventional Power Plants Due to Rising CO2 Prices". 2021. Available online: https://www.ise.fraunhofer.de (accessed on 1 September 2024).
64. The Constructor. Energy Efficiency in Buildings—Need, Benefits, and Methods. 2023. Available online: https://theconstructor.org/building/buildings/energy-efficiency-building/562100/ (accessed on 1 September 2024).
65. Energy Theory. Net Billing vs Net Metering: Key Differences and Benefits. 2024. Available online: https://energytheory.com/net-billing-vs-net-metering/ (accessed on 1 September 2024).
66. Pachta, V.; Giourou, V. Comparative Life Cycle Assessment of a Historic and a Modern School Building, Located in the City of Naoussa, Greece. *Sustainability* **2022**, *14*, 4216. [CrossRef]
67. Bilir, L.; Yildirim, N. Photovoltaic System Assessment for a School Building. *Int. J. Hydrogen Energy* **2017**, *42*, 17856–17868. [CrossRef]
68. European Commission. Energy School—Guardians of Smart Energy Schools. Available online: https://ec.europa.eu/regional_policy/projects/projects-database/energy-school-guardians-of-smart-energy-schools_en (accessed on 1 September 2024).

applied sciences

MDPI

Article

Advanced Energy Management for Residential Buildings Optimizing Costs and Efficiency Through Thermal Energy Storage and Predictive Control

Ricardo Felez [1,2] and Jesus Felez [1,*]

[1] Mechanical Engineering Department, Universidad Politécnica de Madrid, 28006 Madrid, Spain; ricardo.felez@gmail.com

[2] IGNIS Energía S.L., 28016 Madrid, Spain

* Correspondence: jesus.felez@upm.es

Abstract: This paper addresses the challenge of decarbonizing residential energy consumption by developing an advanced energy management system (EMS) optimized for cost reduction and energy efficiency. By leveraging the thermal inertia of building envelopes as a form of thermal energy storage (TES), the proposed EMS dynamically balances energy inputs from the electrical grid, photovoltaic (PV) systems, and battery storage, while regulating HVAC operations. Employing model predictive control (MPC) and integrating real-time data on energy prices, weather conditions, and consumption patterns, the system minimizes energy costs, while maintaining indoor thermal comfort. Simulations conducted for both summer and winter scenarios demonstrate significant reductions in energy costs. Laboratory and real-world testing confirm the EMS's efficacy in enhancing energy efficiency, increasing PV self-consumption, and reducing reliance on grid energy, marking a step forward in sustainable energy solutions for residential applications.

Keywords: energy management system; thermal energy storage; model predictive control; renewable energy integration; residential heating and cooling; HVAC optimization; sustainable energy systems; real-time energy management; smart homes

1. Introduction

Residential heating is a significant source of greenhouse gas emissions in Europe, where cold winters and high standards of comfort drive a substantial demand for heating. Across the continent, homes are primarily heated by burning fossil fuels, with natural gas, oil, and coal accounting for the majority of the energy used in household heating systems. This widespread reliance on fossil fuels has made residential heating one of the leading contributors to carbon dioxide (CO_2) and other greenhouse gas emissions in Europe, directly impacting air quality and contributing to climate change. According to recent estimates, the residential sector is responsible for nearly 30% of Europe's total energy consumption, with heating comprising the vast majority of this demand, and residential heating alone produces approximately 12% of the European Union's total CO_2 emissions [1], a figure that fluctuates depending on seasonal temperatures and energy sources. Countries with colder climates, such as those in Northern and Eastern Europe, often exhibit even higher emissions due to longer heating periods and lower average temperatures.

In the United States, buildings account for approximately 39% of all primary energy consumption and 74% of all electricity consumption [2]. Thermal end uses, including space conditioning, water heating, and refrigeration, account for approximately 50% of building energy demand and are projected to increase in the years ahead.

The adoption of aerothermal and geothermal heating systems in Europe is proving to be an effective strategy to reduce greenhouse gas emissions from the residential sector. Unlike conventional fossil fuel-based heating, aerothermal and geothermal systems extract renewable heat from the air or ground, greatly reducing direct CO_2 emissions. Studies estimate that replacing fossil fuel heating systems with heat pumps (which include both aerothermal and geothermal technologies) could cut emissions by up to 70–80% per household. Across Europe, approximately 20 million heat pumps were installed by 2022 [3], helping to replace a portion of the emissions-heavy natural gas and oil heating systems.

The European Heat Pump Association (EHPA) reports that these installations are already saving about 50 million tons of CO_2 annually [4]. As part of the European Green Deal, the EU aims to accelerate this shift, with projections indicating that heat pumps could power up to 40% of Europe's residential heating demand by 2030 [5]. This large-scale transition is essential for Europe to reach its net-zero emissions targets by 2050, as heat pumps not only reduce emissions directly but also operate more efficiently when paired with renewable electricity sources

As energy systems evolve toward greater sustainability, there is growing interest in leveraging the thermal storage capacity of buildings to reduce energy consumption and shift demand patterns. Thermal energy storage (TES) is a crucial enabling technology for the large-scale deployment of renewable energy, facilitating the decarbonization of thermal end uses, including refrigeration, water heating, and space heating and cooling, and the transition to a decarbonized building stock and energy system by 2050.

A variety of thermal storage solutions are available, which permit the storage of excess thermal energy or heat for subsequent utilization at a more opportune time, whether that is hours or even days later. Thus, there are several investigations in the field of advanced thermal storage materials. For example, reference [6] investigates new composites materials, and reference [7] presents a novel approach to encapsulate salt hydrate. In the field of thermal energy storage projects related to envelope systems, for example, reference [8] presents an active thermally anisotropic building envelope that redistributes thermal loads in response to weather conditions and building energy demand. However, while all of these ideas are promising, they are all in the early research phase and far from real application.

The objective of this work is to develop an energy management system that makes decisions on how to distribute the electrical demand within the house among the possible sources that supply it, such as the electrical grid, self-consumption (for example, with solar panels), and energy storage with electrical batteries with the ultimate goal of reducing the cost of electrical energy.

To this end, this energy management system considers a residential dwelling as TES in which, taking advantage of the thermal and insulation characteristics of its structure and envelopes, excess heat can be stored in the structure and materials of the dwelling and released, when necessary, while maintaining comfortable interior temperatures. The proposed use of the building envelope as TES can be considered as a passive solution for sensible heat storage system for moderate temperature storage applications in buildings, increasing or decreasing the temperature of a given material with high heat capacity. In this case, the envelope includes solid materials, such as concrete and bricks as main components, which are materials that can be used in buildings to improve the TES capacity. Their properties have been exploited to use the thermal inertia of the building structure to attenuate the indoor temperature oscillation.

Therefore, no new envelope materials will be developed, but rather, using the existing ones, an advanced energy management system will be designed with a strategy that effectively regulates the energy management of the building with an efficient use of the HVAC system. In this way, the target temperature inside the house will be slightly modified, always within the thermal comfort margins of the occupants. This is intended to supplement the energy storage capacity of the house as a whole, optimizing the use of batteries, since part of the energy that could provide the electric battery, will provide the thermal energy stored in the structure of the house. This will improve the efficiency of the system, reducing the energy consumption of heating or cooling and, consequently, reducing the cost of energy.

This strategy is especially valuable in areas where renewable energy sources, such as solar and wind, produce intermittent power, as it enables energy use to be adjusted based on availability and price fluctuations.

The remainder of the paper is organized as follows. Section 2 contains a comprehensive literature review and the main objectives of this paper. Section 3 describes the methods used and a general overview of the proposed solution. Section 4 includes the models developed for the thermal behavior and the HVAC control. Section 5 describes the energy consumption scheduler. Section 6 explains the implementation of the proposed solution. Section 7 presents and discusses the results obtained from the simulations performed for the different use cases indicated in a single-family house. Section 8 includes the validation in a test environment. Finally, Section 9 presents the main conclusions of this paper.

2. Literature Review

The phenomenon of global warming is leading to the emergence of a novel habitat scenario, characterized by a heightened prevalence of environmental degradation, sea level rise, and ecosystem extinction. National and interregional governments are developing legislation to reduce greenhouse gas (GHG) emissions. High energy consumption in the building sector is identified as a significant contributor to this phenomenon [9]. The poor energy efficiency of the majority of buildings is a primary factor driving high energy consumption. This is particularly prevalent in countries, where the majority of buildings were designed and constructed prior to the implementation of the first energy efficiency regulations.

A substantial body of research has already demonstrated the necessity of considering climate change when determining building energy retrofitting measures. This is evidenced by numerous studies, including those referenced in [10–17]. In more detail, the paper [18] examined the impact of global warming on the significance of passive design strategies for the heating and cooling of European single-family residences.

A further consideration is the thermal comfort or thermal wellbeing of buildings, which is defined as people's subjective assessment of their thermal sensation. It is typically defined as the feeling of satisfaction with the thermal environment expressed by people. There are also numerous references that have studied and discussed this topic, such as [19–22]. These references indicate that thermal control of buildings is essential to ensure comfort.

The objective of [23] was to review the topics related to the optimized operation of intelligent buildings with respect to occupant comfort and energy consumption. The results demonstrate that energy management systems have a beneficial effect on occupants' overall satisfaction within an enclosed space. By regulating indoor environmental parameters, energy management systems in intelligent buildings can provide thermal and visual comfort for occupants, as well as satisfactory indoor air quality.

There are also relevant publications in the field of thermal control of buildings. The specific control methodology employed is contingent upon the type of controller and algorithm utilized [24]. The reference [25] offers a comprehensive review of recent literature on systems and methodologies for the control of thermal comfort in buildings. The papers analyzed exhibited discernible trends with regard to both thermal comfort analysis and control strategies. With respect to control aspects, the methodologies employed for calculating control settings encompass a range of techniques, from expert rules to sophisticated modelling approaches, such as machine learning and model predictive control (MPC).

In relation to methodologies such as machine learning (ML) or, more generally, artificial intelligence (AI), reference [26] included a systematic review of artificial intelligence-assisted techniques for thermal comfort and energy efficiency in intelligent building control systems. This review demonstrated that the application of AI/ML technology in the construction industry is still in its relatively early stages of development. Furthermore, the study identified several open challenges and potential areas for future research, including the security, privacy, and data sensitivity issues associated with smart buildings, the lack of sufficient data for modelling in buildings, and the need for more AI-based modelling in this field. However, there is a lot of recent work in this field, as shown by the references [27–30].

In relation to control methodologies, reference [31] undertook a review of research works concerning occupancy-based control systems. These were classified according to the integration of occupancy information with control systems, and the strengths and limitations of the research, as well as research gaps in this field, were discussed from different aspects, including performance evaluation metrics, control methods, occupancy models, and building types. Predictive control strategies were proposed in the literature as a solution to this issue by enhancing thermal comfort of occupants, and the review revealed that MPC was the most frequently employed approach for occupancy-based optimal control in the literature.

Model predictive control (MPC) is a control scheme that employs a model to anticipate the future behavior of a system over a finite time horizon. Based on these predictions and the current measured or estimated state of the system, the optimal control inputs with respect to a defined control objective and subject to system constraints are calculated. Subsequently, after a defined time interval, the aforementioned measurement, estimation, and computation process is repeated with a shifted horizon.

The principal benefits of MPC in comparison to conventional reactive control methodologies, such as PID, are as follows. Firstly, the characteristic of proactive control action, whereby the controller anticipates future disturbances and setpoints, is worthy of note. Secondly, MPC can explicitly consider nonlinear systems. Thirdly, MPC can accommodate arbitrary control objectives, including traditional set-point tracking and regulation, as well as economic MPC, where control actions are optimized to satisfy generic economic or performance cost functions. Finally, MPC explicitly considers physical, safety, or operational system constraints.

MPC was first introduced in the 1980s for the regulation of processes in the chemical industry. However, it was not until the 2010s that its use experienced significant growth in the field of thermal regulation for buildings, due to the high computational requirements needed [32].

A first implementation of the present model predictive controller applied to the temperature control of a real building was presented in [33]. The paper [34] presented a unified framework for model predictive control (MPC) technology in building management, with a particular focus on real-world applications. From a theoretical perspective, it provided an overview of MPC formulations for building control, modelling paradigms, and model types, together with the algorithms necessary for real-life implementation. Similarly, an overview of the applications of MPC in building HVAC systems has been provided in [35]. Furthermore, the article discusses modelling techniques and optimization algorithms in greater detail.

The reference [36] developed a machine learning-based MPC with an instantaneous linearization scheme, linear MPC, to control the indoor climate of a hospital. Although the approach is advantageous for computation time, the cooling energy consumption of MPC without linearization is still lower than that of linear MPC. In reference [37], temperature and humidity are considered but not for renewable energy systems. The objective of [38] was to examine advanced control strategies and their influence on buildings and technical systems with regard to energy and cost savings, addressing research gaps in advanced control strategies for buildings and HVAC systems. Their findings indicate that, despite the prevalence of MPC in building applications, it is not optimally suited to systems characterized by uncertainties and unpredictable data, and significant advancements have been made, especially as more intelligent technologies, such as cloud computing and IoT, are being developed.

The principal objective of the paper [39] was to define the MPC formulation framework and undertake a critical analysis of the results produced by a range of existing MPC algorithms for the management of building and HVAC systems. The potential benefits of the application of MPC in improving energy efficiency in buildings were highlighted.

The paper [40] presented a novel control approach for indoor temperature regulation in tertiary buildings, based on predictive functional control. The primary control objective is to reduce energy consumption, while maintaining the desired thermal comfort of the occupants. The proposed approach demonstrates the potential to perform effectively in the presence of disturbances, including changes in the occupancy profile, electrical equipment usage, indoor temperature, surrounding temperatures, and weather data.

A model predictive control system with adaptive machine learning-based building models for building automation and control applications was proposed in [41]. The system incorporates an adaptive machine learning-based building dynamics modelling scheme that updates the building model on a regular basis, using online building operation data through a dynamic artificial neural network. Furthermore, the system utilizes a multi-objective function that may optimize both energy efficiency and indoor thermal comfort, two frequently contradictory demands. The proposed model predictive control system is implemented to regulate the HVAC and mechanical ventilation systems in two single-zone test beds, an office and a lecture theatre, situated in Singapore, for the purpose of experimental evaluation of its control performance. The implementation of a model predictive control system has been demonstrated to result in a reduction of 58.5% in cooling thermal energy consumption in the office environment and 36.7% in cooling electricity consumption in the lecture theatre, in comparison to the levels of consumption observed in the original control system. Furthermore, the introduction of the model predictive control system has also been shown to significantly enhance the indoor thermal comfort experienced by users in both testbeds.

In the field of thermal energy storage systems and MPC, several references can also be considered. A preliminary study on the control of thermal energy storage in building cooling systems was presented in [42]. The focus of this study was on buildings equipped with a water tank used for actively storing cold water produced by a series of chillers.

The paper [43] presented a systematic review of the application of model predictive control strategies to active thermal energy storage systems. This paper presented a discussion of the strengths and weaknesses of this technology, with particular emphasis on the difficulty of operating with complicated physical models as the key limitation to overcome.

In reference [44], an MPC strategy was proposed as a means of improving the operation of a space-heating system that is coupled with renewable resources. This model employs a dynamic approach based on forecasting all energy inputs into the system over a specified period of time in advance, prior to the implementation of any operational decisions. The

MPC strategy was implemented with the objective of reducing the annual energy costs associated with the heating system of a detached house, which was equipped with a heat pump, a thermal energy storage unit, and photovoltaic panels. The results demonstrated the efficacy of the model predictive control with a 24 h time horizon, indicating that energy cost savings of 58% could be achieved in comparison to the same heating system without smart control.

The objective of the study [45] was to ascertain the feasibility of MPC strategy for commercial buildings in response to variations in occupancy and time-variant electricity prices, in comparison to a conventional, rule-based control (RBC) strategy.

In reference [46], an MPC optimization framework that integrates a data-driven cooling load prediction model, a system physical model, and an advanced optimization algorithm was proposed and applied to a district cooling system (DCS) coupled with an ice-based TES in Beijing, China. The operational strategy of the DCS was optimized based on the predicted cooling load, with the objective of minimizing operating costs under the current time-of-use (ToU) tariff. The superior economic performance of the proposed optimization framework was verified by comparing it with two conventional operational strategies. The optimal strategy demonstrated a reduction in the operating costs of approximately 8% over a two-month cooling period.

Reference [47] evaluated a price-responsive MPC strategy for a solar thermal heating system integrated with TES for buildings with high occupancy variability. The coupled system supplies the building with heating through a low-temperature underfloor heating system. A case study was conducted on a university building in Nottingham, UK, with the objective of evaluating the feasibility of the proposed heating system, which is controlled by an MPC strategy. The MPC controller was designed to optimize the solar heating system's operation by dynamically adjusting to forecasted weather, occupancy, and solar availability, with the aim of balancing indoor comfort with energy efficiency.

Finally, economic model predictive control (EMPC) [48] represents a variant of model predictive control used for optimizing economic revenues arising from controlled dynamical processes. It has established itself as a variant of standard MPC, departing from the latter in that arbitrary cost functions are allowed in the formulation of the stage cost. The reference [49] illustrated how the EMPC can be adapted to accommodate time-varying or parameter-varying costs.

Several references have dealt with EMPC for energy management in residential buildings. In [50], an MPC utilizing artificial neural networks (ANNs) was implemented in a residential building. The ANN-MPC was successfully tested and demonstrated satisfactory performance in predicting the building's energy consumption. Subsequently, the controller was modified to function as an EMPC, with the objective of optimizing demand flexibility, defined as the ability to adapt energy demands in response to fluctuations in supply. The operational costs of energy usage were associated with this demand flexibility, which was represented by three flexibility indicators: the flexibility factor, the supply cover factor, and the load cover factor. The results of the day-long test demonstrated that these flexibility indicators were optimized when the EMPC controller's demand flexibility was compared to that of a conventional proportional–integral (PI) controller. Specifically, the flexibility factor ranged from -0.88 to 0.67, the supply cover factor from 0.04 to 0.13, and the load cover factor from 0.07 to 0.16.

To conclude, reference [51] examined the potential of the EMPC of complex residential energy systems with electric coupling to the public grid. The system under examination incorporates a battery energy storage system, photovoltaic power generation, an air-to-water heat pump, thermal energy storage, and a building model. The aforementioned power generation system provided energy for electric loads, as well as domestic hot water and space heating. Two algorithms for distributed MPC were developed, and their efficacy was evaluated in comparison to a common reference control concept and a centralized model predictive control approach. In order to facilitate a comparison of the annual operational costs, the current German energy prices and subsidies were incorporated into the economic calculation. The results demonstrated an enhanced performance of the developed approaches, with an 11.6% reduction in cost in comparison to the reference. This was achieved by increasing the heat pump seasonal performance factor by 3.4%, reducing the curtailment of electrical photovoltaic energy to 21.5% of the reference value and preventing the operation of the auxiliary heater. Furthermore, increased photovoltaic self-consumption by the heat pump resulted in a slight reduction in battery storage operation. In conjunction with the aforementioned monetary and energetic advantages, MPC increased comfort with regard to the violation of minimum limits for the comfort criteria.

Finally, the issue of resilience is of particular concern. The global effects of climate change are projected to increase the frequency and intensity of extreme weather events, including heatwaves and power outages. These events can have significant consequences for buildings and their cooling systems, underscoring the need for resilient design and operational strategies. Reference [52] conducted a comprehensive review of existing cooling strategies, with a particular focus on their performance under heatwaves and power outages. The study advanced a definition of resilient cooling and delineated four criteria for resilience: absorptive capacity, adaptive capacity, restorative capacity, and recovery speed. The literature review and qualitative analyses indicated that, to achieve resilient cooling, the four resilience criteria should be considered during the design phase of a building or during the planning of retrofits. It is essential to take into account the characteristics of both the building and its cooling system to ensure the ability to withstand extreme events. A combination of strategies with different resilience capacities, such as a passive envelope strategy coupled with a low-energy, space-cooling solution, may be needed to obtain resilient cooling.

This paper introduces a novel approach to residential energy management by integrating advanced TES and MPC concepts into a unified system. Unlike existing studies that often focus solely on optimizing energy usage in isolation (e.g., HVAC systems or battery storage), this work leverages the thermal inertia of building envelopes as an additional storage mechanism, contributing to resilient design and operational strategies. By dynamically adjusting the target indoor temperatures within comfort margins, the system effectively utilizes the building's thermal mass as a low-cost and sustainable alternative to conventional energy storage solutions. This strategy not only enhances energy efficiency but also significantly reduces reliance on high-cost electricity during peak demand periods, a feature rarely explored in comparable studies.

Furthermore, the developed system incorporates a multilayered optimization process, combining a seven-day prediction horizon for energy planning with real-time control adjustments. This approach allows the EMS to anticipate fluctuations in energy prices, weather conditions, and household demand, enabling more accurate and cost-effective decision making. Unlike traditional energy management frameworks, which often rely on simpler rule-based or reactive strategies, the proposed system employs MPC to achieve a proactive and robust response to varying conditions. The integration of these capabilities into real-world testing environments—validating performance improvements both in laboratory prototypes and residential installations—sets this work apart by demonstrating its practical applicability and scalability. This comprehensive evaluation highlights the potential of the system to contribute meaningfully to the decarbonization of the residential sector.

3. Methods

The objective of this work is to develop an energy management system that, using the basic principles of EMPC, minimizes the cost of electricity consumed in a house by optimizing its electrical energy consumption, of which air conditioning is an important part but not the only one. It has been developed for use with a regulated electricity market, in which the price of energy is updated in real time every hour.

As starting and design conditions, the analysis is carried out for a house that has an energy input from the grid, a self-consumption system (solar panels), and an energy storage system (batteries). In addition, the structural characteristics of the house envelope are also used as a thermal energy storage system (also called thermal battery).

The proposed system has two layers: a planning layer that proposes the optimal distribution of energy sources over a given time horizon that selects when it is best to use grid or stored energy based on energy costs, and a second layer with a real-time control system that regulates the HVAC system for efficient use.

First, a system for planning the electrical energy consumption of the house was designed. As inputs to the planning system, predictions of the electrical consumption of the different appliances present in the house are needed. In addition, it is also necessary to forecast the intended use and occupancy of the residence in order to ascertain the necessity for HVAC. HVAC consumption is obtained from the predictions of outdoor atmospheric conditions, such as temperature and solar radiation, for example. Finally, it is also necessary to forecast the variation of energy prices.

The data provided by OMIE [53] were used to forecast electricity market prices. OMIE is the Iberian Market Operator that manages the day-ahead and intraday markets based on the criteria of independence, transparency, and objectivity. The forecast model for longer periods used for this work is based on the reference [54].

In this way, the planning system estimates the electricity consumption depending on the time of day based on predictions of occupancy, the consumption of appliances, lighting, etc., along with the prediction of energy demand of the HVAC system. Statistical data provided by Eurostat [55] have been used to predict the consumption of the different electrical appliances, lighting, etc.

For the photovoltaic (PV) production estimates, the PVGIS platform [56], widely used in the sector, has been chosen, thanks to which, by entering the orientation and inclination data of the panels, a production curve is obtained for the desired location.

Finally, an algorithm based on model predictive control (MPC) has been used for the prediction of the HVAC system consumption, which aims to accurately calculate the energy consumption as a function of time. This control system is based on a thermal model of the house and is fed with predictions based on the time of day that affect the energy demand, such as outdoor temperature, solar radiation, etc.

When choosing a source of weather forecasts, there are different sources of information, as well as agencies that provide reliable weather forecasts. In this case, Weatherbit [57] has been chosen, since it contains information on different time scales and weather parameters, such as temperature, humidity, etc.

Then, the objective of the system planner is to make decisions on how to distribute the electrical demand within the house among the possible sources that supply it, such as the electrical grid, self-consumption (for example, with solar panels), electric batteries, and the control of the HVAC. The system planner is schematically represented in Figure 1. In the planner's operating scheme, the communication with the grid is bidirectional, taking energy from the grid when necessary or returning energy to the grid in case of surplus. The consumption of electrical appliances and lighting is always an input. It also manages the use of the energy storage system, taking energy from it, or recharging it (prioritizing recharging over returning energy to the grid). Finally, the planner sets the HVAC system operating guidelines to make efficient use of it, giving directives for the indoor temperature that it should try to reach in the rooms to supplement the energy storage capacity of the house as a whole considered as a TES, optimizing the use of batteries.

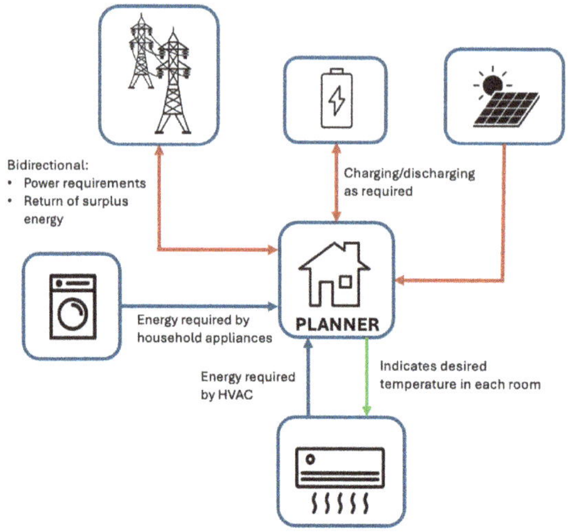

Figure 1. General overview of the system.

In order to minimize the cost of electric energy, once the need for electric energy to meet the demand at a given time has been obtained, priority will be given to the use of self-consumption energy that has no cost. The remaining energy will be distributed between the energy supplied by the grid and the energy supplied by the storage system, prioritizing the use of storage when energy prices are high, and grid consumption when prices are low and then recharging the batteries.

Additionally, the house is considered as a TES when it takes advantage of the thermal and insulation characteristics of its structure and envelopes. Heat can be stored in the structure and materials of the house and released, when necessary, by acting on the interior temperature to be reached in the rooms. For this purpose, the materials of the envelope are not modified, but with the existing ones, the planning system proposes slight variations of the target temperature inside the rooms, always maintaining comfortable indoor temperatures that cause energy storage based on the abovementioned TES principles. This approach leverages the thermal inertia of the building to store or release heat strategically, reducing energy demand during peak periods and enhancing overall efficiency. By allowing controlled deviations from the setpoint temperature, the system can shift energy usage to times when it is more cost-effective or environmentally friendly, contributing to a sustainable and optimized climate control strategy. This approach improves the efficiency of the system, reducing the heating or cooling energy consumption and, consequently, reducing the energy cost.

In this way, the planner works by analyzing data on energy demand and indoor temperatures, outdoor weather conditions, energy prices, and occupant schedules to anticipate heating and cooling needs. Based on these data, the optimizer proactively adjusts heating and cooling systems, storing thermal energy when renewable energy is abundant, and/or energy prices are low. During peaks in energy demand or when energy costs are high, the system reduces active heating or cooling by drawing on thermal energy stored in the home's structure. This approach not only improves energy efficiency and reduces greenhouse gas emissions but also offers flexibility to the grid, reducing stress during peak hours and allowing for the greater integration of renewable energy sources. By means of continuous optimization, the home can be made to achieve energy savings, improve comfort levels, and contribute to broader sustainability goals.

At the planning level, the use of a seven-day forecast horizon offers a significant advantage over a one-day horizon by providing the system with a broader perspective on future energy demand, costs, and generation trends. With this extended horizon, the planner can better anticipate fluctuations in energy prices, PV generation, and thermal energy demand, allowing him to make more strategic decisions about when to store energy in the TES and when to consume it. A shorter horizon lacks the ability to forecast and plan for longer-term trends, leading to less efficient energy management and higher cumulative costs. While a one-day horizon can respond reactively to immediate conditions, it misses opportunities to optimize for the longer term, underscoring why a seven-day model achieves better results in both cost savings and operational efficiency. This optimization minimizes energy costs by avoiding reliance on expensive electricity during periods of high demand and using cheaper energy during periods of surplus or low cost.

The results obtained from the planning layer provide the target temperature guidelines within the rooms that the HVAC controller should try to reach, so as far as the real-time HVAC controller is concerned, given that both electricity market prices and weather conditions are changing, the planning layer is updated hourly, considering it as a shifting horizon.

4. HVAC Control

This section includes everything related to the control system used to predict the consumption of the HVAC system. It is based on model predictive control (MPC) and is derived from a thermal model of the house fed with predictions based on the time of day that affect energy demand, such as outdoor temperature, solar radiation, etc.

First of all, the thermal model considered will be described. Next, the characteristics of the MPC control system that has been developed will be described.

4.1. Thermal Model

The first step involves defining a theoretical model for the temperature evolution within an enclosure. Based on the First Law of Thermodynamics, when applied to a control volume, the principle of energy conservation dictates how temperature changes within that space. This relationship can be generally expressed as the following differential Equation (1):

$$\dot{E}_{vol} = \dot{E}_{in/out} + \dot{E}_g + \dot{Q}_{clim} \tag{1}$$

where $\dot{E}_{vol}$ represents the rate of change in energy stored in the volume, expressed in W, $\dot{E}_{in/out}$ denotes the rate of energy exchanged with the outside of the enclosure, expressed in W, and $\dot{E}_g$ represents the rate of energy generated in the volume, expressed in W.

By applying the generic control volume approach to a specific enclosure, we derive the equation that describes its thermal behavior. In this context, several considerations can be made to further break down the previously defined terms.

The $\dot{E}_{vol}$ term (2) represents the sum of the rate of thermal energy change in the air inside the enclosure plus the rate of thermal energy change in the materials within the enclosure. It is assumed that the air temperature is uniform and consistent throughout the entire enclosure. Based on this, the following Equation (2) is obtained:

$$\dot{E}_{vol} = \left(m_{air} \cdot c_{p_{air}} + \sum_j m_{matj} \cdot c_{p_{mat,j}} \right) \dot{T} = M\,\dot{T} \tag{2}$$

In (2), m_{air} and $m_{mat\,j}$ represent the mass of air inside the enclosure and the mass of each material in the enclosure, respectively, both measured in kg. The terms $c_{p_{air}}$ and $c_{p_{mat,\,j}}$ denote the specific heat capacities of the air and the materials, expressed in J/(kg·K). Finally, $\dot{T}$ represents the rate of change in the temperature over time.

The term $\dot{E}_{in/out}$ in (1) includes the thermal power related to the solar radiation received ($\dot{Q}_{rad}$), the heat transferred through the envelope ($\dot{Q}_{enc}$), and the ventilation thermal load ($\dot{Q}_{vent}$) required to maintain indoor air quality in compliance with legal standards. The breakdown of this term is as follows in (3):

$$\dot{E}_{in/out} = \dot{Q}_{rad} + \dot{Q}_{enc} + \dot{Q}_{vent} \tag{3}$$

The heat due to solar radiation $\dot{Q}_{rad}$, expressed in W, is indicated in (4):

$$\dot{Q}_{rad} = G_i \cdot S \cdot F \cdot Or \cdot f \tag{4}$$

In (4), G_i represents the global irradiance, expressed in W/m^2, while S denotes the area of the window on which it is incident, expressed in m^2. F signifies the modified solar factor, Or denotes the solar radiation coefficient according to the orientation of the window, and f represents the attenuation correction factor for elements such as curtains or awnings.

The heat transmitted through a wall j, denoted $\dot{Q}_{enc\ j}$, expressed in W, is calculated as follows in (5):

$$\dot{Q}_{enc\ j} = U_j \cdot S_j \cdot (T_{ext\ j} - T) \tag{5}$$

In (5), the heat transmission coefficient for a wall, U_j, is expressed in W/(K m^2), while the area, S_j, is expressed in m^2. The indoor temperature, T, is expressed in K, as is the temperature of the external environment with which the enclosure area contacts, $T_{ext\ j}$. It is important to note that this external environment includes both the external environment and other possible enclosures.

In consideration of the fact that an enclosure may comprise multiple walls, the general expression is represented in (6):

$$\dot{Q}_{enc} = \sum_j U_j \cdot S_j \cdot (T_{ext\ j} - T) \tag{6}$$

where the subscript j indicates the number of walls present in the entire enclosure.

The ventilation heat load, $\dot{Q}_{vent}$, is expressed in W and has the form indicated in (7):

$$\dot{Q}_{vent} = \dot{m}_{vent} \cdot c_p \cdot (T_{ext} - T) \tag{7}$$

where $\dot{m}_{vent}$ is the mass flow rate of ventilation air required for the enclosure, expressed in kg/s; c_p is the specific heat of the ventilation air, expressed in J/(kg K); T is the temperature of the enclosure, expressed in K; and T_{ext} is the outdoor temperature, expressed in K. The flow rate $\dot{m}_{vent}$ can be calculated as the product of the outdoor air density ρ_{ext}, expressed in kg/m^3 by the ventilation flow rate $\dot{V}_{vent}$ introduced into the enclosure, expressed in m^3/s. This term $\dot{m}_{vent}$ represents the mass flow rate of ventilation air required for the enclosure, expressed in kg/s. Therefore, this can be expressed as shown in (8):

$$\dot{Q}_{vent} = \rho_{ext} \cdot \dot{V}_{vent} \cdot c_p \cdot (T_{ext} - T_{int}) \tag{8}$$

The $\dot{E}_g$ term includes the thermal loads due to lighting $\dot{Q}_{light}$, equipment $\dot{Q}_{equip}$, and occupancy $\dot{Q}_{ocup}$, which has a latent heat component in addition to a sensible heat component.

$$\dot{E}_g = \dot{Q}_{light} + \dot{Q}_{equip} + \dot{Q}_{ocup} \tag{9}$$

Finally, the thermal load introduced by the HVAC system present in the enclosure $\dot{Q}_{clim}$ must be incorporated as shown in (9).

By aggregating all the developments of the terms that have been made, Equation (1) can be reorganized in the following differential equation that determines the time evolution of the temperature:

$$M\,\dot{T} = A + B\,T + \dot{Q}_{clim} \quad (10)$$

Equation (10) represents the thermal model to be solved. The terms A and B represent the grouping of terms obtained in the different Equations (3) to (8) that do not depend on T (term A) and those that multiply the variable T (term B), as well as those that represent all the actions on the enclosure outside the HVAC system to be controlled. In general terms, expressions (3) to (8), and, therefore, A and B, are time-dependent functions, since they depend on the time of day and the thermal loads not associated with HVAC in each time phase.

Finally, the term $\dot{Q}_{clim}$ represents the action performed by the HVAC system and that will be the variable to be controlled to try to achieve the desired interior temperature T of an enclosure.

4.2. Design of the MPC

The proposal is to implement a control system, with a dedicated unit for each room, based on an MPC approach. The controller is nonlinear, and the constraints are explicitly handled [58].

According to the receding horizon principle, at each time step, the MPC algorithm computes the optimal temperatures solving a finite horizon optimization problem.

For the formulation of the MPC, a prediction horizon $\left[t, t+N_p\right]$ is considered at time t. The notation $x_{t+k|t}$ represents the state vector at time $t+k$, predicted at time t, obtained by starting from the current state $x_{t|t} = x(t) \equiv x_t$, and where $u_{\cdot|t} = \left[u_{t|t}, \ldots, u_{t+N_p-1|t}\right]$ denotes the unknown input variables to be optimized.

The control system aims to achieve the target comfort temperature in each room, while satisfying the state and input constraints. So, for a particular room i, the state variable is the temperature $T^i_{k|t}$ and the unknown input variable to be optimized is the thermal load introduced by the HVAC system $\dot{Q}^i_{k|t}$.

Therefore, the optimization problem for the temperature controller can be formulated in (11):

$$\min_{\dot{Q}_{\cdot|t}} J\left(T^i_{k|t}, \dot{Q}^i_{k|t}, \Delta T^i_{k|t}\right) \quad (11)$$

subject to:

$$T^i_{t|t} = T^i_t \qquad \forall i = 1, \ldots, N_e \quad (12)$$

$$T^i_{k+1|t} = T^i_{k|t} + \sum\nolimits_{j=1}^{K} \omega_j \Delta t f\left(T^i_{k,j|t}, \dot{Q}^i_{k,j|t}\right) \qquad \forall k \in \left[t, t+N_p-1\right]\ \forall i = 1, \ldots, N_e \quad (13)$$

$$f\left(T^i_{k|t}, \dot{Q}^i_{k|t}\right) = \left(A^i + B^i T^i_{k|t} + \dot{Q}^i_{k|t}\right)/M^i \qquad \forall k \in \left[t, t+N_p-1\right]\ \forall i = 1, \ldots, N_e \quad (14)$$

$$\dot{Q}^i_{ref} \leq \dot{Q}^i_{k|t} \leq \dot{Q}^i_{heat} \qquad \forall k \in \left[t, t+N_p-1\right]\ \forall i = 1, \ldots, N_e \quad (15)$$

$$T^i_{k|t} - \Delta T^i_t \leq T^i_{ref} \ \text{with}\ \Delta T^i_t \geq 0 \qquad \forall k \in \left[t, t+N_p-1\right]\ \forall i = 1, \ldots, N_e \quad (16)$$

$$T^i_{N_p|t} - \Delta T^i_t \leq T^i_{ref} \qquad \forall i = 1, \ldots, N_e \quad (17)$$

$$(1-\alpha)M^i \leq M^i \leq (1+\alpha)M^i \qquad \forall i = 1, \ldots, N_e \tag{18}$$

$$-\delta\text{T} \leq w_{T_{ext}} \leq \delta\text{T} \tag{19}$$

In this optimization problem, N_e is the number of air-conditioned rooms in the dwelling, and N_p is the prediction horizon.

The state $T^i_{t|t}$ at time t is initialized with the actual temperature of the room T^i_t, as seen in Equation (12). Then, future states are obtained with Equation (13), which follows a linear discretization approach of Equation (10).

It should be noted that hard constraints are used for the input variables in this optimization problem; whereas, soft constraints are used for the state variables.

A hard constraint from Equation (15) limits the input variable $\dot{Q}^i_{k|t}$, between the maximum refrigeration $\dot{Q}^i_{ref}$ and heat $\dot{Q}^i_{heat}$ thermal loads.

Moreover, Equation (16) bound the temperature of the room, where $T^i_{k|t}$ (K) is the estimated temperature of the controlled room at instant k calculated at time t, and T^i_{ref} is the temperature objective to be reached. This is a soft constraint where Equation (16) penalizes, exceeding the maximum speed through the positive slack variable ΔT^i_t.

The terminal constraint in Equation (17) serves as an operational constraint.

In addition, for the real-time MPC, as it is designed in a robust scenario-based MPC that considers uncertainties in the design parameters and disturbances in the inputs, the following constraints are added.

Equation (18) includes the uncertain in the design parameters considered in the model. In this way, an uncertainty has been considered in the evaluation of the thermal inertia of the building (18), where α is the deviation from the nominal value of thermal inertia expressed as a percent of one. On the other hand, constraint (19) represents the white noise disturbance considered for the outside temperature of the enclosures being δT the amplitude of the uncertainty for the temperature sensor.

Therefore, we define the objective function J to be optimized in (11) as in Equation (20). The objective function has been constructed following the usual and recommended practice of normalizing the terms of the objective function, in order to avoid the terms whose values belong to a higher order of magnitude not presenting more weight than the remaining terms.

$$J = \sum_{k=t}^{t+N_p} \sum_{i=1}^{N_u} \left(K^i_T \cdot \left\| 1 - \frac{T^i_{k|t}}{T^i_{ref}} \right\| + K^i_{\Delta T} \cdot \left\| \frac{\Delta T^i_{k|t}}{T^i_{ref}} \right\| + K^i_{\dot{Q}} \cdot \left\| \frac{\dot{Q}^i_{k|t}}{\dot{Q}^i_{ref}} \right\| \right) \tag{20}$$

The cost function (20) has several dimensionless coefficients: $K^i_T \geq 0$ represents the weight penalizing the output deviation from the maximum temperature of reference T^i_{ref}, $K^i_{\Delta T}$ represents the weight penalizing the positive slack variable ΔT^i_t, and $K^i_{\dot{Q}}$ is a tuning parameter used to smoothen the obtained optimal solution.

The developed HVAC controller is used for two functions. Firstly, as a further component of the planner, it is used for the prediction of the consumption due to the air conditioning of the building. In this case, the MPC is focused on a nominal control design, i.e., we assume perfect model knowledge at the design stage. In this case, the MPC is built based on Equations (11)–(17) and (20).

However, secondly, MPC is also used, once the planning has been completed, as a real-time control system, where the HVAC behavior is regulated. In this case, it is considered a robust controller where uncertainties in the estimation of certain parameters and disturbances in the inputs to the system are then considered, being built based on Equations (11)–(20).

Once the optimization problem has been solved, for each room, the resulting optimal states and inputs are denoted as following:

$$\begin{aligned} T_t^{i*} &= \begin{pmatrix} T_{t|t}^{i*} & T_{t+1|t}^{i*} & \cdots & T_{t+N_p|t}^{i*} \end{pmatrix}^T \\ \dot{Q}_t^{i*} &= \begin{pmatrix} \dot{Q}_{t|t}^{i*} & \dot{Q}_{t+1|t}^{i*} & \cdots & \dot{Q}_{t+N_p-1|t}^{i*} \end{pmatrix}^T \end{aligned} \tag{21}$$

For closing the loop, the first input is applied to the system (10) during the time interval $[t, t+1)$:

$$\dot{Q}_t^{i} = \dot{Q}_{t|t}^{i*} \tag{22}$$

At the next time step, $t+1$, a new optimal problem in the form of (11)–(20), based on the new measurement of the state, is solved over a shifted horizon, yielding a moving receding horizon control strategy with control law.

5. Energy Consumption Planner

The energy consumption planner has been developed to manage and optimize the energy consumption of the house, minimizing the cost of the required electrical energy. For this purpose, an optimization problem has been formulated.

The objective function to be minimized is the cost of electricity over a period of time t_p, so the objective function can be written as:

$$\min_{C_b, D_b, C_{PV}} \sum_{t=t_0}^{t_0+t_p} \left(P_g^t \times \left(Cg^t + C_g^t - D_g^t \right) + P_{PV}^t \times C_{PV}^t \right) \tag{23}$$

where P_g^t represents the price of electricity from the grid at time t. Cg^t is the grid electricity consumption of the house, obtained by adding the predicted consumption of appliances, lighting, and HVAC, subtracting the part already covered by the self-consumption coming from the PV installation.

Related to the batteries, C_g^t represents the charging power coming from the grid and D_g^t represents the discharging power by the batteries. P_{PV}^t is the price of charging the batteries with energy from the PV installation, and C_{PV}^t is the charging from photovoltaic surplus energy. If we assume that the cost of photovoltaic energy is 0, the objective function can be rewritten as follows:

$$\min_{C_b, D_b, C_{PV}} \sum_{t=t_0}^{t_0+t_p} \left(P_g^t \times \left(Cg^t + C_g^t - D_g^t \right) \right) \tag{24}$$

subject to the following constraints:

$$\begin{aligned} (1 - DOD) * STG \le SOC^t \le STG \\ SOC^t = SOC^{t-1} + \alpha^t \cdot \left(C_g^t + C_{PV}^t \right) - \beta^t \cdot D_g^t \end{aligned} \tag{25}$$

$$\alpha^t + \beta^t \le 1 \tag{26}$$

$$0 \le C_g^t + C_{PV}^t \le Pow_C \tag{27}$$

$$0 \le D_g^t \le Pow_D \tag{28}$$

$$Cg^t + \alpha^t \cdot \left(C_g^t + C_{PV}^t \right) - \beta^t \, D_g^t \le Pow_{PS} \tag{29}$$

Constraint (25) refers to the technical limits of the batteries, where DOD is the depth of discharge of the batteries, STG is the capacity in kWh, and SOC^t in kWh is state of charge of the batteries at time t. The model used does not take into account battery degradation.

Constraint (26) ensures that the battery cannot charge and discharge simultaneously. For this reason, α^t and β^t act as binary flags: when $\alpha^t = 1$, charging occurs, and when $\beta^t = 1$, discharging occurs. Their sum being limited to 1 prevents overlapping operations, preserving the logical integrity of the system.

Constraints (27) and (28) are also related to the technical limitations of the batteries, limiting the maximum charging Pow_C and discharging Pow_D power in kWh of the battery.

Additionally, a constraint called "peak shaving" (29) is implemented to avoid power spikes, where Pow_{PS} in kWh represents the maximum power that should not be exceeded at any time. A graphical example of this function can be seen in Figure 2.

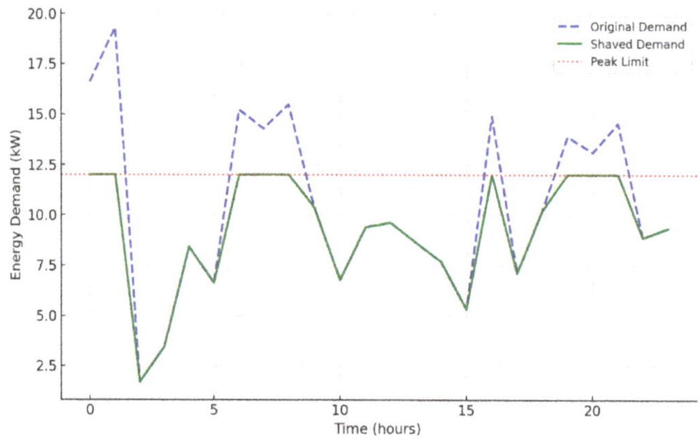

Figure 2. Peak saving constraint.

6. Implementation

To evaluate the results of the proposed system, a mock-up consisting of a residential house, specifically a single-family home for two people, is proposed. Its location was established in downtown Madrid, since this is a location where it is interesting to achieve energy savings and reduce CO_2 emissions. The house has a single floor and a sloping gable roof.

Appendix A includes a plan of the house, and Appendix B includes the parameters considered for the thermal model of the house. The parameters used in the thermal model have been obtained with CYPECAD MEP 2025.c [59]. CYPECAD MEP is a program for the design and sizing of the building envelope, layout, and facilities on a 3D model integrated with the different elements of the building. The program allows us to obtain the necessary parameters for the thermal model, based on the edges detected, depending on the construction solutions adopted and the description of the building from the point of view of thermal analysis (zoning, description of spaces, etc.), according to the standards ANSI/ASHRAE/ACCA Standard 183-2007 [60], CEN/TR 12831 [61], or ISO 6946 [62]. The data for each enclosure include interior conditions, occupancy, illumination, ventilation established according to current regulations, and equipment or other loads. The program also includes libraries of enclosures, partition walls, floors, and roofs, as well as various types of windows.

For the calculations in this paper, the design conditions considered are an operating temperature range inside the enclosures of 22 °C to 26 °C in summer and 18 °C to 23 °C in winter, with a relative humidity of 50% maintained throughout both seasons for all air-conditioned rooms in the house.

For sizing the photovoltaic system, the following factors were taken into account: the available roof surface area for installing solar panels, the absence of shading that could hinder proper radiation reception, the geographical location of the house, and the increased energy demand during winter for heating. Consequently, a photovoltaic setup with 14 solar panels of 450 W each was selected, providing a total peak power of 6.3 kW. Additionally, a 10 kWh electrical battery was included in the design.

The developed system is designed for continuous operation. As previously mentioned, predictions are made with a seven-day horizon, as this represents the optimal balance: it provides the planner with enough foresight to improve accuracy and maximize energy savings, while avoiding unnecessary increases in computation time. This process uses data from prediction models for costs, house usage, atmospheric conditions, and other factors to generate hourly predictions for the entire week ahead. For the planner, the HVAC MPC provides power requirements as well. With these data, the optimizer provides the demand prediction and the battery usage planning with the abovementioned time frequency. The execution process has several steps and is schematically represented in Figure 3.

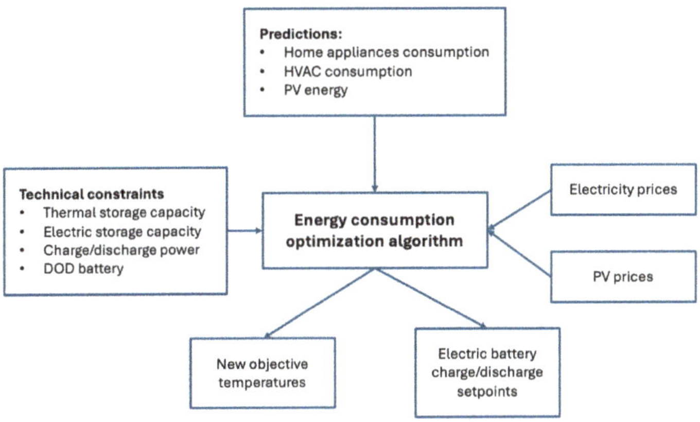

Figure 3. General overview of optimization process.

First, the price prediction models, electrical consumption prediction (excluding HVAC consumption), and solar radiation are executed, providing results with an hourly frequency, and are considered constant during that hour. Parameters and values considered for the planner module are included in Appendix D.

Second, the MPC model is run to obtain an accurate prediction of the HVAC consumption. This model is run by updating the inputs to the model (outdoor temperature, atmospheric conditions, etc.) and target temperature to be reached in each room once every hour and allows us to obtain the temperature of each room at each instant of time.

Subsequently, once all the predictions have been obtained, the energy optimization model is run for two purposes: first, to determine a schedule for running the electric battery and, second, to establish a target temperature profile inside the rooms for the house to act as a thermal battery, proposing operating temperature profiles for weekdays and weekends designed to balance comfort and energy efficiency based on occupancy patterns.

According to an IDAE study [63], the ideal indoor temperature is between 21–23 °C during the day and 16–18 °C at night for the winter months. However, in cold areas, these figures can rise to 22–24 °C during the day and 18–20 °C at night. In the same way, the ideal temperature is between 21–23 °C during the day and 16–18 °C at night for the summer months. For this reason, the capacity of the home's thermal battery has been defined as the energy required to increase the indoor temperature by 2 °C and maintain it for at least 2 h. Based on this calculation, the required energy is 6 kWh thermal, which is equivalent to 2 kWh electrical, considering that the HVAC system has a COP of 3.

Regarding the charging strategy, priority has been given to the use of the electrical battery. This means that the electrical battery will be charged first, and once it is fully charged, the thermal battery will then be charged. Therefore, if the planner determines that the thermal battery needs to be charged, the target temperature of the enclosures will be increased by 2 °C to achieve and maintain the required level.

Finally, once the desired temperature profile for each room is determined, the MPC model is run again to control the HVAC and achieve the desired objectives.

The MPC used for to control the HVAC has been implemented by programming in Python, using the do-mpc toolbox [64]. Do-mpc is a comprehensive, open-source toolbox for robust model predictive control and moving horizon estimation that enables the efficient formulation and solution of control and estimation problems for nonlinear systems, including tools to deal with uncertainty and time discretization. Ipopt [65] has been used as the solver for the optimization MPC problem.

In an MPC controller, the parameter values of the sampling time, the prediction horizon, and the control horizon have to be set. The values assigned to them have a very relevant influence on the results provided by the controller, so it is necessary to select them properly to achieve the best results. However, there are no analytical methods that allow for determining the optimal combination of these parameters, beyond a series of general considerations that allow for starting the iteration of values, finally obtaining, through the trial-and-error method, those values that allow for the generation of adequate results.

In addition to these parameters, it is also necessary to fix the positive values of the coefficients of the different terms that constitute the objective function that the controller tries to minimize. Similarly, these weights have a notable influence on the results provided by the controller and must be carefully chosen by trial and error, there being no analytical methods for this purpose. It is emphasized that for these parameters, as well as for those mentioned above, the values that provide optimum controller behavior vary with the modeling that has been carried out in each case, as well as with the initial conditions that are considered.

In the different cases analyzed, the control horizon is equal to the prediction horizon, a fairly common practice. The choice of sampling time, prediction horizon, and weights are made considering several criteria, such as the accuracy of the temperature results, the energy cost associated with power consumption, or the computation time needed by the controller to solve the system of differential equations.

The parameters used for the MPC model are included in Appendix C.

The time step (1 min) and prediction horizon (60 min) play a crucial role in determining the operation of the HVAC controller and the overall effectiveness of the energy management strategy. The 1 min time step allows for the controller to make frequent updates to its decisions, enabling precise and fine-grained control of the HVAC system. This ensures that the system can quickly adapt to changes in indoor temperature or external factors, like wind or solar radiation variations. While a shorter time step increases computational demand due to more frequent calculations, it ensures that the controller operates with a high level of precision, minimizing energy waste and maintaining thermal comfort. The 60 min prediction horizon, on the other hand, provides the controller with the ability to forecast system behavior, including the building's thermal response and anticipated changes in other factors that influence the controller over a sufficient period.

7. Results and Discussion

This section presents and discusses the results obtained from the simulations performed for the different use cases indicated in the single-family house. The objective of this section is to show the results obtained with the planner and to quantify the economic savings achieved with the proposed solution. Two scenarios have been considered, one corresponding to a week in the winter period and the other to a week in the summer period. With these two cases, it will be possible to see in detail the operation of the planner, and how the use of the batteries is managed, according to the different energy consumption demands of the house.

7.1. Winter

On weekdays, the temperature is set to 18 °C during the night and during working hours when the house is typically unoccupied. It increases to 20 °C in the morning for waking up and preparation and to 21 °C in the evening for comfort during home activities, before dropping to 19 °C for sleeping. On weekends, when people tend to be at home more, the temperature starts at 18 °C during the night, rises to 20 °C in the early morning, and stays at 21 °C throughout the day to ensure comfort during indoor activities, before settling at 19 °C at bedtime. This profile ensures a comfortable environment, while optimizing energy use.

Figures 4 and 5 show the outside temperature profile and the temperature profile inside the enclosures of a typical week during winter, respectively.

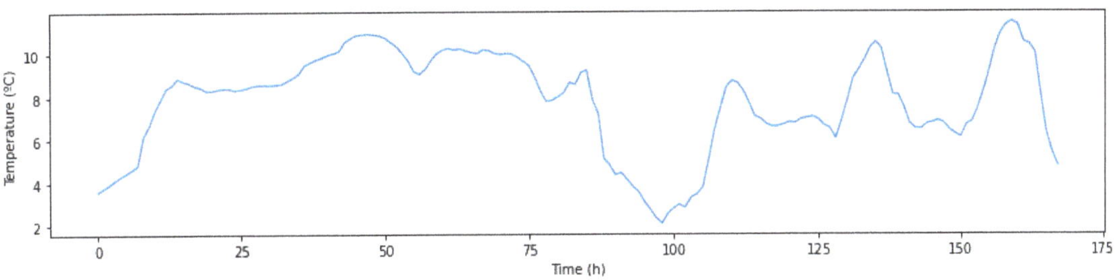

Figure 4. Outside temperature profile of a typical week during winter.

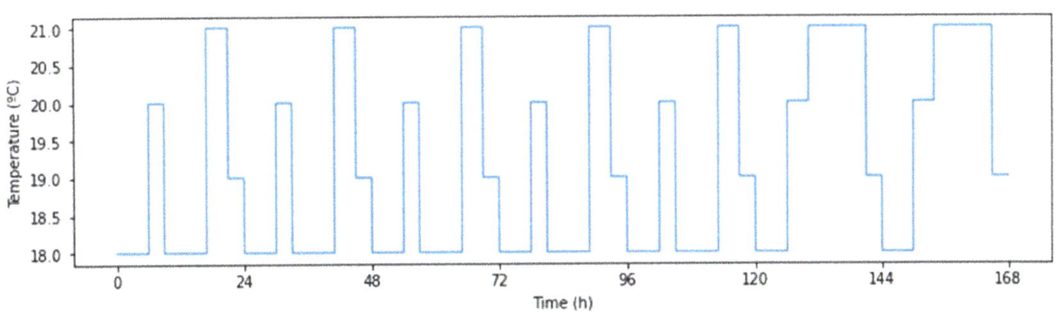

Figure 5. Temperature profile inside the enclosures in a typical week during winter.

As mentioned earlier, the MPC model is first executed to obtain the most reliable possible heating consumption curve as a prediction for the optimization model of the energy consumption.

Figures 6 and 7 show the temperatures obtained and the thermal power required by the HVAC system, respectively.

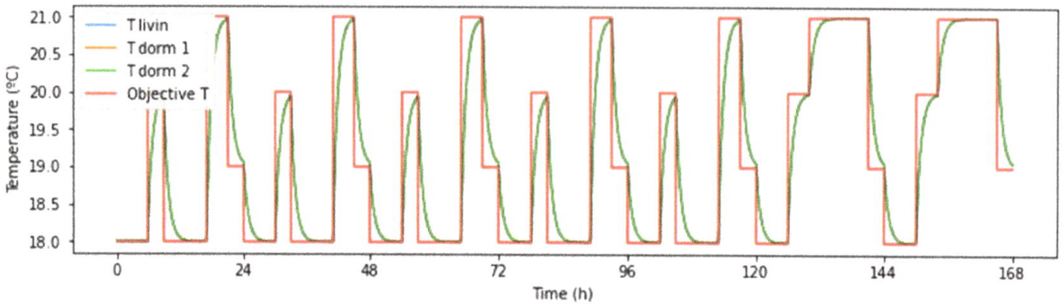

Figure 6. Evolution of the temperature inside the enclosures for a winter week.

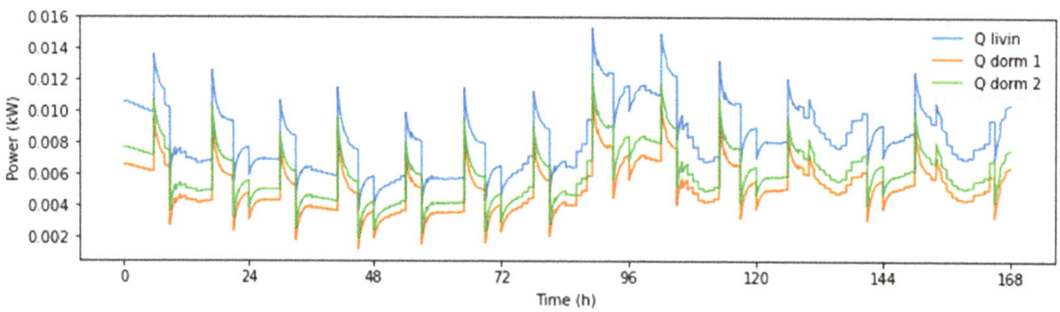

Figure 7. Thermal power required by the HVAC system in each enclosure for a winter week.

In Figure 6, it is evident that the actual temperatures follow the evolution of the target temperature. It can also be observed that there is a delay between the setpoint increase and the system response due to thermal inertia. The blue and yellow lines in Figure 6 are coincident and overlapping with the green line, so that only the green line is visible in this figure.

In Figure 7, it is shown that the thermal power used exhibits periodic peaks, indicating that the system is responding to changes in the target temperature. It is observed that the living room consistently requires more energy than the other two bedrooms, because it has higher thermal losses and higher heating requirements in this zone due to higher outdoor exposure and higher volume. The peaks in the HVAC system's energy consumption shown in Figure 7 coincide with the moments when the indoor target temperatures change, as illustrated in Figure 6. During winter, the system must quickly compensate for the differences between the ambient temperature and the new, higher target temperatures. This leads to increases in power consumption, especially when the system adjusts the heating to reach the higher temperatures. Once the desired temperature is reached, the consumption decreases, stabilizing until the next significant change in the target temperature.

By analyzing both figures together, it can be concluded that the HVAC system is successfully meeting the temperature objectives.

Using the total HVAC consumption curve, combined with the electricity consumption of other household loads, photovoltaic production, and grid electricity prices, the optimization model is run to determine the best use of the different energy sources.

The results are shown in Figure 8, which shows the electricity consumption of HVAC, other household loads, and the photovoltaic production.

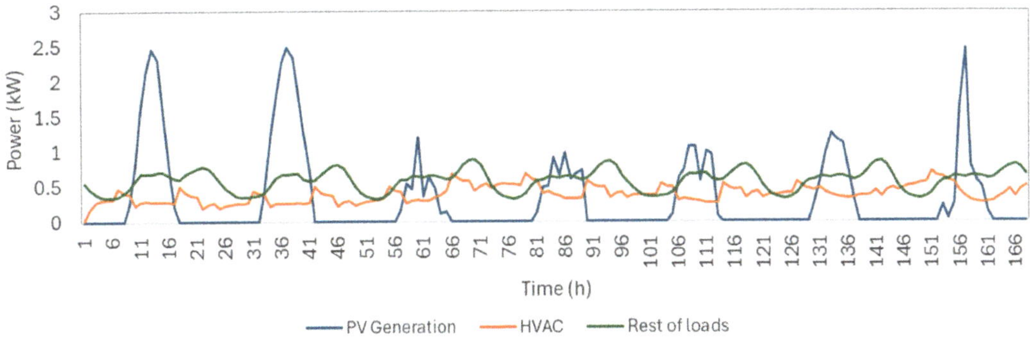

Figure 8. Electricity consumption of HVAC, other household loads, photovoltaic production, and storage elements for a winter week.

Figure 8 shows that, during the daylight hours of the first two days, the PV generation significantly exceeds the combined demand of HVAC and other loads, creating an opportunity for energy storage or export to the grid. However, during nighttime or periods of low sunlight, PV generation drops significantly, and the home relies entirely on external power sources to meet the relatively stable consumption of HVAC and the rest of the loads.

In relation to outdoor temperature variation (Figure 4) and energy use patterns (Figure 8), the indoor temperature profile introduces a pattern in the HVAC system's energy consumption, which interacts with the outdoor temperature, as seen in Figure 9.

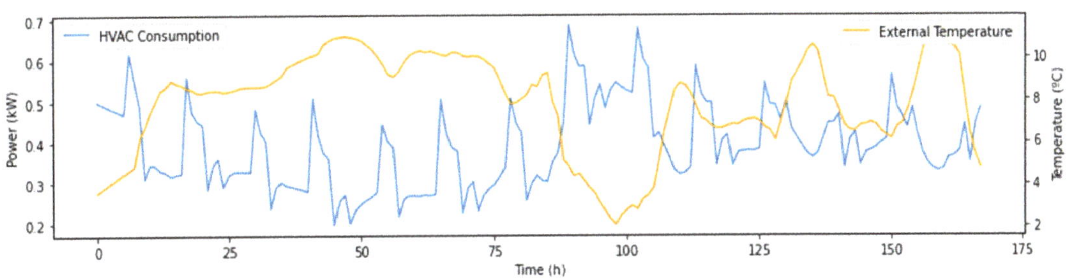

Figure 9. Outdoor temperature variation and HVAC system's energy consumption.

During nighttime and working hours on weekdays, when the indoor temperature is set to 18 °C, the HVAC system uses less energy, which aligns with the observed consumption minimums. In contrast, during the morning, when the temperature rises to 20 °C for preparation, and in the evening, when it increases to 21 °C for comfort, the HVAC consumption rises due to the additional effort required to reach the programmed temperatures, especially on colder days.

On days with lower outdoor temperatures, the HVAC system needs to work harder to heat the house, resulting in higher consumption peaks. Conversely, on days with higher outdoor temperatures, the HVAC requires less energy, since the outdoor temperatures are closer to the desired indoor settings.

On weekends, when the indoor temperature remains at 21 °C for most of the day to ensure comfort, HVAC consumption is likely more constant and higher compared to weekdays, though this pattern depends on outdoor temperatures.

Then, after running the model for a representative winter week, and as a result of the planner, the new target temperature profile is as indicated in Figure 10.

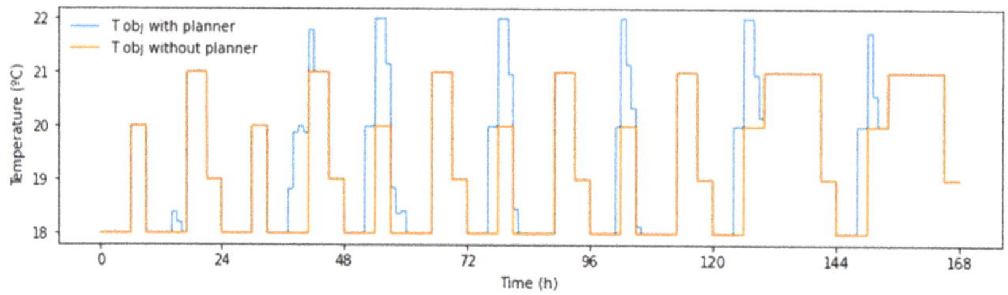

Figure 10. New target temperature profile inside the enclosures for a winter week.

As can be seen in Figure 10, there are certain hours when the target temperature has risen due to the use of the home as a battery. With the new temperature profile, we run the MPC model again, obtaining the results that appear in Figure 11. Again, the blue and yellow lines in Figure 11 are coincident and overlapping with the green line, so that only the green line is visible in this figure.

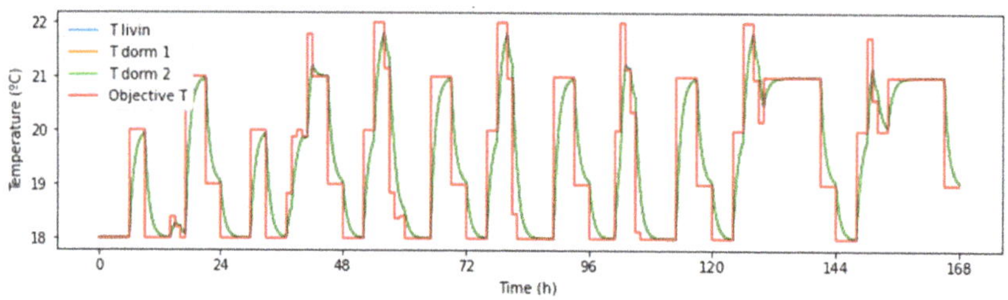

Figure 11. Temperature evolution inside the different enclosures for a winter week.

The temperatures in Figure 11 show greater stability and do not drop as abruptly after reaching the target temperature. This smoother evolution suggests that the TES of the home is functioning, meaning that the heat stored in the walls, floors, or structural elements helps maintain the temperature for a longer period. As a result, the HVAC system does not need to respond as aggressively or as frequently, reducing peak energy demand and improving system efficiency.

The comparison between Figures 6 and 11 highlights the influence of the TES on the temperature evolution of the home. In Figure 6, the lack of thermal storage capacity generates more pronounced temperature fluctuations, forcing the HVAC system to work more intensively. In contrast, in Figure 11, the use of the thermal battery provides greater thermal inertia, reducing heat losses and stabilizing the temperature across different zones of the building. This not only enhances thermal comfort but also decreases energy consumption by optimizing the operation of the HVAC system.

The thermal power required by the HVAC system in the scenario of using the planner and the TES is shown in Figure 12.

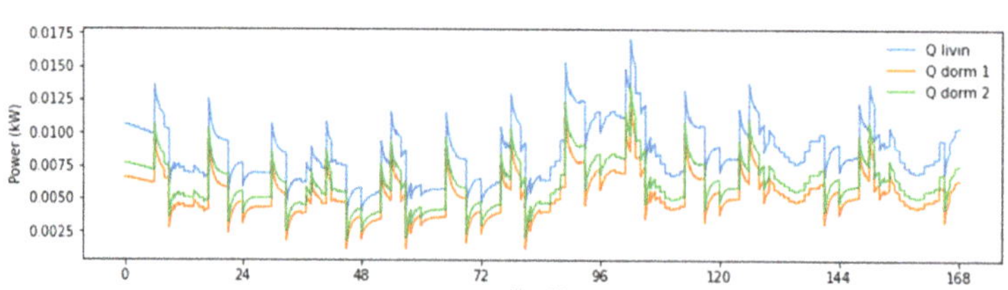

Figure 12. Thermal power required by the HVAC system for a winter week.

Figure 13 highlights the differences in energy consumption between systems with and without a planner. With a planner, energy usage exhibits noticeable peaks at specific times, reflecting the system's strategy of charging the battery and TES during periods of lower electricity demand and cheaper prices. This optimized approach stores energy during these optimal periods, reducing consumption during high-demand or high-cost times. In winter, as shown in Figure 13, this strategy concentrates energy usage exclusively during low-price periods, ensuring no energy is consumed during peak pricing times, thereby maximizing cost efficiency. In contrast, without a planner, energy consumption remains constant yet frequent, with instances of usage even during high-cost periods. This lack of optimization leads to inefficient and more expensive energy use. By shifting consumption to the most cost-effective periods, the planner reduces overall energy costs, ensures sustainable energy management, and achieves a substantial reduction in the final electricity bill.

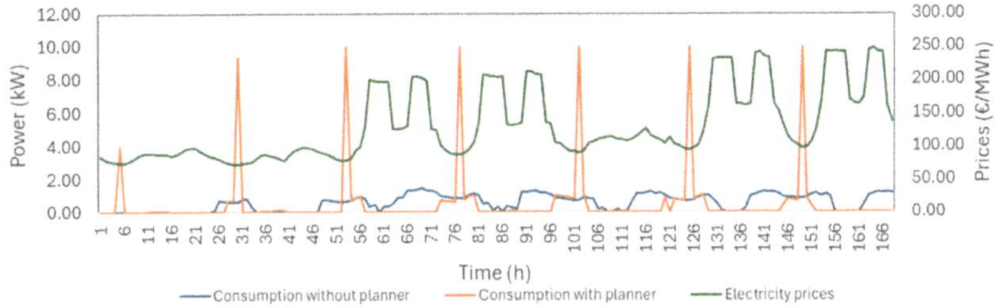

Figure 13. Comparison of electric consumption with and without planner for a winter week.

Figure 14 illustrates the state of charge (SOC) of the system, with and without using the planner. Without the planner, the electric battery only works by storing the excess energy produced by the PV system during periods of surplus generation. When the solar panels generate more electricity than the home requires, the excess energy is used to charge the battery instead of being sent back to the grid. Once the battery is fully charged, it remains on standby, ready to supply power when needed. As soon as the installation starts consuming energy from the grid, the battery automatically discharges to meet this demand, reducing grid reliance.

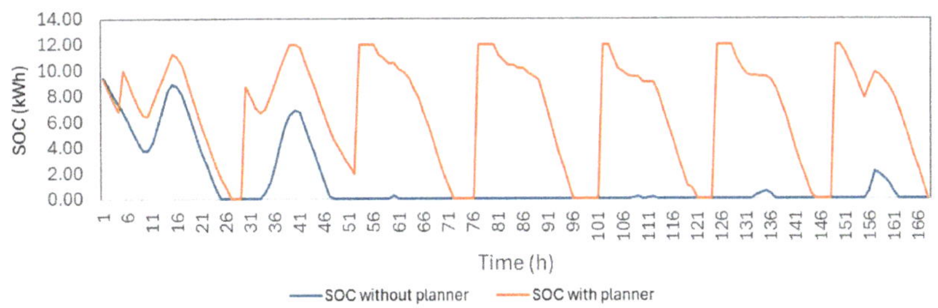

Figure 14. SOC of the system with and without using the planner for a winter week.

In contrast, when using a planner, the thermal storage is efficiently managed, displaying clear charging and discharging cycles. This indicates that energy is stored during periods when it is cheaper or more convenient and discharged later when needed. Without the planner, the SOC remains largely low or empty, highlighting that the storage systems are not effectively utilized, leading to a direct reliance on instantaneous energy consumption from the grid.

Finally, Figure 15 shows the cost of energy, comparing the results of the optimizer without and with indoor temperature control using the TES concept.

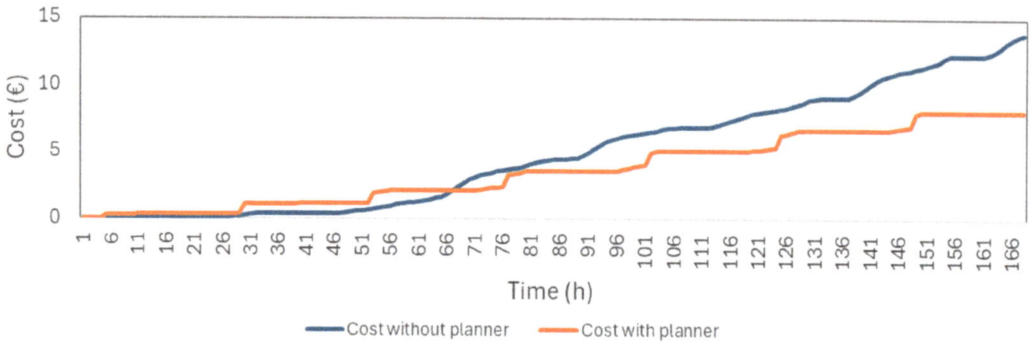

Figure 15. Cost of energy consumption comparing the results of the optimizer without and with indoor temperature control for a winter week.

Figure 15 shows the cumulative cost over time and the impact on total cost. With the planner, the cost increases at a slower rate, indicating the savings achieved by strategically charging and using the TES. This optimization allows the system to avoid using power during expensive periods. In contrast, without the planner, costs accumulate more quickly, reflecting the inefficiencies of using energy immediately without leveraging storage capabilities. The jumps observed in Figure 15 are due to the system where the scheduler strategically concentrates energy consumption during certain periods when electricity prices are at their lowest. These discrete increases in cost correspond to the planner's decision to charge the battery and TES during optimal pricing windows. By doing so, the planner avoids consuming electricity during high-cost periods, resulting in cost savings and fewer incremental increases in the overall cost. In contrast, the system without a planner incurs more frequent and gradual increases in cost, as it directly consumes energy on demand without optimizing for lower-priced periods.

In summary, these results show that the TES acts as an energy reservoir that allows systems to store thermal energy when electricity is cheap or plentiful and use it later when energy demand or costs are higher. The planner optimizes this process by determining the ideal times to charge and discharge the storage, thereby minimizing costs and improving energy efficiency. Without the planner, the system fails to take advantage of thermal storage, resulting in higher costs and more inefficient energy use.

7.2. Summer

The temperature profile for the home, based on the time of the day and day of the week, is designed to optimize comfort and energy efficiency. On weekdays, the target temperature is set at 24 °C overnight, rising to 26 °C in the morning to accommodate daytime activities. In the afternoon, the temperature is set back to 24 °C to provide a comfortable temperature for the end of the day. In the evening, the temperature is raised again to 26 °C to accommodate nighttime preferences. On weekends, the temperature remains at 24 °C from midnight to noon, providing a cooler atmosphere for sleeping and early morning hours. In the afternoon, the target temperature drops slightly to 23 °C for a cooler, more restful environment. Finally, the temperature returns to 24 °C for the rest of the day, providing comfort for evening activities.

Figures 16 and 17 show the outside temperature profile and the temperature profile inside the enclosures of a typical week during summer, respectively.

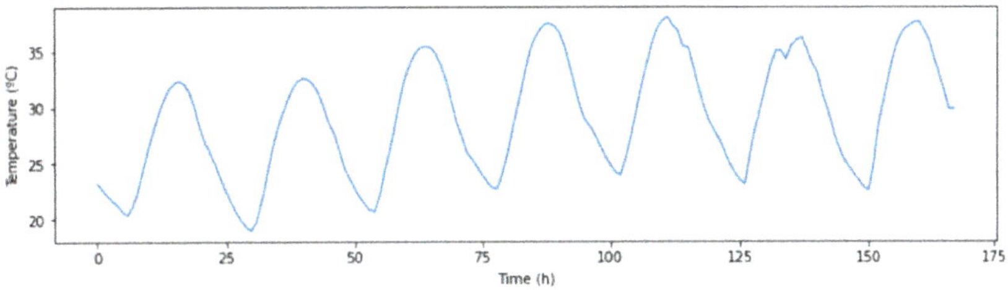

Figure 16. Outside temperature profile of a typical week during summer.

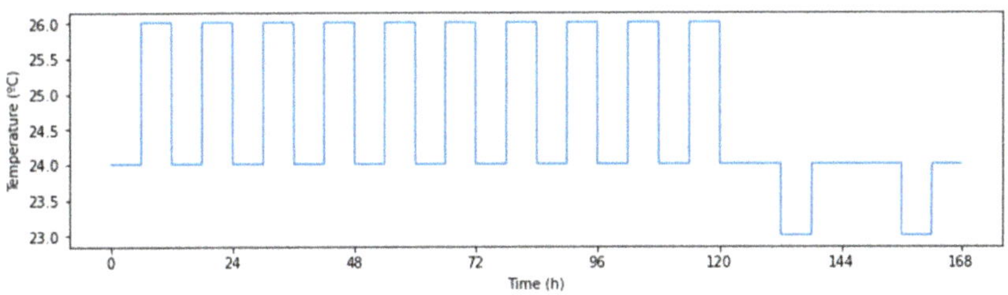

Figure 17. Temperature profile inside the enclosures in a typical week during summer.

Figures 18 and 19 show the temperatures obtained and the thermal power required by the HVAC system, respectively.

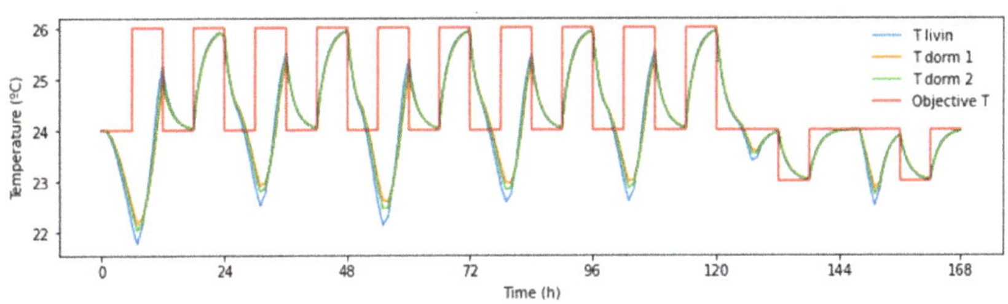

Figure 18. Evolution of the temperature inside the enclosures for a summer week.

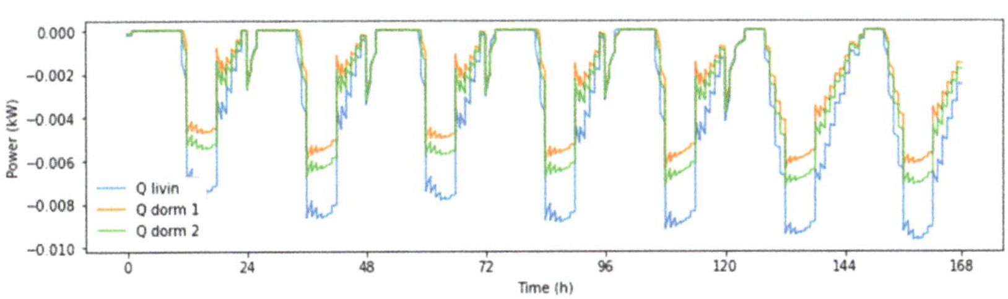

Figure 19. Thermal power required by the HVAC system in each enclosure for a summer week.

The HVAC system exhibits a cyclical dynamic where power consumption and temperature evolution interact to maintain thermal comfort in the different rooms. During the day, internal temperatures rise due to external thermal loads, such as solar radiation, and internal gains generated by occupants and devices. Once temperatures reach the upper limit of the target range, the system activates cooling, reflected in negative power consumption in Figure 19. This controlled cooling gradually reduces temperatures until they approach the lower limit of the target range, at which point the system decreases or stops its operation to save energy. This pattern repeats periodically and shows slight variations between the different zones (living room and bedrooms), which, as mentioned before, is due to differences in solar exposure, size, or thermal characteristics of the spaces. Additionally, the cycles are more pronounced during the day when external thermal loads are higher; whereas, at night, the system requires less intervention due to natural ambient cooling. Overall, this demonstrates an efficient thermal control system that dynamically adjusts power usage to maintain temperatures within the set range, maximizing comfort, while optimizing energy consumption.

We can observe the temperature drops in Figure 18, which can be explained in relation to the outside temperature. When the outside temperature is lower than the target indoor temperature, the house loses heat to the exterior. Since the system is set to cooling mode, it is normal for the temperature in various rooms to decrease. This is because the system cannot switch to heating mode while in cooling mode, and the heat from the interior is transferred outside, causing a drop in the indoor temperature.

Again, using this HVAC consumption curve, combined with the electricity consumption of other household loads, PV production, and grid electricity prices, the optimization model is run to determine the best use of the different energy sources.

The results are shown in Figure 20, which shows the electricity consumption of HVAC, other household loads, photovoltaic production, and storage elements.

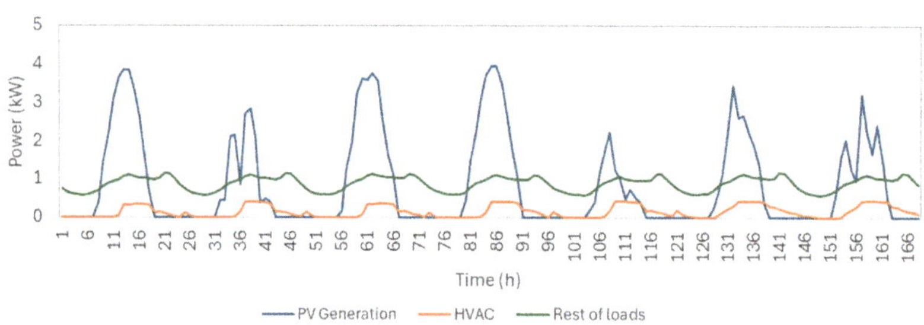

Figure 20. Electricity consumption of HVAC, other household loads, photovoltaic production, and storage elements for a summer week.

In Figure 20, it is evident that, due to the high solar radiation typical of this season, the PV generation reaches significant peaks during the day, exceeding the home's electrical demands, such as the HVAC system and other loads. This results in energy surpluses, visible as the differences between the PV generation curve and the total loads. These surpluses could be stored in batteries or exported to the grid.

After running the model for a representative summer week, and as a result of the planner, the new target temperature profile is as indicated in Figure 21.

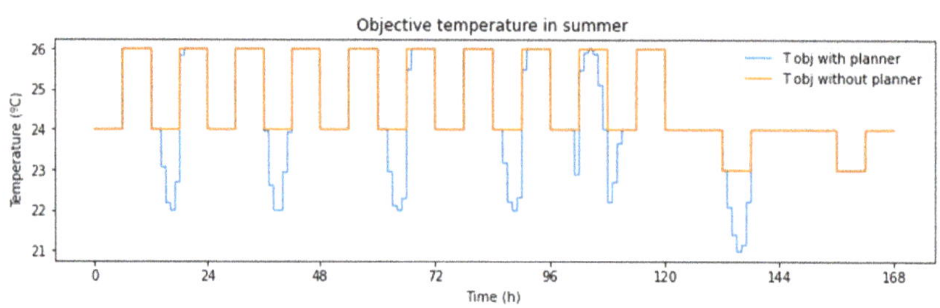

Figure 21. New target temperature profile inside the enclosures for a summer week.

With the new temperature profile, the results appear in Figure 22.

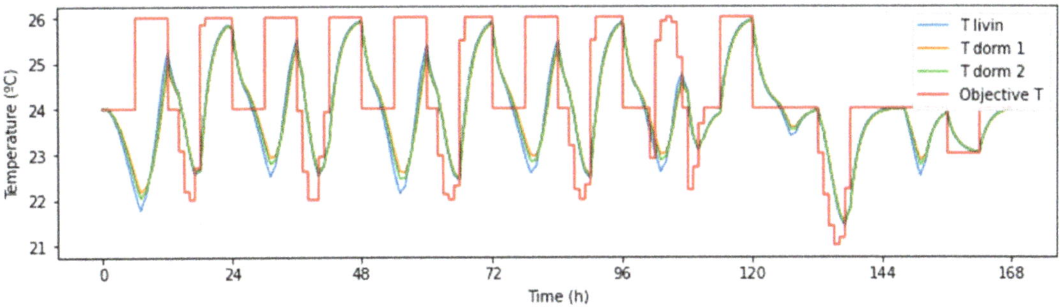

Figure 22. Temperature evolution inside the different enclosures for a summer week.

The thermal power required by the HVAC system is shown in Figure 23.

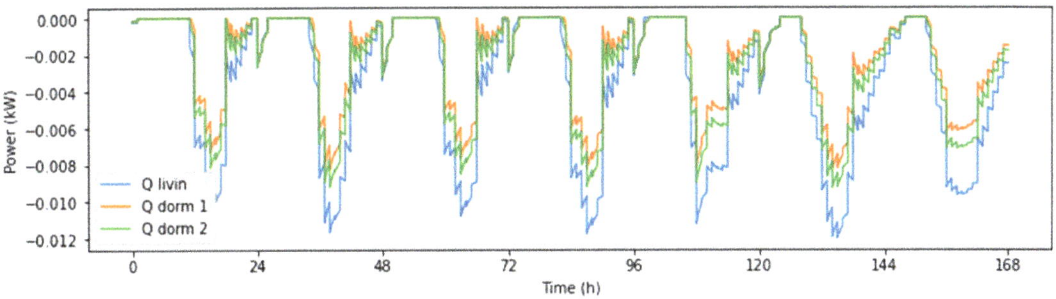

Figure 23. Thermal power required by the HVAC system for a summer week.

When comparing the dynamics of the two systems (with and without a planner), it can be observed that, in the case without a planner, the oscillations in power and temperature are more pronounced. This reflects that the system operates more directly to respond to temperature variations, activating and deactivating cooling in a more reactive manner than in the planner case. On the other hand, in the case with a planner, the graphs reflect a smoother and more uniform control strategy, where the system better anticipates temperature variations, storing or releasing thermal energy more gradually.

This demonstrates a more efficient use of the thermal battery in the case where the planner is used, minimizing power peaks and achieving more stable temperature control. This strategy reduces the dependence on frequent system activations, optimizing energy consumption, while maintaining thermal comfort. In summary, analyzing both systems highlights how the thermal battery is also used to smooth system operation and improve energy efficiency.

Similar to the winter scenario, Figure 24 illustrates variations in energy consumption patterns. With a planner, energy usage shows distinct peaks at specific times, reflecting the system's deliberate strategy to charge the battery and TES during periods of lower electricity prices and reduced demand. This approach allows the system to store energy efficiently, minimizing consumption during high-demand or high-cost periods. In contrast, without a planner, energy consumption is steadier but occurs more frequently, indicating a reliance on power as needed without strategic optimization, leading to less efficient energy use. As observed in winter, during a typical summer week, the planner concentrates energy consumption during low-cost periods, while, in its absence, energy is consumed even during times of higher electricity prices.

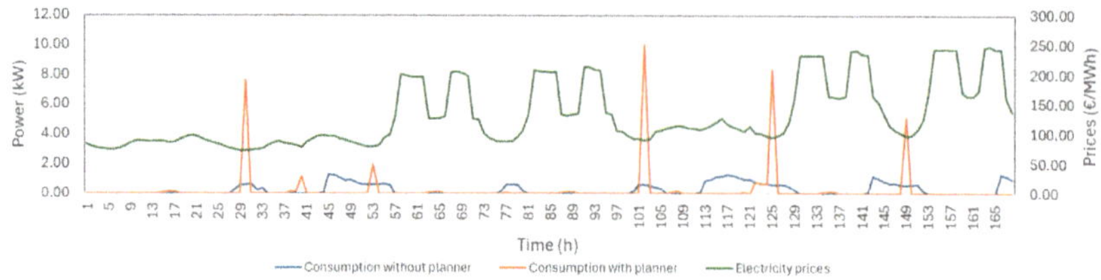

Figure 24. Electric consumption with and without planner for a summer week.

Figure 25 shows that when the planner is implemented, the SOC follows a strategic pattern, with distinct charging peaks during periods of low electricity costs or high PV generation. This allows the TES to store thermal energy efficiently for later use, especially during times of high demand or elevated energy prices. Conversely, without a planner, the SOC remains lower and fluctuates less significantly. This reactive approach leads to less efficient energy storage and utilization strategy, missing opportunities to capitalize on periods of surplus energy or cost advantages.

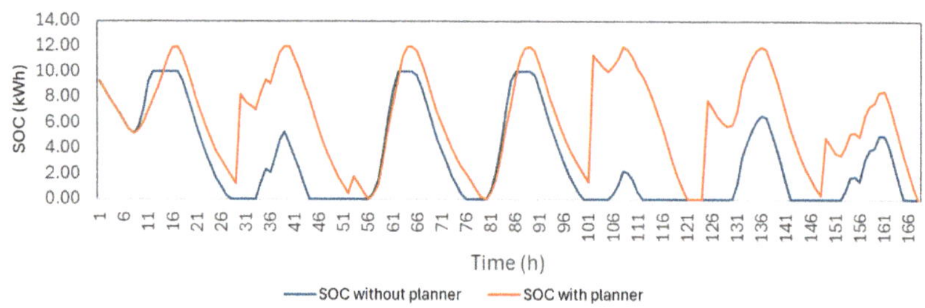

Figure 25. SOC with and without planner for a summer week.

Finally, Figure 26 shows the cost of energy comparing the results of the optimizer without and with indoor temperature control using the TES concept.

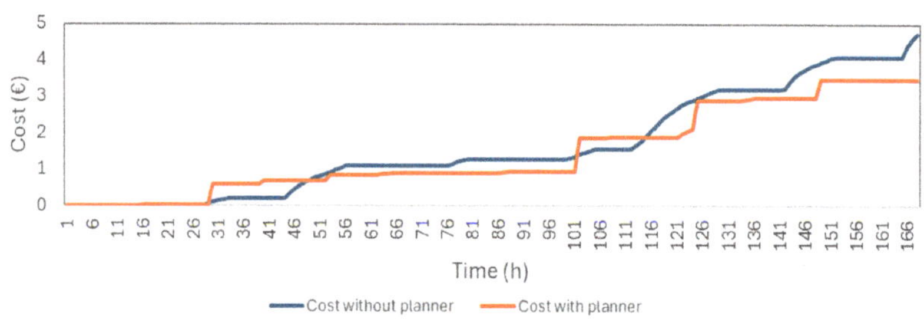

Figure 26. Accumulated cost of energy consumption comparing the results of the optimizer without and with indoor temperature control for a summer week.

Figure 26 illustrates the effect on overall costs during the summer period. Similarly to the winter scenario, using the planner results in a slower cost increase, highlighting the savings obtained by strategically managing the charging and discharging of the TES. This approach enables the system to minimize energy usage during high-cost periods. On the other hand, without the planner, costs rise more rapidly, demonstrating the inefficiencies of relying on immediate energy use without taking advantage of storage to optimize consumption.

7.3. Effect of the Thermal Inertia

To evaluate the effect of thermal inertia and its potential in the battery energy storage system, several simulations of the effect on the HVAC and TES system have been carried out in different houses with different levels of insulation and thermal inertia. To study this effect, we have used different thermal inertia coefficients, which were calculated from that of the studied dwelling, initially reducing it by 25% and 12.5% and then increasing it by 12.5% and 25%. For simplicity, only the analysis carried out for the winter period is included.

In Figure 27, it is observed that higher thermal mass results in less frequent and smaller operational cycles for the HVAC system. This is because the additional thermal mass effectively buffers temperature fluctuations, reducing the regulation effort required and leading to more efficient energy consumption. Conversely, lower thermal mass causes faster temperature fluctuations, forcing the HVAC system to respond more frequently and at higher power levels, resulting in larger energy consumption spikes.

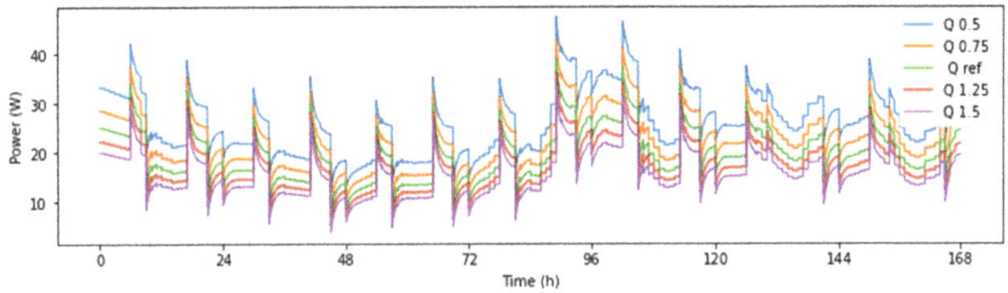

Figure 27. Effect on the HVAC consumption for different thermal inertias.

Table 1 presents the cumulative energy consumption variation in one week for the different thermal masses compared to the case study. It clearly shows that higher thermal mass results in lower cumulative energy consumption over time. This is because the system requires less effort to maintain thermal stability. Conversely, lower thermal mass results in higher cumulative consumption, because the system works harder to compensate for rapid temperature changes.

Table 1. Cumulative energy consumption variation in one week for different thermal masses.

Thermal Mass Variation	Consumption Variation
−25.0%	+33.8%
−12.5%	+14.5%
+12.5%	−11.3%
+25.0%	−20.3%

In summary, higher thermal mass reduces temperature fluctuations and cumulative HVAC energy consumption, improving overall system efficiency. On the other hand, lower thermal mass increases the frequency of HVAC system response and cumulative energy use, as the system's ability to buffer temperature changes is significantly reduced.

Similarly, in less insulated houses, the TES responds quickly to changes in thermal load, which can be beneficial in applications requiring the rapid delivery of heat or cooling. However, this quick response makes the system less stable in the face of demand fluctuations. Additionally, a TES with low thermal mass undergoes more frequent charge and discharge cycles, which increases system wear and reduces its lifespan. Furthermore, its ability to retain thermal energy over extended periods is diminished, making it less effective in long-term storage applications.

Conversely, a TES with high thermal mass better buffers temperature fluctuations, providing a more stable and consistent energy delivery. Although this reduces the system's ability to respond quickly to abrupt changes in demand, it minimizes the frequency of charge and discharge cycles, thereby improving system durability and reducing operational costs. Moreover, greater thermal mass allows for the internal temperature to remain stable over prolonged periods, making it ideal for applications where thermal stability is crucial, such as renewable energy storage or the climate control of large buildings.

8. Validation in a Testing Environment

To study the integration of this solution in a real-world environment, a prototype was first developed in a laboratory in order to test the main functionalities of a household energy system, as shown in Figure 28.

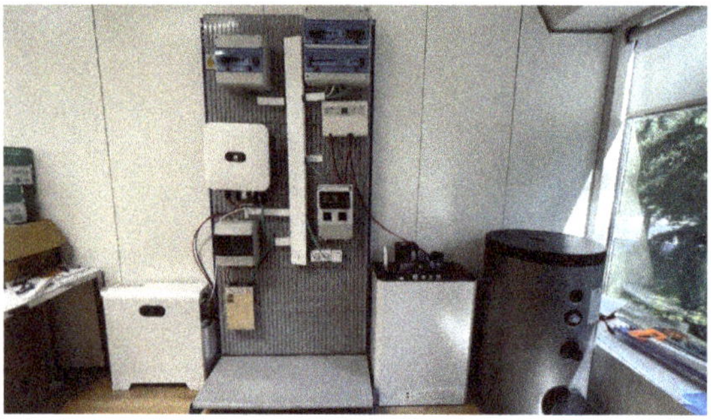

Figure 28. Prototype for laboratory tests.

The developed prototype has been configured to replicate the behavior of real systems, providing a comprehensive and adaptable testing environment. One key element of the prototype is a 3 kW power supply that simulates the operation of solar panels by converting alternating current (AC) to direct current (DC). This power supply acts as a controllable generator, capable of reproducing variations in photovoltaic energy production under different conditions, such as changes in solar radiation or generation capacity throughout the day.

The system also includes an inverter and an electric battery. Meanwhile, the electric battery enables energy storage, allowing excess energy to be stored for use during periods when demand exceeds generation. Together, these components facilitate the evaluation of charge and discharge cycles, system efficiency, and the integration of renewable energy with storage systems.

In addition, an aerothermal system has been used as an HVAC system. The prototype incorporates two water tanks designed to simulate different aspects of thermal energy management. The first tank is a dual-chamber unit configured to replicate the operation of an aerothermal heat pump, heating water on demand. This tank dynamically simulates how an aerothermal system responds to a household's heating requirements. The second tank is a large storage tank that simulates the thermal loads of a household, encompassing both heating and cooling demands. This tank acts as a thermal sink, enabling the evaluation of energy transfer, thermal inertia, and system behavior under various simulated household conditions. Thermal loads are simulated by introducing heat into the large tank to mimic heating consumption or by extracting heat to replicate cooling demands.

The prototype is closely aligned with the final marketable solution and plays a critical role in validating its core components and functionality.

The marketable solution consists of two primary elements: an in-home hardware device and a cloud-based component. The hardware device acts as a controller for various devices, such as the PV inverter and the aerothermal system, while also collecting key data, such as home temperature, battery SOC, and PV power production. The cloud-based component complements this by storing the data in a database, running computational models, and sending commands back to the hardware device to optimize performance.

This prototype laid the groundwork for testing these concepts. These tests ensured that the hardware device could successfully monitor system states and that commands from the cloud could be accurately executed by actuators, even when these actuators were commercial devices not originally designed for this solution.

The prototype also served as a testbed for refining the installation process. This involved ensuring that the system could be easily set up by customers without requiring extensive technical expertise. By simulating household thermal and electrical loads, the prototype allowed for the verification of key functionalities, such as the seamless integration of the PV inverter, electric battery, and aerothermal system.

Although the prototype could not replicate the exact behavior of a TES system, it successfully tested the broader capabilities of energy storage and thermal management in a household context. TES systems in real homes are vital for storing surplus thermal energy generated during peak production and providing it later for heating or hot water. This aspect, while not fully implemented in the prototype, was acknowledged as a critical next step for real-world applications.

Once the validations in the laboratory prototype were completed, including communication between the hardware device and the cloud, the compliance of actuators with issued commands, and the viability of the installation procedure, the system was ready for deployment in a real-world environment.

Then, the next phase involved testing the solution in a home equipped with a PV system, an electric battery, and an aerothermal system, to confirm its performance in a practical setting. This progression from prototype to field testing ensured that the final solution would be robust, reliable, and user-friendly.

The installation where the real-world test will be carried out is located in Madrid and consists of a 6.3 kWp PV system, a 10 kWh electric storage, and an 8 kW aerothermal machine. As can be seen in Figure 29, the hardware device was placed next to the photovoltaic inverter.

Figure 29. Installation in a real environment.

The chosen temperature sensor is an NTC (negative temperature coefficient) thermistor with an uncertainty of ±0.3 °C. This choice is based on the sensor's high precision and reliability in measuring temperature in outdoor and indoor environments. NTC thermistors offer a narrow range of uncertainty, making them ideal for applications where accuracy is crucial, such as in climate control systems within a home. The relatively low uncertainty of ±0.3 °C ensures that temperature readings are highly consistent, providing accurate data for managing heating, ventilation, and air conditioning (HVAC) systems. Additionally, NTC thermistors are cost-effective, durable, and widely available, making them a practical choice for this purpose.

In this test, several important aspects were verified, further demonstrating the advantages of the proposed solution for the customer.

Once this analysis was completed, both the MPC models and the planner were implemented, and a validation test was performed in the real environment for which data were collected during one month of spring.

In order to discuss the validation of the thermal model, the results of the real test can be seen in Figure 30, which, for simplicity, illustrates the behavior of the energy consumption of the air conditioning system during one week of the month in which the test was performed. This analysis was carried out to evaluate the consistency between the theoretical consumption model and the actual measured values obtained from the real installation.

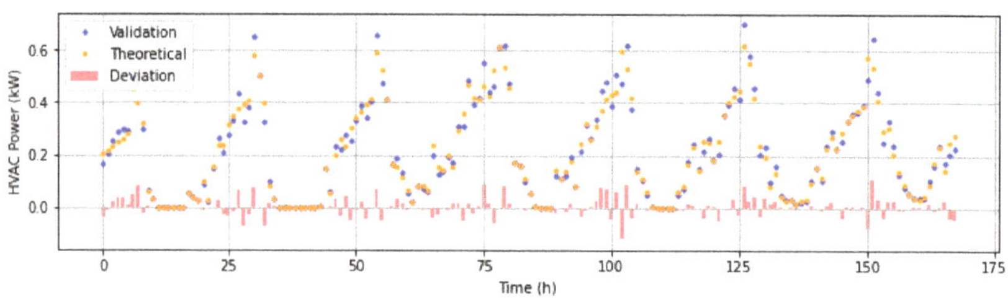

Figure 30. Correlation between theoretical and experimental data for a particular week.

Figure 30 shows a characteristic consumption pattern with daily peaks, reflecting the system's effort to reach the target temperatures during periods of higher thermal demand, such as mornings and evenings. Overall, there is an acceptable correlation between the theoretical and validation curves, suggesting that the theoretical model adequately captures the expected behavior of the HVAC system. The pink bars highlight certain discrepancies, which correspond to the difference between the predictions made by the proposed system and the measurements obtained in real operating conditions, where slight climatic fluctuations or changes in space occupation occur.

Figure 31 shows the results obtained in the test during the whole month.

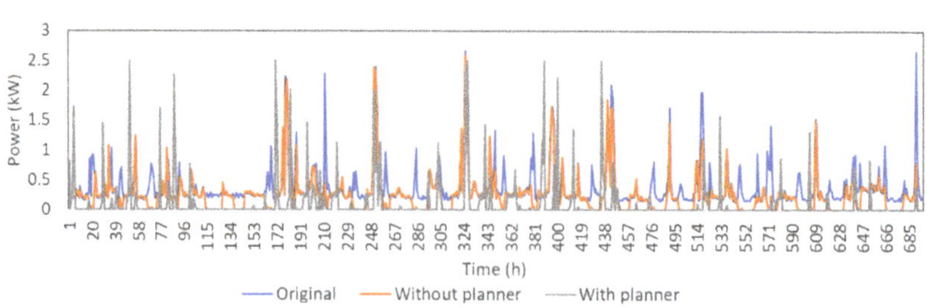

Figure 31. Electrical consumption in the real environment.

The blue curve represents the original consumption without considering the self-consumption system and batteries; the orange curve represents the consumption with the PV system and battery, but without optimization or implementation of the thermal battery; while the gray curve represents the final consumption after optimization and implementation of the thermal battery. As can be seen, the energy savings are significant, with a reduction in consumption of more than 57%, as shown in Figure 32.

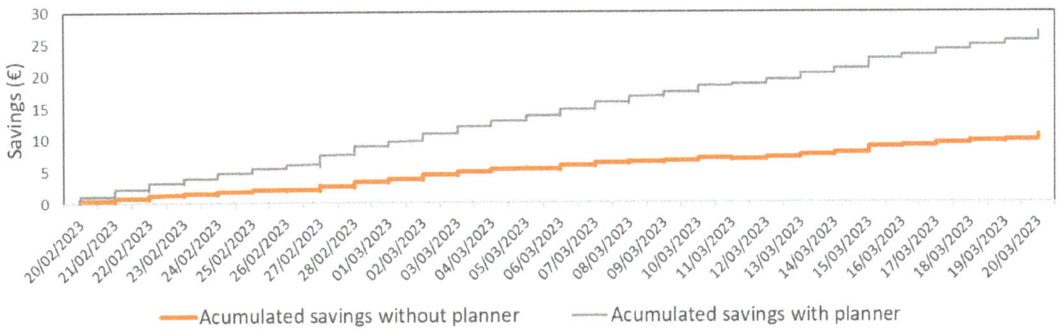

Figure 32. Cost savings with the proposed solution.

Furthermore, if we analyze the energy balance by electricity periods, we can observe a significant reduction in consumption during the most expensive periods, P1 and P2 of the electricity tariff, as shown in the following Figure 33:

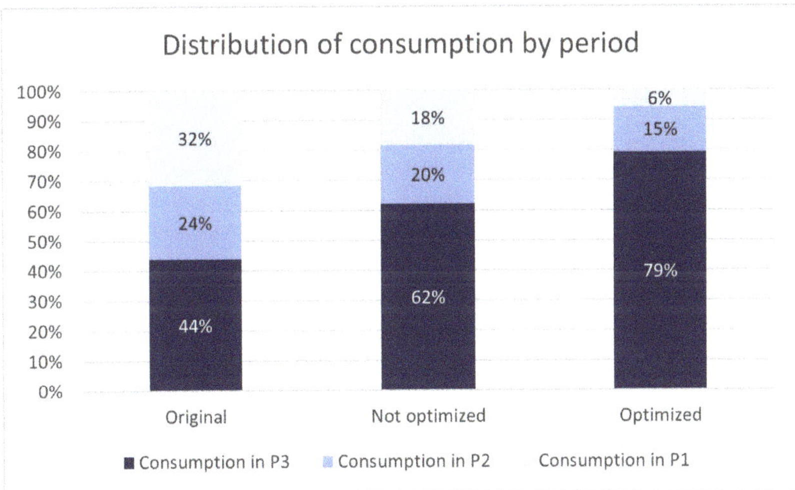

Figure 33. Energy balance by electricity periods.

Both the reduction in electricity consumption and its distribution across periods have resulted in savings of more than EUR 400 on the customer's energy bill over the approximately one-year period during which these tests were conducted in this home.

9. Conclusions

The EMS developed in this study represents a significant advancement in optimizing residential electricity consumption, demonstrating its ability to reduce costs and enhance energy efficiency. During simulations, the EMS achieved substantial cost reductions, highlighting its effectiveness across diverse climatic scenarios. This was accomplished by strategically utilizing thermal energy stored in the building's structure, maximizing the TES capacity, and reducing reliance on expensive electricity sources during peak demand periods.

The EMS significantly increased PV self-consumption, particularly in summer, by prioritizing the use of solar-generated energy to meet the household's thermal and electrical demands. This approach reduced grid dependency, not only lowering costs but also decreasing the carbon footprint of the system. Additionally, the EMS exhibited strong resilience to extreme climatic conditions, such as heatwaves or cold spells, maintaining optimal thermal comfort levels by anticipating and efficiently managing temperature fluctuations and energy demand.

The innovative use of TES allowed the system to leverage the building's thermal inertia for strategic heat storage and release. This resulted in the more efficient operation of the HVAC system, reducing energy demand peaks and improving thermal stability within the home. This approach optimized the use of electric batteries and improved the integration of intermittent renewable sources, like solar energy, by dynamically adjusting the indoor temperature setpoints within defined comfort margins.

Real-world testing confirmed the EMS's applicability in typical residential settings, demonstrating cumulative electricity consumption savings exceeding 57% in some cases. These findings underscore the system's feasibility for deployment in similar housing configurations, promoting the adoption of advanced energy management technologies that contribute to the sustainability and decarbonization of the residential sector.

Environmentally, the EMS proved to be an effective tool for reducing CO_2 emissions by decreasing dependence on grid electricity, particularly during high carbon intensity periods. This aligns with global energy transition goals towards renewable sources and enhances the flexibility of energy systems by enabling greater integration of renewables into existing infrastructure.

Finally, the EMS design features a scalable and flexible architecture, making it adaptable to various residential configurations and existing systems. This opens new opportunities for large-scale implementation, supporting energy sustainability and achieving emission reduction targets in the residential sector. This study establishes a milestone in integrating predictive control and thermal storage solutions into residential settings, paving the way for future developments in this field.

Author Contributions: Conceptualization, R.F. and J.F.; methodology, R.F. and J.F.; formal analysis, R.F. and J.F.; investigation, R.F. and J.F.; validation, R.F.; writing—original draft preparation, R.F.; writing—review and editing, R.F. and J.F.; supervision, J.F. All authors have read and agreed to the published version of the manuscript.

Funding: This research received no external funding.

Institutional Review Board Statement: Not applicable.

Informed Consent Statement: Not applicable.

Data Availability Statement: Data used in this article can be found in [66].

Acknowledgments: The authors would like to thank IGNIS Energía for their support of Ricardo Felez's Industrial PhD Thesis and for the facilities and equipment used in the validation of the results.

Conflicts of Interest: Author Ricardo Felez was employed by the company IGNIS Energía S.L. The remaining authors declare that the research was conducted in the absence of any commercial or financial relationships that could be construed as a potential conflict of interest.

Appendix A. Drawing of the House Used in the Analysis Model

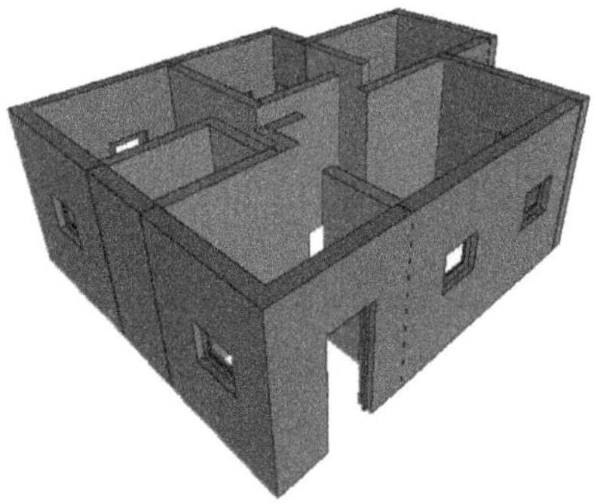

Figure A1. 3D model for the hose used for the study in Section 6.

Appendix B. Thermal Model Parameters

For the thermal model, the following parameters and units have been considered. Their specific values can be found in the section of reference [66], included in the Data Availability Statement.

Table A1. Thermal model parameters.

Parameter	Meaning	Units
V_{air}	Volume of air inside the enclosure	m^3
ρ_{air}	Density of air inside the enclosure	m^3
$c_{p_{air}}$	Specific heat capacities of the air	J/(kg·K)
$m_{mat\,j}$	Mass of material in the enclosure	kg
$c_{p_{mat,j}}$	Specific heat capacities of the materials	J/(kg·K)
G_i	Global irradiance	W/m^2
S	Area of the window	m^2
F	Modified solar factor	-
Or	Solar radiation coefficient	-
f	Attenuation correction factor	-
U_j	Heat transmission coefficient	$W/(K\,m^2)$
S_j	Area	m^2
$\dot{m}_{vent}$	Mass flow rate of ventilation air	kg/s
ρ_{ext}	Air density	kg/m^3
$\dot{V}_{vent}$	Ventilation flow rate	m^3/s

Appendix C. HVAC Controller Parameters

For the HVAC Controller, the following parameters and units have been considered. Their specific values can be found in the section of reference [66], included in the Data Availability Statement.

Table A2. HVAC controller parameters.

Parameter	Meaning	Units
N_e	Number of air-conditioned rooms	-
N_p	Prediction horizon	-
Δt	Time step	min
$\dot{Q}^{i}_{ref}$	Maximum refrigeration thermal load	kW
$\dot{Q}^{i}_{heat}$	Maximum heat thermal load	kW
K^i_T	Weight for the output deviation from the maximum temperature of reference	-
$K^i_{\Delta T}$	Weight for the positive slack variable	-
T^i_{ref}	Reference temperature	°C
α	Uncertainty for the thermal inertia	-
δT	Disturbance for the outside temperature of the enclosures	°C

Appendix D. Planner Module Parameters

For the Planner module, the following parameters and units have been considered. Their specific values can be found in the section of reference [66], included in the Data Availability Statement.

Table A3. Planner module parameters.

Parameter	Meaning	Units
P^t_g	Electric prices of the grid	€/MWh
Cg^t	Grid electricity consumption	kWh
C^t_g	Charging power coming from the grid	kW
D^t_g	Discharging power	kW
P^t_{PV}	Price of the energy coming from the PV	€/MWh
C^t_{PV}	Charging power coming from the PV	kW
DOD	Depth of discharge	-
STG	Storage Capacity	kWh
SOC^t	State of charge	kWh
Pow_C	Maximum charging power	kW
Pow_D	Maximum discharging power	kW
Pow_{PS}	Peak shaving power	kW

References

1. Eurostats. Energy Consumption in Households. 2024. Available online: https://ec.europa.eu/eurostat/statistics-explained/index.php?title=Energy_consumption_in_households (accessed on 18 November 2024).
2. Office of Energy Efficiency & Renewable Energy. Thermal Energy Storage Technologies Subprogram Area. Available online: https://www.energy.gov/eere/buildings/thermal-energy-storage (accessed on 18 November 2024).
3. European Environment Agency. Decarbonising Heating and Cooling—A Climate Imperative. 2024. Available online: https://www.eea.europa.eu/publications/decarbonisation-heating-and-cooling (accessed on 18 November 2024).
4. Thomas Nowak and Pascal Westring. European Heat Pump Market and Statistics Report. 2023. Available online: https://www.ehpa.org/wp-content/uploads/2023/06/EHPA_market_report_2023_Executive-Summary.pdf (accessed on 18 November 2024).

5. European Heat Pump Association. REPowerEU: Heat Pump Strategy Required to Help Sector Deliver. 2024. Available online: https://www.ehpa.org/news-and-resources/press-releases/repowereu-heat-pump-strategy-required-to-help-sector-deliver-2/ (accessed on 18 November 2024).
6. Gladen, A.C.; Bajwa, D. An Improved Thermochemical Energy Storage Material Using Nanocellulose to Stabilize Calcium Chloride Salt. *J. Sol. Energy Eng.* **2022**, *144*, 030905. [CrossRef]
7. Huo, X.; Xie, D.; Zhao, Z.; Wang, S.; Meng, F. Novel method for microencapsulation of eutectic hydrated salt as a phase change material for thermal energy storage. *Int. J. Low-Carbon Technol.* **2022**, *17*, 760–767. [CrossRef]
8. Howard, D.; Shrestha, S.S.; Shen, Z.; Feng, T.; Hun, D. Thermally anisotropic building envelope for thermal management: Finite element model calibration using field evaluation data. *J. Build. Perform. Simul.* **2024**, *17*, 756–775. [CrossRef]
9. Pérez-Lombard, L.; Ortiz, J.; Pout, C. A review on buildings energy consumption information. *Energy Build.* **2008**, *40*, 394–398. [CrossRef]
10. Pajek, L.; Potočnik, J.; Košir, M. The effect of a warming climate on the relevance of passive design measures for heating and cooling of European single-family detached buildings. *Energy Build.* **2022**, *261*, 111947. [CrossRef]
11. Guo, S.; Yan, D.; Hu, S.; An, J. Global comparison of building energy use data within the context of climate change. *Energy Build.* **2020**, *226*, 110362. [CrossRef]
12. Andrade, C.; Mourato, S.; Ramos, J. Heating and Cooling Degree-Days Climate Change Projections for Portugal. *Atmosphere* **2021**, *12*, 715. [CrossRef]
13. Li, Y.; Wang, W.; Wang, Y.; Xin, Y.; He, T.; Zhao, G. A Review of Studies Involving the Effects of Climate Change on the Energy Consumption for Building Heating and Cooling. *Int. J. Environ. Res. Public Health* **2020**, *18*, 40. [CrossRef]
14. Larsen, M.; Petrović, S.; Radoszynski, A.; McKenna, R.; Balyk, O. Climate change impacts on trends and extremes in future heating and cooling demands over Europe. *Energy Build.* **2020**, *226*, 110397. [CrossRef]
15. Berardi, U.; Jafarpur, P. Assessing the impact of climate change on building heating and cooling energy demand in Canada. *Renew. Sustain. Energy Rev.* **2020**, *121*, 109681. [CrossRef]
16. Yang, Y.; Javanroodi, K.; Nik, V.M. Climate change and energy performance of European residential building stocks—A comprehensive impact assessment using climate big data from the coordinated regional climate downscaling experiment. *Appl. Energy* **2021**, *298*, 117246. [CrossRef]
17. Bienvenido-Huertas, D. Analysis of the Impact of the Use Profile of HVAC Systems Established by the Spanish Standard to Assess Residential Building Energy Performance. *Sustainability* **2020**, *12*, 7153. [CrossRef]
18. Peeters, L.; Van der Veken, J.; Hens, H.; Helsen, L.; D'haeseleer, W. Control of heating systems in residential buildings: Current practice. *Energy Build.* **2008**, *40*, 1446–1455. [CrossRef]
19. Ma, Z.; Zhao, D.; She, C.; Yang, Y.; Yang, R. Personal thermal management techniques for thermal comfort and building energy saving. *Mater. Today Phys.* **2021**, *20*, 100465. [CrossRef]
20. Li, Q.; Zhang, L.; Zhang, L.; Wu, X. Optimizing energy efficiency and thermal comfort in building green retrofit. *Energy* **2021**, *237*, 121509. [CrossRef]
21. Sansaniwal, S.K.; Mathur, J.; Mathur, S. Review of practices for human thermal comfort in buildings: Present and future perspectives. *Int. J. Ambient. Energy* **2022**, *43*, 2097–2123. [CrossRef]
22. Zheng, P.; Wu, H.; Liu, Y.; Ding, Y.; Yang, L. Thermal comfort in temporary buildings: A review. *Build. Environ.* **2022**, *221*, 109262. [CrossRef]
23. Mofidi, F.; Akbari, H. Intelligent buildings: An overview. *Energy Build.* **2020**, *223*, 110192. [CrossRef]
24. Afram, A.; Janabi-Sharifi, F. Supervisory model predictive controller (MPC) for residential HVAC systems: Implementation and experimentation on archetype sustainable house in Toronto. *Energy Build.* **2017**, *154*, 268–282. [CrossRef]
25. Grassi, B.; Piana, E.A.; Lezzi, A.M.; Pilotelli, M. A Review of Recent Literature on Systems and Methods for the Control of Thermal Comfort in Buildings. *Appl. Sci.* **2022**, *12*, 5473. [CrossRef]
26. Merabet, G.H.; Essaaidi, M.; Ben Haddou, M.; Qolomany, B.; Qadir, J.; Anan, M.; Al-Fuqaha, A.; Abid, M.R.; Benhaddou, D. Intelligent building control systems for thermal comfort and energy-efficiency: A systematic review of artificial intelligence-assisted techniques. *Renew. Sustain. Energy Rev.* **2021**, *144*, 110969. [CrossRef]
27. Gupta, A.; Badr, Y.; Negahban, A.; Qiu, R.G. Energy-efficient heating control for smart buildings with deep reinforcement learning. *J. Build. Eng.* **2020**, *34*, 101739. [CrossRef]
28. Gao, G.; Li, J.; Wen, Y. DeepComfort: Energy-Efficient Thermal Comfort Control in Buildings Via Reinforcement Learning. *IEEE Internet Things J.* **2020**, *7*, 8472–8484. [CrossRef]
29. Ngarambe, J.; Yun, G.Y.; Santamouris, M. The use of artificial intelligence (AI) methods in the prediction of thermal comfort in buildings: Energy implications of AI-based thermal comfort controls. *Energy Build.* **2020**, *211*, 109807. [CrossRef]
30. Valinejadshoubi, M.; Moselhi, O.; Bagchi, A.; Salem, A. Development of an IoT and BIM-based automated alert system for thermal comfort monitoring in buildings. *Sustain. Cities Soc.* **2021**, *66*, 102602. [CrossRef]

31. Esrafilian-Najafabadi, M.; Haghighat, F. Occupancy-based HVAC control systems in buildings: A state-of-the-art review. *Build. Environ.* **2021**, *197*, 107810. [CrossRef]
32. Wang, Y.; Kuckelkorn, J.; Liu, Y. A state of art review on methodologies for control strategies in low energy buildings in the period from 2006 to 2016. *Energy Build.* **2017**, *147*, 27–40. [CrossRef]
33. Prívara, S.; Široký, J.; Ferkl, L.; Cigler, J. Model predictive control of a building heating system: The first experience. *Energy Build.* **2011**, *43*, 564–572. [CrossRef]
34. Drgoňa, J.; Arroyo, J.; Figueroa, I.C.; Blum, D.; Arendt, K.; Kim, D.; Ollé, E.P.; Oravec, J.; Wetter, M.; Vrabie, D.L.; et al. All you need to know about model predictive control for buildings. *Annu. Rev. Control* **2020**, *50*, 190–232. [CrossRef]
35. Yao, Y.; Shekhar, D.K. State of the art review on model predictive control (MPC) in Heating Ventilation and Air-conditioning (HVAC) field. *Build. Environ.* **2021**, *200*, 107952. [CrossRef]
36. Yang, S.; Wan, M.P. Machine-learning-based model predictive control with instantaneous linearization – A case study on an air-conditioning and mechanical ventilation system. *Appl. Energy* **2022**, *306*, 118041. [CrossRef]
37. Oldewurtel, F.; Ulbig, A.; Parisio, A.; Andersson, G.; Morari, M. Reducing peak electricity demand in building climate control using real-time pricing and model predictive control. In Proceedings of the 49th IEEE Conference on Decision and Control (CDC), Atlanta, GA, USA, 15–17 December 2010; pp. 1927–1932. [CrossRef]
38. Gholamzadehmir, M.; Del Pero, C.; Buffa, S.; Fedrizzi, R.; Aste, N. Adaptive-predictive control strategy for HVAC systems in smart buildings—A review. *Sustain. Cities Soc.* **2020**, *63*, 102480. [CrossRef]
39. Serale, G.; Fiorentini, M.; Capozzoli, A.; Bernardini, D.; Bemporad, A. Model Predictive Control (MPC) for Enhancing Building and HVAC System Energy Efficiency: Problem Formulation, Applications and Opportunities. *Energies* **2018**, *11*, 631. [CrossRef]
40. Ouali, N.; Lehouche, H.; Belkhier, Y.; Achour, A. Indoor temperature regulation and energy consumption inside a working office in building system using a predictive functional control. *Energy Sources Part A Recover. Util. Environ. Eff.* **2024**, *46*, 10473–10493. [CrossRef]
41. Yang, S.; Wan, M.P.; Chen, W.; Ng, B.F.; Dubey, S. Model predictive control with adaptive machine-learning-based model for building energy efficiency and comfort optimization. *Appl. Energy* **2020**, *271*, 115147. [CrossRef]
42. Ma, Y.; Borrelli, F.; Hencey, B.; Packard, A.; Bortoff, S. Model Predictive Control of thermal energy storage in building cooling systems. In Proceedings of the 2009 Joint 48th IEEE Conference on Decision and Control (CDC) and 28th Chinese Control Conference (CCC), Shanghai, China, 15–18 December 2009; pp. 392–397.
43. Tarragona, J.; Pisello, A.L.; Fernández, C.; de Gracia, A.; Cabeza, L.F. Systematic review on model predictive control strategies applied to active thermal energy storage systems. *Renew. Sustain. Energy Rev.* **2021**, *149*, 111385. [CrossRef]
44. Tarragona, J.; Fernández, C.; de Gracia, A. Model predictive control applied to a heating system with PV panels and thermal energy storage. *Energy* **2020**, *197*, 117229. [CrossRef]
45. Lee, D.; Ooka, R.; Ikeda, S.; Choi, W.; Kwak, Y. Model predictive control of building energy systems with thermal energy storage in response to occupancy variations and time-variant electricity prices. *Energy Build.* **2020**, *225*, 110291. [CrossRef]
46. Tang, H.; Yu, J.; Geng, Y.; Liu, X.; Lin, B. Optimization of operational strategy for ice thermal energy storage in a district cooling system based on model predictive control. *J. Energy Storage* **2023**, *62*, 106872. [CrossRef]
47. Wei, Z.; Calautit, J. Evaluation of model predictive control (MPC) of solar thermal heating system with thermal energy storage for buildings with highly variable occupancy levels. *Build. Simul.* **2023**, *16*, 1915–1931. [CrossRef]
48. Angeli, D. Economic Model Predictive Control. In *Encyclopedia of Systems and Control*; Springer: Berlin/Heidelberg, Germany, 2021; pp. 665–671. [CrossRef]
49. Angeli, D.; Casavola, A.; Tedesco, F. Theoretical advances on Economic Model Predictive Control with time-varying costs. *Annu. Rev. Control.* **2016**, *41*, 218–224. [CrossRef]
50. Finck, C.; Li, R.; Zeiler, W. Economic model predictive control for demand flexibility of a residential building. *Energy* **2019**, *176*, 365–379. [CrossRef]
51. Kuboth, S.; Heberle, F.; König-Haagen, A.; Brüggemann, D. Economic model predictive control of combined thermal and electric residential building energy systems. *Appl. Energy* **2019**, *240*, 372–385. [CrossRef]
52. Zhang, C.; Kazanci, O.B.; Levinson, R.; Heiselberg, P.; Olesen, B.W.; Chiesa, G.; Sodagar, B.; Ai, Z.; Selkowitz, S.; Zinzi, M.; et al. Resilient cooling strategies—A critical review and qualitative assessment. *Energy Build.* **2021**, *251*, 111312. [CrossRef]
53. OMIE. Prices and Volumes. Day-Ahead Market Price. Available online: https://www.omie.es/ (accessed on 13 December 2024).
54. Garcia-Martos, C.; Rodriguez, J.; Sanchez, M.J. Mixed Models for Short-Run Forecasting of Electricity Prices: Application for the Spanish Market. *IEEE Trans. Power Syst.* **2007**, *22*, 544–552. [CrossRef]
55. Eurostats. Energy Consumption in Households. 2024. Available online: https://ec.europa.eu/eurostat/statistics-explained/SEPDF/cache/58200.pdf (accessed on 13 December 2024).
56. EU Science Hub. Photovoltaic Geographical Information System (PVGIS). Available online: https://joint-research-centre.ec.europa.eu/photovoltaic-geographical-information-system-pvgis_en (accessed on 13 December 2024).
57. Weatherbit. Weather API. Available online: https://www.weatherbit.io/ (accessed on 13 December 2024).

58. Mayne, D.Q.; Rawlings, J.B.; Rao, C.V.; Scokaert, P.O.M. Constrained model predictive control: Stability and optimality. *Automatica* **2000**, *36*, 789–814. [CrossRef]
59. CYPE. Cypecad Mep. Available online: https://info.cype.com/es/producto/cypecad-mep/ (accessed on 2 January 2025).
60. *ANSI/ASHRAE/ACCA Standard 183-2007 (RA 2020)*; Peak Cooling and Heating Load Calculations in Buildings Except Low-Rise Residential Buildings. ASHRAE: Peachtree Corners, GA, USA, 2020.
61. *CEN/TR 12831*; Energy Performance of Buildings—Method for Calculation of the Design Heat Load. 2017. Available online: https://standards.iteh.ai/catalog/standards/cen/4c907284-01a1-4023-afef-c55a6c9ebc21/en-12831-1-2017 (accessed on 13 December 2024).
62. *ISO 6946*; Building Components and Building Elements—Thermal Resistance and Thermal Transmittance—Calculation Methods. ISO: Geneva, Switzerland, 2021.
63. IDEA. Thermal Comfort in an Air-Conditioned Space. 2020. Available online: https://www.idae.es/uploads/documentos/documentos_BIENESTAR_TERMICO_EN_UN_ESPACIO_CLIMATIZADO_2_articulo_ASV_3725727c.pdf (accessed on 16 December 2024).
64. Fiedler, F.; Karg, B.; Lüken, L.; Brandner, D.; Heinlein, M.; Brabender, F.; Lucia, S. do-mpc: Towards FAIR nonlinear and robust model predictive control. *Control Eng. Pract.* **2023**, *140*, 105676. [CrossRef]
65. Wächter, A.; Biegler, L.T. On the implementation of an interior-point filter line-search algorithm for large-scale nonlinear programming. *Math. Program.* **2006**, *106*, 25–57. [CrossRef]
66. Felez, R.; Felez, J. Dataset for the paper Advanced energy management for residential buildings optimizing costs and efficiency. *Zenodo* **2024**. [CrossRef]

applied sciences

MDPI

Article

Exploiting the Value of Active and Multifunctional Façade Technology through the IoT and AI

Matteo Giovanardi *, Alessia Baietta, Francesco Belletti, Sara Magnani, Oscar Casadei and Alessandro Pracucci

Focchi Spa, 47824 Poggio Torriana, Italy
* Correspondence: m.giovanardi@focchi.it or innovation@focchi.it

Abstract: In recent years, the interest in multifunctional façade (MF) technology has increased significantly. Recent advances in the integration of active and passive technologies have led to a new concept of building skins with highly flexible and decentralized control. Such an approach is considered capable of tackling environmental challenges and enhancing indoor environmental quality (IEQ). Integrated HVAC systems, dynamic blinds, and renewable energy systems can drastically increase façade responsiveness and efficiency. Although the technical feasibility of active and MF technology has already been demonstrated, market applications are still limited. The goal of this paper is to define the state of the art of MFs and clarify how the integration of IoT technologies, supported by AI, can increase market interest by fully exploiting the value of these systems. Indeed, recent advances in the IoT and data analysis tools are opening up attractive scenarios in optimization process. Starting with an overview of the most interesting EU-funded projects, this paper presents a MF case study in which IoT infrastructures are fully integrated. The prototype, realized within the MEZeroE Horizon project, stimulates a debate on future trajectories (and gaps) for the marketability of MF.

Keywords: multifunctional façade; active façade; close cavity façade; IoT; indoor environmental quality

1. Introduction

The building sector is facing a period of deep transformation. On the one hand, it must adhere to the European Union's policies for decarbonization of the built environment [1,2]. On the other hand, it must meet an increasing demand for high standards in indoor environmental quality (IEQ) to ensure the user's well-being and a healthier environment [3,4]. To face both challenges, the EU has pinpointed the building façade sector as a perfect incubator for exploring transformative innovation for the entire construction domain [5,6]. Indeed, the building façade plays a critical role in impacting a building's overall energy consumption [7] and carbon footprint [8], and simultaneously, it is strictly responsible for the IEQ and users' well-being [9,10] by impacting thermal, acoustic, and visual comfort. Thus, several research projects have been funded by the EU for the years with the aim to develop innovative and high-performance building skins [5,11].

In this context, one of the most interesting scenarios in balancing energy efficiency aspects and IEQ refers to the development of active and multifunctional façade (MF) systems [12]. Compared to a "traditional" building skin, a MF expands the façade's role by incorporating additional functionalities and enhancing performance, including aspects of modularity and prefabrication [13]. Various systems such as heating, ventilation, air conditioning (HVAC), automated blinds, solar panels, and photovoltaics technologies can be integrated into the building envelope to optimize energy efficiency, exploit renewable energy, enhance thermal insulation, improve indoor air quality, and assure high and responsive IEQ [14]. By transforming façades into active and adaptive building components, the system's ability to meet environmental and user comfort needs increases significantly. In this perspective, the MF system can adjust the visual, thermal, and acoustic performance of a building according to environmental conditions or specific users' needs.

However, despite the significant technical advances in conceptualizing, designing, and prototyping complex MF systems, widespread adoption in the market is still limited [15–17]. Although the construction market is characterized by a certain latency in integrating innovative technologies and systems, some issues need to be addressed before MF is a ready-to-market technology. Aspects related to cost, the supply chain, design complexity, long-term benefits, system integration, and MF control are identified as the main obstacles that limit product scalability in the market [16–18]. This paper, starting by defining the state of the art of MF systems, aims to clarify how the Internet of Things (IoT) and Artificial Intelligence (AI) technologies can be integrated to address technical barriers and, consequently, foster market deployment. Indeed, recent advances in the field of IoT technologies and AI for the built environment are enabling a new situation-aware paradigm for decision making [19] needed to achieve higher levels of efficiency. The potential contribution of digital innovation is also recognized by the European Commission as a key strategy for enabling sustainable models and promoting the more rational use of resources [20]. In this paper, these benefits are assumed as potential innovative factors to exploit the value of MF systems. Although several theoretical studies have investigated topics of the façade sector [12,21,22], few real experiences have been carried out. This implies a gap in the research on the topic. As reported by Founti et al. [11], further research in recognizing the added value of advanced façades provided by digital technologies is needed to address significant market segments. In this perspective, the development of prototypes is a key action to investigate the advantages and obstacles to be addressed to trigger the technology transfer from research to industry [23].

The novelty of the paper lies in a real case study used to address technical, managerial, and economic issues. This approach is essential to grounding the research efforts and results. This paper is organized as follows: After a methodology section, this paper provides an overview of MF technology, starting from an assessment of European case studies. Subsequently, the MEZeroE MF prototype is presented as the result of an open innovation approach, and finally, the discussion and conclusion parts summarize the main achievements reached and future perspectives on the topic.

2. Materials and Methods

This research is conducted within the framework of the MEZeroE Horizon 2020 project (GA ID 953157). This EU-funded initiative aims to establish a collaborative ecosystem to promote product and process innovation in the building façade sector for the development of highly energy-efficient and healthy nearly Zero Energy Buildings (nZEBs). Embracing an open innovation approach [24], the specific goal of the MEZeroE project is to support the development of innovative high-performance building envelopes through the sharing of knowledge and the creation of a Living Lab and test facilities for cutting-edge solutions [25].

Through a real case study, this paper investigates the potential role of the IoT and AI integrated in a MF as enabling factors to facilitate product marketability. Before presenting the technical features of the innovative building skin, a wider overview of the topic is provided to contextualize the research activities in a theoretical and technological framework. Specifically, a list of relevant MF case studies is collected through the CORDIS database on EU-funded projects [26], white papers, and open-access research papers. This activity is needed to clarify the state of the art and investigate the reasons for a lack of market penetration years after the first prototypes. A qualitative assessment of the results is carried out to confirm (or refute) aspects already described in the literature and investigate how digital technologies can now be integrated for value exploitation and marketability. This paper focuses mainly on aspects of digital/physical integration, digital technologies, and façade control algorithms as the most novel part of the research. Integrated technologies such as heat pumps, mechanical ventilation, automatic blinds, and opening windows are, in fact, products already on the market. The MF prototype, reaching a TLR of 7, has been tested in an operating environment.

3. The Multifunctional Façade Concept

There is no unambiguous definition of the MF. Despite years of investigation, the "multifunctional" term refers to a wide range of functionality and, consequently, technologies that are difficult to fit into closed classifications. For this study, the term MF refers to a fully prefabricated modular building envelope that includes passive and active components to provide a set of non-traditional and customizable functions (Figure 1) [13]. Passive components are generally no-energy-using technologies to assure basic functions, and active components represent additional sub-systems powered by energy for improving or expanding façade functionality. On the one hand, passive components refer to insulation panels, internal or external solar shadings, and other technologies needed to ensure the main envelope functionality, such as mechanical behavior, thermal resistance, air and water tightness, and noise reduction.

On the other hand, a wide range of technologies can be considered active components, able to improve façade performances or provide new customized services such as mechanical ventilation or renewable energy systems (RES) [16,27].

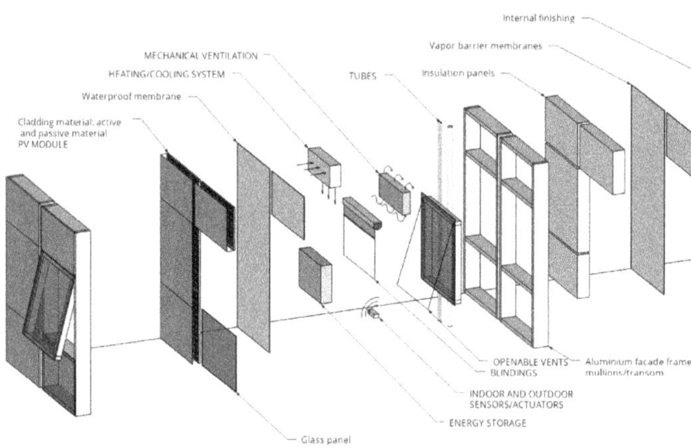

Figure 1. MEZeroE multifunctional façade theoretical scheme (credit: L. Vandi).

In this context, several EU-funded projects have been developed in the last fifteen years to test and validate the technical feasibility of MF systems [5,11], as they are considered impactful technologies for starting deep renovation schemes of the existing building stock [28]. Indeed, most of the EU-funded projects presented in Table 1 aim to develop a fully prefabricated façade for building retrofits.

Table 1. EU-funded projects on multifunctional façades.

EU Project	Project Year	MF Technology Description	TLR	Website Ref.
Meefs	2012–2016	Meefs develops two new energy-efficient modules for retrofitting including PV panels, solar thermal units, and a green façade module.	TLR 9	[29]
Adaptiwall	2013–2017	Adaptive wall is a novel panel consisting of 3 elements: lightweight concrete with nano additives; adaptable polymer for switchable thermal resistance; and a total heat exchanger with nanostructured membrane.	TLR9	[30]

Table 1. *Cont.*

EU Project	Project Year	MF Technology Description	TLR	Website Ref.
Plug-N-Harvest	2017–2022	Plug-N-Harvest develops a plug-and-play module for building retrofit including a PV module, battery, thermal collector, electric heat, and automatic ventilation.	TLR5	[31]
Energy Matching	2017–2022	Energy Matching develops a BIPV module with high aesthetical value and flexibility.	TLR9	[32]
RenoZEB	2017–2022	RenoZEB will provide a cost-effective plug-and-play façade integrating ventilation systems, solar and PV, automated blinds, and sensors.	TLR9	[33]
PowerSkin Plus	2019–2024	PowerSkin Plus combines enhanced insulation and renewable energy technology based on photovoltaics in modular solutions.	TLR8	[34]
ENsnare	2021–2025	The ENSNARE project will strive to increase the implementation of renovation through digitalization. A MF with BIPV, PV, a roll-bond solar collector, thermal batteries, and windows will be developed.	n.a.	[35]
MEZeroE	2021–2026	MEZeroE will focus on developing nZEBs. A MF is developed integrating mechanical ventilation, heating and cooling systems, automated blinds, and window sensors.	TLR8	[25]

From a qualitative assessment of these projects, some considerations can be made. According to the type of technology to be integrated, different types of Technology Readiness Levels (TRLs) have been reached, and different market adoptions have occurred.

Firstly, as demonstrated by Meefs and Energy Matching projects, the integration of RES, including solar and PV modules, has been viable for the construction market for several years [5]. Concerning RES technologies, obstacles to the widespread adoption of MF systems still relate to architectural, aesthetic, and energy efficiency issues. In this context, an intriguing perspective, as explored in the Plug-N-Harvest case, involves incorporating batteries for energy harvesting to transform the façade into active nodes of smart microgrids [36]. Although this perspective is very attractive, the lack of a multi-scalar infrastructure (from the building to the city) limits its application. In addition to this, the integration of advanced materials has gained much interest among researchers. In the AdaptiWall project, phase-change nanomaterials were integrated to enhance energy performance and control temperature, moisture, and indoor health. The integration of such advanced materials is still hampered by higher costs with respect to traditional materials and a lack of knowledge by designers, as well as a lack of communication across the entire supply chain [37].

Again, several EU projects are exploring the benefits of "plug & play" façades capable of integrating HVAC systems. Such an approach holds significant promise for technologically upgrading the existing building stock, which must meet escalating comfort and efficiency demands [38]. Within the framework of RenoZEB and MEZeroE projects, a MF with integrated heating, cooling, and ventilation system technology has been investigated and developed [25,33,39]. If, on the one hand, these projects have shown how technical issues can be managed for real applications [11], on the other, it is evident how far the commercialization of such products has come. Indeed, achieving a fully integrated MF product means redesigning both technologies and the entire supply chain.

All these experiences have demonstrated that MF technology represents a niche of the façade industry. Indeed, despite ongoing efforts toward developing fully integrated systems, practical application experiences remain limited to prototypes and case study demonstrations [11]. As highlighted in the white paper [17], there still is no market for the widespread deployment of active multifunctional systems, as clients and designers do not recognize their value. This implies that the benefits of MF systems in terms of energy savings and comfort provided need to be demonstrated to attract greater market interest.

In the proposed projects, the focus on sensor integration mainly involves monitoring the performance of integrated technologies. Only in RenoZEB and MEZeroE is strict reference made to the use of such sensors for the regulation of active systems. From this perspective, the use of IoT sensors in combination with AI algorithmic control systems remains an open issue with great room for improvement. Although the topic is relevant in the research landscape for building energy optimization and IEQ, this topic has not yet been analyzed in relation to complex envelope systems.

4. An IoT-Based Multifunction Façade Prototype

The MEZeroE MF is a complex advanced system designed to offer a highly adaptive building skin for users' multidomain comfort and to enable energy efficiency actions. Different experts have been involved during the design and manufacturing phase to aid in the development of a non-standard product. This complexity comes from the integration of active and passive technologies put in communication with each other through an IoT infrastructure. Through these technologies, this façade can shift to several configurations based on internal and external environmental stresses and user feedback. This flexibility allows for greater adaptability to changing conditions throughout the day, week, and season. The opening of windows, adjustments to blinds, and control of the heat pump and mechanical ventilation are thus managed through a physical (and digital) intelligence installed in the façade. To investigate the behavior of the façade, a prototype of dimensions 3750 × 3000 mm was designed (Figure 2), built, and tested (Figure 3). The prototype has been prefabricated in a factory.

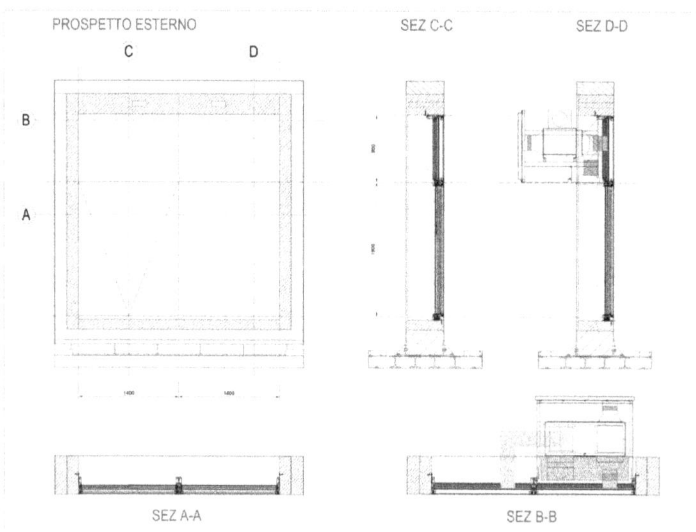

Figure 2. MEZeroE MF technical details.

Precisely, the prototype comprises two unitized curtain wall façade modules with vision and spandrel panels to replicate the real conditions in a building. Specifically, the façade system technology is a close cavity façade (CCF) with structural sealant glazing of the glasses to the aluminum profiles. The CCF system exploits the cavity generated between two layers (approximately 53 mm) to increase the thermal performance of the system. Externally, a low iron heat straightening glass (55.2) with anticondensation coating is used to limit the radiative exchange between the sky and the glass. Internally, the vision panel consists of triple glazing (4-18-4-18-44.2) with argon (90%) and a double low-e coating in face #3 and #5. The spandrel panel is composed of the same external glass, and a metal sheet is installed behind the outer layer to ensure façade uniformity from the outside and cover the technical space and insulation layer. From a thermal point of view, this system has the following performance:

- Thermal transmittance: U = 0.5 W/m^{2}K;
- Solar factor (no blind): g = 0.45;
- Solar factor (with blind): g = 0.04;
- Solar transmission (no blind): Tl = 0.64.

Figure 3. The MEZeroE MF prototype, different stages of the manufacturing phase. From left to right: the installation of the frame, spandrel panels, vision panels (below), and finally, the sub-structure needed for the heat pump machine (behind the spandrel panel).

In the MEZeroE MF, the following components are integrated (Figure 4).

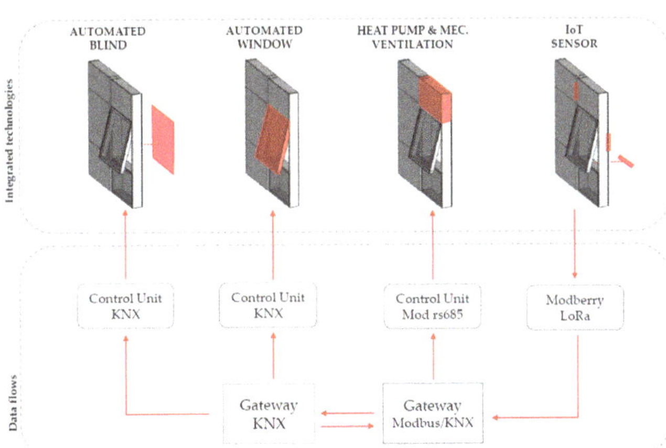

Figure 4. The MEZeroE MF prototype, different stages of installation activities.

4.1. Dynamic Blinds

Solar radiation is regulated by roller blinds installed in the CCF. An adjustable lamella with a high degree of reflection is designed to regulate the thermal load inside the building, reducing the load on summer days and exploiting it on winter days. The lamella system allows for double adjustment: firstly, the blind can be raised and lowered, and secondly, the slat can be oriented to 180°. This allows for a high degree of flexibility in proving shade from sun's rays according to the position of the sun, while at the same time ensuring maximum visual and perceptual comfort. The engine of the lamella and electrical and data cabling have been integrated into the façade frames to guarantee the best architectural integration. For the MEZeroE MF, an on-board control unit for dynamic blinds allows the operation to proceed according to a schedule. The lamella control unit is managed via a KNX signal.

4.2. Automatic Window

The vision façade module is an opening window with an automatic overhang system. The window opening can be adjusted in percentage, and it allows for the possibility of exploiting free heating and free cooling for energy saving or the exchange of indoor air if CO_2 levels are higher than the threshold. The window engine is integrated in the façade frame using a commercial product that allows for a total integration into the façade. The opening is controlled via a KNX signal.

4.3. Heat Pumps and Mechanical Ventilation

A heat pump and mechanical ventilation system were integrated into the prototype to provide a fully autonomous system. A commercial product for nZEB installation has been selected and integrated directly into the façade. This was selected from the market according to several parameters such as cooling/heating power, cost, and the possibility of not installing an external machine. The development of a fully customized solution was not considered, as this would require a long development time. Specifically, a thermodynamic controlled mechanical ventilation is selected to ensure double flow, heating, and cooling dehumidification. Although the physical integration will have to be rethought according to the building type (e.g., the presence of the ceiling, slab), the prototype allows for an evaluation of the potential benefits of a decentralized system with a good degree of approximation. In the MEZeroE MF, the HVAC system takes and ejects air directly from the hidden holes provided in the perimeter of the spandrel frame. Specifically, one of the two modules is responsible for air intake, the other for extraction. From an aesthetic point of view, this solution means that there are no visible air vents in the façade or external units. A smart thermostat is generally used to control and set the indoor air temperature, but for the MEZeroE prototype, the smart thermostat is replaced with a dedicated IoT infrastructure with edge-AI. Data gathered from IoT sensors are communicated to an on-board PCB (printed circuit board) to manage the adjustment of the different machine components (e.g., air velocity, air temperature, etc.). In this case, data are communicated in the Modbus protocol.

4.4. IoT Sensors for Façade

Several sensors are integrated in the building façade to monitor environmental parameters and regulate technologies automatically. Specifically, the IoT system is a proprietary kit [40] composed of three sensor nodes designed for a curtain wall façade for a full integration with an edge-AI to trig data transmission and optimize data collection. The kit is currently energy supplied by wire, and it communicates data collected via LoRaWan to a Modberry gateway. The LoraWan protocol was adopted due to a low energy consumption need and the low data rate for monitoring environmental trends. As a Modbus IoT kit system, as reported in Figure 5, IoT kit nodes are called Master, Slave 1, and Slave 2: the "Master" node is integrated in the lower transom of the vision part (not openable), and it monitors the transom temperature, glass temperature, and internal radiation passed through glass; the "Slave 1" node is integrated in the mullion at 140 cm from the interior floor and monitors indoor parameters and mullion temperature; the "Slave 2" node is installed into the mullion within the vertical and monitors outdoor parameters. The kit was designed to match sensor requirements for dimension, range, precision, and cost so to have a cost-effective, low-cost, and aesthetically appealing solution in line with curtain wall façade market expectations. Bespoke PCB have been designed, manufactured, and integrated within a 3D powder-bed fusion case. The selection of these sensors was established to support the data collection, analysis, and utilization of parameters to enable users' multidomain comfort and energy efficiency. The list of sensors integrated in the kits is shown in Table 2.

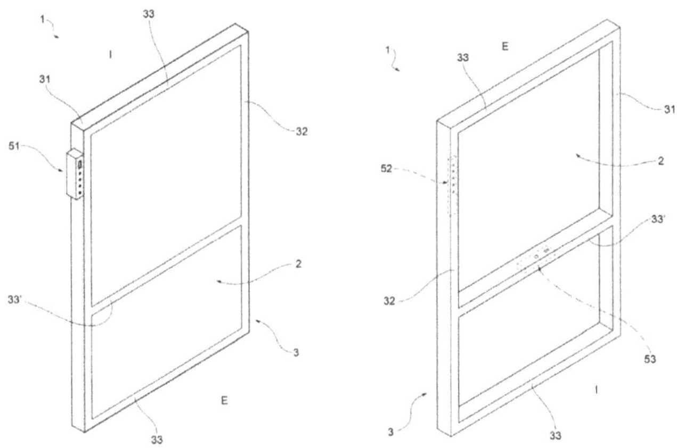

Figure 5. MEZeroE IoT sensor positions outdoors (**left**) and indoors (**right**).

Table 2. List of sensors integrated in MEZeroE MF.

Sensor Kit	Parameter	Units	Range Value
Master (transom)	Glass surface temperature	°C	−40 to 200 °C
	Transom surface temperature	°C	−10 to 50 °C
	Internal solar radiation	W/m^2	0 to 1500 W/m^2
Slave 1 (internal mullion)	Internal air temperature	°C	−55 to 125 °C
	Internal relative humidity	%	0 to 100%
	Internal illuminance	lux	0 to 100 klux
	Internal CO_2	ppm	0 to 1000 ppm
	Internal TVOC [1]	$\mu g/m^3$	0 to 110 kPa
	PIR (presence)	-	0/1
Slave 2 (external mullion)	External air temperature	°C	−55 to 125 °C
	External relative humidity	%	0 to 100%
	External pressure	Pa	0 to 110 kPa
	External illuminance	lux	0 to 100 klux
	External solar radiation	W/m^2	0 to 1500 W/m^2
	External noise	dB	0 to 130 dB

[1] TVOC: total volatile organic compounds.

4.5. Façade Control System

The MF MEZeroE multifunctional façade is managed through a Modberry gateway that collects and analyzes sensors every 5 s and uses ruled-based decision-making AI to define input for the integrated technologies. Thus, environmental data feed energy and IEQ algorithms for adjusting actuation for the heat pump, mechanical ventilation, solar blinds, and opening of windows. These optimization rules also included aspects related to acoustic, perceptual, visual, and air quality to provide multidomain comfort for users. The rules are managed and implemented in Node-RED, an open-source flow-based development tool designed for visual programming. It offers a user-friendly interface for connecting hardware devices, APIs, and online services to IoT sensors, automation workflows, and more.

The main rules for energy and IEQ optimization can be summarized as follows (in order of prioritization of the algorithm functionality):

- Energy and thermal comfort. Depending on the season, different temperature thresholds are set to ensure ideal thermal comfort. These rules provide operating ranges according to the predictive mean vote (PMV) and predicted percentage of dissatisfied (PPD) standards defined for office buildings [3,4]. Implementation of the different technologies will be achieved to enhance the energy performance of the system as much as possible.
- Indoor air quality (IAQ). CO_2, atmospheric particulate, and VOC values are monitored to adjust natural/mechanical ventilation in the room when strictly necessary. This approach drastically reduces heat loss and increases the well-being of users.
- Acoustic comfort. Noise control is carried out externally to promote a healthier and more comfortable working environment. This implies a control on the opening of windows if noise pollution values are too high.
- Visual comfort. Lighting and solar radiation are integrated to guarantee visual and perceptual comfort to allow aspects related to the quality of the working space to be included. For example, a view towards the outside of the building can be favored when there is no glare inside.

These algorithms are designed to adapt dynamically the thresholds of the main comfort parameters (temperature and lighting) based on users' discomfort feedback. Thanks to the adoption of a web app, the users can interact with the algorithms, identifying a status of discomfort. Such an approach overcomes a priori comfort threshold settings and considers the real condition of users. The application of advanced machine-learning models enables the system to provide the best possible regulations in terms of energy efficiency and comfort. In this perspective, the implementation of AI-based rules opens up interesting scenarios in addressing a still unsolved issue such as IEQ.

5. Discussion

To clarify the current state of the art for MF systems and define which aspects could facilitate their market adoption, an analytical assessment is carried out on issues dealt with during the design, manufacturing, and utilization stages. From this practical experience, two main issues clearly emerged.

On the one hand, the technical feasibility of advanced MF systems is confirmed. As demonstrated by several projects, the integration of active and passive technologies is no longer a technical barrier. Although the complexity of the integration largely depends on the type of technology to be embedded, a widespread knowledge of technical aspects and the high level of engineering of the façade companies allow for a wide degree of flexibility. In this regard, it is necessary to point out how the development of a supply chain capable of dialogue with designers and manufacturers could represent a pivotal aspect. This could also have direct effects on the reduction of costs related to MF systems, still considered one of the main barriers to widespread market adoption. The cost of the MEZEroE MF, also due to the ad hoc engineering of some components, is still high (ranging from 1000 to 1500 EUR/sqm), placing it in a narrow market niche and making it difficult to scale up on large building estates.

On the other hand, the novelty of this design lies in the integration of IoT and AI algorithms for façade automatization. Compared to previous MF projects, this study focuses more on new services enabled by data than on physical integration issues. Indeed, moving towards the digitalization of services, an IoT infrastructure for a building façade could increase product value through higher knowledge, flexibility, and customization.

Although these benefits are clarified, further analysis needs to be conducted to address the following issues:

- Design complexity. Developing a fully integrated IoT infrastructure in a building façade requires experts with cross-cutting skills during the design phase. In this context, the "engineer to order" type of manufacturing means that several design aspects have to be redesigned each time, increasing the complexity. Aspects related to the energy consumption, physical integration, aesthetics, maintenance, and cost-effectiveness of IoT infrastructure need to be considered according to the specific building requirements. To this end, further exploration in the development of modular and scalable IoT solutions must be carried out.
- Supply chain. The integration of IoT technologies in the MF systems expands the number of stakeholders involved during the design and manufacturing process. Today, the lack of professionals in the field of IoT automation for building systems causes a major difficulty in the development of non-standard projects. In the case of the MEZeroE prototype, three external developers (hardware developer, software developer and data communication in KNX, software developer and data communication in Mobdus) were involved in tackling a complex challenge.
- Interoperability. The interoperability with different technologies, design tools (e.g., BIM), and existing BMS is an aspect still to be solved. The lack of "plug & play" technologies for MF systems requires specific actors for collecting different data in a common data lake. This results in a constant commitment of IoT practitioners in the manufacturing and installation phases. In the MEZeroE prototype, setting up the communication system required more than a week's work.
- Cost. Although the cost of individual sensors is limited compared to the cost of the façade, the development of a fully integrated IoT infrastructure is still an impactful factor in the market adoption of the technology. Due to a higher design complexity and the lack of ad hoc technologies for building façade products, the global cost of development of a customized infrastructure is still high. To address this issue, it will be necessary to develop modular low-cost technology and quantify the potential savings in economic terms.

6. Conclusions

After years of research and development in the field of MF systems, market adoption still seems far off. Although substantial funding at the European level has led to the validation of complex MF systems proving technical feasibility and high performance in terms of energy and IEQ, further steps need to be taken. To exploit the maximum market value of MFs, this paper has investigated the potential value added by the IoT and AI. As demonstrated in other sectors, IoT technologies are considered potential enablers to exploit the value of a product through greater customization, optimization, and the provision of new services. Through a real case study developed within the MEZeroE research project, technical, economic, and process issues in the design and use of a MF system with an on-board IoT infrastructure were investigated.

This experience showed how a lack of ready-to-market MF technologies is hampered by different issues. First of all, the construction market evolves and includes innovative products (only) if the market demands it. This means that the main barrier in the market deployment of MF lies in customers' and developers' lack of knowledge on advanced building envelopes that can guarantee high levels of comfort and energy efficiency. In this perspective, the development of real IoT-based prototypes is essential to collect data and consolidate knowledge.

To answer the objective of this paper, the authors can state that an integrated IoT infrastructure for MF, both hardware and software, is needed to exploit the maximum value of products. In addition to the product customization provided by digital services, the IoT enables the communication of different technologies, thus overcoming vertical and closed management of them. Such an approach allows for automated, optimal, and highly adaptive performance management of a complex product, providing clear benefits in terms of energy savings and IEQ levels. In this context, the potential widespread and unlimited application of AI for automation appears to be the most important value with the greatest prospects. Rule-based algorithms and machine-learning approaches can drastically revolutionize the way we regulate active and passive technologies in buildings, leading to higher levels of energy efficiency and IEQ. All these aspects could turn into significant value for the market deployment of MF products. However, further demonstrations of the benefits generated by the IoT and AI will need to be validated to make the market ripe for the application of innovative building skins. In conclusion, this research aimed to stimulate a debate on the future of MF systems, identifying a unique opportunity in digital technologies to make further progress in innovative façade systems that are able to assure high energy efficiency and IEQ standards.

Author Contributions: Conceptualization, M.G. and A.P.; methodology, M.G.; validation, M.G. and A.P.; design and case study investigation, F.B., A.B., S.M., O.C., M.G. and A.P.; writing—original draft preparation, M.G.; writing—review and editing, M.G. and A.P. All authors have read and agreed to the published version of the manuscript.

Funding: This research was carried out within the MEZeroE (Measuring Envelope products and systems contributing to the next generation of healthy nearly Zero Energy Buildings) project. This project has received funding from the European Union's Horizon 2020 research and innovation programme under grant agreement No 953157.

Institutional Review Board Statement: Not applicable.

Informed Consent Statement: Not applicable.

Data Availability Statement: Data are contained within the article.

Acknowledgments: The authors would like to thank the entire team involved in the MEZeroE MF project.

Conflicts of Interest: Authors Matteo Giovanardi, Alessia Baietta, Francesco Belletti, Sara Magnani, Oscar Casadei and Alessandro Pracucci were employed by the company Focchi Spa.

References

1. European Commission. The European Green Deal. 2019. Available online: https://eur-lex.europa.eu/legal-content/EN/TXT/?qid=1576150542719&uri=COM:2019:640:FIN (accessed on 11 December 2023).
2. European Commission. A Renovation Wave for Europe—Greening Our Buildings, Creating Jobs, Improving Lives. 2020. Available online: https://eur-lex.europa.eu/legal-content/EN/TXT/?qid=1603122220757&uri=CELEX:52020DC0662 (accessed on 11 December 2023).
3. *ISO 7739:2005*; Ergonomics of the Thermal Environment. Analytical Determination and Interpretation of Thermal Comfort Using Calculation of the PMV and PPD Indices and Local Thermal Comfort Criteria. ISO: Geneva, Switzerland. Available online: https://www.iso.org/standard/39155.html (accessed on 11 December 2023).
4. ASHRAE. Standard 55—Thermal Environmental Conditions for Human Occupancy. Available online: https://www.ashrae.org/technical-resources/bookstore/standard-55-thermal-environmental-conditions-for-human-occupancy (accessed on 11 December 2023).
5. D'Oca, S.; Ferrante, A.; Ferrer, C.; Pernetti, R.; Gralka, A.; Sebastian, R.; Veld, P. Technical, Financial, and Social Barriers and Challenges in Deep Building Renovation: Integration of Lessons Learned from the H2020 Cluster Projects. *Buildings* **2018**, *8*, 174. [CrossRef]
6. Gasparri, E.; Brambilla, A.; Lobaccaro, G.; Goia, F.; Andaloro, A.; Sangiorgio, A. *Rethinking Building Skins: Transformative Technologies and Research Trajectories (Preface)*; Elsevier: Amsterdam, The Netherlands, 2022.
7. Giordano, R.; Giovanardi, M.; Guglielmo, G.; Micono, C. Embodied energy and operational energy evaluation in tall buildings according to different typologies of façade. *Energy Procedia* **2017**, *134*, 224–233. [CrossRef]
8. ARUP. Carbon Footprint of Facades: Significance of Glass. 2022. Available online: https://www.arup.com/perspectives/publications/research/section/carbon-footprint-of-facades-significance-of-glass (accessed on 11 December 2023).
9. Navarro, A.L.; Lori, G.; Callewaert, D.; Overend, M. Semi-automated vs manually controlled dynamic facades: Assessment through a field study on multi-domain occupant satisfaction. *Energy Build.* **2023**, *286*, 112912. [CrossRef]
10. Pastore, L.; Andersen, M. The influence of façade and space design on building occupants' indoor experience. *J. Build. Eng.* **2022**, *46*, 103663. [CrossRef]

11. Founti, M.; Avesani, S.; Esnarrizaga, P.E. Multifunctional façades for renovation through industrialization. *J. Facade Des. Eng.* **2023**, *11*, V–VI. [CrossRef]
12. Negar, H.M.; Ali, E. Technologies used in responsive facade systems: A comparative study. *Intell. Build. Int.* **2022**, *14*, 54–73.
13. Knaack, U.; Klein, T.; Bilow, M.; Auer, T. *Façades: Principles of Construction*; Birkhäuser: Boston, MA, USA, 2014.
14. Biloria, N.; Makki, M.; Abdollahzadeh, N. Multi-performative façade systems: The case of real-time adaptive BIPV shading systems to enhance energy generation potential and visual comfort. *Front. Built Environ.* **2023**, *9*, 1119696. [CrossRef]
15. Du, H.; Huang, P.; Jone, P. Modular façade retrofit with renewable energy technologies: The definition and current status in Europe. *Energy Build.* **2019**, *205*, 109543. [CrossRef]
16. Pracucci, A.; Magnani, S.; Vandi, L.; Casadei, O.; Uriarte, A.; Bueno, B.; Vavallo, M. An Analytical Approach for the Selection of Technologies to Be Integrated in a Plug&play Façade Unit: The RenoZEB Case Study. *Proceedings* **2020**, *65*, 29.
17. Pracucci, A.; Dugè, A.; Richet, N.; Abdullah, T.; Baser, E.; Caneva, S.; Deneyer, A.; Dias, P.; Diygu, C.; Fakhari, M.; et al. IWG5 White Paper. Available online: https://backend.orbit.dtu.dk/ws/portalfiles/portal/327410776/IWG5_Active_Module_white_paper_v1.pdf (accessed on 11 December 2023).
18. Prieto, A.; Klein, T.; Knaack, U.; Auer, T. Main perceived barriers for the development of building service integrated facades: Results from an exploratory expert survey. *J. Build. Eng.* **2017**, *13*, 90–106. [CrossRef]
19. Cheng, B.; Zhu, D.; Zhao, S.; Junliang, C. Situation-Aware IoT Services Coordination Platform Using Event Driven SOA Paradigm. *IEEE Trans. Netw. Serv. Manag.* **2016**, *85–86*, 44–55. [CrossRef]
20. Katsigiannis, E.; Gerogiannis, P.; Atsonios, I.A.; Manolitsis, A.; Founti, M. Smart Wall. *J. Facade Des. Eng.* **2023**, *11*, 29–50. [CrossRef]
21. Boke, J.; Denz, P.R.; Suwannapruk, N.; Vongsingha, P. Active, Passive and Cyber-Physical Adaptive Façade Strategies: A Comparative Analysis through Case Studies. *J. Façade Des. Eng.* **2022**, *10*, 1–18. [CrossRef]
22. Giovanardi, M.; Konstantinou, T.; Pollo, R.; Klein, T. Internet of Things for building façade traceability: A theoretical framework to enable circular economy through life-cycle information flows. *J. Clean. Prod.* **2023**, *382*, 135261. [CrossRef]
23. Giordano, R.; Montacchini, E.; Tedesco, S. Experiment, develop and provoke: The prototype as an instrument of design. *TECHNE* **2019**, *18*, 89–98.
24. Salvalai, G.; Sesana, M.M.; Dell'Oro, P.; Brutti, D. Open Innovation for the Construction Sector: Concept Overview and Test Bed Development to Boost Energy-Efficient Solutions. *Energies* **2023**, *16*, 5522. [CrossRef]
25. MEZeroE. Available online: https://www.mezeroe.eu/ (accessed on 11 December 2023).
26. European Commission. CORDIS Databases. Available online: https://cordis.europa.eu/ (accessed on 11 December 2023).
27. Favoino, F.; Perino, M.; Serra, V. Advanced Integrated Façades: Concept evolution and new challenges. In Proceedings of the Thermal Performance of the Exterior Envelopes of Whole Buildings XIV International Conference, ASHRAE, Clearwater Beach, FL, USA, 5–8 December 2019.
28. Attia, S.; Eleftheriou, P.; Xeni, F.; Morlot, R.; Ménézo, C.; Kostopoulos, V.; Betsi, M.; Kalaitzoglou, I.; Pagliano, L.; Cellura, M.; et al. Overview and future challenges of nearly zero energy buildings (nZEB) design in Southern Europe. *Energy Build.* **2017**, *155*, 439–458. [CrossRef]
29. Meefs. Available online: https://cordis.europa.eu/project/id/285411 (accessed on 11 December 2023).
30. Adaptiwall. Available online: https://cordis.europa.eu/project/id/608808 (accessed on 11 December 2023).
31. Plug-N-Harvest. Available online: https://www.plug-n-harvest.eu/ (accessed on 11 December 2023).
32. Energy Matching. Available online: https://www.energymatching.eu/ (accessed on 11 December 2023).
33. RenoZEB. Available online: https://renozeb.eu (accessed on 11 December 2023).
34. PowerinSkin Plus. Available online: https://www.powerskinplus.eu (accessed on 11 December 2023).
35. Ensnare. Available online: https://www.ensnare.eu (accessed on 11 December 2023).
36. Dannapfel, V. The PLUG-N-HARVEST Facade: A Second Skin with Active and Passive Components. *J. Facade Des. Eng.* **2019**, *7*, 101–114. [CrossRef]
37. Vilà, D.M.; Marrone, G.; Ribas, I.R. Implementation of a multifunctional Plug-and-Play façade using a set-based design approach. *J. Facade Des. Eng.* **2023**, *11*, 71–96.
38. Gubert, M.; Ngoyaro, J.A.; Guiterrez, M.J.; Pinotti, R.; Brandolini, D.; Avesani, S. Comparative cost analysis of traditional and industrialised deep retrofit scenarios for a residential building. *J. Facade Des. Eng.* **2023**, *11*, 145–168. [CrossRef]
39. Torres, J.; Garay-Martinez, R.; Oregi, X.; Torrens-Galdiz, J.I.; Uriarte-Arrien, A.; Pracucci, A.; Casadei, O.; Magnani, S.; Arroyo, N.; Cea, A.M. Plug and Play Modular Façade Construction System for Renovation for Residential Buildings. *Buildings* **2021**, *11*, 419. [CrossRef]
40. Focchi Spa. Elysia. Available online: https://www.focchi.it/ww/elisya (accessed on 11 December 2023).

applied sciences

MDPI

Article

Double-Skin Façades for Building Retrofitting and Climate Change: A Case Study in Central Italy

Camilla Lops *, Samantha Di Loreto, Mariano Pierantozzi and Sergio Montelpare

Department of Engineering and Geology, University G. d'Annunzio of Chieti-Pescara, 65122 Pescara, Italy; samantha.diloreto@unich.it (S.D.L.); mariano.pierantozzi@unich.it (M.P.); s.montelpare@unich.it (S.M.)
* Correspondence: camilla.lops@unich.it; Tel.: +39-085-453-7258

Featured Application: Forecast procedures for building energy renovation and design.

Abstract: In recent years, the need to make the built environment more resilient and adaptable to climate change has become essential. In Europe, this aspect concerns most existing buildings with several deficiencies from the energy efficiency point of view, considering they were designed before the introduction of modern codes. Among the various strategies for building energy retrofitting, Double-Skin Façades (DSFs) have gained attention due to their potential to improve the building performance and inhabitants' comfort. This research aims to evaluate the use of adequately designed DSFs for the energy restoration of buildings. In detail, various DSF configurations are applied to a residential building located in Central Italy and investigated under present and future climate conditions, estimated through regional climate models. The installation of multi-layered façades, particularly the Multi-Storey typology, greatly reduces energy consumption and increases the expected comfort rates. When the selected configuration was considered, the results underline a decrease in the annual building energy requirement of about 37–56% up to 42–59%, respectively, for 2030 and 2070. Moreover, using multi-layer façades can increase indoor minimum operative temperatures up to 3.8% during the coldest months and reduce the maximum summer ones by 1.9–3.8%, raising comfort levels.

Keywords: building energy retrofitting; Double-Skin Façades; climate change; regional climate model; dynamic energy modelling; MM5; CORDEX

1. Introduction

In recent decades, the economy centred on the general reduction in energy consumption and CO_2 emissions has dictated important changes in every sector, especially in the construction world. Buildings are, in fact, key consumers of energy in Europe, and a rising tendency in energy use has been globally recorded in the last twenty years [1]. According to the European Environment Agency (EEA), in 2017 the transport sector accounted for 31% of the total final energy consumption in the European Member States, followed by households (27%), industry (25%) and services (15%) sectors [2].

The energy use in households is mainly related to space and water heating, which together are responsible for 80% of the total building energy consumption [3]. This high percentage is mostly due to poorly insulated envelopes and less-efficient heating equipment, which are mainly fossil-fuel-based and traditionally present in existing buildings.

Similar trends are also estimated for greenhouse gas emissions. Buildings and construction together account for 39% of energy-related CO_2 emissions when upstream power generation is included [4]. Considering the high potential for cost-effective energy savings, the building sector has become a priority area for the European Commission, which has sponsored various actions to reduce the building requirement and promote their renovation. Prominent examples of this effort are the Directive 2002/91/EC [5] and the Directive

2010/31/EU [6], commonly known as the EPBD (from its full name the Energy Performance of Buildings Directive) and its recasting. The first is mainly centred on defining a standardised methodology more oriented to new buildings. The second, on the other hand, aims to deal with existing buildings not only when they are subjected to a significant renovation but also in replacing and retrofitting a few elements or technical systems. After the EBPDs, the European Member States have shown a growing interest in building energy improvement, and, as a main result, a specific article centred on building renovation was introduced in the new Energy Efficiency Directive 2012/27/EU [7].

The great interest in the existing building stock is due to the fact that most European buildings currently in use were erected before the 1960s when the sustainability problem was in a preliminary phase, and energy building regulations were minimal. According to the Building Performance Institute Europe (BPIE) survey, in fact, 35–42% of these buildings date back to before the 1960s, and another consistent part was erected between 1961 and 1990 due to the massive boom in the construction sector [8]. The statistics underline the need for energy-efficient retrofit solutions, which are essential under present and, in particular, future conditions, considering that the construction age of a building is among the main features that affect its climate change vulnerability. Various innovations for improving the energy performance of buildings and, generally, cities, to reduce their environmental impact involve smart and intelligent devices, which are mostly renewable-based and installed directly on building surfaces [9,10] or at an urban scale [11–13].

In this way, the present research aims to evaluate the climate change vulnerability of an existing building in Central Italy by estimating its energy needs and comfort rates under future meteorological conditions and considering both the current and energy-improved versions. In detail, the investigated retrofit solutions take into account the installation of Double-Skin Façades (DSFs) on the building's main elevations, and we analyse multiple configurations to assess the best typology for the specific climatic area. The novelty of the work lies, in particular, in locating the multi-layer façade on more than one elevation. Generally, in fact, DSF systems are placed on only one façade, the north one, and designed to act as a filter zone [14–18]. Instead, a so-conceived configuration could also assume a structural purpose after the installation of seismic devices, in pursuit of the building's holistic renovation. A second innovative aspect of the work is related to the selection of the climatic parameters adopted for the analyses, which are usually performed using historical or current datasets [19,20]. Studying DSF systems under future weather conditions allows the estimation of their sensitivity to climate change phenomena, underlining their potentiality and weakness. The above-mentioned features are analysed using performing dynamic energy simulations in EnergyPlus (Version 8.9) with its DesignBuilder interface (Version 6.1.3).

In detail, the paper is structured as follows. After this brief introduction, considerations of Double-Skin Façades and their use in new and retrofitted buildings are reported in Section 2. Section 3 describes the case study, the main energy modelling features and the preliminary simulations carried out for testing the model's capability in predicting the real energy behaviour of the reference building. Section 4 focuses on the presentation of the methodology adopted for generating future climatic files and shows the obtained results expressed in energy consumption and comfort rates. A discussion of the outputs is described in Section 5. Finally, Section 6 draws the main conclusions.

2. Double-Skin Façades for New and Retrofitted Buildings

The Double-Skin Façade is an architectural trend mainly driven by the aesthetic desire for all-glass façade buildings, the need for a consistent reduction in energy consumption and the parallel increase in indoor comfort conditions, both acoustic and thermic [21]. Described as a pair of glass skins separated by an air corridor ranging in width from 0.20 m to several meters by Uuttu [22], the DSF concept appeared in the early 20th century in northern European countries as a way to reduce heating consumption thanks to the air buffer enclosed in the cavity, which acts as a barrier for heat loss and as a filter for exchanges through the external envelope. The extreme variety of possible DSF typologies can be classified according to different criteria, independent of one another and based on the geometric characteristics of the façade and its operation mode [23]. The criteria mentioned above are the following:

- The ventilation mode (natural, mechanical and hybrid ventilation).
- The compartmentalisation of the façade (Multi-Storey, Shaft-Box, Corridor and Box-Window DSF).
- The airflow type (Exhaust Air, Supply Air, Static Air Buffer, External Air Curtain, Internal Air Curtain.

In the second decade of the 21st century, more articulate configurations and advanced materials started to be combined with multi-layer systems to enhance their energy performance, and more accurate simulation approaches were adopted to overcome modelling limitations and difficulties [24–26].

Nowadays, the analysis of solutions for improving the energy behaviour of Double-Skin Façades is a topic frequently explored by researchers all over the world. Detailed investigations are conducted to evaluate DSFs' efficiency when different glazing options and shading systems are used. According to the literature review [27], using single clear glass for the inner pane and double reflective glazing for the outer surface is the best option, with annual cooling savings of up to 26% compared to a traditional external wall and single absorptive window glazing. Another essential and widely analysed aspect involves the shading system and the main parameters responsible for better solar protection. According to Gratia et al. [28], in fact, building cooling energy consumption can be decreased by up to 23% by paying attention to three main aspects: the location of the blinds, the colour of the blinds and the opening of the Double-Skin. Other investigations involve possible connections of DSFs with the Heating, Ventilation and Air Conditioning (HVAC) system, its use as a solar chimney or the installation of new elements like Photovoltaic (PV) cells, vegetation, or Phase-Changing Materials (PCMs) [29–39].

DSF innovation does not involve only energy efficiency aspects. A novel area of focus is the installation of structural elements able to confer an extra function, a seismic one, to the so-called Engineered Double-Skin Façade [40–45]. Such a façade becomes a mass damper system or an exoskeleton, which, in the case of restoration, can reduce the seismic vulnerability of existing buildings.

Based on the considerations mentioned above, multi-layer façades represent interesting options for both new and retrofitted buildings. In fact, several works study the improvements achievable by installing DSF systems in existing buildings, underlining their key role in cutting building energy consumption and CO_2 emissions if properly designed and oriented [46–49].

3. The Case Study

The reference building is located in a residential suburb of Pescara, Central Italy, and belongs to the social housing stock (Figure 1). Selected for being representative of an existing building without architectural quality but mostly present on the territory, it reflects the typical construction practice of the past, with no consideration of energy problems. The area is known for its Mediterranean climate, with mild winters and hot, sunny summers. According to the climate classification system developed by Köppen–Geiger [50], the city has a humid subtropical climate denominated Cfa. The C group corresponds to "Warm Temperate" climates; the small letter f means "fully humid" and indicates the lack of a dry season, whereas the letter a corresponds to a "hot summer".

The case study is a seven-storey reinforced concrete building characterised by a rectangular shape with 60.00 m and 12.00 m as its main dimensions. The interstorey height is 2.70 m, except for the ground floor, which is higher than the others (3.50 m). The garages and the entrance are on the ground floor, while residential apartments are on the upper levels. The building envelope does not comply with the current standard in the sector. The perimeter walls (0.33 m depth), in fact, do not present any insulation materials but only an air gap enclosed by two brickwork layers, with a total transmittance value (U) equal to

1.46 W/m^{2}K. There are also 0.23 m thick walls placed to delimit the apartment from the staircase, and partitions (with 11 cm of depth) to subdivide each apartment into various thermal zones. Moreover, single glass windows with 3.78 W/m^{2}K transmittance contribute to the building's inadequate performance, especially during the coldest months. Table 1 summarises the thermal properties of the building envelope.

Figure 1. The localisation of Pescara (**left** and **middle**) and the reference building (**right**).

Table 1. Description of building envelope components.

Building Component	Material (Outer to Inner)	s (m)	l (W/mK)	U (W/m^{2}K)
0.33 m External Wall	Lime plaster	0.02	0.80	1.46
	Brickwork	0.13	0.84	
	Air gap	0.08	0.30	
	Brickwork	0.08	0.62	
	Lime plaster	0.02	0.80	
0.23 m Semi-External Wall	Lime plaster	0.02	0.80	1.60
	Brickwork	0.13	0.84	
	Air gap	0.03	0.30	
	Brickwork	0.08	0.62	
	Lime plaster	0.02	0.80	
0.11 m Partitions	Plaster	0.02	0.16	1.27
	Plasterboard	0.07	0.25	
	Plaster	0.02	0.16	
Glazing	Single clear glazing	0.006	-	3.78
Interstorey roof	Cast concrete	0.02	1.13	0.78
	Concrete slab	0.16	0.16	
	Plaster	0.02	0.16	
External roof	Ceiling tiles	0.02	0.06	0.61
	Cast concrete	0.02	1.13	
	Concrete slab	0.16	0.16	
	Plaster	0.02	0.16	
Pavement	Cast concrete	0.02	1.40	3.37
	Ceramic	0.02	1.30	

The building heating system involves the use of radiators with a seasonal Coefficient of Performance (COP) equal to 0.84 and natural gas as the source. Splits are installed to cool the inner spaces and improve the inhabitants' comfort condition. The cooling COP is set equal to 1.40, and the source is electricity.

The temperature setpoints for heating and cooling are, respectively, 22 °C (with 20 °C as the set-back temperature) and 28 °C in all thermal zones. The metabolic factor is set to 0.90 for the whole building, as well as for the clothing insulations, which is 0.50 clo for the summer season and 1.00 for the winter period. The model infiltration is considered to be constant and fixed at 0.70 ac/h. Table 2 depicts the input parameters, which vary according to the activity of the thermal zone.

Table 2. The main input parameters assigned to each thermal zone.

Thermal Zone	Occupation (People/m^2)	Fresh Air (l/s per Person)	Illuminance (Lux)	Schedule (-)
Bedroom	0.0229	10	100	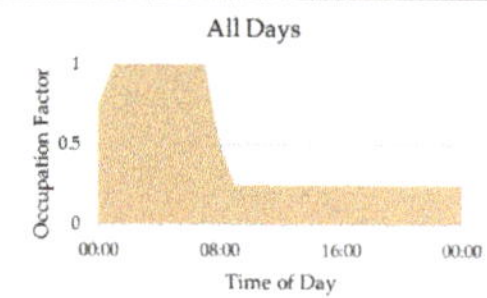
Bathroom	0.0187	12	150	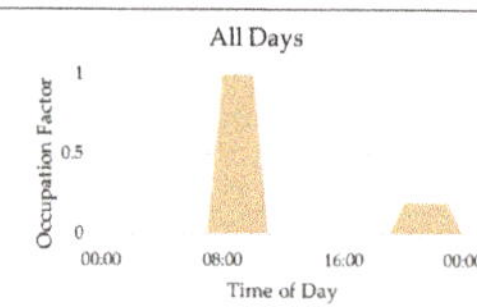
Kitchen	0.0237	12	300	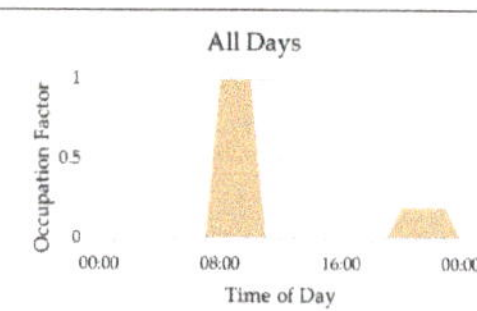
Storage Room	0.0243	12	100	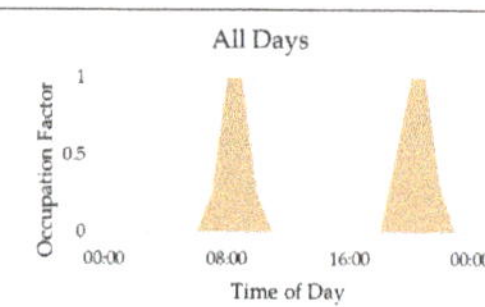
Dining Room	0.0169	10	150	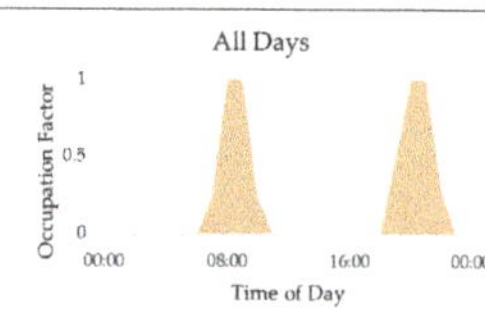
Circulation Area	0.0196	10	100	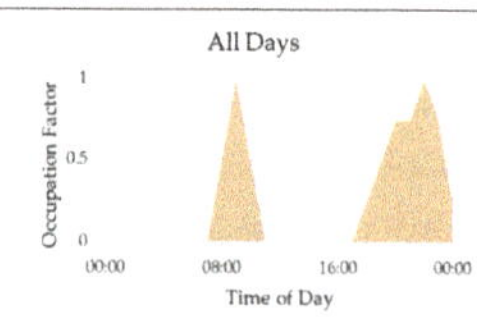

The influence of neighbouring buildings is taken into account by inserting, as block components, the closest structures that could shade the case study and, for the same reason, the balconies on the northwest façade are considered by placing external profiles. Figure 2 shows the main elevations of the elaborated energy model.

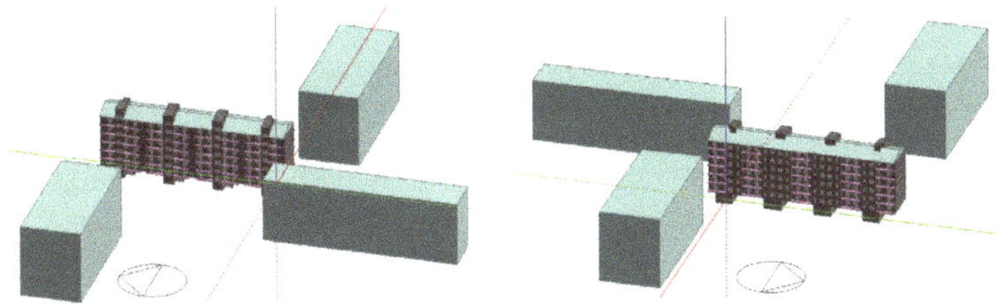

Figure 2. Northwest (**left**) and southeast (**right**) views of the energy model generated in DesignBuilder.

Once the case study had been modelled, preliminary dynamic energy simulations were carried out, and the building's natural gas consumption was estimated and compared with data from building bills. The collected bills refer to 2017, and the heating period (1 October–31 March) was considered for the comparisons. Thus, monthly simulations were performed, setting natural gas consumption (heating + cooking) as a key output, expressed in kWh/m^2. Following the recorded data, the consumption was determined for a two-month range, except for January and December. The analyses were simulated using measured weather data collected by the Climate Network, a private and professional network of urban meteorological stations that record hourly values for all climatic parameters [51]. Error bars, set at 5%, were inserted as inferior and superior limits.

Figure 3 shows the general accordance between estimated and measured values. The model, in fact, can predict within an acceptable range the energy consumption for January and the February/March bimester. More significant differences were calculated for December. Such behaviour can be explained by considering that the available bill refers to December 2017–January 2018, and the expected consumption for a single month is calculated by dividing the total amount by two. For this reason, the December data were not strictly taken into account for this step. The same inaccuracy can also be seen for the October/November bimester, but, in this case, the principal reason is the limitation of predictions for transitory months because these are deeply influenced by the inhabitants' comfort condition and their use of occupied spaces. In fact, as underlined by various works previously carried on in energy simulations, the building requirement/consumption for mild periods could vary more than the rest of the year, according to human behaviour (inhabitants' age, habits, etc.). It is essential to bear in mind that, considering the cumulative value (bimonthly) of measured data, which are related to a specific apartment located inside the building under analysis, a global comparison, rather than a monthly calibration, was performed. Nevertheless, by applying the ASHRAE guideline 14 criteria only to the accurate data, the Normalised Mean Bias Error (NMBE) and the Coefficient of Variation of the Root Mean Square Error (CVRMSE) were determined, respectively, to be 1% and 0% for January and −3% and 3% for the February/March bimester, which are both inside the acceptable range established by the standard. Based on these comparisons, the model can be considered able to describe the real energy behaviour of the case study.

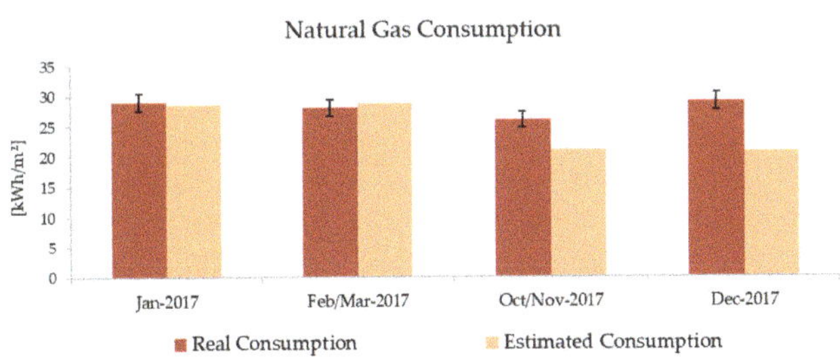

Figure 3. Comparisons among estimated and measured data.

DSF Energy Modelling

A Double-Skin Façade with a 1.00 m cavity depth is installed on the northwest and southeast building elevations, and external blinds with low reflectivity slots are installed and designed to be operable only during the hottest months, thus preventing the risk of overheating the inner spaces. The DSF is designed to be naturally ventilated. External and internal grills are located, respectively, on the outer skin and the inner layer, represented by the existing building envelope. External grills are active, allowing the air to enter the Double Façade during the hottest months, whereas they are closed for the winter period, creating a buffer area. The internal grills, instead, are designed to be adjustable according to the inner temperature distribution. The natural ventilation of the building is controlled by conducting an outdoor maximum temperature check. This prevents overheated air inside the cavity from entering the building and having adverse effects on the cooled side. The outer skin of the DSF is made of a steel structure and high-performance windows composed of triple glazing (a 13 mm cavity filled with Argon gas sealed by two layers of 3 mm single clear glass) with a low U value and solar heat gain coefficient equal to, respectively, 0.78 W/m^{2}K and 0.47. The DSF inner partitions, both horizontal and vertical, present steel frame and single-pane windows made of 6 mm clear glass with a transmittance and solar heat gain coefficient of 3.78 W/m^{2}K and 0.72.

The energy modelling of the DSF was achieved by setting the cavity as an unoccupied zone with no HVAC or lighting template data. Moreover, the internal convection mechanism was activated to model the cavity air space correctly, and a complete interior and exterior solar distribution algorithm was switched on, allowing solar radiation to be accurately transmitted through the interior glazing in the partition.

Various configurations were modelled and tested to evaluate the effectiveness of DSF systems. Multi-Storey, Shaft-Box, Corridor and Box-Window DSFs were investigated in terms of energy consumption and thermal comfort conditions. Two different cases were studied for the Corridor type: the outer grills were inserted on the principal elevation of the façade in one case and on the lateral envelope in the other. The inner partitions, which were eventually inserted inside the cavity according to the selected category, present single-glazed windows enclosed in a steel structure. Figure 4 schematises the chosen options, identifying the air fluxes which enter/exit each DSF typology.

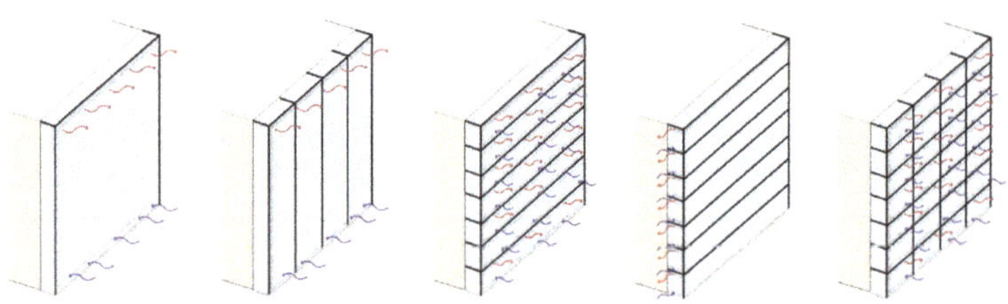

(a) (b) (c) (d) (e)

Figure 4. Schematisation of the investigated Multi-Storey (**a**), Shaft-Box (**b**), Corridor (**c**,**d**) and Box-Window (**e**) DSFs.

4. Dynamic Energy Simulations under Future Climate Conditions

4.1. Generation of Future Climatic Files for Dynamic Energy Simulations

Future climatic files are generated by using the Coordinated Regional Climate Downscaling Experiment (CORDEX). Sponsored by the World Climate Research Program (WCRP) to develop a coordinated framework for evaluating and improving Regional Climate Downscaling (RCD) techniques, CORDEX produces worldwide fine-scale climate data [52]. The CORDEX results are assumed as a baseline by the Intergovernmental Panel on Climate Change (IPCC) to define climate change impact and adaptation studies. Its various domains allow the estimation of meteorological variables all over the world, and the Euro-CORDEX, reported in Figure 5, is its European branch.

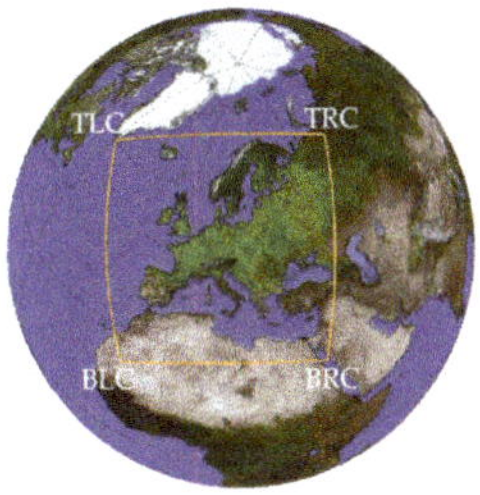

Non-Rotated Coordinates of Domain Corners

Top Left Corner (TLC): 315.86; 60.21

Top Right Corner (TRC): 64.4; 66.65

Bottom Left Corner (BLC): 350.01; 22.20

Bottom Right Corner (BRC): 36.30; 25.36

Figure 5. Euro-CORDEX Domain and the coordinates of its corners.

The Aire Limitée Adaptation dynamique Développement InterNational (ALADIN), used at the Centre National de Recherches Météorologiques (CNRM) with the name CNRM-ALADIN and available in Euro-CORDEX was selected in the present work [53,54]. This limited-area bi-spectral model was chosen due to its capability to predict with good approximation climatic parameters both spatial and temporal scales inside the European area [55,56]. Thus, data with a 3 h time frequency were extracted from the nearest grid point to Pescara, which is 7.50 km from the case study, at 42°28′57″ northern latitude and 14°08′07″ eastern longitude. The selected RCP was 4.5, which is consistent with a future with relatively ambitious emission reductions, and this scenario was chosen for being in accordance with national policies centred on reducing greenhouse gas emissions.

The available weather variables were relative humidity, atmospheric pressure, global solar radiation, wind velocity and temperature. The definition of the reference year was conducted according to the method described in the technical standard EN ISO 15927-4:2005 [57] and following its suggestions [52,58]. Multiple years were extracted to obtain various typical years.

The first step involved the calculation of the daily averaged value for each climatic parameter (p), month (mt) and year (y) of the datasets. Then, the averaged values for a specific month in all the available years were sorted in increasing order to calculate the cumulative function $\phi_{(p,mt,i)}$ for each parameter and ith day using Equation (1).

$$\phi_{(p,mt,i)} = {}^{k(i)}/_{N+1} \tag{1}$$

where $k_{(i)}$ is the rank order of the ith day and N is the total number of days in a month across all years.

The following step consisted of sorting the averaged values for a specific month and year in increasing order to obtain the cumulative distribution function $F_{(p,y,mt,i)}$ for each parameter and ith day (Equation (2)).

$$F_{(p,y,mt,i)} = {}^{J(i)}/_{n+1} \tag{2}$$

Then, for each month and year, the statistics by Finkelstein–Schafer were defined according to Equation (3). The last two steps involved the sorting of months, for which the rank was calculated for every parameter and summed to obtain the total ranking and for each month. Among the first three months with the lowest ranking sum, the one with the lower absolute deviation was chosen as a representative for the TMY generation.

$$Fs_{(p,y,mt)} = \sum_{i=1}^{n} \left| F_{(p,y,mt,i)} - \phi_{(p,mt,i)} \right| \tag{3}$$

To improve the quality of the generated TMY, weighted meteorological parameters were inserted in the Finkelstein–Schafer, as suggested by Cebecauer et al. [58]. The stronger influence of some variables than others is taken into account by increasing the weight of those parameters. Higher impact factors are attributed to surface temperature and solar radiation (8/24), whereas a lower value (4/24) is assigned to relative humidity and wind speed, which slightly affect energy simulations. A twenty-year range was selected to generate a typical year. Thus, data from 2020 to 2040, 2040 to 2060 and 2060 to 2080 were used to elaborate, respectively, 2030, 2050 and 2070 TMYs. Once the selected years were generated, a cubic spline function was used to obtain interpolated values and transform the time frequency of the investigated parameters from 3 h to 1 h.

Figures 6 and 7 depict comparisons of monthly mean values for temperature and global solar radiation between actual (Climate Network database) and future (2030, 2050 and 2070 TMYs) climate data. The future conditions (dotted lines) tended, in general, to assume similar values and trends to those measured in 2017. According to the predictions, higher temperatures are expected for the coldest months (from November to February), while a slight reduction is estimated for summer. More profound variations are expected, instead, for global solar radiation. In this case, the values estimated by CORDEX TMYs are much higher than those recorded in the present, reaching up to 330 W/m^2.

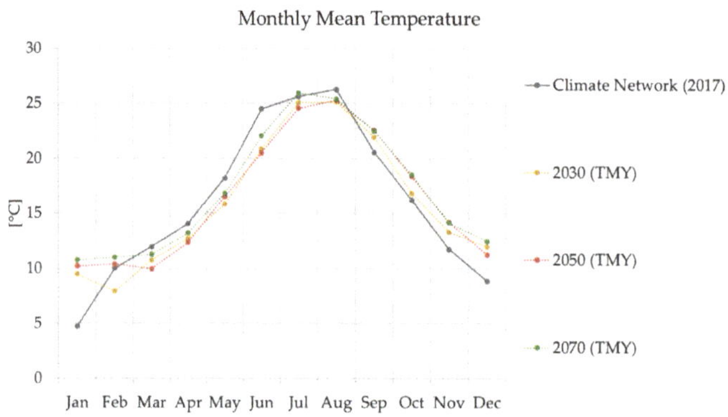

Figure 6. Monthly mean values referring to the temperature.

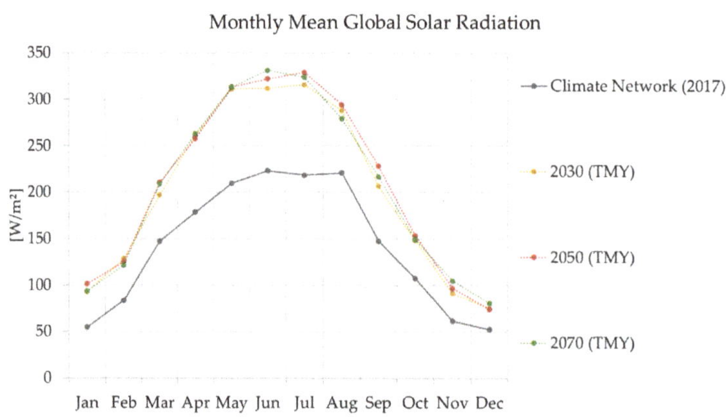

Figure 7. Monthly mean values referring to global solar radiation.

The same trend was confirmed by comparing these parameters to early averaged values (Figure 8). On the temperature side, from actual to future conditions, the increase is gradual, reaching a 2 °C delta between 2017 and 2070, whereas higher variation is expected for solar radiation. It is essential to underline that these comparisons involve projections related to a specific scenario which considers the adaptation of policies for CO_2 emission containment.

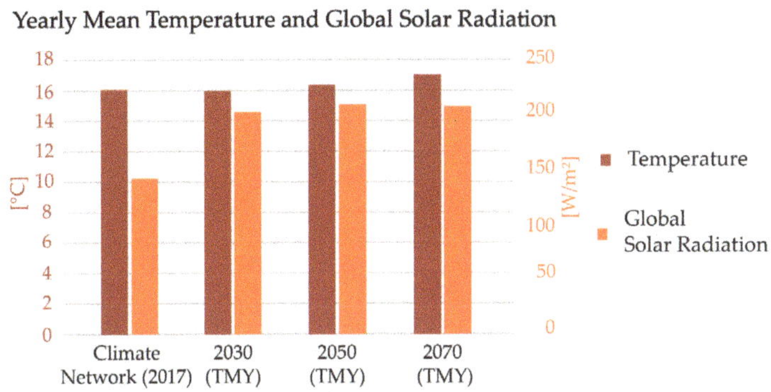

Figure 8. Yearly mean values referring to temperature and global solar radiation.

4.2. Building Energy Modelling under Future Climate Conditions

The impact of climate change was analysed, evaluating the effect that future outdoor conditions could have on the reference building in its existing and improved version. The case study's energy performance was assessed in terms of energy consumption and thermal comfort rates. In the first case, the results refer to the annual building heating and cooling use, whereas the thermal comfort was investigated considering a weekly timestep, simulating alternatively a typical summer (August, from the 17th to 23rd) and winter (January, from the 20th to 26th) conditions. Moreover, to define comfort/discomfort rates, the operative temperature was estimated for two different building thermal zones, selected for their orientation. North- and south-exposed thermal zones were chosen to investigate winter and summer comfort, respectively. This choice was made considering the building dimensions and to prevent averaged values that did not represent the real energy behaviour.

The energy consumptions, depicted from Figures 9–11, underline a constant increase in the building cooling load and a similar decrement in the heating need, and this happens for both the original state and that with DSF configurations. Among the investigated typologies, the Multi-Storey DSF performed better than the others, allowing, at the same time, a good cooling reduction and a more intense heating decrement. Additionally, the Shaft-Box, Corridor and Box-Window configurations exhibited a general and quasi-identical energy reduction under future conditions. Moreover, significant differences were not observed for the two analysed Corridor DSFs.

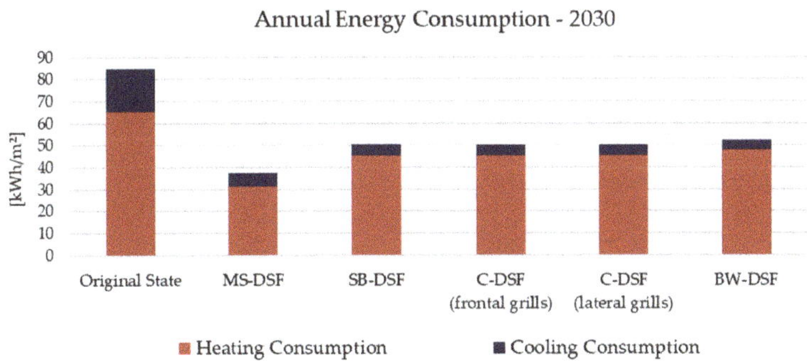

Figure 9. Building annual energy consumption estimated for the year 2030.

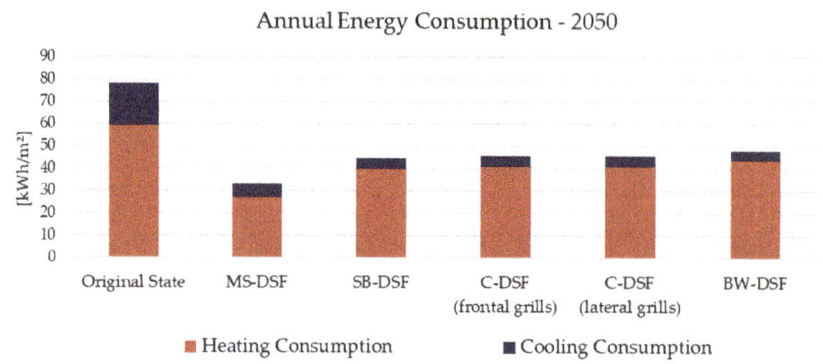

Figure 10. Building annual energy consumption estimated for the year 2050.

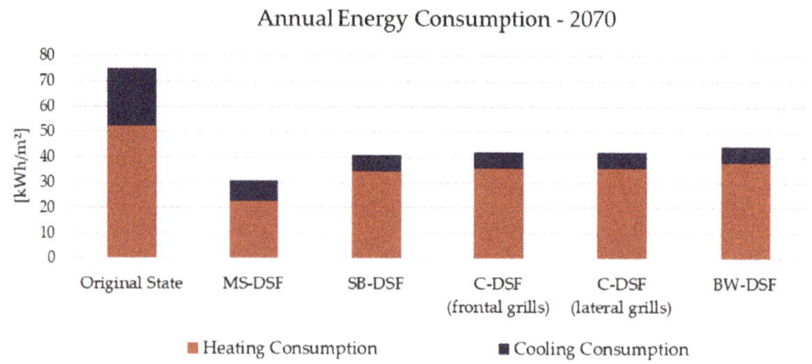

Figure 11. Building annual energy consumption estimated for the year 2070.

The evaluation of indoor thermal comfort was carried out according to the UNI EN 16798-1:2019 (CEN 2019) [59], which associates comfort/discomfort rates with four categories. In detail, Category I has a high level of expectation, and it is recommended for spaces occupied by very sensitive people. Category II is for normal levels of expectations and is suitable in the case of new buildings or renovations. Category III represents a moderate, acceptable level of expectation and can be used for existing buildings, whereas the last, Category IV, involves other conditions not covered by the previous cases. Moreover, the standard establishes for each comfort category a temperature interval, suggesting recommended indoor operative temperatures for the design of heating and cooling systems. These values are used to define each category's upper and lower limits, as summarised in Table 3. It is essential to underline that the here-investigated parameter is a building with a heating and cooling system. Thus, the performed simulations aimed to evaluate its effectiveness in guaranteeing acceptable comfort rates inside the occupied spaces.

Table 3. Indoor operative temperature ranges for comfort categories according to the UNI EN 16798-1:2019 [59] for residential buildings.

Category	Winter Operative Temperature Range	Summer Operative Temperature Range
I	$T_{operative} \geq 21$ °C	$T_{operative} \leq 25.5$ °C
II	20 °C $\leq T_{operative} < 21$ °C	25.5 °C $< T_{operative} \leq 26$ °C
III	18 °C $\leq T_{operative} < 20$ °C	26 °C $< T_{operative} \leq 27$ °C
IV	$T_{operative} < 18$ °C	$T_{operative} > 27$ °C

The thermal comfort analysis was used to evaluate the indoor operative temperature expected for the thermal zones under study before and after the installation of the DSFs. In detail, the operative temperature intervals for both the summer and winter periods were adopted to establish the percentage of the corresponding indoor environmental quality categories associated with each building configuration. In this specific case, only Categories II, III and IV were considered as the simulations were carried out for an existing building, and a normal level of expectation would be "Medium" (Category II), as recommended by the standard. Furthermore, the mean, minimum and maximum expected operative temperature was extracted for better comparisons. The indoor temperatures estimated for the building in its original state were assumed as baseline, and percentages of increments/decrements were calculated and are reported in the tables. The obtained categories are plotted from Figures 12–17 and the temperatures are summarised in Tables 4–9.

The comparisons underline several variations which should be taken into account for further consideration of the effectiveness of DSF systems. The estimated summer performance of Double Façades could lead to a different distribution of future comfort levels due to the high levels of predicted global solar radiation. As highlighted in the previous section, the amount of monthly mean radiation expected for the years 2030, 2050 and 2070 is much more significant than actual conditions, and this could generate the worst comfort rates for various DSF typologies. Due to a high solar load, the Double Façade, in fact, becomes a heated element which continually emits accumulated heat. This element is much more sensitive to this phenomenon than the original building envelope because it is entirely made of glazed surfaces. While on the one hand the DSF acts as a heat damper, reducing the effect of solar load on the inner operative temperature, on the other it also tends to increase the number of hours for which higher temperatures are estimated and, consequently, leads to a lower level of comfort. A mechanical system should be introduced to improve the summer indoor environment and to enhance natural ventilation in case of extra loads, converting the DSF into a hybrid technology.

In evaluating winter comfort rates (Figures 15–17), the general temperature increase predicted for the coldest months allows higher levels of expectation, and this was estimated for both the building's original state and its state after the DSF's installation. The discomfort rates under future conditions tend, in fact, to decrease, and this reduction assumes more significant values for DSF systems, ensuring better performances than those reachable by the reference building in its original state. Additionally, in this case, the DSF that allows the best indoor environment is the Multi-Storey variety, which is capable of guaranteeing the highest percentage with a better level of expectations. Moreover, it is interesting to note that the winter comfort level predicted for the year 2050 shows better values than 2070. This phenomenon is explained by the higher temperatures and solar radiation predicted for the investigated period (January), which positively affect the heating side, reducing its load.

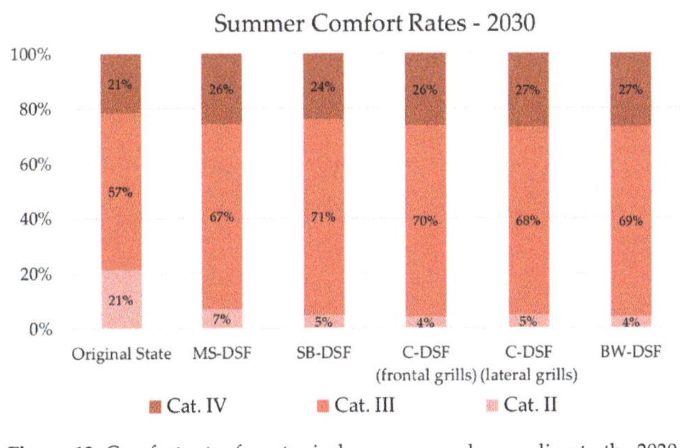

Figure 12. Comfort rates for a typical summer week according to the 2030 climatic file.

Table 4. Indoor operative mean, minimum and maximum temperature expected for a typical summer week in 2030.

Configuration	Mean Temperature	Minimum Temperature	Maximum Temperature
Original State	26.69	24.81	29.04
MS-DSF	26.72 (+0.1%)	25.44 (+2.5%)	28.00 (−3.6%)
SB-DSF	26.73 (+0.1%)	25.47(+2.7%)	27.93 (−3.8%)
C-DSF (frontal grills)	26.74 (+0.2%)	25.48(+2.7%)	28.02 (−3.5%)
C-DSF (lateral grills)	26.75 (+0.2%)	25.48 (+2.7%)	28.02 (−3.5%)
BW-DSF	26.76 (+0.3%)	25.52(+2.9%)	27.95 (−3.8%)

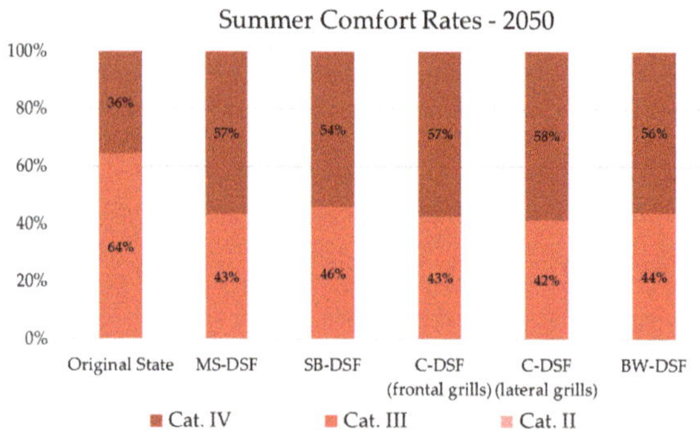

Figure 13. Comfort rates for a typical summer week according to the 2050 climatic file.

Table 5. Indoor operative mean, minimum and maximum temperature expected for a typical summer week in 2050.

Configuration	Mean Temperature	Minimum Temperature	Maximum Temperature
Original State	26.94	26.02	28.39
MS-DSF	27.07 (+0.5%)	26.54 (+2.0%)	27.78 (−2.1%)
SB-DSF	27.05 (+0.4%)	26.54 (+2.0%)	27.73 (−2.3%)
C-DSF (frontal grills)	27.08 (+0.5%)	26.58 (+2.2%)	27.79 (−2.1%)
C-DSF (lateral grills)	27.08 (+0.5%)	26.57 (+2.1%)	27.79 (−2.1%)
BW-DSF	27.06 (+0.4%)	26.57 (+2.1%)	27.74 (−2.3%)

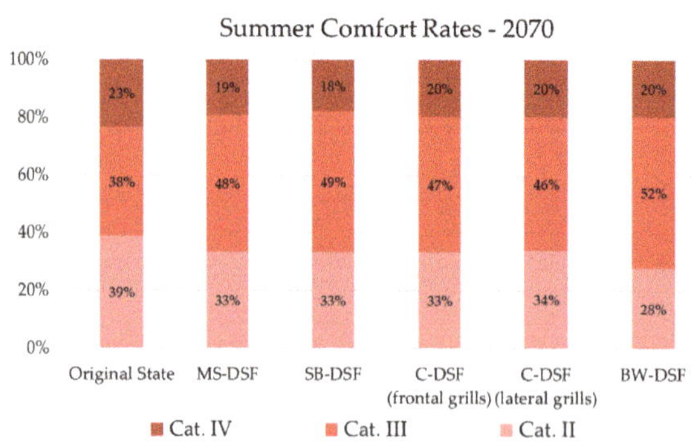

Figure 14. Comfort rates for a typical summer week according to the 2070 climatic file.

Table 6. Indoor operative mean, minimum and maximum temperature expected for a typical summer week in 2070.

Configuration	Mean Temperature	Minimum Temperature	Maximum Temperature
Original State	26.36	24.53	28.37
MS-DSF	26.41 (+0.2%)	25.05 (+2.1%)	27.82 (−1.9%)
SB-DSF	26.41 (+0.2%)	25.07 (+2.2%)	27.75 (−2.2%)
C-DSF (frontal grills)	26.43 (+0.3%)	25.06 (+2.2%)	27.82 (−1.9%)
C-DSF (lateral grills)	26.42 (+0.2%)	25.05 (+2.1%)	27.81 (−2.0%)
BW-DSF	26.44 (+0.3%)	25.11 (+2.4%)	27.76 (−2.2%)

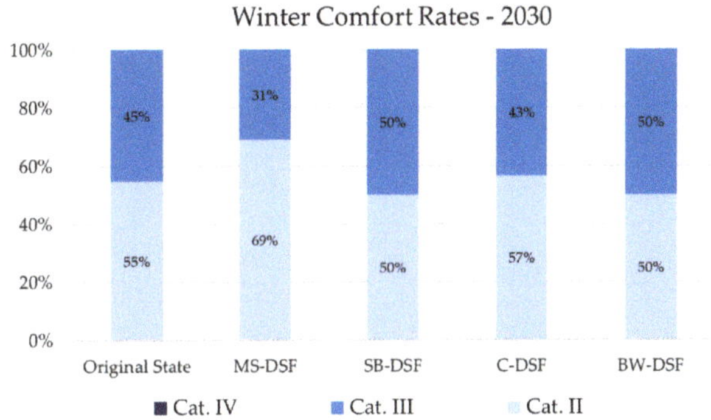

Figure 15. Comfort rates for a typical winter week according to the 2030 climatic file.

Table 7. Indoor operative mean, minimum and maximum temperature expected for a typical winter week in 2030.

Configuration	Mean Temperature	Minimum Temperature	Maximum Temperature
Original State	20.22	18.68	24.62
MS-DSF	20.50 (+1.4%)	18.92 (+1.3%)	23.83 (−3.2%)
SB-DSF	20.05 (−0.8%)	18.86 (+1.0%)	21.67 (−12.0%)
C-DSF	20.21 (+0.0%)	18.93 (+1.3%)	21.86 (−11.2%)
BW-DSF	20.05 (−0.8%)	18.84 (+0.9%)	21.59 (−12.3%)

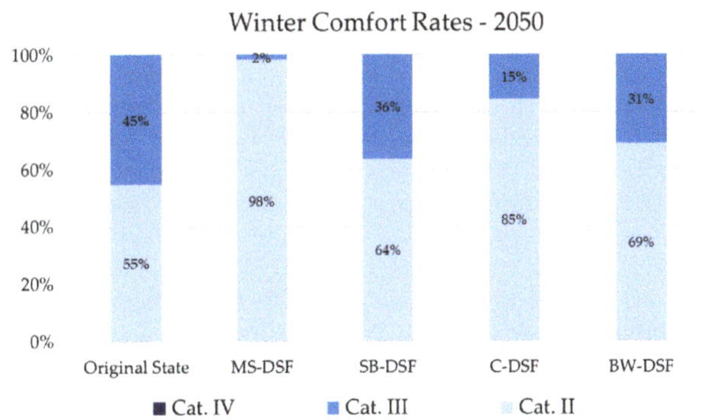

Figure 16. Comfort rates for a typical winter week according to the 2050 climatic file.

Table 8. Indoor operative mean, minimum and maximum temperature expected for a typical winter week in 2050.

Configuration	Mean Temperature	Minimum Temperature	Maximum Temperature
Original State	20.13	19.15	20.92
MS-DSF	21.72 (+7.9%)	19.88 (+3.8%)	25.61 (+22.4%)
SB-DSF	20.56 (+2.1%)	19.45 (+1.6%)	22.11 (+5.7%)
C-DSF	20.86 (+3.6%)	19.65 (+2.6%)	22.65 (+8.3%)
BW-DSF	20.59 (+2.3%)	19.49 (+1.8%)	21.98 (+5.1%)

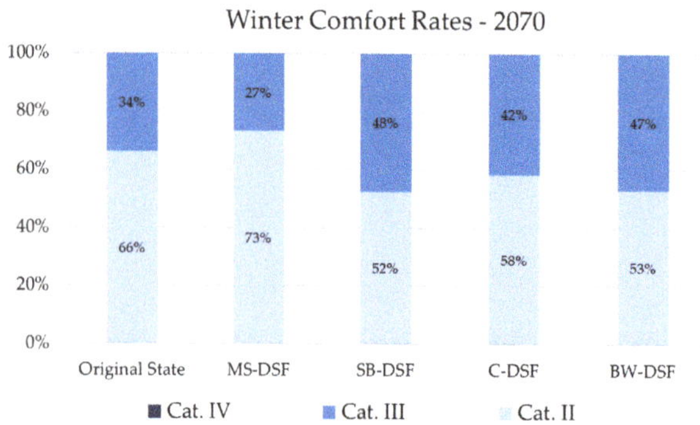

Figure 17. Comfort rates for a typical winter week according to the 2070 climatic file.

Table 9. Indoor operative mean, minimum and maximum temperature expected for a typical winter week in 2070.

Configuration	Mean Temperature	Minimum Temperature	Maximum Temperature
Original State	20.48	18.94	25.67
MS-DSF	20.74 (+1.3%)	19.14 (+1.1%)	24.27 (−5.5%)
SB-DSF	20.17 (−1.5%)	19.01 (+0.4%)	27.72 (+8.0%)
C-DSF	20.36 (−0.6%)	19.12 (+1.0%)	21.95 (−14.5%)
BW-DSF	20.18 (−1.5%)	19.01 (+0.4%)	21.64 (−15.7%)

5. Discussion

As underlined by the performed simulations, the installation of multiple DSFs in existing buildings leads to a general improvement in the energy performance with regard to both consumption and comfort. Intense heating and cooling savings are expected when comparing the building's original state and the Multi-Storey configuration. The annual heating and cooling consumption, in fact, can be cut down, respectively, to 52–57% and 65–68% for the considered year. Similar but less intense trends, especially regarding the heating requirement, were estimated for the other DSFs. The presence of horizontal or vertical partitions and the subdivision of the cavity into multiple chambers, in fact, causes a higher heating requirement and, even if they are associated with slightly better cooling needs (a reduction of 63–74% for the Shaft-Box and 68–79% for the Box-Window), the global annual energy is higher than when the Multi-Storey option is considered. The air channel increases the natural ventilation of the occupied spaces, thus reducing their temperature and, at the same time, creates a more effective barrier against heat loss throughout the building envelope during the wintertime. Finally, the two Corridor DSF versions exhibit negligible differences, underlining that the system performance is not affected by the air grills' location but is influenced by their dimensions.

The analysis of the expected thermal comfort conditions provides a few challenging aspects that should be considered in the case of DSFs. The implementation of an additional insulation layer represented by the Double-Skin Façade reduces heat transfer and ensures more stable indoor temperature fluctuations. This improvement becomes particularly noticeable during extreme meteorological conditions, such as hot summers and cold winters. As a main consequence of more uniform indoor temperatures, the discomfort conditions last longer because heat dissipation occurs slowly. For this reason, the summer comfort rate evaluation highlights higher discomfort, reaching the worst comfort categories, predicted for DSF solutions when compared to the original building state. Among the investigated options, the Multi-Storey configuration exhibits slightly better values than the others, decreasing from Category II, or 21% of the time, to 7% for 2030 and from 39% to 33% for 2070, respectively, for the existing configuration and the retrofitting one. The analysis of the indoor operative temperature confirms the above-mentioned considerations. In fact, even if the expected mean temperature is quasi-identical (a variation of about 0.1–0.5% is predicted for the selected years and configurations) among the original and energy-improved version, lower maximum temperatures were estimated in the retrofitted model (−1.9% for the Multi-Storey DSF in 2070 and up to −3.8% for the Shaft-Box DSF in 2030), confirming the effectiveness of DSFs in reducing extreme conditions.

The winter week, instead, highlights the benefits achievable by installing the multi-layer façade. In particular, the Multi-Storey configuration ensures the highest percentage of Category II, reaching 98% for 2050 but also performing well for the other years. Acceptable comfort rates are also estimated with the other configurations, especially with the Corridor DSF. Additionally, in this case, a minor temperature fluctuation is evidenced by the analyses after the installation of the Double-Skin Façade. The expected minimum operative temperature is always higher in the case of the retrofitted building than its current version, and this increment rises from 0.4% for the Shaft-Box and Box-Window configurations in 2070 up to 3.8% for the Multi-Storey configuration in 2030.

While the thermal comfort assessment results and energy savings in buildings retrofitted with Double-Skin Façades are promising, there are some limitations to consider. The study focused on a specific building type, which is largely present in existing Italian building and located in a climatic context quite representative of the Mediterranean area. Nevertheless, further research is needed to explore different climate zones, building typologies and variations in façade design to understand the broader applicability and potential challenges associated with Double-Skin Façades. Moreover, more advanced and smart ventilation systems should be introduced and analysed to tackle climate change phenomena and reduce the vulnerability of DSFs to global warming and the overheating risk of the cavity.

6. Conclusions

The present research aimed to assess the climate change vulnerability of a residential building located in Pescara, Central Italy. The energy performance of the case study was evaluated by considering its current and improved version through dynamic energy simulations. The building energy requirement, subdivided into cooling and heating needs, and the indoor thermal comfort rates were estimated under future climate conditions in order to establish the best DSF retrofit solution. Thus, typical meteorological years were generated using regional climate models, and the main weather parameters expected for 2030, 2050 and 2070 were defined.

The results underline the impact that the continuous increase in temperature and solar radiation have on building energy consumption, analysing both the heating and cooling needs. Lower winter energy loads were, in fact, estimated for the investigated years in comparison to current weather conditions. In contrast, higher cooling consumption will be necessary to ensure satisfactory comfort rates inside the occupied spaces. By inserting a DSF on the northwest and southeast elevations, energy savings can be expected, with benefits throughout the year. The installation of multi-layer façades, particularly in the Multi-Storey configuration, ensures significant reductions of up to about 52–57% and 65–68% for the considered year, respectively, for the heating and cooling needs. Similar performances, even though they were slightly less effective overall, can be expected with the other DSFs, for which higher cooling savings were calculated for the summertime (up to 63–74% for the Shaft-Box and 68–79% for the Box-Window) and lower savings were predicated for the coldest months (with an average of 32%, 13% and 27%, for the Shaft-Box, Corridor and Box-Window configurations).

In addition to the improvements achievable on the energy side, using DSFs to restore existing buildings positively affects the comfort rates, especially during the wintertime. By reducing extreme weather peaks and, in general, temperature fluctuation, the air cavity allows more uniform conditions to be maintained than the building's current configuration. The capability of DSFs in enhancing heat dissipation directly impacts the expected indoor operative temperatures, which were calculated to be ±3.8% higher and lower, respectively, than the indoor minimum and maximum operative temperatures calculated for the original state. Moreover, mechanical ventilation should be introduced to prevent the summer overheating risk of the air cavity due to elevated solar radiation and temperature levels with consequently high discomfort rates expected inside the inner spaces, thus enhancing natural convection in the case of extreme thermal loads. Additionally, in this case, the configuration with the best performance was the Multi-Storey configuration.

Furthermore, Double-Skin Façades not only provide thermal benefits but also contribute to improved indoor air quality. The additional layer acts as a buffer zone and reduces outdoor air pollutants and noise infiltration, leading to a healthier and more comfortable indoor environment for occupants. Moreover, DSFs maintain adequate natural daylight penetration while minimising glare and excessive solar heat gain. This balance ensures occupants have access to natural light, reducing the reliance on artificial lighting during the day and promoting a more pleasant visual environment.

In view of the achieved results, the use of DSFs for the energy retrofit of existing buildings proves to be a suitable option, with savings in terms of energy requirements and comfort rates. Despite these achievements, there are still some knowledge gaps the remain to be filled. Among them, long-term monitoring and evaluation to understand their performance over extended periods is still required. Factors such as maintenance requirements, durability and ageing effects on thermal and optical properties need to be investigated to ensure that the expected benefits are sustained over the lifespan of the building. More research is needed to assess DSF performance variations and optimise the design for different climatic contexts and to understand the impact of Double-Skin Façades on occupant comfort during different seasons, including winter. Factors such as thermal comfort, visual comfort, acoustic performance and user preferences should be thoroughly investigated to create designs that meet both energy and comfort objectives.

Author Contributions: Conceptualization, S.M. and C.L.; methodology, S.M. and M.P.; investigation C.L.; writing—original draft preparation, C.L.; writing—review and editing, C.L., S.D.L., M.P. and S.M.; supervision, S.M. and M.P. All authors have read and agreed to the published version of the manuscript.

Funding: This research was developed within the framework of the Italian Research Project "Soluzioni integrate energetiche e antisismiche per costruzioni edili", under the PON action "Dottorati Innovativi con caratterizzazione Industriale", funded by the Italian Ministry for the Economic Development—CUP: D92F16000360003.

Institutional Review Board Statement: Not applicable.

Informed Consent Statement: Not applicable.

Data Availability Statement: Not applicable.

Conflicts of Interest: The authors declare no conflict of interest.

References

1. Mazzarella, L. Energy Retrofit of Historic and Existing Buildings. The legislative and Regularity Point of View. *Energy Build* **2015**, *95*, 23–31. [CrossRef]
2. European Environment Agency. *The European Environment—State and Outlook 2020. Knowledge for the Transition to a Sustainable Europe*; European Union: Brussel, Belgium, 2019; pp. 1–499.
3. Buildings Performance Institute Europe (BPIE). *Europe's Building under the Microscope. A Country-by-Country Review of the Energy Performance of Buildings*; Buildings Performance Institute: Brussel, Belgium, 2011; pp. 1–132.
4. UN Environment and International Energy Agency. *Towards a Zero-Emission, Efficient, and Resilient Buildings and Construction Sector. Global Status Report 2017*; UN Environment: Nairobi, Kenya, 2017; pp. 1–48.
5. Directive 2002/91/EC of the European Parlament and of the Council of 16 December 2002 on the Energy Performance of Buildings. *Off. J. Eur. Communities*. 2003. Available online: http://data.europa.eu/eli/dir/2002/91/oj (accessed on 23 January 2023).
6. Directive 2010/31/EU of the European Parlament and of the Council of 19 May 2010 on the Energy Performance of Buildings (Recast). *Off. J. Eur. Union*. 2010. Available online: http://data.europa.eu/eli/dir/2010/31/oj (accessed on 23 January 2023).
7. Directive 2012/27/EU of the European Parlament and of the Council of 25 October 2012 on Energy Efficiency, Amending Directives 2009/125/EC and 2010/30/EU and Replaing Directives 2004/8/EC and 2006/32/EC. *Off. J. Eur. Union*. Available online: http://data.europa.eu/eli/dir/2012/27/oj (accessed on 23 January 2023).
8. Buildings Performance Institute Europe (BPIE). *State of the Building Stock-Briefing*; Buildings Performance Institute: Brussel, Belgium, 2012; pp. 1–2.
9. Oh, M.; Lee, C.; Park, J.; Lee, K.; Tae, S. Evaluation of Energy and Daylight Performance of Old Office Buildings in South Korea with Curtain Walls Remodeled Using Polymer Dispersed Liquid Crystal (PDLC) Films. *Energies* **2019**, *12*, 3679. [CrossRef]
10. Moreno, Á.; Chemisana, D.; Vaillon, R.; Riverola, A.; Solans, A. Energy and Luminous Performance Investigation of an OPV/ETFE Glazing Element for Building Integration. *Energies* **2019**, *12*, 1870. [CrossRef]
11. Sethi, M.; Lamb, W.; Minx, J.; Creutzig, F. Climate change mitigation in cities: A systematic scoping of case studies. *Environ. Res. Lett.* **2020**, *15*, 1748–9326. Available online: https://iopscience.iop.org/article/10.1088/1748-9326/ab99ff (accessed on 27 May 2023). [CrossRef]
12. Kabisch, N.; Frantzeskaki, N.; Pauleit, S.; Naumann, S.; McKenna, D.; Artmann, M.; Haase, D.; Knapp, S.; Korn, H.; Stadler, J.; et al. Nature-based solutions to climate change mitigation and adaptation in urban areas: Perspectives on indicators, knowledge gaps, barriers, and opportunities for action. *Ecol. Soc.* **2016**, *21*, 239. [CrossRef]
13. Ricci, R.; Vitali, D.; Montelpare, S. An innovative wind-solar hybrid street light: Development and early testing of a prototype. *Int. J. Low-Carbon Technol.* **2015**, *10*, 420–429. [CrossRef]
14. Kim, D.; Cox, S.J.; Cho, H.; Yoon, J. Comparative investigation on building energy performance of double skin façade (DSF) with interior or exterior slat blinds. *J. Build. Eng.* **2018**, *20*, 411–423. [CrossRef]
15. Lucchino, E.C.; Gennaro, G.; Favoino, F.; Goia, F. Modelling and validation of a single-storey flexible double-skin façade system with a building energy simulation tool. *Build. Environ.* **2022**, *226*, 109704. [CrossRef]
16. Jankovic, A.; Goia, F. Characterization of a naturally ventilated double-skin façade through the design of experiments (DOE) methodology in a controlled environment. *Energy Build.* **2022**, *263*, 112024. [CrossRef]
17. Matour, S.; Garcia-Hansen, V.; Omrani, S.; Hassanli, S.; Drogemuller, R. Thermal performance and airflow analysis of a new type of Double Skin Façade for warm climates: An experimental study. *J. Build. Eng.* **2022**, *62*, 105323. [CrossRef]
18. Bugenings, L.A.; Schaffer, M.; Larsen, O.K.; Zhang, C. A novel solution for school renovations: Combining diffuse ceiling ventilation with double skin façade. *J. Build. Eng.* **2022**, *49*, 104026. [CrossRef]
19. Ballestini, G.; De Carli, M.; Masiero, N.; Tombola, G. Possibilities and limitations of natural ventilation in restored industrial archaeology buildings with a double-skin façade in Mediterranean climates. *Build. Environ.* **2005**, *40*, 983–995. [CrossRef]
20. Catto Lucchino, E.; Gelesz, A.; Skeie, K.; Gennaro, G.; Reith, A.; Serra, V.; Goia, F. Modelling double skin façades (DSFs) in whole-building energy simulation tools: Validation and inter-software comparison of a mechanically ventilated single-story DSF. *Build. Environ.* **2021**, *199*, 107906. [CrossRef]
21. Poirazis, H. *Double Skin Façades: A Literature Review*; Lund University, Lund Institute of Technology Department of Architecture and Built Environment: Lund, Sweden, 2006; pp. 1–252.
22. Uuttu, S. *Study of Current Structures in Double-Skin Façades*; Helsinki University of Technology: Espoo, Finland, 2001.
23. Heimrath, R.; Hengsberger, H.; Mach, T.; Streicher, W.; Waldner, R.; Flamant, G.; Loncour, X.; Guarracino, G.; Erhorn, H.; Erhorn-Kluttig, H.; et al. BESTFAÇADE. Best Practice for Double Skin Façades. WP 1 Report "State of the Art". 2005, pp. 1–151. Available online: http://www.bestfacade.com/pdf/downloads/Bestfacade_WP1_Report.pdf (accessed on 27 May 2023).
24. Zhai, Z.; Chen, Q.; Haves, P.; Klems, J.H. On Approaches to Couple Energy Simulation and Computational Fluid Dynamics Programs. *Build Environ.* **2002**, *37*, 857–864. [CrossRef]
25. Lops, C.; Germano, N.; Ricciutelli, A.; D'Alessandro, V.; Montelpare, S. Naturally Ventilated Double Skin Façades: Comparisons Between Different CFD Models. *Math. Model. Eng. Probl.* **2021**, *8*, 837–846. [CrossRef]
26. Germano, N.; Lops, C.; Montelpare, S.; Camata, G.; Ricci, R. Determination of Wind Pattern Inside an Urban Area Through a Mesoscale-Microscale Approach. *Math. Model. Eng. Probl.* **2020**, *7*, 515–519. [CrossRef]
27. Chan, A.L.S.; Chow, T.T.; Fong, K.F.; Lin, Z. Investigation on Energy Performance of Double Skin Façade in Hong Kong. *Energy Build.* **2009**, *41*, 1135–1142. [CrossRef]

28. Gratia, E.; De Herde, A. The Most Efficient Position of Shading Devices in a Double-Skin Façade. *Energy Build.* **2007**, *39*, 364–373. [CrossRef]
29. Joe, J.W.; Choi, W.J.; Huh, J.H. Operation Strategies for an Office Building Integrated with Multi-Story Double Skin Façades in the Heating Season. In Proceedings of the Building Simulation 2011: 12th Conference of International Building Performance Simulation Association, Sydney, Austrilia, 14–16 November 2011.
30. Ding, W.; Hasemi, Y.; Yamada, T. Natural Ventilation Performance of a Double-Skin Façade With a Solar Chimney. *Energy Build.* **2005**, *37*, 411–418. [CrossRef]
31. Gontikaki, M. Optimization of a Solar Chimney Design to Enhance Natural Ventilation in a Multi-Storey Office Building. In Proceedings of the 10th International Conference for Enhanced Building Operations, Kuwait, 26–28 October 2010.
32. Abraham, S.B.; Ming, T.Z. Numerical Analysis on The Thermal Performance of a Building With Solar Chimney and Double Skin Façade in Tropical Country. *IOP Conf. Ser. Mater. Sci. Eng.* **2018**, *453*, 012030. [CrossRef]
33. Stec, W.J.; van Paassen, A.H.C.; Maziarz, A. Modelling the Double Skin Façade with Plants. *Energy Build.* **2005**, *37*, 419–427. [CrossRef]
34. De Gracia, A.; Navarro, L.; Castell, A.; Cabeza, L.F. Energy Performance of Ventilated Double Skin Façade with PCM Under Different Climates. *Energy Build.* **2015**, *91*, 37–42. [CrossRef]
35. Johny, A.E.; Shanks, K. Optimization of Double-Skin Facades for High-Rise Buildings in Hot Arid Climates. *Int. J. Environ. Sustain.* **2018**, *7*, 88–100. [CrossRef]
36. Gaillard, L.; Giroux-Julien, S.; Ménézo, C.; Pabiou, H. Experimental Evaluation of a Naturally Ventilated PV Double-Skin Building Envelope in Real Operating Conditions. *Sol Energy.* **2014**, *10*, 223–241. [CrossRef]
37. Athienitis, A.K.; Buonomano, A.; Ioannidis, Z.; Kapsis, K.; Stathopoulos, T. Double Skin Façades Integrating Photovoltaics and Active Shadings: A Case Study for Different Climates. In Proceedings of the 1st International Conference on Building Integrated Renewable Energy Systems, Dublin, Ireland, 6–9 March 2017.
38. Luo, Y.; Zhang, L.; Wang, X.; Xie, L.; Liu, Z.; Wu, J.; Zhang, Y.; He, X. A comparative study on thermal performance evaluation of a new double skin façade system integrated with photovoltaic blinds. *Appl. Energy* **2017**, *199*, 281–293. [CrossRef]
39. Jayathissa, P.; Luzzatto, M.; Schmidli, J.; Hofer, J.; Nagy, Z.; Schlueter, A. Optimising Building Net Energy Demand With Dynamic BIPV Shading. *Appl. Energy* **2017**, *202*, 726–735. [CrossRef]
40. Moon, K.S. Structural Design of Double Skin Facades as Damping Devices for Tall Buildings. *Procedia. Eng.* **2011**, *14*, 1351–1358. [CrossRef]
41. Zhang, R. SMART BUILDINGS: An Integrative Double Skin Façade Damper System For Structural Safety and Energy Efficiency. Ph.D. Dissertation, University of New Hampshire, Durham, UK, 2017.
42. Takeuchi, T.; Yasuda, K.; Iwata, M. Studies on Integrated Building Facade Engineering with High-Performance Structural Elements. *IABSE Symp. Rep.* **2006**, *92*, 33–40. [CrossRef]
43. Scuderi, G. Adaptative Exoskeleton for the Integrated Retrofit of Social Housing Buildings. Ph.D. Dissertation, University of Trento, Trento, Italy, February 2016.
44. Passoni, C.; Belleri, A.; Marini, A.; Riva, P. Sustainable Restoration of Post-WWII European Reinforced Concrete Buildings. In Proceedings of the 16th World Conference on Earthquake, Santiago, Chile, 9–13 January 2017.
45. Labò, S. Holistic Sustainable Renovation of Post-World War II Reinforced Concrete Building Under A Life Cycle Perspective by Means Diagrid Exoskeletons. Ph.D. Dissertation, University of Bergamo, Bergamo, Italy, March 2019.
46. Khabir, S.; Vakilinezhad, R. Energy and thermal analysis of DSF in the retrofit design of office buildings in hot climates. *Archit. Eng. Des. Manag.* **2022**, 1–23. [CrossRef]
47. Ascione, F.; Bianco, N.; Iovane, T.; Mastellone, M.; Mauro, G.M. The evolution of building energy retrofit via double-skin and responsive façades: A review. *Solar Energy* **2021**, *224*, 703–717. [CrossRef]
48. Sarihi, S.; Mehdizadeh Saradj, F.; Faizi, M. A Critical Review of Façade Retrofit Measures for Minimizing Heating and Cooling Demand in Existing Buildings. *Sustain. Cities Soc.* **2021**, *64*, 102525. [CrossRef]
49. Montelpare, S.; D'Alessandro, V.; Lops, C.; Costanzo, E.; Ricci, R. A Mesoscale-Microscale approach for the energy analysis of buildings. *J. Phys. Conf. Ser.* **2019**, *1224*, 012022. [CrossRef]
50. Peel, M.C.; Finlayson, B.L.; McMahon, T.A. Updated World Map of the Köppen-Geiger Climate Classification. *Hydrol. Earth Syst. Sci.* **2007**, *11*, 1633–1644. [CrossRef]
51. Fondazione Osservatorio Meteorologico Milano Duomo ETS. Available online: https://www.fondazioneomd.it/climate-network (accessed on 20 January 2019).
52. Giorgi, F.; Jones, C.; Asrar, G.R. Addressing climate information needs at the regional level: The CORDEX framework. *World Meteorol. Organ. Bull.* **2019**, *58*, 175.
53. Bubnovà, R. Integration of the Fully Elastic Equations Cast in the Hydrostatic Pressure Terrain-Following Coordinate in the Framework of the ARPAGE/Aladin NWP SYstem. *Am. Meteorol. Soc.* **1995**, *123*, 515–535. [CrossRef]
54. Farda, A.; Déué, M.; Somot, S.; Horányi, A.; Spiridonov, V.; Tóth, H. Model ALADIN as Regional Climate Model for Central and Eastern Europe. *Stud. Geophys. Geod.* **2010**, *54*, 313–332. [CrossRef]
55. Horanyi, A.; Ihasz, I.; Radnoti, G. ARPEGE/ALADIN: A numerical Weather Prediction Model for Central-Europe with the Participation of the Hungarian Meteorological Service. *IdoJárás* **1996**, *100*, 277–301.

56. Huth, R.; Mládek, R.; Metelka, L.; Sedlák, P.; Huthová, Z.; Kliegrová, S.; Kyselý, J.; Pokorna, L.; Helenka, T.; Janoušek, M. On the Integrability of Limited-Area Numerical Weather Prediction Model ALADIN over Extended Time Periods. *Stud. Geophys. Geod.* **2003**, *47*, 863–873. [CrossRef]
57. *EN ISO 15927-4:2005*; Hygrothermal Performance of Buildings—Calculation and Presentation of Climatic Data—Part 4: Hourly Data for Assessing the Annual Energy Use for Heating and Cooling. ISO: Geneva, Switzerland, 2005.
58. Cebecauer, T.; Suri, M. Typical Meteorological Year Data: SolarGIS Approach. *Energy Procedia* **2015**, *69*, 1958–1969. [CrossRef]
59. *UNI EN 16798-1:2019*; Energy Performance of Buildings–Ventilation for Buildings-Part 1: Indoor Environmental Input Parameters for Design and Assessment of Energy Performance of Buildings Addressing Indoor Air Quality, Thermal Environment, Lighting and Acoustics—Module M1-6. Ente Nazionale Italiano di Unificazione (UNI): Milano, Italy, 2019.

applied sciences

MDPI

Article

Building Energy Flexibility Assessment in Mediterranean Climatic Conditions: The Case of a Greek Office Building

Georgios Chantzis *, Effrosyni Giama and Agis M. Papadopoulos

Process Equipment Design Laboratory, Department of Mechanical Engineering, Aristotle University of Thessaloniki, 54124 Thessaloniki, Greece; fgiama@auth.gr (E.G.); agis@auth.gr (A.M.P.)
* Correspondence: chantzis@auth.gr

Abstract: The EU energy and climate policy has set quantitative goals for decarbonization based on the energy efficiency and the evolution of energy systems. The utilization of demand side flexibility can help towards this direction and achieve the target of higher levels of penetration in regard to intermittent renewable energy production and carbon emission reduction. This paper presents a simulation-based assessment of thermal flexibility in a typical office building in Greece, which is a representative Mediterranean country. The use of variable speed heat pumps coupled with hydronic terminal units was evaluated. The research focused mainly on the evaluation of energy flexibility offered by energy stored in the form of thermal energy by utilizing the building's thermal mass. The demand response potential under hourly CO_{2eq} intensity and energy prices was investigated. The flexibility potential was evaluated under different demand response strategies, and the effect of demand response on energy consumption, operational costs, CO_{2eq} emissions and thermal comfort was analyzed and discussed. The results showed that both control strategies based on both the CO_{2eq}intensity signal and spot price signal have, in some cases, the potential for cost and emission savings, and in other cases, the potential to depreciate in terms of emissions and cost the increase of energy consumption due to load shifting.

Keywords: energy efficiency; CO_{2eq} intensity; energy prices; energy flexibility; predictive rule-based control; office building; energy simulation; TRNSYS

1. Introduction

In recent decades, consumer demand for greater living standards has increased along with global electricity use [1]. This rise in global energy consumption, forecasted declines in fossil fuel supply and signs of escalating consequences of global warming have all contributed to an increased focus in RES research. Because of the intrinsic unpredictability that characterizes RES, instability risk in the energy network grows as RES become more prevalent in covering demand [2]. The necessity to control demand and guarantee flexible energy usage is reinforced by the requirement for network stability. Buildings are anticipated to be a key player in this change by supplying the power grid with their energy flexibility to help meet the demands of energy networks [2]. To this end, and in line with the updated European regulations requiring a low-carbon built environment [3], it is imperative that buildings are designed with energy efficiency and decarbonization in mind. To this end, flexible operation and smart energy use in buildings are seen as key strategies for mitigating greenhouse gas emissions and ensuring resilient energy system operations [4].

Buildings can offer their energy flexibility in multiple ways. One of the main methods for offering flexibility is thermal energy storage. The thermal mass of a building can be used for load shifting when an energy deficit is detected since it can store thermal energy for short periods of time [5].

Office buildings are seen as the best option for implementing DR programs due to their high energy usage and generally superior automation infrastructure compared to residential buildings [6]. Commercial HVAC, lighting and other electrical equipment in office buildings can adjust their operational profile to support the energy grid.

Several researchers have come to appreciate the importance of buildings as a source of energy flexibility to the energy grid and tried to quantify the demand side flexibility of the building sector by implementing DR in buildings and their energy systems. In their research [7], they tried to quantify the flexibility that a building with CHP and energy storage can offer, by considering the magnitude of a CHP operation's delay or forced operation time. Arteconi et al., 2019 [8] introduced a multicriteria indicator and tried to quantify the energy flexibility of a residential building with AWHP and TES. De Coninck and Helsen in 2016 [9] introduced cost curves as a metric for measuring flexibility that is applicable to a wide variety of building typologies and energy systems. In [10], power shifting indexes were employed to evaluate an office building's flexibility under model predictive control. Masy et al., 2015 [11], developed a generic flexibility assessment methodology that was applied to various building typologies equipped with heat pumps. In [12,13], indicators that cover time, size and cost were used to quantify the energy storage in building thermal mass as a mean of flexibility in different building typologies. D'Ettorre et al., 2019 [14], implemented different DR scenarios to assess the flexibility of a building equipped with a hybrid heating system with thermal energy storage located in Italy. Several temperature regulation strategies were implemented in detailed resistance–capacitance models of various buildings in order to quantify their demand response potential in [15].

Several researches studied the impact of DR control on equivalent carbon dioxide emissions. The authors of [16–18] tested the implementation of MPC in residential buildings. In every case, a reduction in dioxide emissions was observed. Other researchers studied the implementation of two different CO_2 MPCs to a large commercial building with two different HVAC settings and found decreases in emissions, especially in the case of exploiting renewable energy [19]. Vigna et al., 2021 [20], assessed building clusters with different thermal mass characteristics, which consisted of residential buildings with ideal heating systems, under CO_2 forcing factor smart control. Only two papers were found in the literature that test the implementation of RBC, either by considering CO_2 forcing factors [21] or just studying the impact of economic RBC on carbon emissions [22].

The purpose of this research was to evaluate the energy flexibility of an office building in the Mediterranean area. The study used electricity market data and examined two types of PRBCs: economic PRBC and environmental PRBC. To achieve this goal, classic energy, and environmental and economic indicators were used, as well as energy flexibility indicators. The key innovations of this paper are the initial assessment of energy flexibility in terms of environmental performance, which is quite limited in the literature, especially for office buildings. In addition, PRBC controls were implemented with the aim of reducing energy costs and the carbon footprint using data from the energy market. Another contribution is the detailed modelling and simulation of both the building and its heating system. In most cases when a detailed building is simulated, the HVAC systems are simplified, and in cases when a detailed model of HVAC system is realized, a simplified building model is used, like a resistance–capacitance model. Finally, a holistic analysis was conducted, and the flexibility was evaluated with multiple indicators to produce more detailed results.

The remainder of the paper is structured as follows: The weather and electricity market data, the building, the operating conditions, the installed HVAC systems, and the modelling and simulation methodology, are all described in Section 2. Section 3 includes the description of the implemented control algorithms. Section 4 describes the performance indicators. The findings under the various demand response strategies are presented and discussed in Section 5, before the paper is concluded in Section 6.

2. Methodology

With the target of a building energy flexibility evaluation, a building and its heating system were modeled. As part of the study, an office building in the area of Thessaloniki was selected. Most studies investigate the energy flexibility that residential buildings can offer. This building is building D of the polytechnic school of AUTH. It is a building that is characterized by various uses of spaces with it predominantly being used as an office. The building in its current state meets its heating needs using gas-fired boilers. For the purposes of the study, heating was electrified and the gas-fired boilers were replaced by heat pumps.

2.1. Weather Data

Thessaloniki's mild, warm, subtropical Mediterranean climate (Koppen–Geiger classification: Csa) results in relatively low heating and cooling requirements. Because of the data availability regarding the electricity market, the simulations were conducted using 2022 weather data. The weather for 2022 was not typical weather for Thessaloniki, as it was relatively hot throughout the year with less heating degree days compared to previous years.

In Figure 1, we notice that the average monthly temperature was high throughout the year, with a minimum average monthly temperature of 6 degrees in January. The average monthly global horizontal irradiance remained above 80 W/m^2 throughout the year reaching the maximum value of about 330 W/m^2 in July (Figure 2). High levels of relative humidity were present compared to average Greece levels, with a minimum value of about 12%, a maximum value reaching around 100%, and average around 50%.

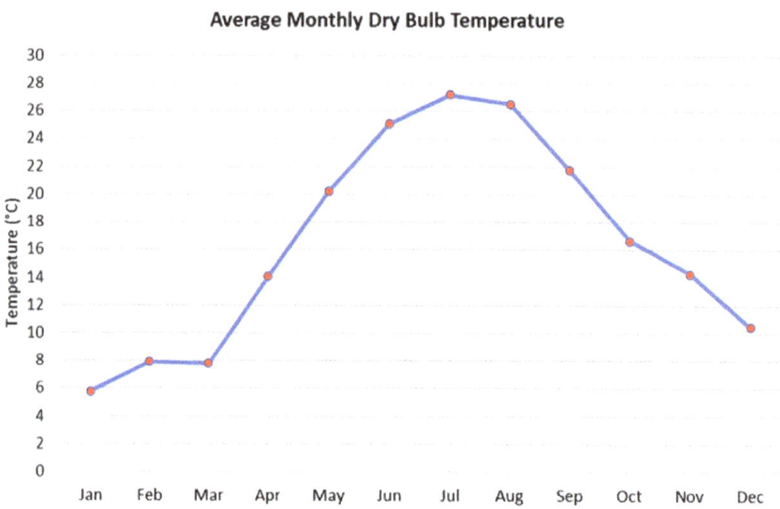

Figure 1. Average monthly dry bulb temperature (2022 weather data).

2.2. Electricity Market Data

For the implementation of the control methods for the heating system, historical data of the Greek energy market were used [23]. The price data refer to hourly wholesale market prices for the whole year. We noticed in Figure 3 that prices varied between approximately 50 EUR/MWh and 900 EUR/MWh. The lowest prices were observed during the winter months and especially in the months of October, January and February, while the highest prices were observed during the summer period and especially in the month of August, due to the particularly increased demand because of tourism. It was also evident that the lowest prices occurred during nighttime (24:00–6:00), while the price seemed to be maximized during the early morning hours (7:00–10:00) and during the afternoon hours (18:00–22:00). On the contrary, in terms of dioxide emissions [24], the participation of RES and especially PV was evident since the minimum daily values were observed during the hours of sunshine (9:00–17:00) and the minimum values on an annual basis were observed during the transitional periods (autumn and spring), both due to the increased production of PV energy compared to the winter season and due to the reduced demand compared to the summer season (Figure 4). In this case, the values ranged between 100 and 550 gCO_2/kWh.

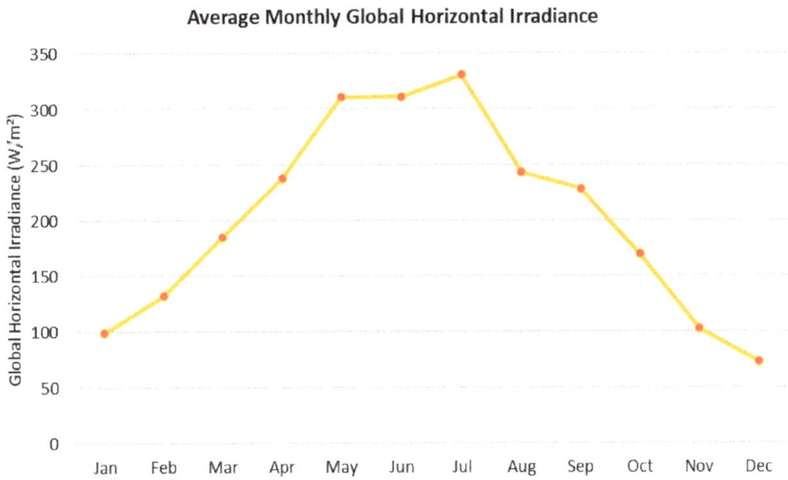

Figure 2. Average monthly global horizontal irradiance (2022 weather data).

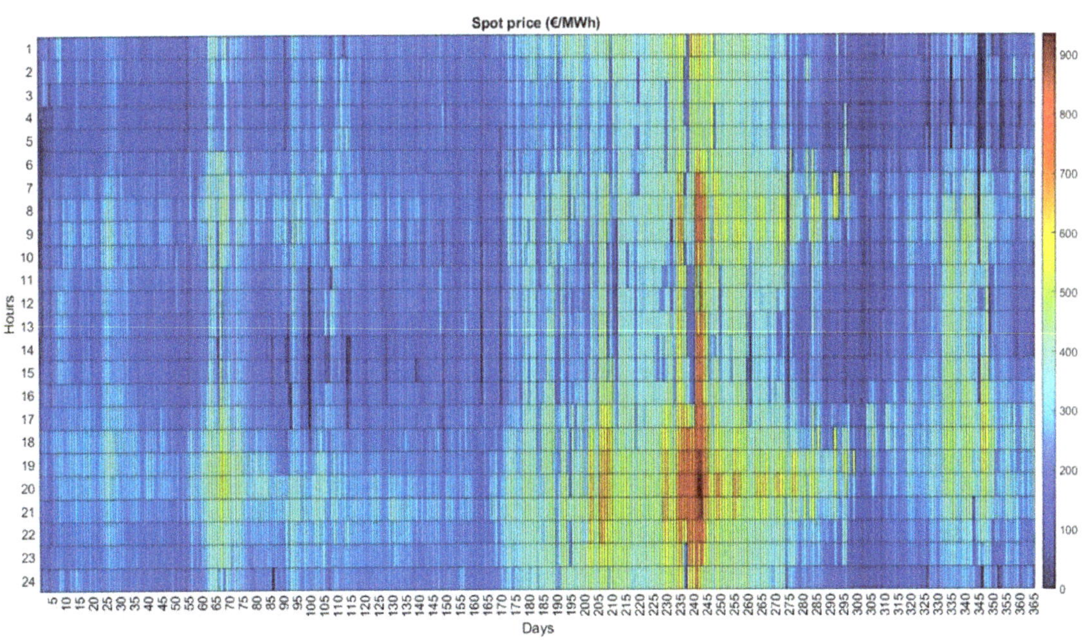

Figure 3. Heatmap plot of the hourly spot price of electricity (Greece 2022, [23]).

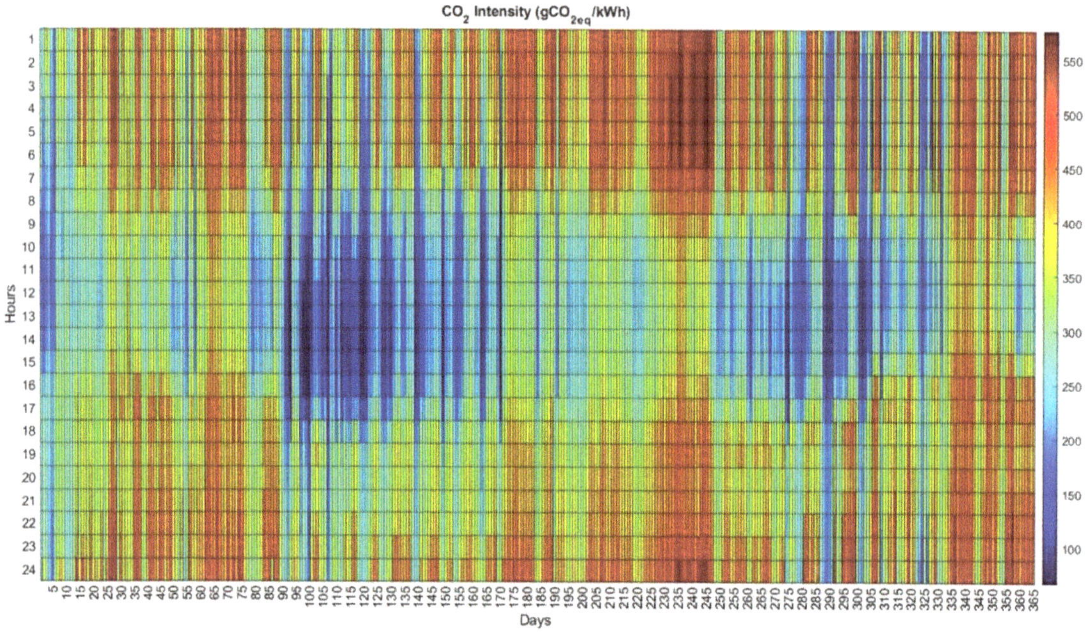

Figure 4. Heatmap plot of the hourly CO_{2eq} intensity of electricity (Greece 2022, [23]).

2.3. Building and Thermal Zoning

Building D′ of the Aristotle University of Thessaloniki's Faculty of Engineering, which is located between Egnatia and 3rd September streets, is the building under the investigation in this study. Building D′, which has ten floors (including the ground floor), houses the secretariats of the various departments as well as the offices of the faculty's teaching and research staff. Some laboratories are also housed in the basement of the building.

The building's overall volume and surface area are 44,671,197 m^3 and 10,595,451 m^2, respectively, of which 29,017,462 m^3 and 8,167,691 m^2 are heated spaces. The structure has not undergone renovation since it was built, and it is thermally insulated in accordance with the 1979 Greek law on thermal insulation [25]. For the building under consideration, the entire building's external surface that is exposed to air is A = 8714.90 m^2. The window-to-wall ratio is roughly 16%, and the overall area of glazing is 1147.78 m^2.

To achieve a more accurate energy analysis, the building under study was divided into 12 thermal zones. Each of the 10 floors comprised a different conditioned thermal zone and the basement and roof were unconditioned thermal zones. On floors 1–9, most of the surface (70%) is occupied by offices, while auxiliary spaces and hallways occupy the remaining area. The zone that corresponds to the ground level of the building is made up of 70% ancillary spaces and 30% offices and secretariats.

2.4. Operating Conditions

There are two different types of uses for this particular building. The main use is as offices, and the second is the use of the hallways and supporting areas. The internal operating conditions of each thermal zone were determined using Technical Chamber of Greece guidelines (T.O.T.E.E. 20701-1), the Greek regulation on buildings' energy performance (KENAK), and the ASHRAE STANDARD 90.1 [26]. In order to comply with the Energy Performance of Buildings Directive (EPBD), the Greek regulation was recently amended [27]. The selection was made with the objective of offering adequate levels of thermal comfort to all inhabitants [28], and the outcomes are outlined in Table 1. The heating setpoint was taken from the ASHRAE STANDARD, as the Greek regulation's setpoint of 20 °C was judged insufficient.

Table 1. Internal operating conditions of thermal zones (imported from [29]).

	Offices	Ancillary Spaces
Working hours	10 h/day	
Days of operation	5 days/week	
Operating period	12 months	
Heating setpoint	21 °C	18 °C
Number of users	10 people/100 m^2	-
Fresh air required	3 $m^3/h/m^2$	2.6 $m^3/h/m^2$
Lighting level	500 lx	100 lx
Installed lighting power	14 W/m^2	2.8 W/m^2
People heat gain	80 W/person	-
Equipment power	15 W/m^2	-

2.5. Building Shell and HVAC Systems

2.5.1. Building Shell

The reference building has been thermally insulated in accordance with the Building Regulations for Thermal Insulation [25], as indicated in Section 2.1. Total thermal permeability values for all exposed structural components were determined for both the reference and case A building and are presented in Table 2.

2.5.2. HVAC Systems

Currently, the building is heated by two 90%-efficient natural gas boilers, a high-temperature water distribution system, and a moderately inefficient pipe insulation distribution system. A compensatory mechanism built within the distribution network handles partial loads and boosts the network's efficiency to 96%. According to the manufacturer's technical specifications, the total rated power of the boiler systems is 800 kW. Water radiators (AKAN type), complemented with thermostatic controls, make up the terminal heating units, which operate at an efficiency of 89% [28]. In order to be able to implement the DR scenarios and assess the building's demand response potential, during the modeling process, the existing heating system was replaced with an electric system. The existing system was replaced with a system consisting of a VSHP, a low temperature distribution network and an emission system consisting of hot water radiators.

2.6. Modeling and Simulation

Energy modeling was completed in TRNSYS 18 after the building geometry was modeled using the software environment of SketchUp 2017 [30]. The model was made utilizing details on the dimensions, architectural elements, orientation, and shadings of the real structure (Figures 5 and 6). Every structural parameter, as well as operational parameters, setpoints and schedules, and internal loads, were set up following the import of the building model into TRNSYS. Using the weather files, the essential climatic information (temperature, relative humidity, radiation, etc.) was added to complete the building model. Detailed models of both the building and its heating systems were realized in TRNSYS. The TRNSYS HVAC library only includes single and double speed heat pumps. In order to model the VSHP, the methodology proposed by Toffanin et al., 2019 was used [31].

Table 2. Thermal characteristics of the building shell.

<table>
<tr><th></th><th>Component</th><th>Materials</th><th>U-Value (W/(m^{2}K)</th></tr>
<tr><td rowspan="4">Reference building</td><td>External walls</td><td>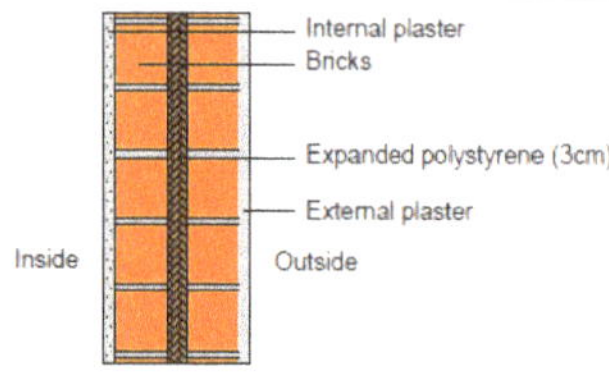
</td><td>0.71</td></tr>
<tr><td>Roof</td><td>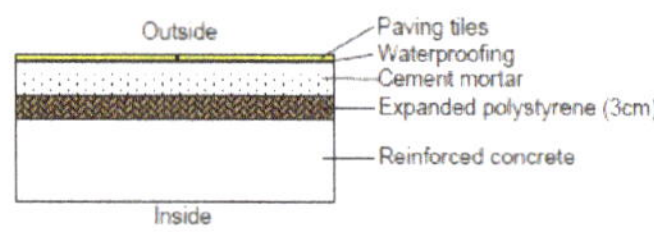
</td><td>0.77</td></tr>
<tr><td>Ground floor</td><td>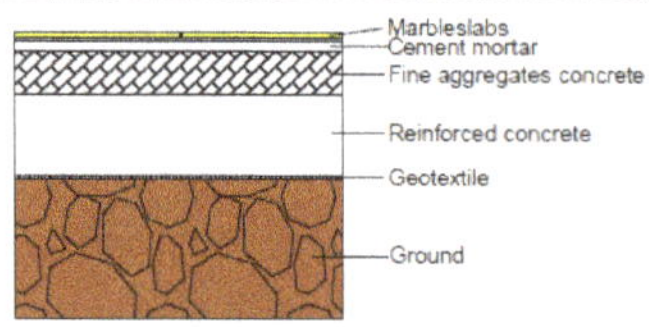
</td><td>3.11</td></tr>
<tr><td>Windows</td><td>double-glazed, aluminum frame</td><td>2.82</td></tr>
<tr><td rowspan="4">Case A building</td><td>External walls</td><td>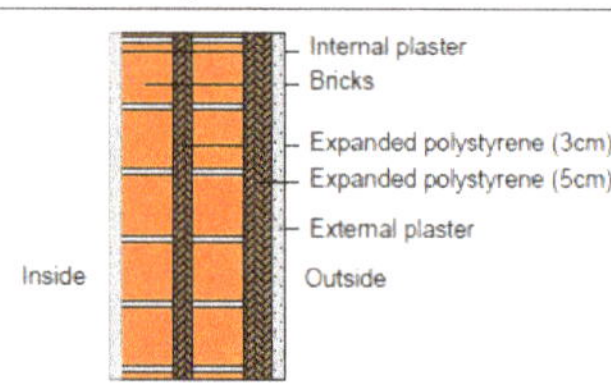
</td><td>0.35</td></tr>
<tr><td>Roof</td><td>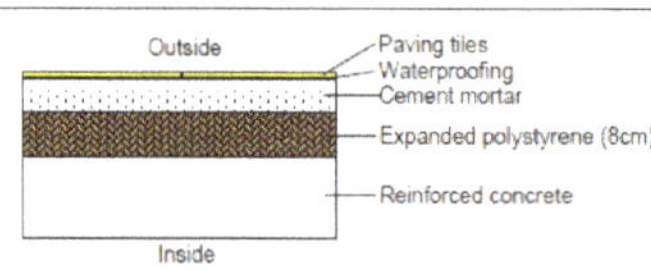
</td><td>0.37</td></tr>
<tr><td>Ground floor</td><td>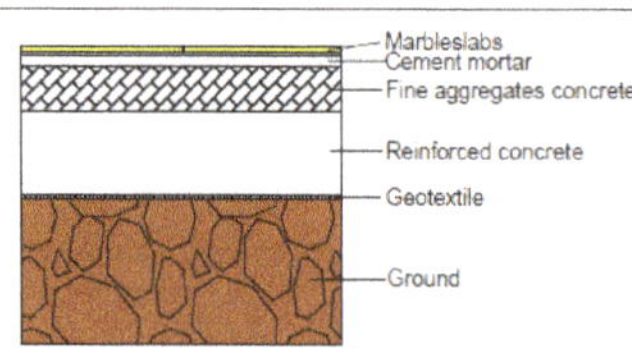
</td><td>3.11</td></tr>
<tr><td>Windows</td><td>Argon filler double-glazing, aluminum frame with thermal break</td><td>1.57</td></tr>
</table>

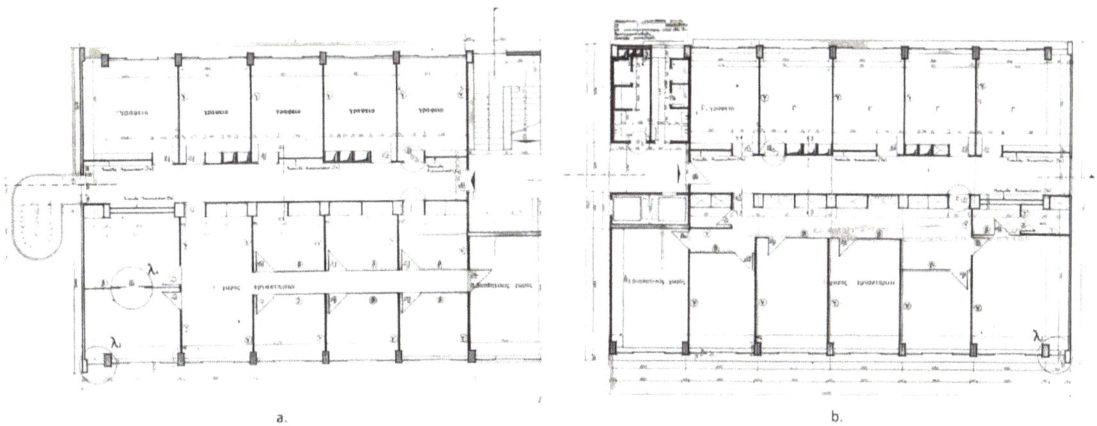

Figure 5. Building D' typical floor plan: (a). left side, and (b). right side.

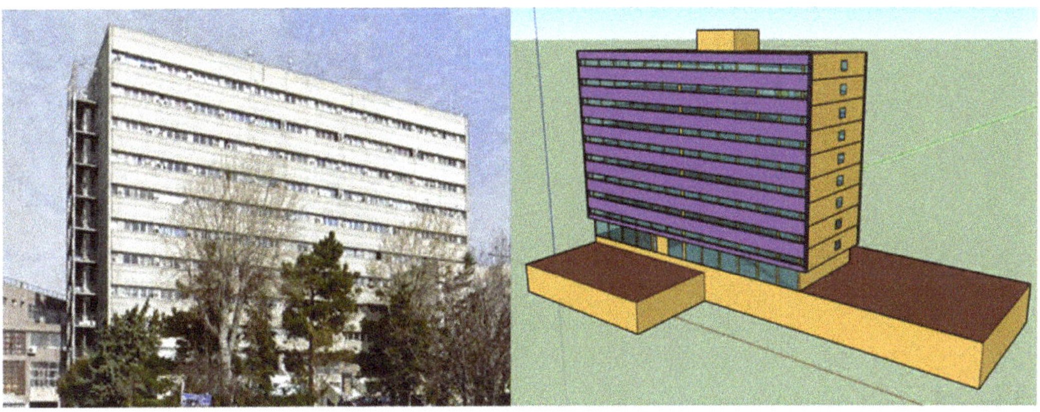

Figure 6. Front (left) and 3D (right) view of the simulated building.

3. Control Methods

3.1. Reference

The reference scenario set a constant temperature setpoint from 7:00 to 17:00 (working hours) of 21 °C and a setback temperature of 15 °C from 17:00 to 7:00. These setpoints are the most commonly used setpoints for office buildings during the heating period in Greece. During the rest control strategies, the setpoints are increased or decreased by 2 °C so that adequate levels of thermal comfort are maintained. The first DR control scenario is a predefined control strategy and is only used to assess the ability of the building to use its thermal mass for thermal energy storage. The second DR case takes into account power price fluctuations over time and tries to increase electricity consumption for heating during low electricity prices and decrease consumption during high electricity prices, respectively. The third DR scenario works in a similar manner as the second scenario but considers CO_{2eq} intensity instead of electricity prices.

3.2. Predefined Control Strategy

The predefined control scenario was only used to evaluate the ability of the building to store energy in its thermal mass, according to the KPIs proposed by Reynders et al., 2017 [12]. During the implementation of this control strategy, the heating setpoint is increased to

23 °C from 10:00 to 12:00, which is considered a low demand daytime period, considering the selected building typology. The setpoint was kept stable at 21 °C for the rest of the working hours and the setback was maintained at the same levels of the reference scenario. The purpose of this scenario was to calculate the heat stored during the increased setpoint period and assess the efficiency of storage by calculating the thermal losses occurring during the thermal mass activation.

3.3. Cost PRBC

The Cost PRBC uses the working hours spot price data for each day to determine the daily low price and high price limits. The limits arise from the calculation of average price and standard deviation for each day. Then, the setpoints were set to 23 °C during the hours when the price was lower than the average + SD; at 19 °C during the hours when the price is higher than the average + SD; and at 21 °C for the rest of the working hours. The setback is kept stable at 15 °C. This scenario tries to shift energy use away from peak times and toward off-peak times.

3.4. CO_2 PRBC

The CO_2 PRBC works in a very similar manner with Cost PRBC. The main difference is that instead of energy prices, the CO_{2eq} emissions were used. Again, a high carbon intensity limit and a low carbon intensity limit were calculated for each day. The temperature was kept constant at 23 °C during the times of day when CO_{2eq} emissions are lower than the average + SD; at 19 °C during the hours when the CO_{2eq} emissions is higher than the average + SD; and at 21 °C for the rest of the working hours.

4. Key Performance Indicators (KPIs)

For the evaluation of the different control strategies, various performance indicators were used for the efficient monitoring and control of the systems studied.

- **Energy use for heating**: The electricity consumption of the heat pump during the heating period.
- **Average hourly heat pump power**: The electricity used for heating averaged to one hour period.
- **Coefficient of performance (COP)**: This metric reflects the heat pump's energy efficiency.
- **CO_{2eq} emissions**: CO_{2eq} intensity is used to measure the level of carbon dioxide equivalent emissions caused by the use of the heating system.
- **Cost of energy**: The cost that occurs during the operation of the heating system.
- **Hours of discomfort**: During the operation of the building, some people are not satisfied with the indoor environment thermal conditions. Hours of discomfort are calculated as number of hours that the PMV index value is lower than −0.5 or higher than 1.
- **Storage capacity**: The heat storage capacity of the thermal mass of a building during a DR event [12].
- **Storage efficiency:** The portion of energy that may be stored in the structure and utilized later for energy flexibility and thermal comfort provision [12].
- **Flexibility:** When all of a building's required heating or cooling energy is utilized at the time of day when energy prices or CO_{2eq} intensity are lowest, the Flexibility index, which assumes values from 0% to 100%, is maximized [11].

5. Results and Discussion

5.1. Demand Response Scenario Implementation

Figures 7 and 8 illustrate the implementation of the control scenarios for 3 consecutive working days. The figures depict from top to bottom the average hourly heat pump power, the indoor air temperature, the PMV index, the COP, the electricity market data and the induced temperature setpoints. The results indicate that the average hourly heat pump power reached around 250 kW for the reference scenario and increases during increased temperature setpoints. The electricity market data indicated a large discrepancy between energy prices and dioxide emissions, since during the hours when the maximum electricity prices were observed, the minimum prices of dioxide emissions appeared and vice versa. This happens because energy production from fossil fuels such as lignite, which are the main source of energy, is characterized by low energy costs but increased emission intensity. Conversely, during the day, due to high demand and production from RES and natural gas base units, prices showed an increase and CO_2 emissions decreased significantly. The indoor air temperature and the PMV index values indicate that, during the reference scenario, the heating system cannot efficiently cover the demand with temperatures reaching the setpoint for a few hours each day. On the other hand, during case A, the high internal gains combined with lower heat losses led to cases of slight overheating even for the reference scenario but mostly for the PRBC scenarios when the 23 °C setpoints were forced.

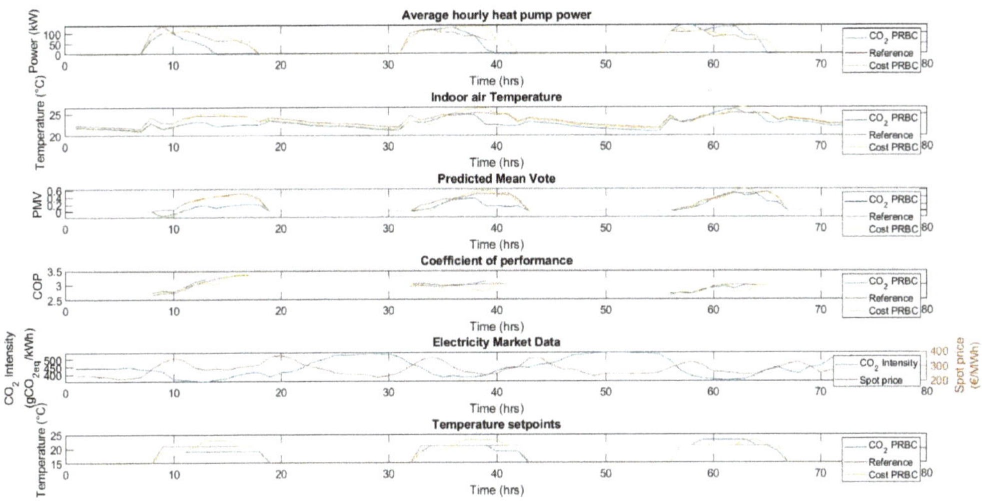

Figure 7. Control scenario implementation for three consecutive days (case A building).

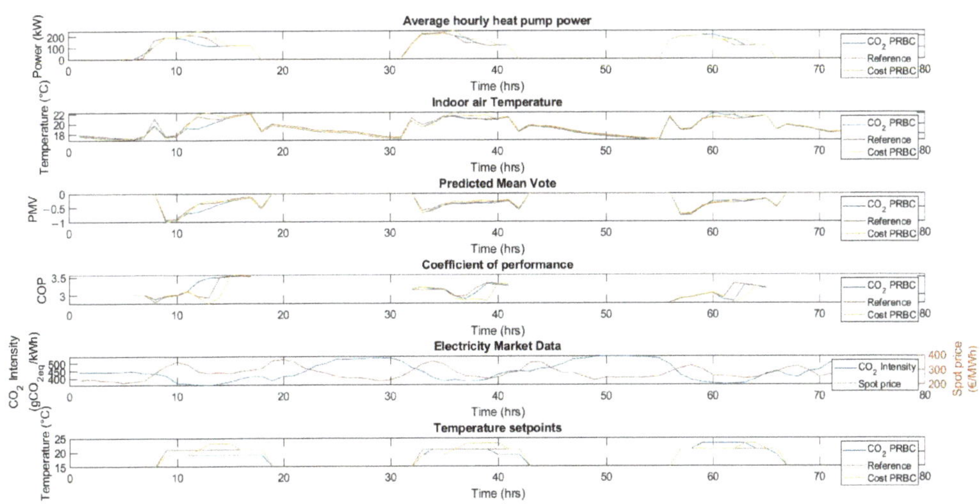

Figure 8. Control scenario implementation for three consecutive days (reference building).

5.2. Evaluation of Thermal Mass Flexibility

In order to evaluate the flexibility that the thermal mass can offer, we used the KPIs proposed by Reynders et al., 2017 [12]. The two main KPIs are storage capacity C_{ADR} and storage efficiency η_{ADR}.

The heat storage capacity of the thermal mass of a building during a DR event can be expressed by the available storage capacity (Equation (1)), while taking into account the building's thermal comfort and varying conditions.

$$C_{ADR} = \int_0^{l_{ADR}} \left(Q_{ADR} - Q_{Ref}\right) dt \quad (1)$$

where $\int Q_{ADR}$ is the amount of energy consumed during the demand response implementation and $\int Q_{REF}$ is the amount of energy consumed in the same period of time, but under the reference control. The difference between the energy consumption is the amount of energy stored in the thermal mass of the building.

The portion of energy that may be stored in the thermal mass and utilized later for energy flexibility and thermal comfort provision is referred to as the storage efficiency (Equation (2)).

$$\eta_{ADR} = 1 - \frac{\int_0^{\infty} \left(Q_{ADR} - Q_{Ref}\right) dt}{\left|\int_0^{l_{ADR}} \left(Q_{ADR} - Q_{Ref}\right) dt\right|} \quad (2)$$

where $\int_0^{l_{ADR}} \left(Q_{ADR} - Q_{Ref}\right) dt$ is the heat storage capacity and $\int_0^{\infty} \left(Q_{ADR} - Q_{Ref}\right) dt$ is the difference between the energy consumption of the DR scenario and the reference scenario until the building returns to steady state conditions. Essentially, this KPI calculates the amount of stored thermal energy that is possible to be recovered and used.

In their research [12], Reynders et al. calculated those two indicators for a residential building and considering stable ambient conditions of 5 °C and zero solar and internal gains. In our case, we evaluated the average value of the same indicators for several days of the year without considering stable outdoor conditions. Reynders et al. used these indicators as a tool to assess a building's flexibility during the design stage. We chose to implement the indexes under real operational conditions in order to better capture the operational flexibility of the building.

The results for each building are presented in Table 3. We can observe the lower capacity of the reference scenario and at the same time the lower storage efficiency compared to the case A. The increased capacity of case A is due to the lower heating power needed before the DR setpoints are implemented, which leaves more power reserve when the heat pump tries to reach the 23 °C setpoint. The increased efficiency is mainly attributed to the decreased heat losses of the upgraded building. Compared to the results of [12], we observed a reduced storage efficiency. This is mainly due to the intermittent schedule of the study building, which leads to cycles of charging and discharging of the heat mass of the building, but also due to the residential building shell's superior thermal properties compared to our building.

Table 3. Thermal mass flexibility evaluation.

	C_{ADR}	η_{ADR}
Unit	kWh	%
Reference	115.6	43.9
Case A	121.9	53.2

5.3. Evaluation of Flexibility in Terms of Cost

The flexibility KPI, proposed by Masy et al., 2015, is used to assess flexibility in terms of cost [11]. The Flexibility index (Equation (3)), which ranges from 0% to 100%, is maximized when all the energy needed is consumed at a time of day when the energy prices are at the lowest levels.

$$Flexibility_{Cost} = \frac{p_{el,max} - p_{el,avg}}{p_{el,max} - p_{el,min}} \quad (3)$$

Figure 9 depicts the results for each building (reference, case A) and each control scenario. We noticed that the reference control scenario presents a cost flexibility of about 47%, both for the reference and the case A building. The flexibility increased in the CO_2 PRBC scenario to about 48% for both building scenarios. As expected, the Cost PRBC led to the highest cost flexibility index, with values around 51%, with the case A building producing a little higher index value. The small increase between scenarios shows the difficulty in shifting high amounts of energy from high prices periods to low price periods and, at the same time, maintaining adequate thermal comfort levels. Still, even this small increase of about 4% during the Cost PRBC implementation shows that this method works as expected and has the potential of increased cost-efficient operation of heating under DR.

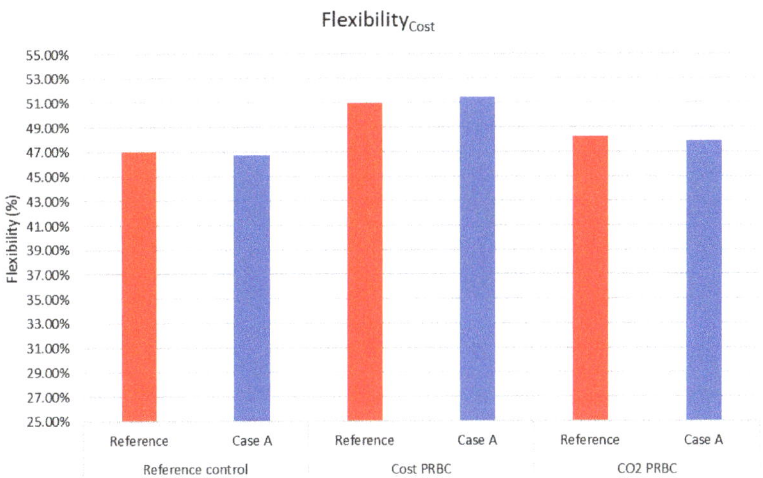

Figure 9. Flexibility index in terms of cost.

In their research, Masy et al. evaluated the flexibility in terms of cost of a residential building with a heat pump for heating, and various levels of building envelope insulation and heat emission systems [11]. The results showed a flexibility index of approximately 78% for the case of hot-water radiators. We noted that the differences with respect to our results are significant. This difference is mainly due to the use of the building, which in our case is a building with limited hours of operation, while the building in Masy et al. had a continuous operation schedule. This fact limits the ability of the building under study to achieve a high value of flexibility since the limited operating hours have an impact both on the ability to charge the thermal mass of the building, but also on the energy cost, which, during the morning hours, showed little variability.

5.4. Evaluation of Flexibility in Terms of CO_{2eq} Emissions

In order to evaluate the flexibility in terms of emissions, we use a modified version of the flexibility KPI proposed by Masy et al., 2015 [11]. The modified Flexibility index (Equation (4)), which ranges from 0% to 100%, is maximized when all the energy needed is consumed at a time of day when the CO_{2eq} emissions are at the lowest levels.

$$Flexibility_{CO2eq} = \frac{em_{el,max} - em_{el,avg}}{em_{el,max} - em_{el,min}} \quad (4)$$

The results for each building (reference, case A) and each control scenario are shown in Figure 10. We noticed that the reference control scenario presented a $CO_{2eq.}$ emission flexibility of about 65% for the reference building and about 64% for the case A building. The flexibility increased in the CO_2 PRBC scenario to about 70% for both buildings. Surprisingly, the cost PRBC presented the highest value of CO_{2eq} emission flexibility (71%), but with a small difference compared to the CO_2 PRBC. The moderate increase between scenarios (about 6%) was attributed to the relatively high variation of CO_{2eq} intensity of electricity throughout the day and showed that PRBC is a promising type of control when the target is reduction of CO_{2eq} emissions.

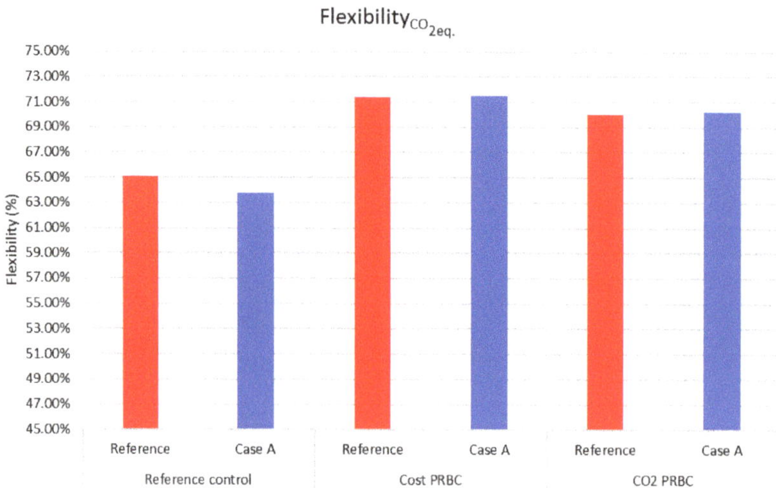

Figure 10. Flexibility index in terms of CO_{2eq}.

5.5. Energy Consumption, Cost of Energy, and Carbon Emission Evaluations

In Figure 11, the monthly energy costs for heating are presented. We can see that the reference building had almost double the energy consumption compared to the case A building, with the highest energy consumption reaching 42 MWh for the reference building and 22 MWh for case A. Additionally, we can notice a larger difference between scenarios regarding monthly energy consumption for the reference building. The reference building under the reference control scenario presented increased energy consumption compared to the PRBC scenarios during January, February and March. The case A building under the reference control presented increased energy consumption compared to the PRBC scenarios only during January. It is evident that implementing DR in the case A building led to increased energy consumption, while implementing DR in the reference building can lead to decreased energy consumption, which probably comes with decreased thermal comfort levels.

Regarding energy costs (Figure 12), we can see that the cost PRBC presented the best overall behavior for the reference building. Specifically, we noticed that during every month, except February and December, the monthly energy costs of the Cost PRBC were lower compared to the other scenarios. In February, the CO_2 PRBC gave the best results regarding cost, which was mainly attributed to the lower energy consumption of CO_2 PRBC during February. In December, the reference control resulted in the lowest costs, which was again attributed to the lower energy consumption of reference scenario. Regarding the case A building, the Cost PRBC proved to be the most efficient scenario during February and March; CO_2 PRBC was the most efficient during January; and the reference control was the most efficient for the rest of the months. This level of difference between the case A building and the reference one, where the CO_2 PRBC was the most cost efficient, can be attributed to the increased energy consumption during DR for the case A building.

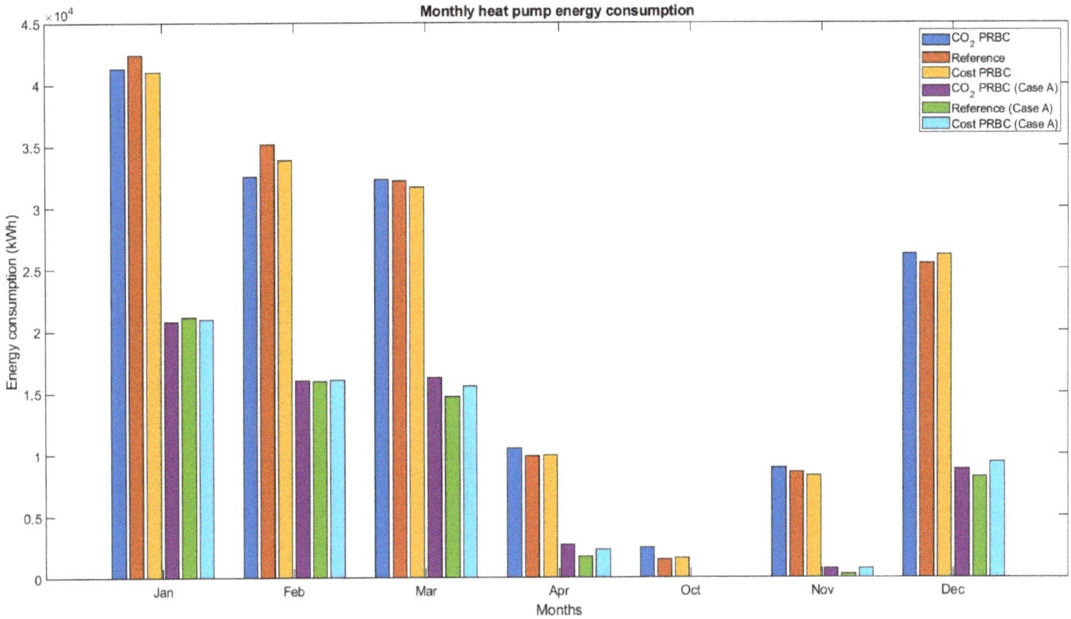

Figure 11. Monthly electricity consumption for heating.

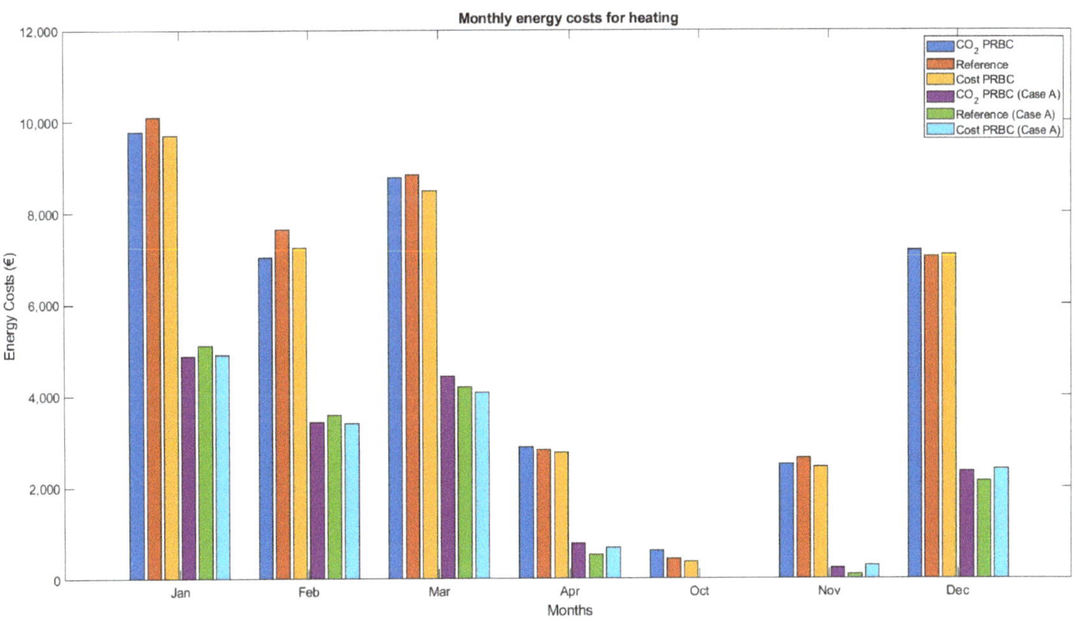

Figure 12. Monthly energy costs for heating.

Finally, regarding CO_{2eq} emissions (Figure 13), the CO_2 PRBC, as expected, seemed to decrease emissions during most months compared to the other two control scenarios.

Specifically for the reference building, the CO_2 PRBC presented the lowest emissions during January and especially February; the cost PRBC gave the greatest results for March, April and November; and the reference control produced the best results for December. The results for the case A building indicate that there were decreased CO_{2eq} emissions during CO_2 PRBC in January and February and the reference control was rated as the most efficient one for the rest of the months.

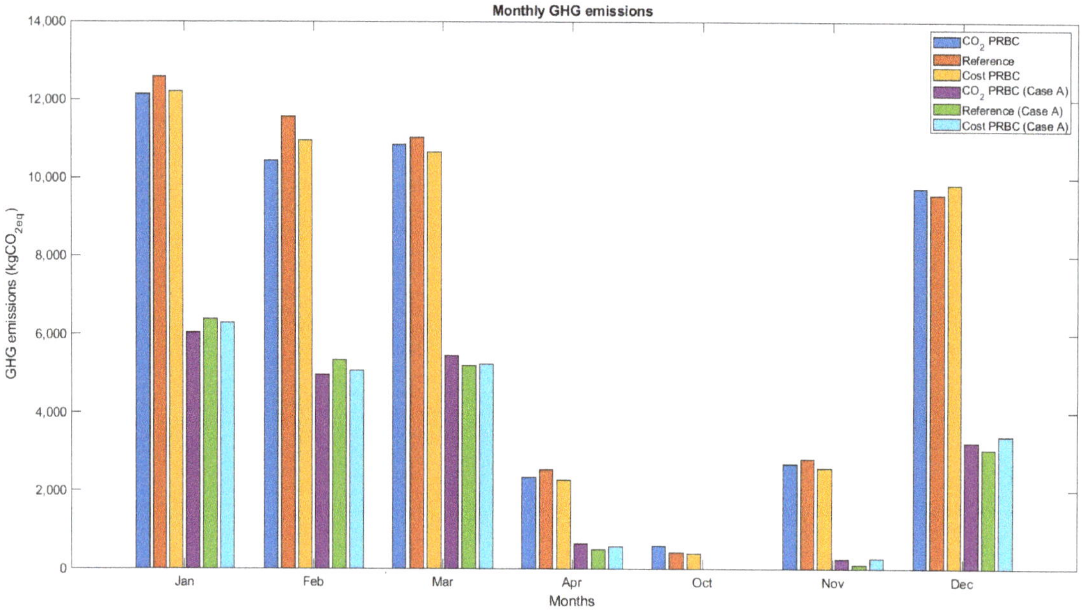

Figure 13. Monthly CO_{2eq} emissions for heating.

Table 4 shows the annual results for both buildings under the different control scenarios. For the reference building, the energy consumption decreased by 1.7% under the Cost PRBC and by 0.6% under the CO_2 PRBC. Both DR scenarios led to decreased costs and emissions compared to the reference control. Specifically, the costs were reduced by 1.9% under the CO_2 PRBC and by 3.6% under the Cost PRBC; emissions were reduced by 3.3% under the cost PRBC and by 3.5% under the CO_2 PRBC. We can notice that a small energy consumption reduction was converted to higher costs and emission reduction, indicating a shifting of consumption to times with a lower cost of energy and emissions intensity. Still, we can notice that DR comes with sacrificing thermal comfort, with increased thermal discomfort hours under both PRBCs (15.4–23.1%).

A similar study was conducted by Clauß et al. in a residential building in Norway [21]. The implemented CO_2 PRBC led to an increase in CO_{2eq} emissions when implemented in the heat pump system. The CO_{2eq} emissions decreased only in the case of CO_2 PRBC implementation in direct electric heating. The increase in CO_{2eq} emissions was primarily attributed to the increasing energy consumption under every control strategy, but also to the relatively low fluctuations of the electricity mix throughout the day, in terms of price and CO_{2eq} intensity. In other studies where cost or CO_2 MPC is applied, a reduction is observed both in energy consumption and in energy costs, and in carbon emissions; in fact, in the majority of cases the percentage reduction exceeds 10% compared to the reference scenario [16–20,22].

Table 4. Annual performance results.

		Energy		Costs		Emissions		Thermal Discomfort	
	Unit	kWh	%	EUR	%	CO_{2eq}	%	h	%
Reference building	Reference	155,220	-	39,483	-	50,548	-	26	-
	Cost PRBC	152,590	−1.7%	38,057	−3.6%	48,897	−3.3%	32	23.1%
	CO_2 PRBC	154,320	−0.6%	38,725	−1.9%	48,795	−3.5%	30	15.4%
Case A building	Reference	62,056	-	15,639	-	20,622	-	0	-
	Cost PRBC	65,119	4.9%	15,747	0.7%	20,839	1.1%	0	0%
	CO_2 PRBC	65,299	5.2%	16,089	2.9%	20,599	−0.1%	0	0%

6. Conclusions and Future Research

Resilient and sustainable planning, construction and management of the built environment are priorities in European energy and environmental policy. The integrated management of the carbon neutrality aim for buildings is a concern that primarily concentrates on energy-intensive operations like heating and cooling. Since distribution grid bottlenecks are anticipated in the near future, reducing energy use during peak load hours is one of the most crucial control objectives in the Greek context. Utilizing buildings' energy flexibility and HVAC systems should also increase a building's and its inhabitants' resilience to climate change and unpredictable conditions [32]. Greek electricity has a relatively high price compared to the other European countries, with prices even reaching 900 €/MWh in 2022. However, the Greece electricity market has modest levels of CO_{2eq} intensity compared to the rest of the European countries, with levels lower than 500 $kgCO_{2eq}$/MWh.

In this paper, two types of PRBC were used to examine the demand response for heating a typical Greek office building. Two cases of a building with different thermal characteristics were evaluated. The actual heating system of the building consists of gas-fired boilers. In order to implement DR, we chose to replace the boilers with a VSHP during the modelling process. In this study, a detailed model of the building and the heating system was realized in TRNSYS. By using hourly spot prices and the hourly CO_{2eq} intensity of the electricity mix as an input, the two PRBC techniques aimed to lower heating energy costs and CO_{2eq} emissions, respectively. A third control scenario was implemented in an effort to quantify energy storage in the thermal mass of the building and assess its storage efficiency.

The price-based PRBC led to both decreased heating costs (−3.6%) and CO_{2eq} emissions (−3.3%) when implemented in the reference building. The implementation of the cost PRBC in the case A building led to increased costs of energy (0.7%) and CO_{2eq} emissions (1.1%) compared to the reference control, which were mainly attributed to the large increase in energy consumption (4.9%) compared to the reference control. The CO_2 PRBC led to reduced emissions compared to the reference control for both the reference building (−3.5%) and the case A building (−0.2%). Again, the slight reduction of CO_{2eq} emissions for the case A building was attributed to the large increase in energy consumption. The carbon-based PRBC also led to a reduction of costs (−1.9%) compared to the reference control for the reference building.

Generally, both PRBC controls managed to achieve the goal of their application when implemented in the reference building. When implemented in the case A building, the increase in the energy consumption due to DR left no room for reducing costs and carbon emissions. Future research includes the implementation of model-predictive control and the investigation of its possible benefits, especially in the case A building where PRBC was found to fail. Other aspects worth investigating is the testing of DR controls on various heat emission systems to further clarify the role of thermal mass in the energy flexibility provided and further assess the impact of demand response to indoor environment quality [33].

Author Contributions: Conceptualization, G.C.; methodology, E.G. and G.C.; validation, G.C.; data curation, G.C.; writing—original draft preparation, G.C.; writing—review and editing, E.G. and G.C.; supervision, A.M.P. All authors have read and agreed to the published version of the manuscript.

Funding: The research work was supported by the Hellenic Foundation for Research and Innovation (H.F.R.I.) under the "First Call for H.F.R.I. Research Project to support Faculty members and Researchers and the procurement of high-cost research equipment grant" (Project Number: 4104).

Institutional Review Board Statement: Not applicable.

Informed Consent Statement: Not applicable.

Data Availability Statement: Not applicable.

Conflicts of Interest: The authors declare no conflict of interest.

Nomenclature

RES	Renewable Energy Sources
HVAC	Heating, Ventilation, and Air-Conditioning System
DR	Demand Response
CHP	Combined Heat and Power
AWHP	Air-to-Water Heat Pump
TES	Thermal Energy Storage
TRNSYS	Transient System Simulation Tool
MPC	Model-Predictive Control
PRBC	Predictive Rule-Based Control
KPI	Key Performance Indicator
COP	Coefficient of Performance
PMV	Predicted Mean Vote

References

1. Heylen, E.; Deconinck, G.; Van Hertem, D. Review and Classification of Reliability Indicators for Power Systems with a High Share of Renewable Energy Sources. *Renew. Sustain. Energy Rev.* **2018**, *97*, 554–568. [CrossRef]
2. Jensen, S.Ø.; Marszal-Pomianowska, A.; Lollini, R.; Pasut, W.; Knotzer, A.; Engelmann, P.; Stafford, A.; Reynders, G. IEA EBC Annex 67 Energy Flexible Buildings. *Energy Build.* **2017**, *155*, 25–34. [CrossRef]
3. European Commission. Proposal for a Directive of the European Parliament and of the Council on the Energy Performance of Buildings (Recast). *Off. J. Eur. Union* **2021**, *0426*, 10–27.
4. Chantzis, G.; Giama, E.; Nizetic, S.; Papadopoulos, A. The Potential of Demand Response as a Tool for Decarbonization in the Energy Transition. In Proceedings of the 2022 7th International Conference on Smart and Sustainable Technologies (SpliTech), Split/Bol, Croatia, 5–8 July 2022; pp. 1–4.
5. Lu, F.; Yu, Z.; Zou, Y.; Yang, X. Cooling System Energy Flexibility of a Nearly Zero-Energy Office Building Using Building Thermal Mass: Potential Evaluation and Parametric Analysis. *Energy Build.* **2021**, *236*, 110763. [CrossRef]
6. Chen, Y.; Xu, P.; Gu, J.; Schmidt, F.; Li, W. Measures to Improve Energy Demand Flexibility in Buildings for Demand Response (DR): A Review. *Energy Build.* **2018**, *177*, 125–139. [CrossRef]
7. Nuytten, T.; Claessens, B.; Paredis, K.; Van Bael, J.; Six, D. Flexibility of a Combined Heat and Power System with Thermal Energy Storage for District Heating. *Appl. Energy* **2013**, *104*, 583–591. [CrossRef]
8. Arteconi, A.; Mugnini, A.; Polonara, F. Energy Flexible Buildings: A Methodology for Rating the Flexibility Performance of Buildings with Electric Heating and Cooling Systems. *Appl. Energy* **2019**, *251*, 113387. [CrossRef]
9. De Coninck, R.; Helsen, L. Quantification of Flexibility in Buildings by Cost Curves–Methodology and Application. *Appl. Energy* **2016**, *162*, 653–665. [CrossRef]
10. Oldewurtel, F.; Sturzenegger, D.; Andersson, G.; Morari, M.; Smith, R.S. Towards a Standardized Building Assessment for Demand Response. In Proceedings of the 52nd IEEE Conference on Decision and Control, Firenze, Italy, 10–13 December 2013.
11. Masy, G.; Georges, E.; Verhelst, C.; Lemort, V.; André, P. Smart Grid Energy Flexible Buildings through the Use of Heat Pumps and Building Thermal Mass as Energy Storage in the Belgian Context. *Sci. Technol. Built Environ.* **2015**, *21*, 800–811. [CrossRef]
12. Reynders, G.; Diriken, J.; Saelens, D. Generic Characterization Method for Energy Flexibility: Applied to Structural Thermal Storage in Residential Buildings. *Appl. Energy* **2017**, *198*, 192–202. [CrossRef]
13. Péan, T.Q.; Torres, B.; Salom, J.; Ortiz, J. Representation of Daily Profiles of Building Energy Flexibility. In *eSIM 2018 Conference Proceedings*; IBPSA: Montréal, QC, Canada, 2018; pp. 153–162.
14. D'Ettorre, F.; De Rosa, M.; Conti, P.; Testi, D.; Finn, D. Mapping the Energy Flexibility Potential of Single Buildings Equipped with Optimally-Controlled Heat Pump, Gas Boilers and Thermal Storage. *Sustain. Cities Soc.* **2019**, *50*, 101689. [CrossRef]

15. Yin, R.; Kara, E.C.; Li, Y.; DeForest, N.; Wang, K.; Yong, T.; Stadler, M. Quantifying Flexibility of Commercial and Residential Loads for Demand Response Using Setpoint Changes. *Appl. Energy* **2016**, *177*, 149–164. [CrossRef]
16. Ruusu, R.; Cao, S.; Manrique Delgado, B.; Hasan, A. Direct Quantification of Multiple-Source Energy Flexibility in a Residential Building Using a New Model Predictive High-Level Controller. *Energy Convers. Manag.* **2019**, *180*, 1109–1128. [CrossRef]
17. Vogler-Finck, P.J.C.; Wisniewski, R.; Popovski, P. Reducing the Carbon Footprint of House Heating through Model Predictive Control—A Simulation Study in Danish Conditions. *Sustain. Cities Soc.* **2018**, *42*, 558–573. [CrossRef]
18. Péan, T.; Costa-Castelló, R.; Salom, J. Price and Carbon-Based Energy Flexibility of Residential Heating and Cooling Loads Using Model Predictive Control. *Sustain. Cities Soc.* **2019**, *50*, 101579. [CrossRef]
19. Esmaeilzadeh, A.; Deal, B.; Yousefi-Koma, A.; Zakerzadeh, M.R. How Combination of Control Methods and Renewable Energies Leads a Large Commercial Building to a Zero-Emission Zone—A Case Study in U.S. *Energy* **2023**, *263*, 125944. [CrossRef]
20. Vigna, I.; Lollini, R.; Pernetti, R. Assessing the Energy Flexibility of Building Clusters under Different Forcing Factors. *J. Build. Eng.* **2021**, *44*, 102888. [CrossRef]
21. Clauß, J.; Stinner, S.; Sartori, I.; Georges, L. Predictive Rule-Based Control to Activate the Energy Flexibility of Norwegian Residential Buildings: Case of an Air-Source Heat Pump and Direct Electric Heating. *Appl. Energy* **2019**, *237*, 500–518. [CrossRef]
22. Pallonetto, F.; Oxizidis, S.; Milano, F.; Finn, D. The Effect of Time-of-Use Tariffs on the Demand Response Flexibility of an All-Electric Smart-Grid-Ready Dwelling. *Energy Build.* **2016**, *128*, 56–67. [CrossRef]
23. EnExGroup Webpage. Available online: https://www.enexgroup.gr/web/guest/home (accessed on 27 February 2021).
24. *IPCC Climate Change 2014: Synthesis Report. Contribution of Working Groups I, II and III to the Fifth Assessment Report of the Intergovernmental Panel on Climate Change 2014*; IPCC: Geneva, Switzerland, 2014.
25. *Greek Government Gazette on the Approval of a Regulation for the Thermal Insulation of Buildings*; National Printing House of Greece: Athens, Greek, 1979; p. 362.
26. *ANSI/ASHRAE Standard 90*; 1-2007 Energy Standard for Buildings Except Low-Rise Residential Buildings. ASHRAE: Peachtree Corners, GA, USA, 2007.
27. European Parliament. *Directive 2010/31/EU*; Official Journal of the European Union; European Union: Brussels, Belgium, 2010; p. L153/13.
28. Chantzis, G.; Antoniadou, P.; Symeonidou, M.; Giama, E.; Oxizidis, S.; Kolokotsa, D.; Papadopoulos, A. Demand Side Flexibility Potential and Comfort Performance of Non-Residential Buildings. *J. Phys. Conf. Ser.* **2021**, *2069*, 12151. [CrossRef]
29. Giama, E.; Chantzis, G.; Kontos, S.; Keppas, S.; Poupkou, A.; Liora, N.; Melas, D. Building Energy Simulations Based on Weather Forecast Meteorological Model: The Case of an Institutional Building in Greece. *Energies* **2023**, *16*, 191. [CrossRef]
30. *SCL TRNSYS 18*; Thermal Energy System Specialists, LLC: Madison, WI, USA, 2018; Volume 3.
31. Toffanin, R.; Péan, T.; Ortiz, J.; Salom, J. Development and Implementation of a Reversible Variable Speed Heat Pump Model for Model Predictive Control Strategies. In Proceedings of the 16th IBPSA Conference, Rome, Italy, 2–4 September 2019.
32. Chantzis, G.; Antoniadou, P.; Symeonidou, M.; Kyriaki, E.; Giama, E.; Oxyzidis, S.; Kolokotsa, D.; Papadopoulos, A. Evaluation of Methods for Determining Energy Flexibility of Buildings. *Green Energy Sustain.* **2022**, *2*, 6. [CrossRef]
33. Giama, E. Review on Ventilation Systems for Building Applications in Terms of Energy Efficiency and Environmental Impact Assessment. *Energies* **2022**, *15*, 98. [CrossRef]

applied sciences

MDPI

Article

Kinetic Photovoltaic Facade System Based on a Parametric Design for Application in Signal Box Buildings in Switzerland

Ho Soon Choi

Department of Architecture, Gachon University, 1342 Seongnamdaero, Sujeong-gu, Seongnam-si 13120, Gyeonggi-do, Republic of Korea; hosoon@gachon.ac.kr; Tel.: +82-31-750-5519

Abstract: This study aims to produce renewable energy by applying a solar-energy-harvesting architectural design using solar panels on the facade of a building. To install as many solar panels as possible on the building elevation, the Signal Box auf dem Wolf, located in Basel, Switzerland, was selected as the research target. The solar panels to be installed on the facade of the Signal Box auf dem Wolf are planned such that they are able to move according to the optimal tilt angle every month to allow maximal energy generation. The kinetic photovoltaic facade system and the simulation of renewable energy generation were implemented using a parametric design. The novelty of this study is the development of a kinetic photovoltaic facade system using a parametric design algorithm. From the perspective of renewable energy in the field of architecture, the kinetic photovoltaic facade system developed in this study has the advantage of producing maximal renewable energy according to the optimal tilt angle of the solar panels. Additionally, building facades that move according to the optimal tilt angle will contribute to the expansion of the field of sustainable architectural design.

Keywords: solar energy harvesting; kinetic photovoltaic facade system; parametric design; sustainable architectural design

1. Introduction

The purpose of this research is to study building facades and the use of sunlight to increase the energy independence of buildings. The reasons for selecting building facades as a research subject are, firstly, the limitation that existing studies have been conducted mainly on the rooftops of buildings [1–4] and, secondly, to utilize the wide facade of the building to produce the maximum amount of renewable energy. This study considered the fact that a solar panel can be moved according to its optimal tilt angle every month in order to maximize the generation of renewable energy. By moving on the facade of the building, the angle of the solar panel can be controlled, also representing an element of new architectural design by being placed on the facade of the building. The kinetic solar panel developed in this study and the simulation of renewable energy generation produced by the solar panel were implemented with digital technology.

Digital technologies are being increasingly applied in the architectural field, wherein architectural designs are created using various computer programs. Designers implement virtual buildings using computer programs during the initial stages of architectural design. Designing a building in virtual reality aims to consider different variables, such as the complex elements required for constructing a building and the given conditions. The method of designing by establishing relationships, dependencies, and interactions with data is called parametric design [5–7]. Parametric architectural design is used not only in the external design of buildings, but also to solve serious contemporary environmental problems.

The building sector consumes a significant amount of energy, accounting for 30% of global energy consumption [8]. From this perspective, architectural design to increase energy independence is considered from the initial stages of design, and parametric design is used for this purpose [9–12]. To this end, parametric design first utilizes the available natural energy, which is a passive architectural design method. Parametric design is also used to derive the optimal building shape, such that sufficient solar energy can reach the interior of the building [13,14]. These approaches should be used in the early stages of architectural design, so that an optimized building form that can utilize the maximum natural energy can be formulated. Additionally, passive design actively converts natural energy into usable energy through special devices, which is a voluntary architectural design method. Energy harvesting is a technology that actively produces renewable energy from natural sources [15,16], mainly leveraging solar and wind power to produce renewable energy. Solar energy is utilized through solar panels and wind power is utilized via wind turbines and fine vibrations generated by the wind [17–20]. When such energy harvesting technology is applied to buildings, parametric design is utilized in all architectural design processes. The variables for the parametric design of representative buildings are the surrounding building types, building site location, and climate conditions. Thus, parametric design in the field of architecture refers to the process of realizing a single building in virtual reality through the interaction of different variables related to the building.

In this study, we aim to realize an architectural design using parametric design to increase the energy independence of buildings. A special device was developed on the facade of a building to incorporate solar-based energy harvesting. The special device is designed such that it moves according to the optimal solar panel angle every month to maximize the production of renewable energy. Such flexible solar panels can be differentiated from solar panels fixed at a certain angle.

Additionally, because building facades are an important factor in determining the appearance of a building, a building facade incorporating energy harvesting will not only increase the energy independence of the building, but also play an important role in creating a new urban landscape.

2. Background

2.1. Energy Harvesting

Energy harvesting was first introduced in 1954 when Bell Laboratories in the United States conducted research on solar cells that convert sunlight into energy. Because the only way to produce electricity in a city is through a generator, natural energy, such as solar heat, is recognized as wasted energy, and the word 'harvest' is used to represent the generation of electricity by collecting wasted energy [21,22].

The energy harvesting field is divided into two categories according to the scale of the energy source (Table 1). The first energy harvesting field is a macro-energy source that produces energy from natural energy sources such as sunlight, vibration, heat, and wind power. The second field of energy harvesting is technology that regenerates energy by harvesting or using wasted energy, such as pressure, thermoelectric, and electromagnetic energy with micro-energy sources [23]. In particular, energy harvesting technology that harvests and utilizes energy that is discarded based on the increase in demand for low-carbon and eco-friendly energy and the rapid use of mobile devices was developed.

Table 1. Energy harvesting field according to the natural energy scale.

Macro-Energy Source	Micro-Energy Source
Solar, wind, and geothermal	Piezoelectric, thermoelectric, photoelectric, and electromagnetic

Energy harvesting, which produces energy from macro-energy sources, primarily utilizes solar, wind, and geothermal sources. Because energy harvesting originally started with the use of sunlight, energy harvesting technology using sunlight has become the most widely used. Solar-based energy harvesting uses solar panels to generate renewable energy. In today's solar-based energy harvesting, research on solar panel materials and batteries for effectively storing the energy produced from solar panels is being conducted to maximize the efficiency of energy production [24,25]. In particular, the development of solar panel modules has resulted in their active use in architectural designs. The diversification of photovoltaic modules, such as multicolored and transparent photovoltaic panels, plays a significant role in the development of architectural designs, as photovoltaic panels are used as architectural design materials.

Energy harvesting from wind power converts wind into electricity using a wind turbine. Wind power is currently used in many countries to produce relatively inexpensive and zero-emission energy sources [26]. Recently, the number of wind-power-generation-linked buildings in the field of architecture has been increasing worldwide [27]. Wind turbines in buildings are used not only to produce renewable energy through wind power generation, but also as eco-friendly design elements of the buildings themselves. Depending on how wind power is connected to buildings, it is divided into building-mounted wind turbines (BMWTs) and building-integrated wind turbines (BIWTs) [28]. BMWTs were first introduced in Germany in 2000, and they have been built worldwide since 2006. BIWTs were first constructed in 2008, later than BMWTs. Since the introduction of BIWTs, wind-power-linked buildings have been built using BIWTs rather than BMWTs [29,30]. This is because BIWTs, which actively utilize wind turbines as an element of their architectural design, are preferred over BMWTs, which involve wind turbines being placed on the roofs of buildings.

Energy harvesting, which generates energy from micro-energy sources, produces energy from piezoelectric, thermoelectric, photoelectric, and electromagnetic energy sources [31]. Piezoelectric energy harvesting has been developed focusing on the development of new materials for synthetic fibers that utilize the piezoelectric effect. Thermoelectric energy harvesting uses the temperature difference that occurs at the parts in contact with both ends of two types of metals or semiconductors and this is converted into voltage. Electromagnetic energy harvesting is a technology that uses an antenna to collect radio waves floating on the ground and convert them into electrical energy [32]. Currently, energy harvesting based on micro-energy sources has low industrialization and commercialization compared with energy harvesting based on macro-energy sources, because the power and voltage of the produced energy are low. However, energy harvesting based on micro-energy sources has the potential to solve the inconsistency of energy production from solar and wind power caused by weather conditions.

2.2. Parametric Design

Parametric design has been increasingly applied to architectural design as digital technology has been applied to the field of architecture. It is a digital design method that can be characterized by rule algorithmic design and multiple solution generation. Parametric design originated from generative design, which is a typical computational design approach based on rules or algorithms [33]. Parametric design has historically evolved from simple architectural models created by inputting parameters into computer scripts with highly sophisticated structures based on the hierarchical dependencies of complex algorithms [34]. Currently, computer software developed for parametric designs can transform architectural designs in real time, providing sophisticated interfaces that provide designers with more control and immediate feedback when the parameters change.

Parametric designs in architecture are architectural designs that consider different variables for architectural design, and this process is expressed according to parameters and regularities based on algorithms. Typical parameters for parametric architectural design include natural factors, such as location, climate, direction, and sunlight, and design concept factors, such as building shape and material selection. Parametric design in architecture focuses on the process of organically establishing conventions among various parameters and implementing them in the form of architecture.

Parametric modeling for architectural design is divided into propagation- and constraint-based systems (Table 2). Propagation-based systems compute from known to unknowns with a dataflow model, whereas constraint-based systems solve sets of continuous and discrete constraints [35]. In architectural design, propagation-based systems derive unknown final architectural shapes by inputting different variables into a parametric algorithm. However, in constraint-based systems, the final building is completed using constraints between the geometric components of the building. In this study, the constraint-based parametric design was oriented such that the interrelationships between different variables and constraints were established through the constructed algorithm.

Table 2. Parametric modeling for architectural design.

Propagation-Based Systems	Constraint-Based Systems
A system that computes from known to unknown using data flow models	A system that solves sets of continuous and discontinuous constraints

Parametric design solutions are rapid and effective for addressing design problems using a simplified and rational technique that requires considerable effort and time when using conventional methods [36]. Compared with conventional methods, parametric design efficiently solves complex geometrical problems by applying mathematical parameters and creating advanced design models in an automated process. In addition, parametric design is based on virtual reality; therefore, problems after the completion of the project can be minimized. Parametric design in architecture first implements a three-dimensional building in a virtual space with a design program and then applies different variables to the building using an algorithm plugged into the design program. The mathematical process of parametric design has evolved through computation. Currently, the representative parametric design algorithms include Grasshopper (a plug-in for Rhino) and Dynamo (a plug-in for Revit). In this study, a parametric design was employed using the Revit program and Revit plug-in Dynamo.

3. Methodology

The study comprised five stages. First, the building to be studied was selected. The research target was a building in which the maximum number of solar panels could be installed on the building's facade. Additionally, to avoid shadow interference, there should be no other buildings around the building. Second, the solar-energy-harvesting-based facade system was modeled and simulated virtually using building information modeling (BIM). BIM integrates various data into the architectural design process and implements them from the architectural planning and design stage to construction and operation [37]. The modeling and simulation in this study were conducted using Autodesk Revit 2020 software [38], a representative BIM. Dynamo [39], a Revit 2020 software plug-in, was used to implement flexible elevation of the solar-energy-harvesting-based facade system. Third, for maximal solar energy generation, the optimal angle of the solar panel for each month was derived, and the solar panel movement according to the monthly angle was simulated. Fourth, Insight [40], a Revit 2020 software plug-in, was used to estimate the amount of energy generated monthly. Fifth, the amount of renewable energy generated in one year was calculated, and the overall appearance of the building with the attached kinetic facade system was evaluated (Figure 1).

3.1. Introduction to Solar Energy Harvesting Based on Parametric Design in Architecture

The objective of this study is to produce renewable energy by utilizing solar energy through architectural design. Generally, solar panels are installed on the roofs of buildings to ensure maximum exposure to sunlight and facilitate maintenance [41]. This study seeks to actively utilize solar panels as an element of architectural design by installing them on the facade of a building, addressing the limitation that solar panels installed on rooftops cannot be used as elements of architectural design.

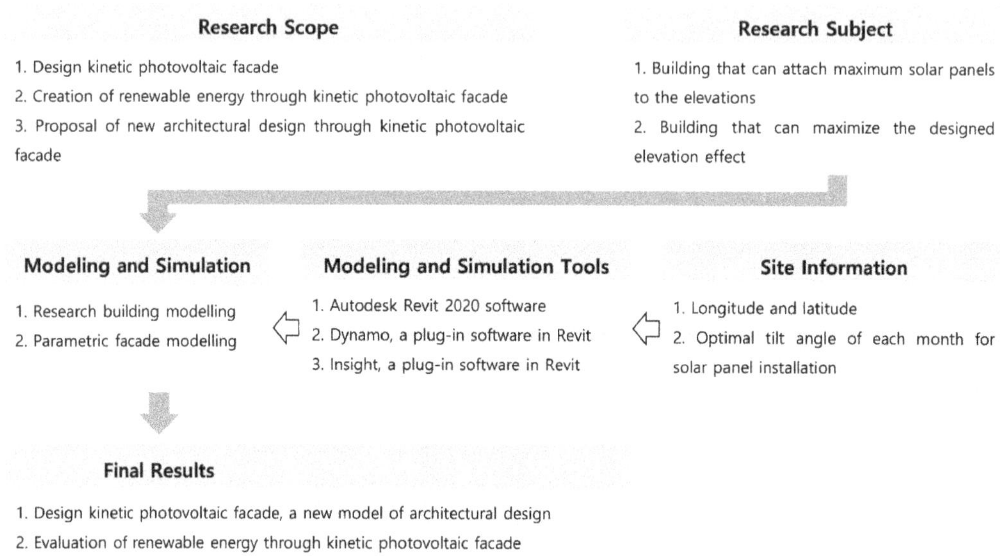

Figure 1. Processes in the proposed method.

Representative studies on solar panels in architectural design are classified according to the size of the solar panel module attached to a building. First, if the photovoltaic module used in the study is macro-sized, the photovoltaic panel is directly installed on the building elevation. To derive the maximum energy production using solar panels attached to the building elevation, a study was conducted to determine the size of the building's outer wall using a parametric design [42]. This study can play an important role in determining the basic form of a building suitable for a given site. There has also been an empirical study on the production of building elevations of a certain size comprising solar panels. A solar panel was manufactured for movement, and the energy produced by the solar panel was measured at certain angles. The optimal angle of the solar panel was derived based on the maximum amount of energy generated by the solar panel [43–45]. Secondly, considering micro-sized solar modules, a study using sunshades installed on windows has been reported previously. When a solar panel is integrated with a sunshade, it is planned for the size of the solar panel to be small. A study was conducted to measure energy production using the size and spacing of the solar panels to be placed on the sunshade and at various angles exposed to sunlight [46,47]. Additionally, research on the development of sunshades with attached solar panels is characterized by the fact that the subject of research is not limited to solar panels but expands to the interior space of buildings. As the area of the solar panel attached to the sunshade is small, increasing the energy utilization of the interior space of buildings using sunshades is considered in addition to optimal energy production [48].

3.2. Research Subject and Site Analysis

The building to be studied was the Signal Box auf dem Wolf (47.57° N and 7.35° E), surrounded by the railway tracks of Basel Station, Switzerland. The Signal Box auf dem Wolf was completed in 1994 and comprises five floors with a height of 24 m (the total floor area is 935 m^{-2}). The width of the building is 11.2 m and the length is 17.2 m. Machinery and equipment occupy most of the building for train services. A characteristic of this building is that all facades are surrounded by copper bands, so the windows are not visible. This is because the main purpose of the building was to store machinery, thus windows were not required. However, there are windows in the space where the building managers reside; therefore, when windows exist on the outer wall of the building, the angle of the copper bars is changed to bring natural light from the outside to the interior (Figure 2 left). Additionally, the external environment of the building is characterized by the absence of other surrounding buildings (Figure 2 right). Because of these characteristics, we selected the Signal Box auf dem Wolf as the research subject. Moreover, the metallic material comprising the elevation of the Signal Box auf dem Wolf can be replaced by a solar panel. Furthermore, the Signal Box auf dem Wolf can maximize the energy production effect of the solar panels, because there is no shadow interference from the surrounding buildings. If the moveable solar panel pursued in this study can be installed in the Signal Box auf dem Wolf to produce renewable energy, it will not only increase the energy independence of the building, but also maximize the current mechanical image.

Figure 2. Research subject. Signal Box auf dem Wolf in Basel (**left**) [49] and master plan (**right**) [50].

4. Design

4.1. Selection of Building Elevations for Energy Generation via Solar Panel Facade Systems

The Signal Box auf dem Wolf has a south-west-facing facade, and the main facade of the building is tilted 30° clockwise from the south. To attach the solar panel facade system to a building elevation, a suitable elevation of the building must be selected. Insight, a Revit software plug-in optimized for solar energy simulation, was used to select the elevation of the building where the solar panels would be installed. The building was implemented using Revit software, and solar energy simulation for the building elevation was conducted according to the latitude and longitude values of the city of Basel, based on data from 1 January to 31 December 2022. Figure 3 shows the results of the solar radiation simulation using the Insight program. The building elevation with the largest cumulative solar radiation appeared in the order of south, east, west, and north. A prerequisite for solar panel installation is securing peak sun hours, which is the time at which sunlight begins to shine directly on the solar panel. While the amount of sunlight received by the panels is important, a more accurate representation of the amount of energy the panels can produce is the peak sun hours. It is important to note that "peak sun hours" are different from "hours of daylight". Peak sun hours refer specifically to the amount of solar energy available in an area during a typical day. A peak sun hour is an hour during which the intensity of sunlight is 1000 Wh·m^{-2} [51,52]. Under these preconditions, the solar panel facade systems were installed at the south, east, and west elevations, except for the north side, where peak sun hours cannot be secured.

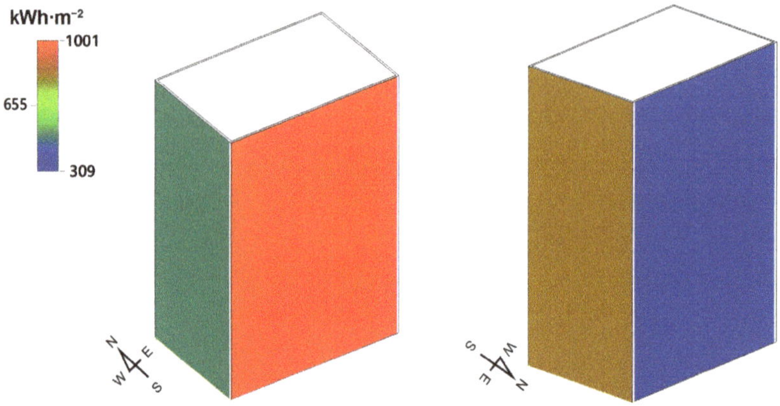

Figure 3. Cumulative solar radiation on the facades of the Signal Box auf dem Wolf for one year in 2022.

4.2. Design of Kinetic Solar Panel Facade System

The solar panel facade system to be applied to the Signal Box auf dem Wolf's elevation is designed to be movable. The solar panels to be installed on the building elevation were arranged according to the optimal tilt angle calculated to maximize solar energy generation efficiency every month. The advantage of such a kinetic solar panel facade system is that more solar energy is generated compared with existing solar panels installed at a fixed angle. Second, the kinetic solar panel facade also functions as a sunshade to control sunlight entering the building. The solar panel, which could be moved according to the angle desired by the user, controlled the amount of sunlight entering the building. The size of the solar panel followed the standard for the copper material attached to the existing building elevation. As a result, a solar panel with a length of 2 m and a width of 1 m was planned (Figure 4).

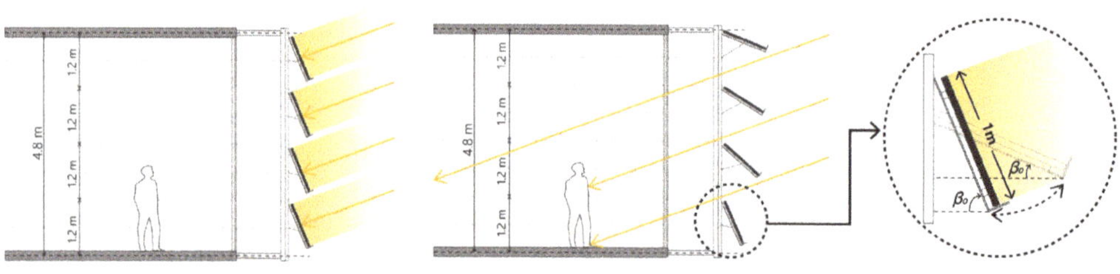

Figure 4. Principle of solar panel facade system and optimal tilt angle (βo).

The solar panel used in this study was the Hanwha Q CELLS Q. PEAK DUO model. Hanwha Q CELLS is among the world's largest and most recognized photovoltaic panel manufacturers. They are renowned for their reliability and the state-of-the-art technologies they use in their modules and cell designs [53]. The Q. PEAK DUO ML-G11 solar module had a power generation efficiency of up to 21.5% and a power capacity of 500 Wp. It is 2054 mm long and 1134 mm wide. The standard size of this model was easily applied to the conceptual design of the elevational solar module in this study (Table 3).

Table 3. Hanwha Q. CELLS' Q. PEAK DUO ML-G11 solar module.

Type	Parameter	Value
Solar panel module	Model	Q. PEAK DUO ML-G11
	Length (mm)	2054
	Width (mm)	1134
	Power capacity (kW/unit)	500 Wp
	Efficiency (%)	21.5%

To increase the energy generation efficiency of the solar panel, it must be tilted at a certain angle. We calculated the optimal tilt angle of the solar panel every month to maximize its energy generation of the solar panel using the following equation [54].

$$\beta o = a_1 + a_2\varphi \quad (1)$$

Here, βo is the optimal tilt angle, φ is the latitude of Basel (47.57° N), and a_1 and a_2 are coefficients derived from solar declination values. Table 4 lists the optimal tilt angle values by function from January to December 2022.

Table 4. Optimal tilt angles of Basel from January to December 2022 (47.57° N, 7.35° E) (βo).

N.	Month	Coefficients		Optimal Tilt Angle (°)
		a_1	a_2	βo
1	January	31.33	0.68	64°
2	February	16.25	0.86	57°
3	March	6.80	0.84	47°
4	April	−6.07	0.87	35°
5	May	−14.95	0.87	26°
6	June	−19.27	0.87	22°
7	July	−15.65	0.83	24°
8	August	−4.23	0.75	31°
9	September	6.42	0.77	43°
10	October	15.84	0.83	55°
11	November	23.61	0.84	64°
12	December	30.56	0.76	67°

Kinetic solar panel facade systems were placed according to the monthly optimal tilt angle values on the south, east, and west facades of the Signal Box auf dem Wolf for the solar energy simulation described in Section 3.1. The solar panels were changed according to the different angles of each month. We used Dynamo, a Revit software plug-in, to simulate a solar panel facade according to the angle change.

5. Analysis

The solar energy simulation for three faces of the Signal Box auf dem Wolf according to the optimal tilt angle (βo) was conducted from January to December 2022. The solar panel was implemented using Revit software, and the placement of the solar panels according to the optimal tilt angle was controlled using Dynamo's algorithm. The algorithm configuration involves entering the angle of the selected solar panel after selecting it. Designate the solar panel to be moved in 'Code Block' of <Solar panel selection> (Figure 5 left). In <Input optimal tilt angle>, input the parameter name and optimal tilt angle in two 'Code Blocks' (Figure 5 right). Figure 5 shows the algorithm for the solar panel with a 67° optimal tilt angle with the best solar energy generation efficiency in December 2022.

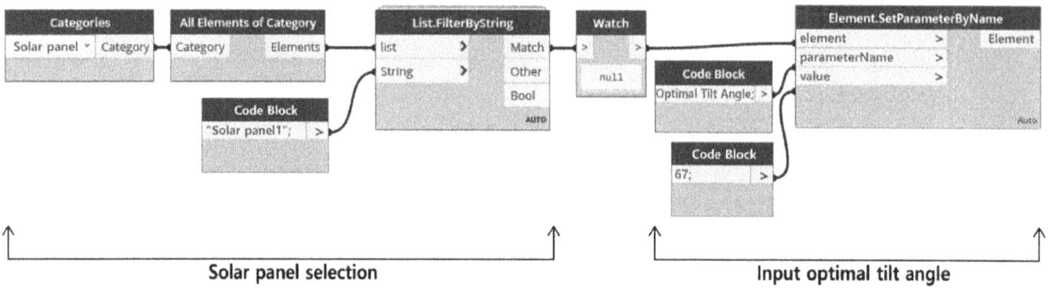

Figure 5. Algorithm input to the optimal tilt angle (67°) in "Solar Panel1".

The Signal Box auf dem Wolf's elevations were simulated using the Dynamo algorithm according to each optimal tilt angle from January to December. This Dynamo algorithm construction has the advantage of being able to simulate changes easily and quickly in architectural design according to different variables. In addition, the fast building simulation results using the Dynamo algorithm are a prerequisite for the subsequent parametric design for the measurement of renewable energy generation.

The solar panels to be placed on the facade of the Signal Box auf dem Wolf were simulated with the Revit program and, as a result, solar panels with a total area of 989 m^{-2} were placed on the south, east, and west facades. Figure 6 shows the energy generation simulation of the solar panel using the optimal tilt angle (24°) in July and the optimal tilt angle (67°) in December 2022. The solar energy generation efficiency was in the order of south (red color) > east (yellow color) > west (blue color). The solar panels implemented with the Revit program were placed according to the optimal tilt angle from January to December 2022 using the Dynamo algorithm, and the renewable energy derived using Insight software is shown in Table 5. The energy generation in July (47,407 kWh·$month^{-1}$) was the highest, and the lowest energy generation was in January (1267 kWh·$month^{-1}$).

Table 5. Energy generation according to the optimal tilt angle (βo) from January to December 2022.

N.	Month	Optimal Tilt Angle (°) βo	Total Surface Area m^{-2}	Energy Generation kWh·$Month^{-1}$
1	January	64°	989	1267
2	February	57°	989	15,767
3	March	47°	989	30,503
4	April	35°	989	39,355
5	May	26°	989	46,821
6	June	22°	989	40,526
7	July	24°	989	47,407
8	August	31°	989	39,294
9	September	43°	989	20,312
10	October	55°	989	9318
11	November	64°	989	2236
12	December	67°	989	11,760
		Total		304,566

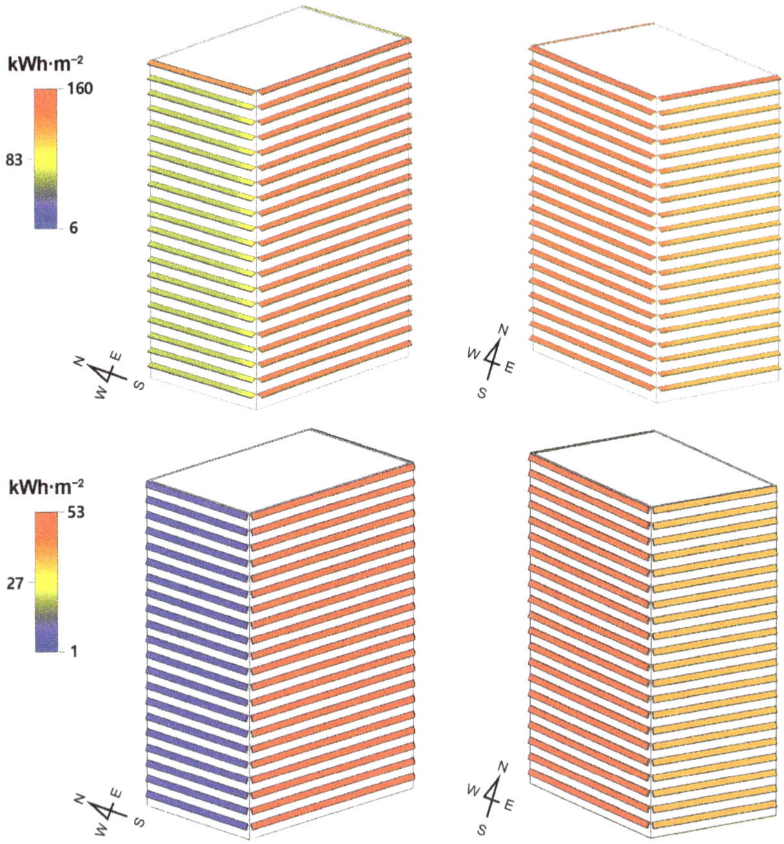

Figure 6. Energy generation simulation of solar panels by optimal tilt angle (24°) in July 2022 (**upper image**) and optimal tilt angle (67°) in December 2022 (**lower image**).

6. Results and Discussion

In this study, solar panels were used to produce renewable energy in the architectural sector. The Signal Box auf dem Wolf, located in Basel, Switzerland, was selected as the target building to produce maximum renewable energy, and solar panels were installed as an architectural design. The optimal angle of the solar panel was derived to maximize the monthly solar energy generation from January to December 2022. The movement of the solar panels was planned according to the optimal monthly tilt angle. This kinetic photovoltaic facade system and the energy generation movement simulation according to the optimal tilt angle were implemented through a parametric design. For the algorithm constituting the parametric design, Dynamo, a Revit software plug-in, was used to control the variables of the solar panel and the optimal tilt angle every month. To measure the optimal amount of renewable energy generated from solar panels, Insight, a Revit software plug-in, was used to control the variables of the site's natural environment every month.

Thus, optimal energy was generated based on the energy generation simulation. The energy generation in July (47,407 kWh·month^{-1}) was the highest, and that in January (1267 kWh·month^{-1}) was the lowest. The total energy generation through the kinetic photovoltaic facade system for the year 2022 was 304,566 kWh·year^{-1}. The total floor area of the study building was 935 m^{-2} and the annual solar energy generation per unit area (m^{-2}) was 325.7 kWh.

The average energy consumption of general office buildings ranges from 100 to 1000 kWh per unit area [55]. Based on these data, the 325.7 kWh of solar energy generation derived from this study accounted for 33% of the average energy consumed by office buildings. Thus, the kinetic photovoltaic facade system had a significant effect on the increase in energy independence in the construction field. Additionally, the solar panel developed in this study, which can be moved according to the monthly optimal tilt angle, can be used as a new architectural facade design for the kinetic facade concept to produce optimal solar energy. In particular, the Dynamo algorithm can be used to propose various building designs by simulating solar panels of various colors and textures that are currently developed for building elevation.

7. Conclusions

This study has the novelty of presenting a 'kinetic facade' architectural design based on the production of renewable energy in the field of architecture. On the other hand, there is a disadvantage in that an economic analysis of the driving principle of the solar panel facade device and the renewable energy produced through this study has not been performed. In relation to previous studies and this study, the intelligent solar tracking control systems [56,57] developed to move solar panels according to the trajectory of the sun can be easily controlled according to the angle desired by the user through the Dynamo algorithm proposed in this study.

Follow-up research will be conducted on the driving principle of the kinetic photovoltaic facade system, system production costs, and economic analysis of the generation of renewable energy compared with the cost of manufacturing the device. We plan to further expand this research through collaboration with the fields of electronic engineering and computer engineering. Additionally, the method in this study can be extended to other buildings. Furthermore, our approach can be considered an extension of architectural remodeling because it targets the design stage of not only new buildings, but also existing buildings.

Funding: This research was supported by the National Research Foundation of Korea (NRF) grant funded by the Korean Government, Ministry of Science and ICT (MSIT) (No. 2022R1F1A1065442).

Institutional Review Board Statement: Not applicable.

Informed Consent Statement: Not applicable.

Data Availability Statement: Not applicable.

Conflicts of Interest: The authors declare no conflict of interest.

References

1. Alboaouh, K.A.; Mohagheghi, S. Impact of rooftop photovoltaics on the distribution system. *J. Renew. Energy* **2020**, *2020*, 4831434. [CrossRef]
2. Hong, T.; Lee, M.; Koo, C.; Jeong, K.; Kim, J. Development of a method for estimating the rooftop solar photovoltaic (PV) po-tential by analyzing the available rooftop area using Hillshade analysis. *Appl. Energy* **2017**, *194*, 320–332. [CrossRef]
3. Gernaat, D.E.; de Boer, H.S.; Dammeier, L.C.; van Vuuren, D.P. The role of residential rooftop photovoltaic in long-term energy and climate scenarios. *Appl. Energy* **2020**, *279*, 115705. [CrossRef]
4. Muñoz-Rodríguez, F.; Jiménez-Castillo, G.; Rus-Casas, C. *Photovoltaic Rooftops in Smart Energy Systems in Handbook of Smart Energy Systems*; Springer: New York, NY, USA, 2022; pp. 1–28.
5. Jabi, W. *Parametric Design for Architecture*; Hachette: London, UK, 2013; pp. 15–16.
6. Panya, D.S.; Kim, T.H.; Choo, S.H. A methodology of interactive motion facades design through parametric strategies. *Appl. Sci.* **2020**, *10*, 1218. [CrossRef]
7. Assasi, R. Parametric Design, A Historical and Theoretical Overview. In Proceedings of the International Conference on Emerging Technologies in Architectural Design (ICETAD 2019), Toronto, ON, Canada, 17–18 October 2019.
8. International Energy Agency. Available online: https://www.iea.org/topics/buildings (accessed on 15 August 2022).
9. Bushra, N. A comprehensive analysis of parametric design approaches for solar integration with buildings: A literature review. *Renew. Sustain. Energy Rev.* **2022**, *168*, 112849. [CrossRef]

10. Eltaweel, A.; Su, Y. Controlling venetian blinds based on parametric design; via implementing Grasshopper's plugins: A case study of an office building in Cairo. *Energy Build.* **2017**, *139*, 31–43. [CrossRef]
11. Bande, L.; Hamad, H.; Alqahtani, D.; Alnahdi, N.; Ghunaim, A.; Fikry, F.; Alkhatib, O. Design of Innovative Paramet-ric/Dynamic Facade Integrated in the Library Extension Building on UAEU Campus. *Buildings* **2022**, *12*, 1101. [CrossRef]
12. Bott, H.; Grassl, G.C.; Anders, S. *Sustainable Urban Planning*; Edition Detail: Berlin, Germany, 2019; p. 201.
13. Kim, H.J.; Choi, W.S.; Kim, J.W. A study on parametric design tool for residential buildings securing valid sunlight hours on the winter solstice. *J. Asian Archit. Build. Eng.* **2022**, *21*, 1657–1676. [CrossRef]
14. Toutou, A.; Fikry, M.; Mohamed, W. The parametric based optimization framework daylighting and energy performance in residential buildings in hot arid zone. *Alex. Eng. J.* **2018**, *57*, 3595–3608. [CrossRef]
15. Liu, H.; Fu, H.; Sun, L.; Lee, C.; Yeatman, E.M. Hybrid energy harvesting technology: From materials, structural design, system integration to applications. *Renew. Sustain. Energy Rev.* **2021**, *137*, 110473. [CrossRef]
16. Ryu, H.; Yoon, H.J.; Kim, S.W. Hybrid energy harvesters: Toward sustainable energy harvesting. *Adv. Mater.* **2019**, *31*, 1802898. [CrossRef]
17. Icaza, D.; Borge-Diez, D.; Pulla Galindo, S.; Flores-Vázquez, C. Modeling and simulation of a hybrid system of solar panels and wind turbines for the supply of autonomous electrical energy to organic architectures. *Energies* **2020**, *13*, 4649. [CrossRef]
18. Škvorc, P.; Kozmar, H. Wind energy harnessing on tall buildings in urban environments. *Renew. Sustain. Energy Rev.* **2021**, *152*, 111662. [CrossRef]
19. Lucchi, E.; Lopez, C.S.; Franco, G. A conceptual framework on the integration of solar energy systems in heritage sites and buildings. *IOP Conf. Ser. Mater. Sci. Eng.* **2020**, *949*, 012113. [CrossRef]
20. Hao, D.; Qi, L.; Tairab, A.M.; Ahmed, A.; Azam, A.; Luo, D.; Pan, Y.; Zhang, Z.; Yan, J. Solar energy harvesting technologies for PV self-powered applications: A comprehensive review. *Renew. Energy* **2022**, *188*, 678–697. [CrossRef]
21. Loferski, J.J. The first forty years: A brief history of the modern photovoltaic age. *Prog. Photovolt. Res. Appl.* **1993**, *1*, 67–78. [CrossRef]
22. Bogue, R. Solar-powered sensors: A review of products and applications. *Sens. Rev.* **2012**, *32*, 95–100. [CrossRef]
23. Harb, A. Energy harvesting: State-of-the-art. *Renew. Energy* **2011**, *36*, 2641–2654. [CrossRef]
24. Yu, C.; Park, J.; Youn, J.R.; Song, Y.S. Sustainable solar energy harvesting using phase change material (PCM) embedded pyroelectric system. *Energy Convers. Manag.* **2022**, *253*, 115145. [CrossRef]
25. Páez-Montoro, A.; García-Valderas, M.; Olías-Ruíz, E.; López-Ongil, C. Solar energy harvesting to improve capabilities of wearable devices. *Sensors* **2022**, *22*, 3950. [CrossRef]
26. Dutton, A.; Halliday, J.; Blanch, M. *The Feasibility of Building-Mounted/Integrated Wind Turbines (BUWTs): Achieving Their Potential for Carbon Emission Reductions*; Energy Research Unit, CCLRC: Didcot, UK, 2005; pp. 77–83.
27. Juan, Y.H.; Wen, C.Y.; Li, Z.; Yang, A.S. Impacts of urban morphology on improving urban wind energy potential for generic high-rise building arrays. *Appl. Energy* **2021**, *299*, 117304. [CrossRef]
28. Grant, A.; Johnstone, C.; Kelly, N. Urban wind energy conversion: The potential of ducted turbines. *Renew. Energy* **2008**, *33*, 1157–1163. [CrossRef]
29. Henderson, A.R.; Morgan, C.; Smith, B.; Sørensen, H.C.; Barthelmie, R.J.; Boesmans, B. Offshore wind energy in Europe—A review of the state-of-the-art. *Wind Energy Int. J. Prog. Appl. Wind Power Convers. Technol.* **2003**, *6*, 35–52. [CrossRef]
30. Nazir, M.S.; Mahdi, A.J.; Bilal, M.; Sohail, H.M.; Ali, N.; Iqbal, H.M. Environmental impact and pollution-related challenges of renewable wind energy paradigm—A review. *Sci. Total Environ.* **2019**, *683*, 436–444. [CrossRef] [PubMed]
31. Briand, D.; Yeatman, E.; Roundy, S.; Brand, O.; Fedder, G.K.; Hierold, C.; Korvink, J.G.; Tabata, O. *Micro Energy Harvesting*; Wiley, Online Library: Hoboken, NJ, USA, 2015.
32. Gould, C.; Edwards, R. Review on micro-energy harvesting technologies. In Proceedings of the 51st International Universities Power Engineering Conference (UPEC), Coimbra, Portugal, 6–9 September 2016.
33. Monedero, J. Parametric design: A review and some experiences. *Autom. Constr.* **2000**, *9*, 369–377. [CrossRef]
34. Oxman, R. Thinking difference: Theories and models of parametric design thinking. *Des. Stud.* **2017**, *52*, 4–39. [CrossRef]
35. Woodbury, R.; Williamson, S.; Beesley, P. Parametric modelling as a design representation in architecture:a process account. In Proceedings of the Canadian Design Engineering Network (CDEN) Conference, Toronto, ON, Canada, 24–26 July 2006.
36. Hernandez, C.R.B. Thinking parametric design: Introducing parametric Gaudi. *Des. Stud.* **2006**, *27*, 309–324. [CrossRef]
37. Smith, P. BIM implementation–global strategies. *Procedia Eng.* **2014**, *85*, 482–492. [CrossRef]
38. Revit Software. Available online: https://www.autodesk.com/products/revit/overview?plc=RVT&term=1-YEAR&support=ADVANCED&quantity=1 (accessed on 18 June 2022).
39. Dynamo, a Revit Software Plug-In. Available online: http://dynamobim.org/download (accessed on 25 June 2022).
40. Insight, a Revit Software Plug-In. Available online: https://www.autodesk.com/products/insight/overview (accessed on 28 June 2022).
41. Reinberg, G.W. *Architecture for a Solar Future*; Birkhauser: Basel, Switzerland, 2021; pp. 13–17.
42. Salimzadeh, N.; Vahdatikhaki, F.; Hammad, A. Parametric modeling and surface-specific sensitivity analysis of PV module layout on building skin using BIM. *Energy Build.* **2020**, *216*, 109953. [CrossRef]
43. Jayathissa, P.; Caranovic, S.; Hofer, J.; Nagy, Z.; Schlueter, A. Performative design environment for kinetic photovoltaic architecture. *Autom. Constr.* **2018**, *93*, 339–347. [CrossRef]

44. Hofer, J.; Groenewolt, A.; Jayathissa, P.; Nagy, Z.; Schlueter, A. Parametric analysis and systems design of dynamic photovoltaic shading modules. *Energy Sci. Eng.* **2016**, *4*, 134–152. [CrossRef]
45. Guo, M.; Zang, H.; Gao, S.; Chen, T.; Xiao, J.; Cheng, L.; Wei, Z.; Sun, G. Optimal tilt angle and orientation of photovoltaic modules using HS algorithm in different climates of China. *Appl. Sci.* **2017**, *7*, 1028. [CrossRef]
46. Park, H.S.; Koo, C.; Hong, T.; Oh, J.; Jeong, K. A finite element model for estimating the techno-economic performance of the building-integrated photovoltaic blind. *Appl. Energy* **2016**, *179*, 211–227. [CrossRef]
47. Lim, H.; Cho, S.H.; Moon, J.; Jun, D.Y.; Kim, S.H. Effects of Reflectance of Backsheets and Spacing between Cells on Photovoltaic Modules. *Appl. Sci.* **2022**, *12*, 443. [CrossRef]
48. Taveres-Cachat, E.; Lobaccaro, G.; Goia, F.; Chaudhary, G. A methodology to improve the performance of PV integrated shading devices using multi-objective optimization. *Appl. Energy* **2019**, *247*, 731–744. [CrossRef]
49. Flickr. Available online: https://www.flickr.com/photos/marcteer/5459146616/in/photolist-9jpy6w-9jpxSq-eXht8m-tp5HC7-t3BDLg-3aa28q-c39zY-UC1oD6-2n1KK8N-PgqX4-pnHCmr-sDqSpT-hAem6g-2jVyreW-c39jo9-owDL2h-2hzH4sN-edgf9y-pCG3nP-8iRbr6-2kQJ1n3-dobqhE-ef31RN-62ruG7-656ooz-mcxWLj-q6vfvT-2hZwNjR-fUWrGV-e27N36-o9dPcj-eX63CD-fHiHyD-ddJqZp-36CmsU-ggwZ2j-Sc8sEo-fUVwbd-AqkZ92-vfsiyR-JrP5bX-4GPSQX-5DRXcL-3B2B53-8szJN-3B2xiE-ja67WH-3B2EHy-g53787-hj6qzv (accessed on 12 January 2023).
50. Google Maps. Available online: https://www.google.com/maps/place/Walkeweg+61,+4053+Basel,+%EC%8A%A4%EC%9C%84%EC%8A%A4/@47.5416222,7.6094535,17z/data=!3m1!4b1!4m6!3m5!1s0x4791b830721fb075:0x4226c7a9bed34898!8m2!3d47.5416222!4d7.6094535!16s%2Fg%2F11c1y3qg3z (accessed on 28 January 2023).
51. Baker, A. How to Calculate Your Peak Sun-Hours. Available online: http://waterfrontballparkdistrict.com.s3.amazonaws.com/10.%20Remainder/AR%200078729-%20AR%200078739.pdf (accessed on 10 January 2023).
52. Foster, R.; Ghassemi, M.; Cota, A. *Solar Energy: Renewable Energy and the Environment*; CRC Press: Boca Raton, FL, USA, 2009; p. 45.
53. Solarhub. Available online: https://solarhub.net.au/product/q-cells-q-peak-duo/ (accessed on 3 February 2023).
54. Calabrò, E. An algorithm to determine the optimum tilt angle of a solar panel from global horizontal solar radiation. *J. Renew. Energy* **2013**, *2013*, 307547. [CrossRef]
55. Nguyen, D.M.; Grace, D.; Runeson, G. Energy and economic analysis of environmental upgrading of existing office buildings. *Constr. Econ. Build.* **2022**, *20*, 82–102. [CrossRef]
56. Nadia, A.R.; Isa, N.A.M.; Desa, M.K.M. Efficient single and dual axis solar tracking system controllers based on adaptive neural fuzzy inference system. *J. King Saud Univ.-Eng. Sci.* **2020**, *32*, 459–469.
57. Rambhowan, Y.; Oree, V. Improving the dual-axis solar tracking system efficiency via drive power consumption optimization. *Appl. Sol. Energy* **2014**, *50*, 74–80. [CrossRef]

applied sciences

MDPI

Article

Methodology to Identify and Prioritise the Sustainability Aspects to Be Considered in the Design of Brazilian Healthcare Buildings

Marília Costa, Maria de Fátima Castro and Ricardo Mateus *

University of Minho, ISISE, ARISE, Department of Civil Engineering, 4800 058 Guimarães, Portugal
* Correspondence: ricardomateus@civil.uminho.pt

Abstract: The evidence of climate change has increased the necessity for actions to avoid severe consequences for future generations. Recognising the benefits of sustainable and energy-efficient buildings has led to the development of several methods to estimate and rate the performance of building. Healthcare buildings are functional and dynamic structures that support their occupants' healing processes and comfort. They have a significant role in society and specific goals in merging their strategic planning requirements (cost reduction, regulatory compliance, social responsibility, and performance improvement) with Sustainable Development Goals. Therefore, this paper addresses critical issues regarding the sustainability of the Brazilian healthcare sector by analysing the suitability of the most common international healthcare building sustainability assessment methods to the specific social, economic, and environmental contexts of the Brazilian healthcare sector. The paper analyses the HBSAtool-PT, a sustainability assessment method developed for this type of building, and identifies the possibilities for adapting to the Brazilian context. As a result, this research proposes a framework to assess the sustainability of healthcare buildings in Brazil. In the framework's definition, a survey was conducted to determine the opinion of building stakeholders regarding the importance of indicators. A weighing system for the proposed sustainability indicators was developed using the AHP method.

Keywords: healthcare buildings; sustainable construction; building sustainability assessment method; sustainable certification; rating systems

1. Introduction

The building industry is one of the largest sectors involving people (services) and materials. It significantly impacts the environment, economy, and society because it consumes large amounts of natural resources [1]. As reported by the World Business Council for Sustainable Development, the building sector accounts for about 40% of total energy use, 30% of greenhouse gas (GHG) emissions, 17% of freshwater consumption, and 25% of harvested wood and is responsible for the production of 45% to 65% of disposed waste in landfills [2].

Consequently, mitigating the environmental impacts of the building sector has become a significant issue, and projections for the future global population and economic growth predict a steady increase in the worldwide energy demand [2]. Energy use in the built environment is the main factor causing climate change. Therefore, promoting energy-efficient buildings is crucial to ensure a more sustainable future [3].

According to Happio and Viietaniemi [4], numerous building sustainability assessment (BSA) methods have been developed over the last decades to evaluate building performance and to analyse the buildings' life cycle impacts on the environment, economy, public health, and well-being in cities. These methods are based on a list of indicators organised in different sustainability dimensions.

BSA methods are tools for evaluating building performance and supporting sustainable design. Therefore, they are essential to support decision making towards implementing sustainable principles in existing and new buildings. It allows us to obtain a sustainability rate after a detailed report [5] and serves as a management tool or guideline to address sustainability targets during the design, construction, and operation phases. Considering the effect these methods can have in fostering the implementation of sustainable buildings, some governments offer tax incentives for buildings that fulfil certain sustainability rating thresholds [6].

Globally, there are different BSA methods developed by private and public institutions. The approaches share the general aim of promoting sustainable buildings and are based on a similar list of core sustainability indicators. Still, they differ in adjusting to the specific territory's environmental, economic, and sociocultural contexts [6].

Healthcare buildings are very complex due to the different types of users, the dynamic demand of spaces, the integration of constantly evolving technologies and systems and their social contribution to the community. Incorporating sustainable development criteria in the design strategy is considered the main factor for the success of sustainable hospital projects. Other essential factors are promoting sustainable initiatives, measuring risks and impacts (goals, metrics, monitoring, and evaluation), management programs (water and energy efficiency), low-impact maintenance products, actions towards transparency, educational programs, communication, and response teams.

In Brazil, the healthcare sector is starting to look at sustainability as an opportunity, and efforts are being made to define the sustainability indicators for different healthcare buildings and locations. One of the limitations of developing a method to support the practical implementation of sustainability goals in the Brazilian healthcare sector is the need for comprehensive information about its performance. In addition, the Brazilian territory is very vast. Therefore, diverse economic, environmental, and social goals make it challenging to develop an effective method to be applied nationwide [7]. The analysis and interpretation of the current situation in Brazil are crucial in developing a sustainability assessment framework that fits the market needs and stakeholders' expectations.

1.1. Challenges for Healthcare Buildings in Brazil

Healthcare is a water- and energy-intensive service, mainly due to the continuous operations that require, among others, light, heat, intensive ventilation, sterilisation, and food preparation. It is responsible for generating significant waste and producing polluting emissions. The built environment accounts for about 40% of all CO_2 emissions, and hospitals alone account for about 4% of the total emissions of the built environment. Therefore, it is necessary that hospitals' stakeholders feel the urgency to undertake actions to reduce these greenhouse gas emissions. CO_2 emissions are prone to cause respiratory diseases and other illnesses. Thus, hospitals are undermining the communities' health [8]. Since hospitals provide patient care for people within a community, they can be characterised as inherent to social responsibility by taking care of patients as their core business. In sharp contrast to this, hospitals are among the most significant contributors to climate change [9].

Brazil's health sector has undergone many changes in the last few years. Concerns regarding sustainability are becoming more evident, mainly because of society's awareness of the growing environmental issues and how the poor quality of indoor environments affects the health and comfort of patients and employees. Additionally, constant technological changes and the need to incorporate them into hospitals lead to a plurality of design issues that must be considered, such as building flexibility and adaptability.

The need to apply modern concepts in hospital management influences the development of suitable techniques and tools that can support decision making towards integrating sustainability principles that contribute to economic growth, social satisfaction, and the preservation of the natural environment. The design professional can improve the healthcare environment, from being cold, sober, and intimidating to a humanised, ludic, and secure space, increasing the healing ambience of the patient and reducing the hospitalisation period. In addition, designing rest areas enables a better working environment for the employees, resulting in better performance and safety. Besides this set of concerns, merging all specific technical requirements with sustainable development goals is necessary.

1.2. Healthcare Building Sustainability Assessment Methods and Their Application in the Brazilian Context

With regard to implementing the sustainable healthcare building concept, several assessment methods define the criteria that can guide professionals to meet the standards and develop high-performance solutions. Nevertheless, some issues still need to be considered for better accuracy. For example, comprehensive databases that gather life cycle information about buildings and materials are required, as are more and better correlations between construction and operational costs. In the construction sector, traditional and outdated processes lead to multiple construction processes and considerable heterogeneity of the final products [10].

According to Bellen [11], comparing existing methods allows for identifying the main advantages and limitations of the different existing evaluation processes. Comparative analysis will enable groups with various objectives to choose the most appropriate method for achieving their goals. Many countries either have or are developing sustainability assessment methods, which makes the need for coordination increasingly relevant. In Brazil, the assessment tool for healthcare buildings is mainly LEED, but there needs to be a prior adaptation to the country's specific context.

2. Objectives

Underestimated decisions at the early design stage are often tricky and expensive to remediate during the operation stage. In addition, the environmental impact and cost can be massive for large-scale projects, such as healthcare buildings, if they are not completed well.

This paper aims to identify and analyse the sustainability assessment methods for healthcare buildings, studying their suitability and the need to be adapted to the Brazilian context. This analysis focuses on assessing the potentiality of adaptation of the Portuguese method for healthcare buildings (HSBAtool-PT) to the Brazilian context. The objective is to define a list of sustainability criteria, adapted to the Brazilian environmental, societal, and economic contexts that can support decision making in the design of more sustainable healthcare buildings in Brazilian. Additionally, it will propose a system of weights for the defined list of indicators, based on the opinion of different Brazilian stakeholders of the healthcare sector. The proposed system of weight enables the identification, from the preliminary design stages, of the priorities to consider in the sustainable design of healthcare buildings. In addition, it allows comparing the sustainability priorities of different countries in the context of these type of buildings.

The HBSAtool-PT, LEED v4 for Building Design and Construction, and BREEAM International New Constructions 2016 were the BSA methods analysed to evaluate the potentialities and possibilities for adaptation. The lists of sustainability indicators were studied comparatively, along with the weights and assessment methods used.

A list of indicators was defined based on the abovementioned assessment methods and validated for the Brazilian context, considering the opinion of a group of experts. Afterwards, the sustainability priorities (weighting system) of the proposed list of indicators were outlined using a survey involving the main stakeholders on the design, management, and maintenance of healthcare buildings. This list would define the priorities of a sustainable healthcare building in Brazil.

3. Methodology

This study seeks to analyse different sustainability assessment methods for healthcare buildings to highlight the key similarities and differences and propose a list of sustainability indicators and a system of weights for use in the Brazilian context. Two international methods have been selected to attain these goals: BREEAM International New Constructions

2016 and LEED v4 for Building Design and Construction, the leading systems operated by well-known organisations (BRE and USGBC) with a proven record in the domain of sustainability development. A method developed for a specific country, i.e., HBSAtool-PT, is also considered.

3.1. Comparative Matrix of Healthcare Assessment Methods

The comparative analysis aims at contrasting the selected rating systems for assessing the performance of healthcare buildings. Considering the criteria requirements and guidelines of the different methods used, it is possible to identify, categorise, and standardise the critical building performance [12] criteria that will influence the decision-making process the most.

Analysing the categories of the considered HBSA methods, it is possible to conclude that there are, on one hand, different categories referring to the same sustainability criteria and, on the other, categories with similar denominations that assess completely different criteria. Therefore, thirteen core sustainability criteria were identified, and the categories of the different methods analysed were grouped in this framework. According to this organisation, the categories most commonly assessed within the analysed methods are Energy (21%) and Indoor Environmental Quality (17%). Other important categories are Materials and Resources (13%), followed by Site Quality, Management, Water, Waste, and Transport and Pollution, which are assessed by the majority of the methods. Nevertheless, there are some categories only considered by a few or one of the BSA methods. For example, outdoor quality and the economy are two categories that are only covered by the HBSAtool-PT method.

The HBSAtool-PT method was designed specifically for Portuguese healthcare buildings, considering the national context, European sustainability standards and policies, and other existing and recognised sustainability methods for this type of building [13]. It can evaluate new, existing, and renovated buildings, allowing for future adaptations to new guidelines, standards, or national laws. Thus, it aims to be practical, easily understandable, and flexible to be adapted. This is important to promote sustainable development, construction, operation, and maintenance. The structure of the method is based on 52 indicators, divided into 22 categories, which are integrated into five areas: Environmental; Sociocultural and functional; Economic; Technical; and Site. The sustainability label delivered communicates the performance at the level of each sustainability area and category, allowing a better comparison between different design scenarios.

The proposed research methodology is based on the exact steps of development of the HBSA method for the specific context of Portugal but considers the Brazilian context. The following research steps are considered:

1. To define the preliminary list of sustainability indicators based on internationally recognised HBSA methods and international goals for sustainability;
2. To pre-validate the preliminary list of indicators through interviews with the most recognised experts and policymakers in the context of the Brazilian healthcare sector;
3. To propose the final list of indicators based on a survey (Appendix A) distributed to the main Brazilian healthcare buildings' stakeholders;
4. To develop the system of weights of the proposed list of indicators by applying the Analytical Hierarchy Process (AHP) method.

3.2. Development of the List of Indicators and Questionnaire

The definition of the initial set of sustainable indicators comprehended all criteria identified in existing methods for the healthcare context. It was based on the literature and the analysis of the following data:

- Two most used assessment methods at the international level to assess the sustainability of healthcare buildings, namely: BREEAM International New Constructions 2016 [14] and LEED v4 for Building Design and Construction [15];
- The list of areas, categories, and indicators of the HBSAtool-PT method [16];

- The Global Agenda for Green and Healthy Hospitals objectives [17].

The initial set of indicators included all different environmental, social, economic, technical, and site indicators identified in the BSA methods mentioned before [18,19]. The methodology was similar to the one used by other authors to develop BSA methods for specific contexts [20,21].

Afterwards, three experts (well-known Brazilian academics and practitioners in sustainability and healthcare buildings) were invited to discuss the results and validate the preliminary list of sustainable categories and indicators. These experts include an ex-president from the Brazilian Association for the Development of Hospital Buildings; a member of the International Academy for Design and Health–South American Chapter; and a professional consultant in building sustainability assessment.

The discussions were carried out through online interviews, which allowed the definition of the list of sustainability categories and related indicators. Some were highlighted and considered a priority according to the Brazilian context. Following the experts' recommendations, one indicator was added to the preliminary list of indicators (Table 1).

Table 1. Indicators suggested by the experts that were not included in the preliminary list of indicators.

Suggested Indicators	Area	Justification
Corruption avoiding plan	Economy	Considering the Brazilian situation, the Economy area should have sustainable purchase policies in order to prevent extortions.
Olfactory Comfort	Sociocultural and functional	To enhance the comfort of users in hospitals by the minimisation of olfactory discomfort and improvement of indoor air quality.
Contingency Plan	Sociocultural and functional	A very punctual situation caused by emergencies in cases of serious claims (e.g., pandemics, floods).
Universal Design	Technical	Composition of an environment that can be accessed, understood, and used by all people, regardless of their age, size, ability, or disability.
Space flexibility and Space adaptability	Sociocultural and functional	Merge flexibility and adaptability into just one indicator because it makes more sense in this type of building.

Experts recommended adding the corruption avoidance plan to the preliminary list of indicators because of its relation to the political context. It is necessary to understand how each healthcare institution implements anti-corruption measures.

On the other hand, the criteria of olfactory comfort are used in the Haute Qualité Environnementale (HQE) and sub-targeted into objectives, such as the provision of efficient ventilation, the minimisation of olfactory discomfort, and the improvement of indoor air quality. These items are already embedded into the User's health and comfort category. Therefore, they will not be considered as indicators to be added.

The Contingency Plan is another item that would not be interesting to include in the list presented because it is only used in odd circumstances and is challenging to measure. In addition, Brazil has no occurrences of natural disasters (e.g., hurricanes, earthquakes, and tsunamis). The Universal Design was not considered because the research is based on the added value, meaning that only the improvements compared to the standard Brazilian design practices will be considered. Universal Design is already mandatory in the design of healthcare buildings (according to, e.g., ANVISA–RDC 50/2002 [22] and ABNT NBR 9050 [23]).

It was also suggested to merge two items into Space flexibility and Space adaptability. Still, it describes the changes that can happen during the period of a day, using furniture elements, doors, panels, and curtains, while the other describes the adaptive capacity of the room in case of renovation.

As a result of the analysis of the existing assessment methods and the expert's opinions, the draft list of indicators (Table 2) was used to develop the survey to produce the final list of indicators. The draft list presents fifty-seven indicators, organised into twenty-three categories: C1 Environmental life cycle impacts assessment; C2 Energy; C3 Soil use and biodiversity; C4 Materials and Solid Waste; C5 Water; C6 Users' health and Comfort; C7 Controllability by the user; C8 Landscaping; C9 Passive design; C10 Mobility plan; C11 Space flexibility and adaptability; C12 Life cycle costs; C13 Local economy; C14 Corruption avoidance plan; C15 Environmental management systems; C16 Technical systems; C17 Security; C18 Durability; C19 Awareness and education for sustainability; C20 Skills in sustainability; C21 Local community; C22 Cultural value; and C23 Conveniences. This list was included in the survey (Appendix A) conducted to healthcare buildings stakeholders.

During the survey, a descriptive methodology was used. Descriptive research involves gathering data that describe events which are then organised, recorded, and analysed. The survey consisted of fifty-seven questions, organised into six sections. The first section was a cover letter that explained the research goals and stated the privacy policies of the collected data. The following section allowed for identifying how involved in the life cycle of a healthcare building the stakeholder was, and the remaining sections recognised the relative importance of each indicator according to each stakeholder's perspective. Respondents were asked to give their opinion regarding the matter of each indicator, using a Likert scale organised into five levels of importance: not at all important (1); slightly important (2); moderately important (3); very important (4); and extremely important (5).

Table 2. List of sustainability areas, categories, and indicators included in the survey.

Areas	Categories	Indicators
A1 Environmental	C1 Environmental life cycle impacts assessment	I1 Assessment of the buildings' life cycle impacts
	C2 Energy	I2 Primary energy consumption
		I3 Onsite energy production
		I4 Energy Performance
	C3 Soil use and biodiversity	I5 Layout optimisation
		I6 Soil sealing
		I7 Reuse of previously built or contaminated areas
		I8 Ecological protection of the site
		I9 Rehabilitation of the surrounding
		I10 Use of native plants
		I11 Heat Island effect
	C4 Materials and Solid Waste	I12 Construction waste
		I13 Reused and recycled products and materials
		I14 Waste separation and storage
		I15 Responsible sourcing of materials
	C5 Water	I16 Potable water consumption
		I17 Recycling and reuse of effluents
		I18 Treatment of contaminated effluents
		I19 Water efficient equipment

Table 2. *Cont.*

Areas	Categories	Indicators
A2 Sociocultural and functional	C6 Occupants' health and comfort	I20 Natural ventilation
		I21 Toxicity of finishing materials
		I22 Thermal comfort
		I23 Visual comfort
		I24 Acoustic comfort
		I25 Indoor air quality
	C7 Controllability by the user	I26 Ventilation and temperature control
		I27 Natural light adjustment
	C8 Landscaping	I28 Visual link with the surrounding landscape
	C9 Passive design	I29 Layout and orientation
		I30 Passive systems
	C10 Mobility plan	I31 Accessibilities
	C11 Space flexibility and adaptability	I32 Availability and accessibility of social areas
		I33 Space optimisation
		I34 Space flexibility
		I35 Space adaptability
A3 Economy	C12 Life cycle costs	I36 Initial cost
		I37 Operational costs
	C13 Local economy	I38 Hiring local goods and services
	C14 Corruption avoidance plan	I39 Sustainable purchase policies
A4 Technical	C15 Environmental management systems	I40 Commissioning
		I41 Environmental management plan
		I42 Infection control
		I43 Reducing noise pollution
	C16 Technical systems	I44 Efficiency of lighting and air conditioning systems
	C17 Security	I45 Occupants safety
		I46 Responsible construction practices
	C18 Durability	I47 Materials of high strength and durability
		I48 Proper selection of furniture
	C19 Awareness and education for sustainability	I49 Education of occupants
		I50 Education of service providers
		I51 Satisfaction surveys
	C20 Skills in sustainability	I52 Integration in the team of a qualified sustainability expert
A5 Site	C21 Local community	I53 Local community development
	C22 Cultural value	I54 Heritage framework
	C23 Conveniences	I55 Accessibility to public transport
		I56 Low-impact mobility
		I57 Local amenities

In the final part of the questionnaire, the respondents were asked to provide their opinion on the list of indicators by stating if, according to their experience, it was necessary or not to amend the list of sustainability indicators.

3.3. Sampling Process

The survey was intended for professionals developing a professional activity in the healthcare sector or sustainable development. It included healthcare building designers, managers, and stakeholders with sustainable development experience. To achieve broad participation, an internet-based questionnaire was sent by e-mail to the stakeholders with a cover letter describing the study's aims and stating the data privacy issues.

The overall sample (Table 3) consisted of fifty-five respondents, 40% of which were architects, followed by civil engineers (38%), hospital managers (18%), and sustainable construction consultants or experts (4%).

Table 3. Feedback numbers from each field.

Field		Frequency	Percentage
Architect	Less than 5 years of experience	2	3.6%
	More than 5 years of experience	8	14.5%
	Specialists in the hospital sector	12	21.8%
Civil Engineer	Less than 5 years of experience	2	3.6%
	More than 5 years of experience	13	23.6%
	Specialists in the hospital sector	6	10.9%
Sustainable Construction Consultant/Expert		2	3.6%
Hospital Manager	Facilities and Equipment Services	2	3.6%
	Superintendence/Management	4	7.3%
	Others	4	7.3%
Total		55	100%

Figure 1 presents the geographic distribution of the Brazilian locations where the stakeholders typically develop their activity. As expected, the majority of people who answered are from the Southeast region of Brazil (59%), where most hospitals are concentrated, followed by the South (15%) and Northeast (11%). A further 9% of the participants develop activities all over the country.

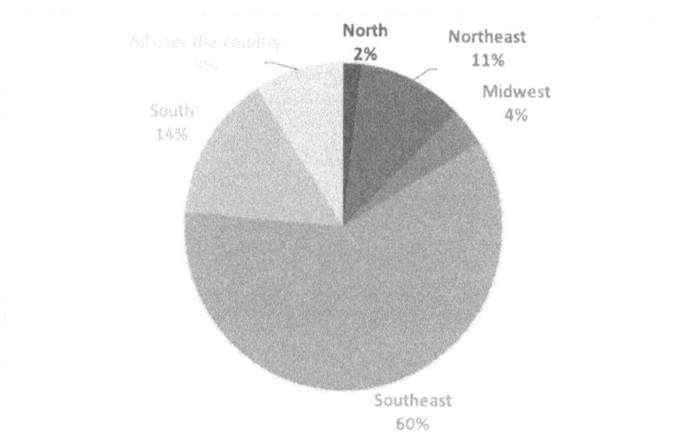

Figure 1. Geographic distribution of the activities developed by the participants in the survey.

3.4. *Development of the Weighting System*

Panel weighing is considered the most accurate method for defining the weighting system of a BSA method [24]. The more prominent and representative the panel, the more precise the result is. In this study, the Analytical Hierarchy Process (AHP) panellist's method was applied to evaluate the relative weight of each indicator. The AHP is an organisational and analytical mathematical method for complex priorities and decisions. It was developed in the 1970s and prioritised human decision making in different fields such as policies, business, project selection, healthcare, and education [25].

The AHP is a simple technique that can translate the decision-makers' qualitative and quantitative data evaluations into multi-criteria rankings. In addition, the AHP includes a valuable tool for checking the consistency of the decision-makers' evaluations, thus reducing bias in the decision-making process [26].

The AHP involves the following steps to hierarchise the priorities of a project [25]:

- Modelling the problem as a hierarchy containing the goal decision, alternatives for reaching it, and criteria for evaluating;
- Establishing priorities among the elements of the hierarchy by making a series of judgments based on pairwise comparisons between them;
- Synthesising these judgments to yield a set of overall priorities for the hierarchy;
- Checking consistency;
- Achieve a final decision based on the results of this process.

The professionals were asked to provide their opinion on the importance of the five sustainability areas (environmental, sociocultural, functional, economy, technical, and site) in the overall building sustainability level. Inside each sustainability area, the respondents ranked the importance of each indicator using the five-point Likert scale presented before. Based on the recommendations of the experts interviewed. On the methodology used in other similar studies, e.g., Mateus and Bragança [18], for the practical use of the method under development, the final list of indicators should be as compact as possible while including the most critical indicators. Therefore, the last step was to reduce the list of indicators by removing those that, according to the results, had a shallow contribution to overall sustainability. At this level, it was decided to remove the indicators with a weight lower than 0.2% from the list. The reasoning for this cut-off rule comes from analysing the abovementioned BSA methods. The lowest weight in BREEAM is 0.4%, 0.9% in LEED, and 0.2% in HBSATool-PT.

Therefore, a list considering only the indicators that contributed 0.2% or more to the overall sustainability level was proposed, which meant that the list presented in Table 2 was reduced to 48 indicators. In this process, the following indicators were excluded from the final list: Layout optimisation; Reuse of previously built or contaminated areas; Use of native plants; Heat Island effect; Visual comfort; Commissioning; Proper selection of furniture; Education of occupants; and Satisfaction surveys. Using the AHP method again, the weight of the final list of indicators was calculated.

4. Presentation of Results and Discussion

After removing the indicators with a weight below the cut-off boundary, the weights of the remaining indicators, categories, and sustainability areas were recalculated using the AHP method.

Figure 2 presents the relative importance of the sustainability areas and shows that according to the professionals' opinion, the Environmental (A1), Sociocultural and Functional (A2), and Technical (A4) areas have similar relative importance. Subsequently, the Economy (A3) area's importance was 17%, and Site (A5) was appointed as the least relevant, corresponding to a weight of 8%.

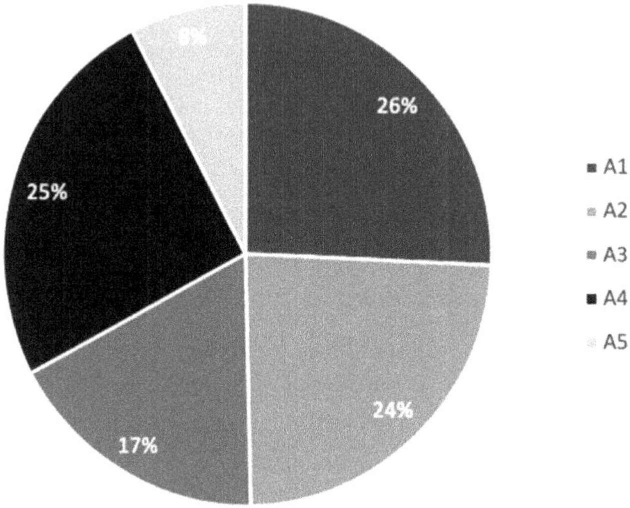

Figure 2. Final weight of the sustainability areas.

Analysing the final weights of the indicators, the "technical systems" and "corruption avoidance plan" were both ranked as the most important ones in the "economic" and "technical" sustainability areas, respectively, accounting for almost about 10% of the total weight in each category. The third most prioritised category in sustainable healthcare buildings was water, with potable water consumption and treatment of contaminated effluents as the most critical indicators.

The "Sociocultural and functional" area was ranked the third most important in healthcare buildings. The most important categories of this area are "mobility plan" and "controllability by the user", both with a weight of 33% inside the area. Furthermore, the fourth-ranked category was "environmental life cycle impacts assessment", closely followed by the "life cycle costs" category. The "Operational costs within the life cycle costs" category scored highest. Finally, in the "Sociocultural and functional" area, the less important categories are "landscaping" and "passive design".

The categories proposed for the "Environmental" area were defined to allow, in a holistic way, the assessment of the most common life-cycle environmental impacts according to national priorities. The "Social and Functional" area presented a list of indicators divided into six categories, including the critical aspects of building occupants' health and comfort, considering the importance of mobility and space design quality.

The "Economic" area was defined to consider the most relevant building life-cycle costs and the impact on the local economy. On the other hand, the "Technical" area reflected issues about security, durability, management of systems and their efficiency and the occupants' education and awareness regarding sustainability concerns. In contrast, the "Site" areas' main issues regarded community and cultural values.

The proposed list of sustainability areas, categories and indicators is presented in Table 4, together with the weights assigned by respondents to each sustainability category and area.

Analysing Table 4, it is essential to highlight that the categories' weight was equal to that of the indicator in categories with only one indicator. Analysing each category made it possible to understand the most important Indicators according to the respondents' opinions.

Table 4. The weighing system of the proposed list of sustainability indicators.

Area	Category	Indicator	Weight	
			Indicator	Category
A1. Environmental	C1. Environmental life cycle impacts assessment	I1. Assessment of the buildings' life cycle impacts	100%	24%
	C2. Energy	I2. Primary energy consumption	40%	13%
		I3. Onsite energy production	40%	
		I4. Energy Performance	20%	
	C3. Soil use and biodiversity	I5. Soil sealing	26%	11%
		I6. Ecological protection of the site	41%	
		I7. Rehabilitation of the surrounding	33%	
	C4. Materials and solid waste	I8. Construction waste	24%	19%
		I9. Reused and recycled products and materials	19%	
		I10. Waste separation and storage	47%	
		I11. Responsible sourcing of materials	10%	
	C5. Water	I12. Potable water consumption	33%	33%
		I13. Recycling and recovery of effluents	17%	
		I14. Treatment of contaminated effluents	33%	
		I15. Water efficient equipment	17%	
A2. Sociocultural and functional	C6. Occupant's health and comfort	I16. Natural ventilation	12%	16%
		I17. Toxicity of finishing materials	10%	
		I18. Thermal comfort	24%	
		I19. Acoustic comfort	11%	
		I20. Indoor air quality	42%	
	C7. Controllability by the user	I21. Ventilation and temperature control	50%	33%
		I22. Natural light adjustment	50%	
	C8. Landscaping	I23. Visual link with the surrounding landscape	100%	3%
	C9. Passive design	I24. Layout and orientation	67%	4%
		I25. Passive systems	33%	
	C10. Mobility plan	I26. Accessibilities	100%	33%
	C11. Space flexibility and adaptability	I27. Availability and accessibility of social areas	36%	10%
		I28. Space optimisation	33%	
		I29. Space flexibility	15%	
		I30. Space adaptability	16%	
A3. Economic	C12. Life cycle costs	I31. Initial cost	20%	33%
		I32. Operational costs	80%	
	C13. Local economy	I33. Hiring local goods and services	100%	10%
	C14. Corruption avoidance plan	I34. Sustainable purchase policies	100%	57%

Table 4. *Cont.*

Area	Category	Indicator	Weight	
			Indicator	Category
A4. Technical	C15. Environmental management systems	I35. Environmental management plan	24%	19%
		I36. Infection control	62%	
		I37. Reducing noise pollution	14%	
	C16. Technical systems	I38. Efficiency of lighting and air conditioning systems	100%	39%
	C17. Security	I39. Occupants' safety	75%	18%
		I40. Responsible construction practices	25%	
	C18. Durability	I41. Materials of high strength and durability	100%	9%
	C19. Education	I42. Education of service providers	100%	10%
	C20. Skills in sustainability	I43. Integration of a qualified sustainability expert	100%	5%
A5. Site	C21. Local community	I44. Local community development	100%	17%
	C22. Cultural value	I45. Heritage framework	100%	17%
	C23. Conveniences	I46. Accessibility to public transport	69%	67%
		I47. Low impact mobility	24%	
		I48. Local amenities	7%	

In Category C2 (Energy), the weight of the two first Indicators (I2—Primary energy consumption; I3—Onsite energy production) was the same, whereas Indicator I4 (Minimum energy performance) had the lowest weight. In Category C3 (Soil use and biodiversity), the most important was I6 (Ecological protection of the site), closely followed by I7 (Rehabilitation of the surrounding). In Category C4 (Materials and Solid Waste), I10 (Waste separation and storage) achieved the highest weight, and the other three indicators had a balanced weight. In Category C5 (Water), I12 (Potable water consumption) and I14 (Treatment of contaminated effluents) were the most important.

At Category C6 (User's health and comfort), I20 (Indoor air quality) was disclosed as a priority, and the other three indicators had a balanced weight. The two indicators had the same weight in Category C7 (Controllability by the user). In Category C9 (Passive design), I24 (Layout and Orientation) was considered more significant than I25 (Passive Systems). In Category C11 (Space flexibility and adaptability), the most important was I27 (Availability and accessibility to social areas), closely followed by I28 (Space optimisation).

Within Category C12 (Life cycle costs), I32 (Operational costs) was considered more important than indicator I31 (Initial cost). In what concerns Category C14 (Corruption avoiding plan), Indicator I34 (Sustainable purchase policies) was proposed by the experts, and the results disclosed it as the most important category within the Economy Area. In Category C15 (Environmental management systems), the most important was I36 (Infection control), and in Category C17 (Security), I39 (Occupants' safety) was more important than indicator I40 (Responsible construction practices). Finally, the most crucial indicator in Category C23 (Conveniences) was I46 (Accessibility to public transport).

Finally, the weight of each category within the areas is presented in Figure 3. The category with the most significant impact on an area was C23 (Conveniences), representing 67% of the A5 (Site) total. Nevertheless, the importance of the Site area in the overall sustainability level is only 8%. The categories with the lower influence were C8 (Landscaping) and C9 (Passive design), which belong to the "Sociocultural and functional" area, having weights of 3% and 4%, respectively.

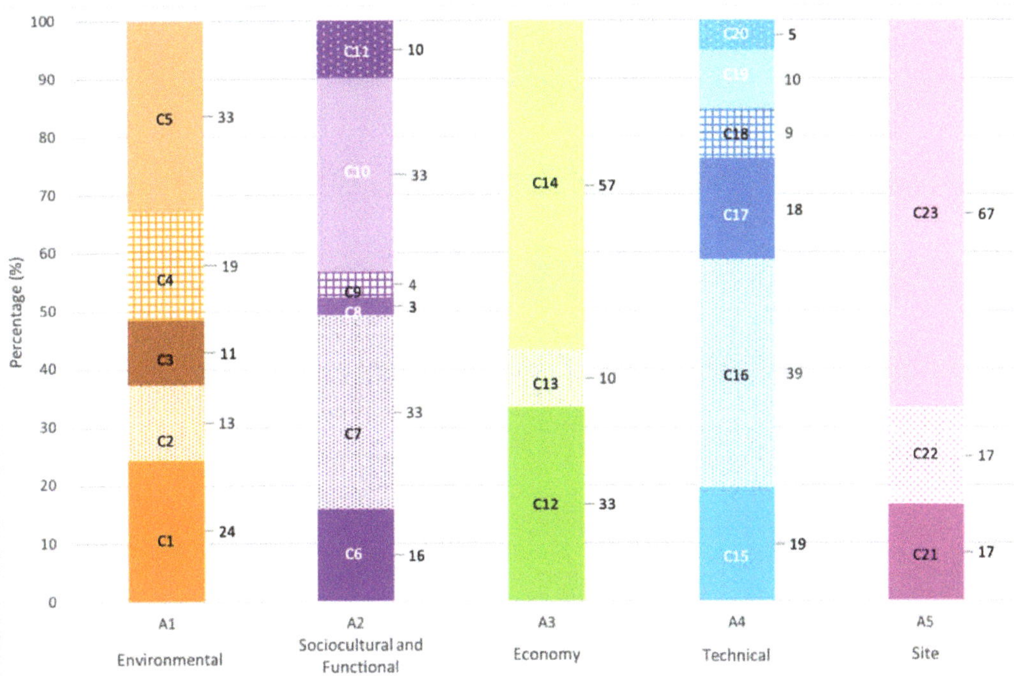

Figure 3. Final weight of categories within each area.

From comparing the weights between the proposed assessment framework and other HBSA methods, it is possible to draw some conclusions regarding the differences in the sustainability priorities and the level of the structure of each method.

Analysing the different HBSA methods addressed in this study, it is possible to conclude that they share common core sustainability categories, such as energy use; water efficiency; indoor and outdoor environmental quality; resources and materials; service quality; and site strategies. The proposed framework in this study also considered those priorities. Nevertheless, the weight of each category in the overall assessment varies among the analysed methods, as presented in Figure 4. This difference is because the methods were developed for different locations; therefore, each system of weights reflects specific sustainability properties.

It is important to note that one of the studied methods gives more importance to one core sustainability category than others. The energy-related indicators in LEED BD + C weigh 33% of the global sustainability score. On the other hand, BREEAM International New Construction has a more balanced weighing system between all core criteria.

The "Economic" category was one of the three leading sustainability categories in the proposed framework, and this is not considered in the two most used BSA methods (LEED and BREEAM). Additionally, in the Brazilian context, the importance of the economy category is more suited to the healthcare sector's role in the Brazilian economy. It also reflects the stakeholders' opinions gathered in the survey.

Based on these differences, it is possible to conclude that this study is an essential contribution to developing a new HBSA method in the Brazilian context because it sets a more comprehensive list of sustainability priorities suitable for the specific environmental, societal, and economic contexts.

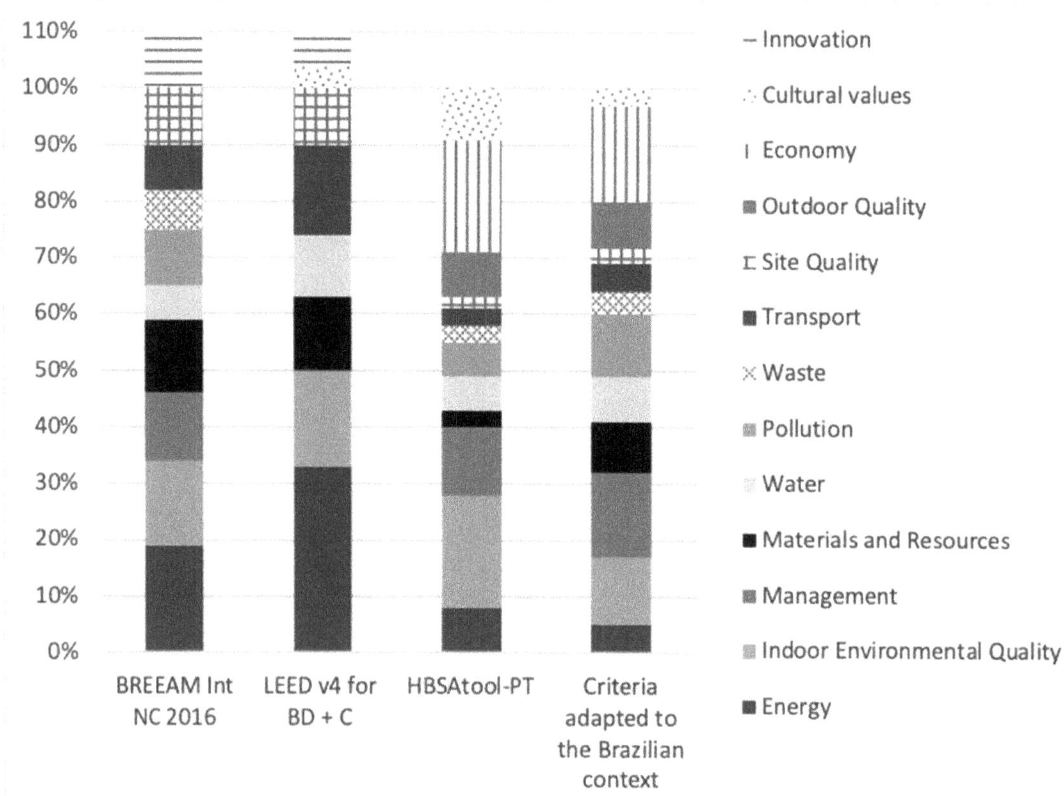

Figure 4. Comparison between the weights of each sustainability category of different HBSA methods and the ones from the proposed framework adapted to the Brazilian context.

5. Conclusions

Healthcare buildings are complex systems due to specific, higher technical, and functional requirements. The number of health services covered and the population served to mean different types of healthcare buildings. Many initiatives related to environmental sustainability in the healthcare sector present necessary steps for hospitals to deal with the eminent global crisis resulting from excessive consumption of natural resources and production of wastes.

Buildings do not have the inherent capacity for regeneration. However, the built environment can be designed to contribute to and support this goal. It also offers the opportunity to align its ecological profile with the health sector's core mission—to heal—by providing essential health services while preserving the natural environment.

If decisions are made at the early design stage, using comprehensive and systematic approaches, it is possible to integrate sustainability principles with a greater probability of success and reduced costs. It is also important to highlight that those principles must be aligned with the environmental, societal, and economic contexts of the country/region where the approach will be applied.

This study's limitations are related to the reduced number of Brazilian experts in the fields of hospital building design, construction, and operation. Sustainable construction applied to this type of building has yet to be made a priority. This situation may cause some need for more understanding by the respondents about what is included in each sustainability category and indicator, thus causing biases in the presented weighting system. An additional limitation may be that it was impossible to involve a more significant number of participants, a situation that is common to other studies that used the same methodology conducted in Brazil.

Healthcare building sustainability assessment methods can support the implementation of sustainability concerns in standard design and building management practices by integrating more comprehensive social and economic aspects and minimizing environmental impacts. These methods will allow identifying the design priorities from the early design stages.

Since there is no common international understanding regarding the weight of each sustainability indicator and it depends overall on the specific context and priorities of the location, the proposed framework was based on the results of a survey involving the main Brazilian healthcare stakeholders. This concept is relevant since it considers the knowledge and experience of different experts in the design process, validates the proposed framework, and evaluates the relative importance of each sustainability area, category, and indicator in the overall sustainability level of a project. This will facilitate priority setting when designing more sustainable healthcare buildings in Brazil.

This kind of initiative brings a significant advantage when seeking improvement towards the performance of healthcare buildings. The assessment method can raise awareness, promote sustainable practices, and decrease consumption rates and costs, thus reducing environmental and economic impacts. Additionally, considering the stakeholders' opinions in the proposed list of indicators, the weighing system becomes more aligned with their expectations, increasing its potential effectiveness.

As a recommendation, environmental issues and strategies must be the focus. However, considering other vital issues, such as societal needs and requirements and economic and technical aspects, is also essential.

The sustainability framework presented in this study differs from other existing HBSA methods since, in this case, the context of the Brazilian construction and healthcare sector was addressed.

To plan a sustainable healthcare building in Brazil, the designer must focus on the most pertinent sustainability categories disclosed in this study, such as efficiency in technical systems, sustainable purchase practices, and accessibility. Additionally, life cycle impacts and costs must be considered, as well as security, environmental management, and aspects related to the controllability of the indoor environmental quality by the user.

Author Contributions: Conceptualisation, M.C., M.d.F.C. and R.M.; Data curation, M.C.; Investigation, M.C.; Methodology, M.d.F.C. and R.M.; Project administration, M.d.F.C. and R.M.; Resources, M.C., M.d.F.C. and R.M.; Supervision, M.d.F.C. and R.M.; Writing—original draft, M.C.; Writing—review and editing, M.d.F.C. and R.M. All authors have read and agreed to the published version of the manuscript.

Funding: This research received no external funding.

Institutional Review Board Statement: Ethical review and approval were waived for this study since it does not disclose personal data nor individual answers of the participants in the survey.

Informed Consent Statement: Not applicable.

Data Availability Statement: Additional data supporting reported results can be found in the following link: https://hdl.handle.net/1822/67176.

Conflicts of Interest: The authors declare no conflict of interest.

Appendix A

Appendix A.1 Survey

In the context of the Master's Dissertation entitled "Building Sustainability Assessment Methods for Healthcare–critical comparative analysis to apply in Brazilian context", the questionnaire was formulated representing a proposal to develop an indicators structure based on international rating systems (LEED v4 for Building Design and Construction, BREEAM New Construction, and Healthcare Building Sustainability Assessment tool–Portugal).

The survey aims to check the sustainability level of hospitals in Brazil. The objective of the evaluation is to contribute to the applicability of the indicator to a broader context, making it more reliable and robust for practical applications and contributing to the development of the measurement of the Sustainable Assessment Method for Healthcare Buildings. This evaluation is carried out through the expert's judgment using specific criteria divided into categories: Environmental, Sociocultural and Functional, Economy, Technical, and Site.

The answers will be critical to the credibility of the final results and should be given the most genuine opinions on all the questions presented. All data processing will be performed for the Dissertation mentioned, respecting their anonymity and confidentiality.

The estimated time to complete the survey is 20 min.

Thank you in advance for your cooperation!

---

1. To identify the individual expectations of each involved, identify the group to which they belong. If you belong to more than one of the groups listed, please select as many options as necessary.

1.1. Occupation or position currently held (choose one or more of the following):

- Architect
  - ☐ With less than five years of experience
  - ☐ With over five years of experience
  - ☐ With experience in hospital projects
- Civil engineer
  - ☐ With less than five years of experience
  - ☐ With over five years of experience
  - ☐ With experience in hospital projects
- Sustainable Construction Consultant/Expert
  - ☐ With less than five years of experience
  - ☐ With more than five years of experience
- Qualified Sustainable Construction Appraiser

☐ AQUA-HQE	☐ LEED
☐ Leader	☐ CASBEE
☐ Procel Edifica	☐ DGNB
☐ BREEAM	☐ Other, which one?

- Hospital Manager
  - ☐ Facilities and Equipment Services
  - ☐ Superintendent/Director
  - ☐ Other, which one?
- Other
  - ☐ What?

2. Identification

2.1. Area of expertise and project development:

- Brazil

☐ North ☐ Southeast
☐ Northeast ☐ South
☐ Midwest ☐ All over the country

3. The proposed Hospital Building Method is a Design Support that consists of three levels to position the selected assessment criteria. Thus the fifty-seven Indicators presented are grouped into twenty-two Categories, framed into five Areas.

3.1. Considering the following tables show the indicators of each category, define the relative importance that each INDICATOR should present in the statement scale.

The criteria are in the form of statements; the evaluator should verify if each of the statements is consistent with the indicator that is being evaluated through a five-level scale (Likert scale):

1—Not important
2—Slightly important
3—Neutral
4—Important
5—Very important

Appendix A.2 Environmental Area

This item intends to evaluate the environmental life cycle impact to promote the use, by the design teams, of low environmental impact solutions, associated with the life cycle of various constructive elements and building materials. The database includes, beyond renewable and non-renewable energy embodied, the accounting of the categories: Global warming potential (GWP) Destruction of the ozone layer (ODP); Potential acidification (PA), Photochemical oxidation potential (POP) and Eutrophication potential (EP). To contribute to a reduction in national NOx emissions levels by using low-emission heat sources in the building.

Environmental Life Cycle Impacts Assessment						
ID	**Description**	**Evaluation**				
		Not Important	**Slightly Important**	**Neutral**	**Important**	**Very Important**
I1	Assessment of the building's life cycle impacts	☐	☐	☐	☐	☐

Primary energy consumption

Total primary energy consumption during the use phase to promote the reduction of energy consumption in healthcare buildings.

Local energy production

Amount of energy produced in the building through renewable sources to reward renewable energy consumption by incorporating systems that enable clean energy. The use of renewable energy allows the reduction of greenhouse gas emissions and other pollutants. It contributes to the conservation of global resources of fossil fuels and the development of technologies that will enable their exploitation, additionally resulting in a reduction in the life cycle costs of the building.

Minimum Energy Performance

Demonstrate an improvement in the proposed building performance rating compared with the baseline building performance rating to reduce the environmental and economic harms of excessive energy use by achieving a minimum level of energy efficiency for the building and its systems.

	Energy					
ID	Description	Evaluation				
		Not Important	Slightly Important	Neutral	Important	Very Important
I2	Primary energy consumption	☐	☐	☐	☐	☐
I3	Local energy production	☐	☐	☐	☐	☐
I4	Minimum Energy Performance	☐	☐	☐	☐	☐

Layout optimisation

Several beds are available per square meter of the built area to recognise the efficient use of the built space.

Soil sealing

Waterproofing Index to promote soil permeability in urban areas to ensure aquifer recharge and decrease peak flow in stormwater drainage systems. The impermeable areas can majorly impact ecosystems as the ground covered by constructions, streets, and other occupations reduces the soil surface available to support natural habitats and perform rainwater absorption. The increased area of impermeable soil has a negative effect on sustainable development, and the most common are nature conservation and the absence of flood control.

Reuse of previously built or contaminated areas

Percentage of intervention areas previously contaminated or built to reward the choice of location of these buildings in areas previously contaminated or constructed once. The primary solution to slow the destruction of natural habitats and wildlife they support and prevent the loss of soil suitability is reuse.

Ecological protection of the site

Protection of the site's ecological and natural resources aims to implement measures to preserve the ecological and natural resources of the building construction site. The ecological value is affected by the type of existing flora and fauna and their interactions, the number of different species, vegetation strata, and water courses.

Rehabilitation of the surrounding

Potential development of the surroundings by rewarding the rehabilitation of deteriorated or/and abandoned surrounding areas. The principle to be adopted when studying the implementation of a building is to minimise the impacts on the ecology of the site or, where possible, contribute to its improvement.

Use of native plants

Percentage of the green area occupied by native plants aiming to promote the integration of pre-existing native plants and the planting of local plants in green spaces. Native plants are the ones that belong to a place where they live for many generations, which is different from an introduced plant, which is a specie that resulted from subsequent introductions. The spontaneous plants grow in a community with other species, providing protection and nourishment, but at the same time, they can interfere with the natural habitat, competing with native plants.

Heat island effect

Percentage of coverage area and surrounding paved areas with reflectance to reduce the heat island effect in urban areas by promoting high reflectance materials or vegetation in outdoor spaces and roofs. The heat island effect indicates a higher temperature in urban areas, compared with the forest and rural areas. It is caused mainly due to the removal of vegetation and its replacement by asphalt and concrete buildings and structures that store and release thermal energy, which has high solar absorption due to its low reflectance. The heat island effect results in additional energy needs of buildings in urban areas, resulting in increased emissions of pollutants into the atmosphere.

Soil use and Biodiversity						
		Evaluation				
ID	**Description**	**Not Important**	**Slightly Important**	**Neutral**	**Important**	**Very Important**
I5	Layout optimisation	☐	☐	☐	☐	☐
I6	Soil sealing	☐	☐	☐	☐	☐
I7	Reuse of previously built or contaminated areas	☐	☐	☐	☐	☐
I8	Ecological protection of the site	☐	☐	☐	☐	☐
I9	Rehabilitation of the surrounding	☐	☐	☐	☐	☐
I10	Use of native plants	☐	☐	☐	☐	☐
I11	Heat island effect	☐	☐	☐	☐	☐

Construction waste

Measurements to reduce the production of Solid Construction and Demolition Waste, and a percentage is destined for reuse or recycling to promote the reduction of waste production and reward its recycling.

Reused products and recycled materials

Percentage of the cost of reused products and materials with recycled content to promote its usage from the construction site or outside, specific for each material type within its components. The reuse of building materials or elements that result from the end of a building's life cycle consists of using them incorporated into new materials for construction or rehabilitation.

Waste separation and storage

The building's performance at the level of this parameter is evaluated by the value of the Potential of the Building's Conditions for Promoting the Separation of Solid Waste (PRSU), which results from criteria related to the indoor and outdoor existing conditions for the deposition and storage of household waste and usually composed of organic material, paper, cardboard, plastic, glass, metals, infectious, pathological, sharps, and chemicals, among others.

Responsible for sourcing materials

To recognise and encourage the specification and procurement of responsibly sourced construction products by using timber/timber-based products legally harvested and traded, a documented policy and procedure that sets out procurement requirements for all suppliers and trades to adhere to relating to the responsible sourcing of construction products and to available the responsible sourcing credits awarded where the methodology responsibly sources the applicable construction products.

Materials and Solid Waste						
ID	Description	Evaluation				
		Not Important	Slightly Important	Neutral	Important	Very Important
I12	Construction waste	☐	☐	☐	☐	☐
I13	Reused products and recycled materials	☐	☐	☐	☐	☐
I14	Waste separation and storage	☐	☐	☐	☐	☐
I15	Responsible sourcing of materials	☐	☐	☐	☐	☐

Potable water consumption

The annual volume of water consumed per square meter inside the building aims to promote reducing water consumption, depending on the efficiency of devices and the average consumption patterns. The quality of water supply, drainage and wastewater treatment strongly impact public health. Due to reducing drinking water supplies, as opposed to increasing consumption, it is necessary to take measures to make its use more efficient.

Recycling and recovery of effluents

Percentage reduction of drinking water consumption rewarding the use of effluents and systems that reduce the unnecessary consumption of potable water. As a precious resource and quality of life, potable water should be used only for functions that require all its qualities. However, it is currently used in applications that can be satisfied with recycled or lower-quality water.

Treatment of contaminated effluents

Separation of contaminated effluents and local wastewater treatment gives space for premises in the building for wastewater treatment and an appropriate contaminated effluent drainage system. Hospital effluents can be classified into household effluents (kitchens, laundries and toilets) and specifically hospital effluents (from analyses, patient care and medicines). Hospital wastewater is classified according to the Generic Recommendations for Hospital Wastewater Management into groups that should be treated appropriately and differentiated according to their category.

Water efficient equipment

To reduce the water consumption by encouraging specification of water-efficient equipment by systems or processes identified to reduce the water demand, and demonstrate, through either good practice design or specification, a meaningful reduction in the total water demand of the building.

Water						
ID	Description	Evaluation				
		Not Important	Slightly Important	Neutral	Important	Very Important
I16	Potable water consumption	☐	☐	☐	☐	☐
I17	Recycling and recovery of effluents	☐	☐	☐	☐	☐
I18	Treatment of contaminated effluents	☐	☐	☐	☐	☐
I19	Water efficient equipment	☐	☐	☐	☐	☐

Appendix A.3 Sociocultural and functional

Natural ventilation

The efficiency of natural ventilation indoors promotes the existence of conditions that allow natural ventilation of the interior space of the building to the exclusive detriment of mechanical ventilation. The levels of indoor air renewal must be guaranteed, safeguarding its quality and reducing occupant exposure to indoor pollutants. The primary influence for the natural ventilation is the depth of the floor drawings plans. Also, courtyards and inner courts favour. When natural ventilation strategy is conceived correctly, this can be as effective as a mechanical ventilation system, with all the advantages associated with the fact that there is no power consumption.

Toxicity of finishing materials

Weight the per cent of low Volatile Organic Compounds (VOC) finishing materials aiming the reward of using materials that do not cause occupant health problems. Several studies reveal the connection between high VOC Sick Building Syndrome (SBS) concentrations. Some examples are formaldehyde, benzene, toluene and xylene. Inside the buildings, the primary sources are products derived from wood produced through adhesives and used as solvents in paints mainly based on synthetic, adhesives, carpeting and polyurethane foams. These compounds are often accidentally released into the atmosphere and are responsible for significant environmental impacts.

Thermal comfort

Average annual thermal comfort level to ensure the conditions within the healthcare providers to meet occupant needs. The thermal environment of the interior spaces has physical and psychological effects on its occupants and is of great importance in building design. When designing a building, creating a microclimate in the interior spaces, despite the weather conditions outside, essentially responds to the needs and expectations of occupants. The climate in Brazil is divided into five sub-types – equatorial, semi-arid, highland tropical, and subtropical – and during much of the year, it is necessary to use a cooling system to maintain indoor temperatures within a comfortable range. This situation explains that most buildings produce large amounts of thermal energy.

Visual Comfort

This item measures the contribution of natural lighting to the proper lighting of the interior environment by promoting the adoption of criteria to improve the visual comfort of occupants through the appropriate use of natural lighting, which will contribute to the recovery of patients and the reduction of energy consumption inside the building. Natural lighting is one factor in conditioning the environment's quality. It has to provide a comfortable visual interior environment, through the minimum energy consumption (artificial lighting). The increasing importance of aspects related to the environment, sustainable development and interior comfort, has contributed to natural light's leading role in the healing process.

Acoustic comfort

The average level of sound insulation aims to promote the option for constructive solutions that improve the acoustic comfort of patients and team works. Considering the problems that noise causes in humans, society must be aware and take necessary measures to preserve the health of building occupants. Thus, those responsible for building design must develop techniques to provide acoustic comfort conditions, creating a suitable environment for the activities developed.

Indoor air quality

Evaluating pollutants measured in indoor air aims to recognise and encourage the search for a healthy indoor environment by controlling the airborne concentration of existing pollutants. The indoor air is a spread source of microorganisms, which in healthcare units leads to the origin of hospital infections.

User's Health and Comfort						
ID	**Description**	**Evaluation**				
		Not Important	**Slightly Important**	**Neutral**	**Important**	**Very Important**
I20	Natural ventilation	☐	☐	☐	☐	☐
I21	Toxicity of finishing materials	☐	☐	☐	☐	☐
I22	Thermal comfort	☐	☐	☐	☐	☐
I23	Visual comfort	☐	☐	☐	☐	☐
I24	Acoustic comfort	☐	☐	☐	☐	☐
I25	Indoor air quality	☐	☐	☐	☐	☐

Ventilation and temperature

The possibility of room control of temperature and openings (windows) to encourage the installation of systems that guarantee indoor air quality (IAQ) conditions at a reduced level of energy consumption, while preventing the sprawl of diseases. The increasing efficiency of natural ventilation and preventing the solar incidence are linked with the requirements of IAQ and thermal comfort, as well as to the potential for reducing energy consumption.

Natural light

The possibility of controlling the entrance of natural light inside the building, through elements for this purpose according to the functional needs of the spaces aiming to recognise daylight control systems to reduce energy consumption, allowing users to administer the space' visual comfort.

Controllability by the User						
ID	**Description**	**Evaluation**				
		Not Important	**Slightly Important**	**Neutral**	**Important**	**Very Important**
I26	Ventilation and temperature	☐	☐	☐	☐	☐
I27	Natural light	☐	☐	☐	☐	☐

The next item persuades the visual contact with the exterior from the main compartments of the building by promoting a design that values the relationship between interior/exterior through visual contact with the outside.

Landscaping						
ID	**Description**	**Evaluation**				
		Not Important	**Slightly Important**	**Neutral**	**Important**	**Very Important**
I28	Visual link with the surrounding landscape	☐	☐	☐	☐	☐

Layout and orientation

Proper implementation and orientation of the building, considering the territorial and landscape framing of the place, promoting the quality of the interior environment to promote, and reward a building implementation and orientation that allows the good use of solar radiation in the different heating and cooling stations and the appropriate use of wind for natural ventilation.

Passive systems

Integration of building systems for passive heating, ventilation and cooling, upgrading the indoor air quality by promoting the design of bioclimatic buildings that encourages comfort conditions of its users, reducing energy consumption.

Passive Design						
ID	**Description**	**Evaluation**				
		Not Important	**Slightly Important**	**Neutral**	**Important**	**Very Important**
I29	Layout and orientation	☐	☐	☐	☐	☐
I30	Passive systems	☐	☐	☐	☐	☐

Accessibilities

Accessibility and ease of circulation area for patients, visitors, and service providers by rewarding the existence of efficient accessibility and mobility plan that covers as many people and paths as possible.

Mobility Plan						
ID	**Description**	**Evaluation**				
		Not Important	**Slightly Important**	**Neutral**	**Important**	**Very Important**
I31	Accessibilities	☐	☐	☐	☐	☐

Availability and accessibility to social areas

To evaluate the existence and accessibility, by users, to activities, living, leisure, and outdoor spaces by aiming for living spaces that provide the well-being of patients and work teams.

Space optimisation

The maximisation of the usable floor area inside the building and reduction of the total construction area by promoting the adoption of space design forms and construction solutions that facilitate the optimisation of the construction area, reducing the environmental impacts associated with the floor area, and increasing the efficiency.

Space flexibility

The need for spatial solutions that contributes to the versatility of the area, analysing the level of flexibility, allowing the increase in and adaptation to the continuous need for alteration of spatial functions into rewarding the option for a design that promotes the flexibility of spaces, so that it can adapt to different operations according to the diverse needs of everyday life.

Space adaptability

The assessment of the adaptive capacity of spaces to change functionalities to promote the adoption of design and construction solutions that simplify their adaptation to different uses, in case of need or rehabilitation of the building.

Space Flexibility and Adaptability						
ID	Description	Evaluation				
		Not Important	Slightly Important	Neutral	Important	Very Important
I32	Availability and accessibility to social areas	☐	☐	☐	☐	☐
I33	Space optimisation	☐	☐	☐	☐	☐
I34	Space flexibility	☐	☐	☐	☐	☐
I35	Space adaptability	☐	☐	☐	☐	☐

Appendix A.4 Economy

Initial cost

The value of initial investment cost per square meter of the total construction area to promote the design of sustainable buildings whose initial investment is at least equivalent to conventional buildings.

Operational costs

The value of utilisation costs per square meter of Total construction area aims to appreciate the design of sustainable buildings whose utilisation costs are lower than conventional buildings.

Life Cycle Costs						
ID	Description	Evaluation				
		Not Important	Slightly Important	Neutral	Important	Very Important
I36	Initial cost	☐	☐	☐	☐	☐
I37	Operational costs	☐	☐	☐	☐	☐

The next item goes into the local community promotion by contracting national and local goods and services and addressing the development of the local economy by hiring local goods and services.

Local Economy						
ID	Description	Evaluation				
		Not Important	Slightly Important	Neutral	Important	Very Important
I38	Hiring local goods and services	☐	☐	☐	☐	☐

The risk of corruption is transversal. Good practices management helps to prevent situations of infractions and, therefore, it is essential to identify the risks of deviation from good practices and their consequences in terms of management. Reducing corruption is necessary for strengthening democratic institutions, promoting relations between citizens and public or private administration, economic development and growth, and the regular functioning of markets. The problem of corruption is associated with many situations, which spoil the functioning of institutions and markets, such as abuse of power, bribery, embezzlement, influence peddling, economic participation in business, and concussion. All of these constitute related crimes, and there is an undue advantage or compensation to be obtained. Sustainable purchase policies aim to identify risks, resources, actions, and responsibilities to mitigate them, as well as the implementation, monitoring, evaluation, and reporting process. The risk can be defined as an event that, if it occurs, will have a negative impact on the achievement of the organisation's mission and objectives. Missed opportunities can also be considered a risk.

Corruption Avoiding Plan						
ID	**Description**	**Evaluation**				
		Not Important	**Slightly Important**	**Neutral**	**Important**	**Very Important**
I39	Sustainable purchase policies	☐	☐	☐	☐	☐

Appendix A.5 Technical

Commissioning

Assessment of building systems and components throughout the different life cycle phases to identify a properly planned commissioning process that ensures the proper functioning of all building systems and components.

Environmental management plan

Adopting a Sustainable Management Plan aims to reward the existence of an Environmental Management System to ensure the design and construction phase lasts throughout the use phase.

Infection control

Monitor and evaluate infection control by promoting adequate cleaning, disinfection, decontamination, and sterilisation of all areas, equipment, and instruments of the hospital.

Reducing noise pollution

Mitigation measures of noise production pleasing the reduction within healthcare buildings.

Environmental Management Systems						
ID	**Description**	**Evaluation**				
		Not Important	**Slightly Important**	**Neutral**	**Important**	**Very Important**
I40	Commissioning	☐	☐	☐	☐	☐
I41	Environmental management plan	☐	☐	☐	☐	☐
I42	Infection control	☐	☐	☐	☐	☐
I43	Reducing noise pollution	☐	☐	☐	☐	☐

The next item concerns the maintenance plan's evaluation, ensuring the proper functioning and efficiency of existing or designed mechanical systems enabling the appropriate operation of all mechanical systems and building components.

Technical Systems						
ID	Description	Evaluation				
		Not Important	Slightly Important	Neutral	Important	Very Important
I44	Efficiency of lighting and air conditioning systems	☐	☐	☐	☐	☐

Occupants safety

The indicator below evaluates measures to ensure occupant safety by limiting the risk of hatching and fire hazards, favour the action of firefighters whenever their intervention is necessary, and provide means for users to initiate combat measures before the firefighter's arrival.

Responsible construction practices

To recognise and encourage construction sites which are managed in an environmentally and socially considerate, responsible and accountable manner by using legally harvested and traded timber, considering and implementing health and safety legislation and regulations for construction sites, and monitoring, recording and reporting the energy use, water consumption and transport data resulting from all on-site processes throughout the programme.

Security						
ID	Description	Evaluation				
		Not Important	Slightly Important	Neutral	Important	Very Important
I45	Occupants safety	☐	☐	☐	☐	☐
I46	Responsible construction practices	☐	☐	☐	☐	☐

Materials of high strength and durability

Assessment of the durability and required level of maintenance of finishing materials and other constituents of building elements to benefit the use of durable materials suitable for their intended use, reducing the complexity and periodicity of maintenance.

Proper selection of furniture

Suitability of furniture and general equipment, mobile and fixed to the functions for which they are intended by promoting the highest durability and eligibility.

Durability						
ID	Description	Evaluation				
		Not Important	Slightly Important	Neutral	Important	Very Important
I47	Materials of high strength and durability	☐	☐	☐	☐	☐
I48	Proper selection of furniture	☐	☐	☐	☐	☐

Education of occupants

Availability and content of the Building's User Manual by rewarding guidelines for occupants to use it efficiently. Regardless of the design of a building, its efficiency and operating costs are strongly influenced by the daily behaviour of its users. With guidance and access to information, these can best use the systems. On the other hand, malfunctioning can lead to discomfort and operating, and maintenance costs differ from those estimated, resulting in a waste of resources.

Education of service providers

Availability and content of the Building's Maintenance and Management Manual by ensuring the proper maintenance for better use and preservation and increasing its useful life.

Satisfaction surveys

Existence of periodic surveys distributed to building users to assess their satisfaction with the buildings' performance.

Awareness and Education for Sustainability						
ID	Description	Evaluation				
		Not Important	Slightly Important	Neutral	Important	Very Important
I49	Education of occupants					
I50	Education of service providers					
I51	Satisfaction surveys					

The indicator below evaluates the existence of a qualified evaluator in the sustainability construction field to promote and value the integration of designers and building management or maintenance from a qualification of the experts.

Skills in Sustainability						
ID	Description	Evaluation				
		Not Important	Slightly Important	Neutral	Important	Very Important
I52	Integration of a qualified sustainability expert					

Site

The next item is about the development of the urban and local community by promoting the creation of new public space areas, access, and services.

Local Community						
ID	Description	Evaluation				
		Not Important	Slightly Important	Neutral	Important	Very Important
I53	Local community development					

The indicator below evaluates the urban context and valorisation of the surrounding space by promoting and enhancing the architectural, landscape, and urban design according to the cultural value of the location.

Cultural Value						
ID	Description	Evaluation				
		Not Important	Slightly Important	Neutral	Important	Very Important
I54	Heritage framework	☐	☐	☐	☐	☐

Accessibility to public transport

The public transport accessibility index aims to promote and value solutions that meet most building users' travel needs through urban transportation services.

Low impact mobility

The potential of the building's sustainable mobility conditions is to provide facilities which encourage building users to travel using low-carbon modes of transport, by stimulating the usage of bicycles and pedestrian accessibility, minimising individual journeys, and using vehicles with less environmental impact.

Local amenities

The accessibility to amenities index intends to enhance the existence of sustainable and integrated communities by establishing basic amenities near the healthcare building.

Conveniences						
ID	Description	Evaluation				
		Not Important	Slightly Important	Neutral	Important	Very Important
I55	Accessibility to public transport	☐	☐	☐	☐	☐
I56	Low impact mobility	☐	☐	☐	☐	☐
I57	Local amenities	☐	☐	☐	☐	☐

4. Considering the proposed Method presented above, are there any indicators that still need to be delivered and that you think should be addressed? Which? Justify.

5. Considering the proposed Method presented above, is there any indicator that could be eliminated? Which? Justify.

Comments:

Thanks for the collaboration!

References

1. Li, Y.; Chen, X.; Wang, X.; Xu, Y.; Chen, P. A review of studies on green building assessment methods by comparative analysis. *Energy Build.* **2017**, *146*, 152–159. [CrossRef]
2. Energy Efficiency in Buildings Facts and Trends: Business Realities and Opportunities. Available online: https://www.wbcsd.org/Programs/Cities-and-Mobility/Energy-Efficiency-in-Buildings/Resources/Business-realities-and-opportunities-Summary (accessed on 10 January 2023).
3. Sahamir, S.; Zakaria, R. Green Assessment Criteria for Public Hospital Building Development in Malaysia. *Procedia Environ. Sci.* **2014**, *20*, 106–115. [CrossRef]
4. Happio, A.; Viitaniemi, P. A critical review of building environmental assessment tools. *Environ. Impact Assess. Rev.* **2008**, *28*, 469–482. [CrossRef]
5. Bragança, L.; Koukkari, H.; Mateus, R. Building Sustainability Assessment. *Sustainability* **2010**, *7*, 2010–2023. [CrossRef]
6. Darko, A.; Zhang, C.; Chan, A.P.C. Drivers for green building: A review of empirical studies. *Habitat Int.* **2016**, *60*, 34–49. [CrossRef]
7. The Scenario of Hospitals in Brazil. Available online: http://cnsaude.org.br/publicacoes/cenario-dos-hospitais-no-brasil-2019/ (accessed on 10 January 2023).
8. Kras, I. Sustainable Hospital Building. Master's Thesis, Technical University of Delft, Delft, The Netherlands, 2011.
9. Healthy Hospital, Healthy Planet, Healthy People: Addressing Climate Change in Healthcare Settings. Available online: http://www.who.int/globalchange/publications/healthcare_settings/en/index.html (accessed on 10 January 2023).

10. Barbosa, M.T.G.; Almeida, M. Developing the methodology for determining the relative weight of dimensions employed in sustainable building assessment tools for Brazil. *Ecol. Indic.* **2017**, *73*, 46–51. [CrossRef]
11. Bellen, H.M.V. Sustainable Indicators: A Comparative Analysis. Master's Thesis, Santa Catarina Federal University, Florianópolis, Brazil, 2002.
12. Lu, Y.; Geng, Y.; Liu, Z.; Cote, R.; Yu, X. Measuring sustainability at the community level: An overview of China's indicator system on National Demonstration Sustainable Communities. *J. Clean. Prod.* **2017**, *143*, 326–335. [CrossRef]
13. International Initiative for a Sustainable Built Environment Home. Available online: http://iisbe.org (accessed on 5 August 2020).
14. BREEAM International New Construction 2016. Available online: https://www.breeam.com/discover/technical-standards/newconstruction/ (accessed on 5 August 2020).
15. USGBC LEED v4 for Homes Design and Construction. Available online: https://www.usgbc.org/resources/leed-v4-homes-design-and-construction-checklist (accessed on 5 August 2020).
16. Castro, M. Building Sustainability: Method of Supporting the Design of Hospital Buildings. Doctoral Thesis, Minho University, Guimarães, Portugal, 2018.
17. Agenda for Hospitals and Health. Global Green and Healthy Hospitals. Available online: https://www.hospitalesporlasaludambiental.net/wp-content/uploads/2016/07/Global-Green-and-Healthy-Hospitals-Agenda.pdf (accessed on 10 January 2023).
18. Mateus, R.; Bragança, L. Sustainability assessment and rating of buildings: Developing the methodology SBToolPT–H. *Build. Environ.* **2011**, *46*, 1962–1971. [CrossRef]
19. Castro, M.D.F.; Mateus, R.; Bragança, L. Development of a healthcare building sustainability assessment method—Proposed structure and system of weights for the Portuguese context. *J. Clean. Prod.* **2017**, *148*, 555–570. [CrossRef]
20. Castro, M.D.F.; Mateus, R.; Bragança, L. Healthcare Building Sustainability Assessment tool–Sustainable Effective Design criteria in the Portuguese context. *Environ. Impact Assess. Rev.* **2017**, *67*, 49–60. [CrossRef]
21. Castro, M.D.F.; Mateus, R.; Bragança, L. A critical analysis of building sustainability assessment methods for healthcare buildings. *Environ. Dev. Sustain.* **2015**, *17*, 1381–1412. [CrossRef]
22. Ministry of Health, National Health Surveillance Agency. Resolution RDC N° 50, Last Revised February 2002. Available online: https://bvsms.saude.gov.br/bvs/saudelegis/anvisa/2002/rdc0050_21_02_2002.html (accessed on 11 March 2023).
23. Brazilian Normative. ABNT NBR 9050, Last Revised August 2020. Available online: https://www.canoas.rs.gov.br/wp-content/uploads/2020/07/norma-abnt-NBR-9050-2015-emenda-1-2020.pdf (accessed on 11 March 2023).
24. Ju, C.; Ning, Y.; Pan, W. A review of interdependence of sustainable building. *Environ. Impact Assess. Rev.* **2016**, *56*, 120–127. [CrossRef]
25. Saaty, T. Decision making with the Analytic Hierarchy Process. *Int. J. Serv. Sci.* **2008**, *1*, 83–98. [CrossRef]
26. Ali, H.H.; Al Nsairat, S.F. Developing a green building assessment tool for developing countries—Case of Jordan. *Build. Environ.* **2009**, *44*, 1053–1064. [CrossRef]

applied sciences

MDPI

Article

Sustainable Operation of Active Distribution Networks

Dimitra G. Kyriakou and Fotios D. Kanellos *

School of Electrical and Computer Engineering, Technical University of Crete, GR-73100 Chania, Greece
* Correspondence: fkanellos@tuc.gr

Abstract: The present and future conditions in the energy market impose extremely high standards to the operation of building energy systems. Moreover, distribution networks face new operational and technical challenges as a result of the rapid penetration of renewable energy sources (RES) and other forms of distributed generation. Consequently, active distribution networks (ADNs) will play a crucial role in the exploitation of smart building prosumers, smart grids, and RES. In this paper, an optimization method for the sustainable operation of active distribution networks hosting smart residential building prosumers, plug-in electric vehicle (PEV) aggregators, and RES was developed. The thermal and electrical loads of the residential buildings were modeled in detail and an aggregation method was implemented to the hosted PEVs. Moreover, smart power dispatch techniques were applied at each building prosumer and PEV aggregator hosted by the active distribution network. Simultaneously, all the operational limitations of the active distribution network, building energy systems, and the hosted PEVs were satisfied. The constrained optimal power flow (OPF) algorithm was exploited to keep the voltages of the hosting distribution network between the permissible bounds. A significant operation cost reduction of 17% was achieved. The developed models were verified through detailed simulation results.

Keywords: active distribution network; sustainability; smart residential buildings; electric vehicles; renewable energy sources; optimization; power management; power dispatch

1. Introduction

The increasing penetration of distributed energy resources (DERs) in power systems affects their reliability, efficiency, and operation cost. Therefore, it is essential to promote, optimally operate, and control active distribution networks in order to achieve the qualitative and cost-effective operation of future electric power systems.

The study of residential building power management and control has received a lot of attention. The home energy management systems (EMSs) studied in [1,2] were exclusively used to optimally control only the power of a residence with renewable energy sources and an energy storage system (ESS). Aiming to decrease the power consumption, the EMS presented in [3] exploits the load and generation forecast, in contrast to [4]. In [5], a fuzzy logic controller was designed for the power management of residential microgrids including distributed generation and energy storage units. However, their ability to control the energy demand was not examined, contrary to studies [6,7] that considered controllable house appliances that are optimally scheduled to reduce the amount of grid-supplied energy. The EMSs examined in [8,9] were deployed in smart residential buildings and aimed at the minimization of the electricity cost and peak load shaving by the appropriate operation scheduling of household devices. In [10,11], the researchers optimally managed the power consumption by regulating heating, ventilation and air conditioning (HVAC) systems of residential buildings with suitable load shifting methods and analyzed the HVAC power demand profile for various building characteristics. The maximum energy cost saving was achieved, while a sufficient level of thermal comfort was ensured. A proper modeling of the thermal and electrical loads as well as the energy storage system of a smart home was developed in [12]. A strategy to optimize energy usage through the appropriate scheduling of domestic appliances, while maintaining user comfort levels within a predetermined range was also proposed. Algorithms for the efficient energy management of residential RESs and battery energy storage systems and the optimal sizing of these units were developed in [13]. The main objective of this work was the maximization of the energy use produced by the RESs and the reduction in the system's overall cost. Studies [14,15] have also focused on the analysis of building energy consumption profiles. Specifically, the authors in [14] evaluated the influence of the occupants' age on the energy consumption of the residential sector. Daily energy consumption habits and trends in residential buildings were identified in [15], mainly through statistical methods and smart metering systems. In [16], a method for the coordinated operation of an active distribution network (ADN) comprising a microgrid of large office building prosumers and PEVs was examined. This method focuses on the interaction of the microgrid and the active distribution network and enables them not to disclose to each other their respective internal technical characteristics and information. In contrast, the work presented in this paper examined the optimization of distributed small residential prosumers of a distribution network, which is a completely different problem.

The operation scheduling and control of PEVs has been the subject of extensive research. In [17], a vehicle to grid (V2G) operation was exploited. The PEVs inject power to the electric grid in order to reduce PEV charging cost. Nevertheless, the network constraints were not considered and the proposed method was applied only to each PEV. Numerous research works have focused on the PEVs' charging management, taking into account the distribution network constraints [18–21]. In [21], PEV aggregation techniques were employed for the most effective control of the distribution network. A hierarchical multi-agent system (MAS) was developed in [22] for the efficient real-time control of large electric ports with a high integration of PEVs, while in [23], the optimal operation of distribution networks comprising PEVs was achieved with the implementation of demand response strategies. To optimally dispatch the total active power of a population of PEVs to each PEV, an effective definition of a PEV's flexibility to adjust its exchanged power was developed in [24]. The power management method suggested in [25] is suitable for real-time application to distribution networks hosting large clusters of PEVs. In this work, simple flexibility indicators were proposed for both PEVs and clusters of PEVs. Additionally, the optimal reactive power dispatch to each PEV was accomplished in real-time. Moreover, a strategy for incorporating energy storage units into ADNs was developed in [26,27]. The ESSs were suitably designed to take into account their potential to support the ADN by exploiting both active and reactive power. In [28], the batteries used in the examined ADN were optimally designed and scheduled in order to maximize the penetration of the distributed generation units.

A brief description of this work and its novel aspects are presented in the following.

- The work presented in this paper examines the optimization of distributed small residential prosumers of a distribution network. A large number of operation and technical limitations of the ADN and its components was ensured and the proposed algorithm achieved the highest possible reduction in the daily operation cost of the ADN.
- The implementation of the proposed method was based on the exploitation of a hierarchical MAS that allows for the optimization of the complex examined power system in a single optimization level. The goals of each type of agent (smart residential building prosumers, PEV aggregators, and RES) and how they cooperated to accomplish them were fully outlined. The coordinated optimal operation of all types of agents was achieved and the suggested approach for ADN's optimal operation aligns with the growing need for sustainable development.
- The thermal and electrical loads of residential buildings and an aggregate battery of numerous plug-in electric vehicles were modeled in detail.
- A simple and effective definition of the buildings' cluster flexibility to adjust their power by exploiting fuzzy logic was first proposed in this article.

- Smart power dispatch techniques were applied to buildings and PEV clusters as well as to single residential buildings and PEVs. This kept the needed computation time very low, as the minimum possible number of decision variables was achieved.

The proposed method for the optimal operation of the ADN was aligned with the increasing need for sustainable development. The examined system included RES that generate clean energy. Moreover, the autonomous operation of the ADN was ensured, mainly through the hosted PEVs and RES, and there was no need for extra investment to conventional polluting auxiliary power generation units. This contributes to a reduction in greenhouse gas emissions and enhances the environmental friendliness of the proposed system. Another significant aspect of the suggested method is the minimization of the daily operation costs of the examined ADN, ensuring its sustainability in this way. Another sustainable feature of the method is the qualitative operation of the ADN as it can ensure the maintenance of the ADN voltages within their permissible bounds.

The remainder of the article is structured as follows. Section 2 provides the description of the suggested hierarchical MAS. Section 3 includes the models of the agents and the connection among them. Finally, Section 4 presents the results received by the simulation of the examined active distribution network and evaluates them; while Section 5 draws the general conclusions.

2. Description of the Suggested Multi-Agent System Structure

The layout of the examined ADN and the structure of the hierarchical MAS developed in this work are shown in Figure 1. In the examined MAS, the agents used were classified into three major types: (1) local agents; (2) cluster agents; and (3) the active distribution network management agent. Local agents are accountable to model an individual module of the ADN, and cluster agents are accountable to gather and properly handle the responses of a cluster of local agents and forward information back to them. All cluster agents are supervised and managed by the active distribution network management agent (ADNM/A). ADNM/A is accountable for the optimal coordination of the MAS according to the responses received by cluster agents and possesses a higher place in the MAS hierarchy, as depicted in Figure 1. It can be concluded by Figure 1 that a bidirectional communication system should be available to each agent. A variety of communication technologies (i.e., DSL, Power Line Communication, Wi-Fi, GPRS, etc.) can be exploited. Their pros and cons can be found in [29]. Usual communication protocols include: AMI, SCADA, cellular communications, and other wireless communication technologies. Smart meters and PMUs can be installed at carefully selected points of the network. A low-cost PMU for smart distribution networks was proposed in [30]. Next, a brief explanation of each type of agent is provided, while their detailed operation is presented in the following section.

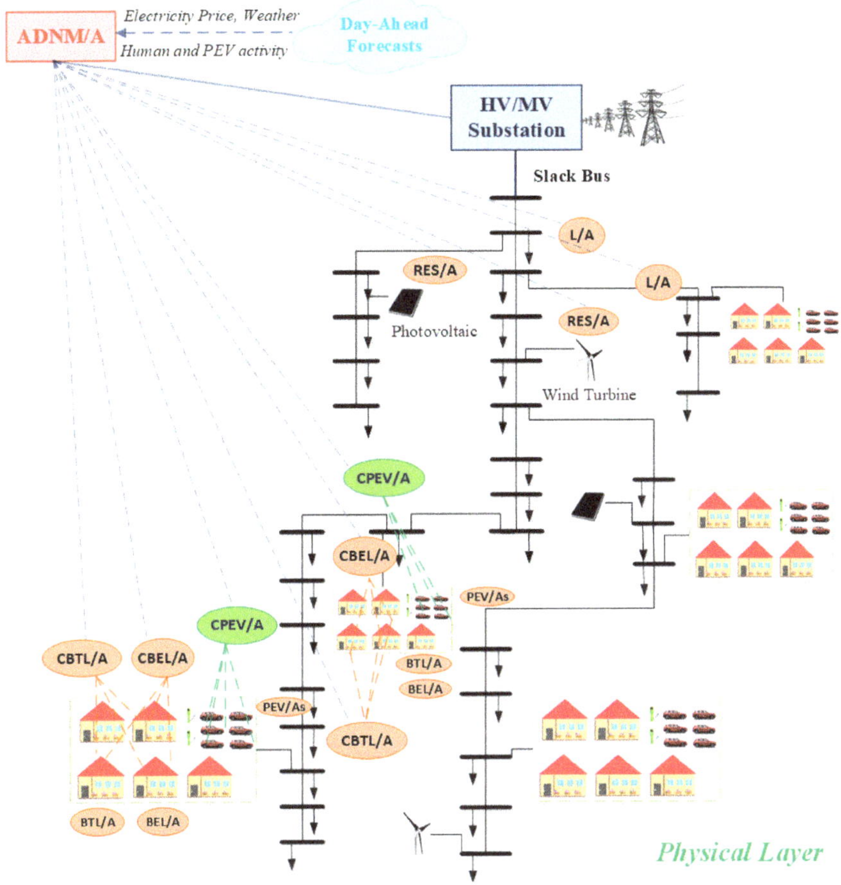

Figure 1. MAS based power management system structure for the active distribution network.

2.1. Local Agents

The building thermal load agent (BTL/A), building electrical load agent (BEL/A), plug-in electric vehicle agent (PEV/A), renewable energy sources agent (RES/A), and load agent (L/A) belong to the type of local agents. The inherent quantities of the distribution network components are calculated by local agents. Specifically, BTL/A and BEL/A were assigned to each building of the network and to model its thermal and electrical loads, respectively. Additionally, BTL/A calculates the indoor temperature of each building over the optimization horizon. The estimated indoor temperature of the building was used with its upper and lower limits to suitably calculate the flexibility of each building to increase or decrease its thermal power demand. Afterward, BTL/A and BEL/A send the necessary data to the supervising cluster agents CBTL/A and CBEL/A, respectively. PEV/A calculates the flexibility of each individual PEV to adjust its power and the bounds of its stored energy and the active and reactive power it exchanges with the distribution network. These quantities are forwarded to the CPEV/A, which supervises the cluster of PEVs. RES/A carries out extremely short-run predictions of the power produced by the renewable energy units it is assigned to. L/A is applied to each node of the distribution network and also estimates the node's active and reactive power requirement by performing very short-run predictions. RES/A and L/A provide their responses to ADNM/A.

2.2. Cluster Agents

Cluster of Building Thermal Loads Agent (CBTL/A), *Cluster of Building Electrical Loads Agent* (CBEL/A), and *Cluster of Plug-in Electric Vehicles Agent* (CPEV/A) belong to the category of cluster agents and they are hierarchically located just above the agents BTL/A, BEL/A, and PEV/A, respectively. They are regularly placed on medium voltage/low voltage (MV/LV) transformers that supply clusters of residential buildings and/or PEVs. Their primary aim is the aggregation of the responses they receive from the BTL/As, BEL/As, and PEV/As in their control areas. Moreover, CBTL/As and CPEV/As define suitable flexibilities exploiting fuzzy logic based on information regarding the respective node voltage and the flexibility of the supervised BTL/As and PEV/As, respectively. Cluster agents are also responsible for the optimal dispatch of their power to the local agents they supervise and control. Finally, they forward the necessary data to the active distribution network management agent.

2.3. Active Distribution Network Management Agent

The *Active Distribution Network Management Agent* (ADNM/A) is at the higher level of the proposed hierarchical MAS. ADNM/A gathers the aggregated responses from the cluster agents and also information for the forecasted price of electricity, the node loads of the network, and the power produced by RESs and it exploits an optimization technique to optimally schedule the operation of the ADN. The proposed method aims to reduce the daily operation cost of the network by shifting the power demand in time periods with a low electricity price. At the same time, all the operational constraints of the ADN, building energy systems, and hosted PEVs should be satisfied. The constrained optimal power flow algorithm (OPF) was exploited to keep the voltages of the ADN within the permissible bounds. Then, the active and reactive power set-points resulted from the OPF exploitation were forwarded back to the respective cluster agents to be optimally dispatched to the local agents they supervise.

3. Agents' Modeling

In this section, the models of local, cluster and ADNM/A are presented.

3.1. Local Agents

3.1.1. Building Thermal Load Agent (BTL/A)

BTL/A calculates the internal temperature of each residential building it supervises over the optimization horizon according to the thermal model described below [16].

$$\rho \cdot C_b \cdot V_b \cdot \frac{dT_{in,b}}{dt} = \dot{Q}_{ex,wall,b} + \dot{Q}_{in,wall,b} + \dot{Q}_{win,b} + Q_{in,b} + \dot{Q}_{sw,b} + \dot{Q}_{sg,b} - Q_{HVAC,b} \quad (1)$$

With

$$\dot{Q}_{ex,wall,b} = \sum_{y \in \varepsilon} U_{wall,y} \cdot F_{wall,y} \cdot (T_{out} - T_{in,b}) \quad (2)$$

$$\dot{Q}_{win,b} = \sum_{y \in \varepsilon} U_{win,y} \cdot F_{win,y} \cdot (T_{out} - T_{in,b}) \quad (3)$$

$$\dot{Q}_{sw,b} = \sum_{y \in \varepsilon} a_w \cdot R_{se} \cdot U_{wall,y} \cdot F_{wall,y} \cdot I_{T,b} \quad (4)$$

$$\dot{Q}_{sg,b} = \sum_{y \in \varepsilon} \tau_{win} \cdot SC \cdot F_{win,y} \cdot I_{T,b} \quad (5)$$

$$\dot{Q}_{in,wall,z} = \sum_{x \in \mathcal{I}} U_{wall,x} \cdot F_{wall,x} \cdot (T_{in,nz} - T_{in,z}) \quad (6)$$

where the b index denotes the building's number; the y index denotes the yth external wall/window; ε is the external walls set τ_{win} is the window glass transmission coefficient; a_w is the external wall absorbance coefficient; C_b is the specific heat capacity; $F_{wall(win)}$

is the wall (window) surface area; $I_{T,b}$ is the total solar radiation on the *b*th building; ρ denotes the air density; $Q_{HVAC,b}$ denotes the power production for the cooling of the *b*th building; $\dot{Q}_{ex,wall,b}$ denotes the external walls' heat exchange of the *b*th building; $Q_{in,b}$ denotes the internal heat gains of the *b*th building; $\dot{Q}_{sg,b}$ denotes the solar radiation through the windows of the *b*th building; $\dot{Q}_{sw,b}$ denotes the heat gain by the external walls' solar radiation of the bth building; $\dot{Q}_{win,b}$ denotes the heat transfer across the windows of the bth building; R_{se} is the heat resistance of the external surface; $T_{in,b}$ denotes the bth building's internal temperature (°C); T_{out} denotes the bth building's outdoor temperature (°C); SC denotes the shading coefficient of the windows; and $U_{wall(win)}$ denotes the factor of the heat exchange of the external wall (window).

The estimated indoor temperature of the building was used with its upper and lower limits to suitably define the flexibility of each building to increase ($FL_b^{\uparrow}$) or decrease ($FL_b^{\downarrow}$) its thermal power demand, as in the following equations.

$$FL_b^{\uparrow}(t) = \frac{T_{in,b}(t) - T_{min,b}}{T_{max,b} - T_{min,b}} \tag{7}$$

$$FL_b^{\downarrow}(t) = \frac{T_{max,b} - T_{in,b}(t)}{T_{max,b} - T_{min,b}} \tag{8}$$

3.1.2. Building Electrical Load Agent (BEL/A)

BEL/A is assigned to each building of the network and the model electrical loads of the building. The modeling of the energy consumption profiles in residential buildings was developed taking into consideration the number, age, and the activity level of the occupants as well as the number, type, and the usage of electric household appliances [14,15].

3.1.3. Plug-in Electric Vehicle Agent (PEV/A)

PEV/A calculates the boundaries of each PEV's state of charge and the active and reactive power it trades with the distribution network. According to these characteristics, the area the PEV is permitted to operate is derived [16], and is shown in Figure 2.

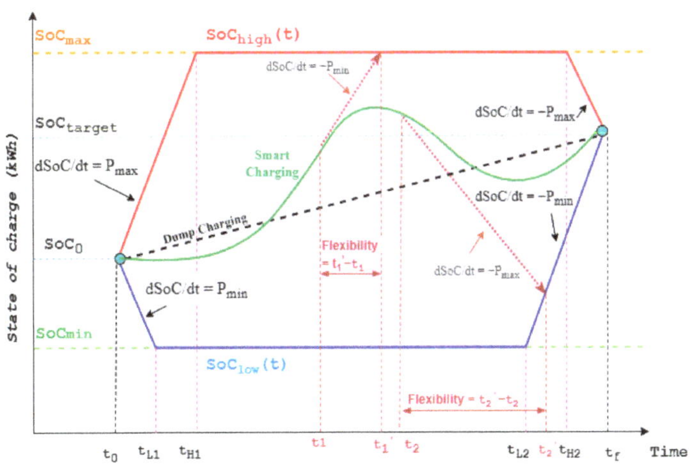

Figure 2. Bounds of the PEV's stored energy.

The time that the EV is plugged to the network and unplugged from it is denoted with t_0 and t_f, respectively. SoC_{max} and SoC_{min} are the maximum and the minimum values of the PEV's battery state of charge. SoC_0 is the initial state of charge (kWh) of each PEV. P_{max}

and P_{min} (kW) are the maximum and minimum power that the PEV's battery can exchange with the network, respectively. SoC_{target} (kWh) is the stored energy target that the driver has defined the PEV should reach at its disconnection time. The dynamic lower and upper bounds of the PEV's SoC, SoC_{low}, and SoC_{high} are generally defined by 4 points where they start to decrease or increase with a constant rate of change P_{min} or P_{max}, respectively. These points are (t_0, SoC_0), (t_{L_1}, SoC_{min}), (t_{L_2}, SoC_{min}), and (t_f, SoC_{target}) for SoC_{low} and (t_0, SoC_0), (t_{H_1}, SoC_{max}), (t_{H_2}, SoC_{max}), (t_f, SoC_{target}) for SoC_{high}.

t_{L_1}, t_{H_1}, t_{L_2} and t_{H_2} are estimated as in the following:

$$t_{L_1}(i) = t_0(i) + \frac{SoC_{min}(i) - SoC_0(i)}{P_{min}(i)} \tag{9}$$

$$t_{H_1}(i) = t_0(i) + \frac{SoC_{max}(i) - SoC_0(i)}{P_{max}(i)} \tag{10}$$

$$t_{L_2}(i) = t_f(i) + \frac{SoC_{target}(i) - SoC_{min}(i)}{P_{min}(i)} \tag{11}$$

$$t_{H_2}(i) = t_f(i) + \frac{SoC_{target}(i) - SoC_{max}(i)}{P_{max}(i)} \tag{12}$$

Moreover, PEV/A determines the flexibility of each individual PEV to adjust its power, $flex_{PEV}(i,t)$ [25]. These quantities are forwarded to the CPEV/A that supervises the cluster of PEVs.

$$flex_{PEV}(i,t) = t'(i,t) - t \tag{13}$$

where $t'(i,t)$ is the time point at which the SoC trajectory crosses the SoC_{high} or SoC_{low} limits provided that the ith PEV absorbs or injects active power $P_{min}(i)$ or $P_{max}(i)$, respectively, as described in [25].

According to Equation (13), the flexibility of the ith PEV to increase (reduce) its active power at a certain time point is the time required to reach the $SoC_{high(low)}$ limit if it absorbs (injects) the maximum active power, respectively.

If $P(i,t) \geq 0$, then $t'(i,t)$ is calculated in Equation (14).

$$t'(i,t) = min\begin{cases} \frac{SoC_{max}(i) - SoC(i,t+\Delta t)}{P_{max}(i) - P_{min}(i)} + \frac{P_{max}(i)\cdot t_{H_2}(i) - P_{min}(i)\cdot t}{P_{max}(i) - P_{min}(i)} \\ t + \frac{SoC(i,t+\Delta t) - SoC_{max}(i)}{P_{min}(i)} \end{cases} \tag{14}$$

If $P(i,t) < 0$, then $t'(i,t)$ is calculated in Equation (15).

$$t'(i,t) = min\begin{cases} \frac{SoC_{max}(i) - SoC(i,t+\Delta t)}{P_{max}(i) - P_{min}(i)} + \frac{P_{max}(i)\cdot t_{H_2}(i) - P_{min}(i)\cdot t}{P_{max}(i) - P_{min}(i)} \\ t + \frac{SoC(i,t+\Delta t) - SoC_{max}(i)}{P_{min}(i)} \end{cases} \tag{15}$$

If $P(i,t)$ is the optimal active power of the ith PEV (generator convention) and the n_{ch}, n_{dis} charging and discharging efficiencies, then the SoC (kWh) of the ith PEV at the end of the next time interval is estimated according to Equation (16).

$$SoC(i,t+\Delta t) = \begin{cases} SoC(i,t) - P(i,t)\cdot n_{ch}\cdot \Delta t, & P^*_{PEV}(i,t) < 0 \\ SoC(i,t) - \frac{P(i,t)}{n_{dis}}\cdot \Delta t, & P^*_{PEV}(i,t) \geq 0 \end{cases} \tag{16}$$

The state of charge calculated in Equation (16) is used in Equations (13)–(15) to estimate the flexibility of the ith PEV, $flex_{PEV}(i, t+\Delta t)$, at time $t+\Delta t$.

3.2. Cluster Agents

3.2.1. Cluster of Building Thermal Loads Agent (CBTL/A), Cluster of Building Electrical Loads Agent (CBEL/A)

The primary goal of CBTL/As is to aggregate the responses they receive from the BTL/As being in their control areas. The flexibility of a cluster of buildings to increase ($FL_{CB}^{\uparrow}$) or decrease ($FL_{CB}^{\downarrow}$) its power is the respective weighted average of the flexibilities of the buildings it supervises, as described in Equations (17) and (18). V_b is the air volume of the bth building.

$$FL_{CB}^{\uparrow}(t) = \frac{\sum_b FL_b^{\uparrow}(t) \cdot V_b}{\sum_b V_b} \tag{17}$$

$$FL_{CB}^{\downarrow}(t) = \frac{\sum_b FL_b^{\downarrow}(t) \cdot V_b}{\sum_b V_b} \tag{18}$$

Moreover, CBTL/As define suitable flexibilities ($FL_{CB,fz}$) by exploiting fuzzy logic based on information regarding the respective node voltage and their primary flexibility to adjust their power, as shown in Figure 3. Initially, the inputs and outputs were mapped into fuzzy sets using the respective membership functions. Both inputs were fuzzified using five sets with linguistic variables for the node voltage deviation from its upper or lower bound and the flexibility of the cluster of buildings to change its power. These variables were defined as very small (VS), small (S), medium (M), large (L), and very large (VL). The output of the fuzzy logic was determined by defining the rules indicated in Table 1.

Figure 3. Fuzzy inference system.

Table 1. Fuzzy rules for the calculation of the building clusters' flexibility ($FL_{CB,fz}$).

		ΔV				
		VS	**S**	**M**	**L**	**VL**
Cluster of buildings flexibility (FL_{CB})	**VS**	VS	VS	VS	VS	VS
	S	VS	S	S	M	L
	M	VS	S	M	L	L
	L	VS	M	L	L	VL
	VL	VS	L	L	VL	VL

BEL/As aggregate the responses they receive from the BEL/As they control in order to model the electrical load profiles of the clusters of residential buildings.

3.2.2. Cluster of Plug-In Electric Vehicles Agent (CPEV/A)

An innovative flexibility of cluster PEVs to adjust its active power at time t, $flex_{CPEV}(c,t)$, is suggested in [25], and was also followed in the present work as defined in Equation (19).

$$flex_{CEV}(c,t) = t'(c,t) - t \tag{19}$$

where $t'(c,t)$ is the moment in time when the SoC trajectory passes SoC_{high} or SoC_{low} bounds, assuming that the cth cluster of PEVs absorbs or injects active power P_{min} or P_{max}, respectively, as shown in Figure 4. The CEV index denotes the cluster of PEVs;

SoC_{high}, SoC_{low} are the maximum and minimum stored energy by CEV; and P_{max}, P_{min} are the active power boundaries of CEV.

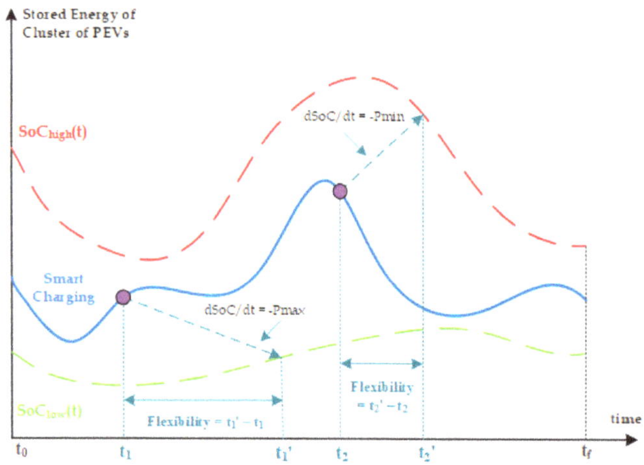

Figure 4. Flexibility of a CPEV/A to adjust its active power.

Moreover, CPEV/As define suitable flexibilities ($FL_{CPEV,fz}$) by exploiting fuzzy logic in a similar way to the flexibilities ($FL_{CB,fz}$) defined by CBTL/As. Both inputs and the output were fuzzified using five sets with linguistic variables. The inputs of the fuzzy logic system are the flexibility of a cluster of PEVs to adjust its active power at time t, $flex_{CPEV}(c,t)$, and the node voltage deviation from its upper or lower bound. The output of the fuzzy logic was determined by defining the rules indicated in Table 1. The obtained flexibility $FL_{CPEV,fz}$ was used by ADNM/A to optimally dispatch the total active power of the aggregate battery of PEVs to each CPEV/A.

3.3. Active Distribution Network Management Agent (ADNM/A)

ADNM/A gathers the aggregated responses from the cluster agents as well as the information for the forecasted price of electricity, the node loads of the network, and the power produced by RESs.

Equations (20)–(27) can be used to estimate the time-varying maximum and minimum bounds of the buildings' electric power consumption.

$$P_{HVAC,total,max}(t) = \sum_{CB} P_{HVAC,CB,max}(t) \quad (20)$$

$$P_{HVAC,total,min}(t) = \sum_{CB} P_{HVAC,CB,min}(t) \quad (21)$$

$$P_{EL,total,max}(t) = \sum_{CB} P_{EL,CB,max}(t) \quad (22)$$

$$P_{EL,total,min}(t) = \sum_{CB} P_{EL,CB,min}(t) \quad (23)$$

with

$$P_{HVAC,CB,max}(t) = \sum_{b} P_{HVAC,b,max}(t) \quad (24)$$

$$P_{HVAC,CB,min}(t) = \sum_{b} P_{HVAC,b,min}(t) \quad (25)$$

$$P_{EL,CB,max}(t) = \sum_{b} P_{EL,b,max}(t) \quad (26)$$

$$P_{EL,CB,min}(t) = \sum_{b} P_{EL,b,min}(t) \tag{27}$$

where $P_{HVAC,b,max}, P_{HVAC,b,min}$ are the bth building's HVAC system power consumption boundaries; $P_{HVAC,CB,max}, P_{HVAC,CB,min}$ are the respective boundaries of CBTL/A; $P_{HVAC,total,max}, P_{HVAC,total,min}$ are buildings' total HVAC power consumption boundaries; $P_{EL,b,max}, P_{EL,b,min}$ are the bth building's electrical load consumption boundaries; $P_{EL,CB,max}, P_{EL,CB,min}$ are the respective boundaries of CBEL/A; and $P_{EL,total,max}, P_{EL,total,min}$ are the total electrical loads' power consumption boundaries.

Equations (28)–(35) can be used to estimate the time-varying maximum and minimum bounds of the power and stored energy of the aggregate battery.

$$P_{PB,max}(t) = \sum_{CEV} P_{CEV,max}(t) \tag{28}$$

$$P_{PB,min}(t) = \sum_{CEV} P_{CEV,min}(t) \tag{29}$$

$$SoC_{PB,high}(t) = \sum_{CEV} SoC_{CEV,high}(t) \tag{30}$$

$$SoC_{PB,low}(t) = \sum_{CEV} SoC_{CEV,low}(t) \tag{31}$$

with

$$P_{CEV,max}(t) = \sum_{i} P_{max}(i,t) \tag{32}$$

$$P_{CEV,min}(t) = \sum_{i} P_{min}(i,t) \tag{33}$$

$$SoC_{CEV,high}(t) = \sum_{i} SoC_{high}(i,t) - SoC_{CEV,diff}(t) \tag{34}$$

$$SoC_{CEV,low}(t) = \sum_{i} SoC_{low}(i,t) - SoC_{CEV,diff}(t) \tag{35}$$

The PEVs' plugging and unplugging over time affects the total stored energy of the dynamic equivalent aggregate battery. The resulting change of the aggregate SoC is indicated by $SoC_{PB,diff}$ and estimated in Equations (36)–(38).

$$SoC_{CEV,diff}(t) = \sum_{T_0:\Delta t:t} \left(SoC_{0,CEV}(t) - SoC_{T,CEV}(t)\right) \tag{36}$$

$$SoC_{0,CEV}(t) = \sum_{\substack{\forall ith\ EV\ plugged\ at\ t, \\ \in CEV}} SoC_0(i) \tag{37}$$

$$SoC_{T,CEV}(t) = \sum_{\substack{\forall ith\ EV\ unplugged\ at\ t, \\ \in CEV}} SoC_{target}(i) \tag{38}$$

Equation (39) indicates the total stored energy of the equivalent PEVs' battery at the next time interval, assuming that $P_{PB,total}(t)$ is the optimal total active power of the aggregate battery and by adopting a generator convention.

$$SoC_{PB}(t+\Delta t) = \begin{cases} SoC_{PB}(t) - P_{PB,total}(t) \cdot n_{ch} \cdot \Delta t, & P_{PB,total}(t) < 0 \\ SoC_{PB}(t) - \frac{P_{PB,total}(t)}{n_{dis}} \cdot \Delta t, & P_{PB,total}(t) \geq 0 \end{cases} \tag{39}$$

where n_{ch}, n_{disch} are the PEVs' charging (discharging) efficiency coefficients; SoC_0 is the level of energy at the beginning of charging; SoC_{target} is the target level of the stored energy at the end of charging; $SoC_{0,CEV}$ is the level of the stored energy of a cluster of PEVs at t_o; $SoC_{T,CEV}$ is the target level of the stored energy of a cluster of PEVs at time t_f; $SoC_{PB,high}, SoC_{PB,low}$ are the maximum/minimum stored energy by the PEVs' aggregate battery, respectively; $SoC_{CEV,high}$, $SoC_{CEV,low}$ are the maximum/minimum stored energy by a cluster of PEVs; $P_{CEV,max}$, $P_{CEV,min}$ are the active power boundaries of a cluster of PEVs; and $P_{PB,max}$, $P_{PB,min}$ are the power boundaries of the PEVs' aggregate battery.

Then, the ADN exploits an optimization technique for the optimal scheduling of its operation. Specifically, ADN uses particle swarm optimization (PSO), which is one of the most highly efficient heuristic methods and is remarkably simple to implement. PSO has been proven to be very robust and efficient for application in complex optimization problems as it does not depend on the selected initial point and leads to a global optimum with a high rate of success. However, it should be noted that no optimization method can guarantee convergence to the global optimum with a 100% rate of success. ADN provides the optimal total electric power consumption of the HVAC systems of all of the hosted residential buildings and the optimal active power of the PEVs' aggregate battery. The structure of each particle of the swarm with its different parts comprising the respective decision variables of the optimization of the ADN is given in Figure 5. Constrained OPF can be exploited to keep the voltages of the active distribution network and the currents of the power lines between the permissible boundaries. Then, the active and reactive power set-points that resulted from the OPF exploitation are forwarded back to the appropriate cluster agents to be suitably dispatched to the local agents they manage.

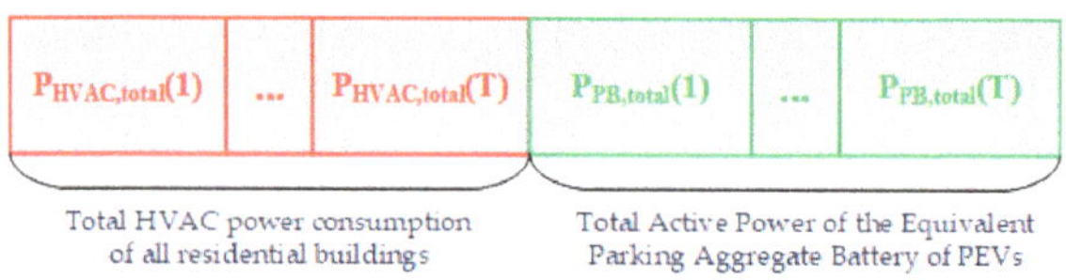

Figure 5. Particle structure used by the ADNM/A.

The augmented cost function used by the PSO incorporating the imposed constraints is provided in Equation (40). The objective of the optimization procedure is to minimize the ADN's overall daily operational costs in accordance with the cost of electricity, while also fully satisfying all of the ADN components' constraints incorporated into the term (Penalty) of the objective function.

$$TC_{ADN} = \min_{P_{PB,total},\, P_{HVAC,total}} \left\{ \left(\sum_t P_{grid}(t) \cdot EP(t) \right) \cdot \Delta t + Penalty \right\} \tag{40}$$

where EP denotes the variable price of electricity (m.u./MWh) and P_{grid} is the power of the main electric grid.

Then, ADNM/A optimally dispatches the total HVAC electric consumption of residential buildings ($P_{HVAC,total}$) to the CBTL/As it supervises according to Equation (41) and the obtained active and reactive power set-points are delivered back to these clusters. Afterward, CBTLAs dispatch their power to the local agents BTL/As they supervise and control, as formulated in Equation (42).

$$P_{HVAC,CB}(t) = \frac{FL_{CB}(t) \cdot \sum_b V_b}{\sum_{CB} \{FL_{CB}(t) \cdot V_{CB}\}} \cdot P_{HVAC,total}(t) \tag{41}$$

$$P_{HVAC,b}(t) = \frac{FL_b(t) \cdot V_b}{\sum_b \{FL_b(t) \cdot V_b\}} \cdot P_{HVAC,CB}(t) \tag{42}$$

where V_{CB} is the sum of the volumes of the buildings belonging to the respective cluster of buildings denoted by CB.

The process of the HVAC system's optimal power dispatch to the CBTL/As and then to the local agents BTL/As and more specifically to each hosted residential building is depicted in Figure 6. The average execution time of this process was 0.035 s for the examined system.

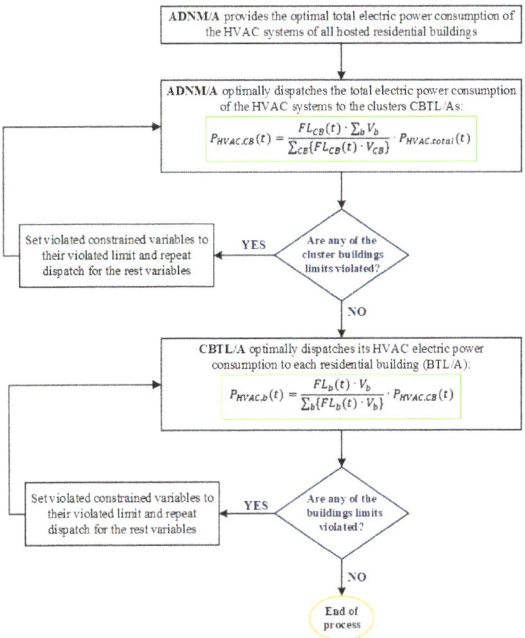

Figure 6. Flowchart of the optimal power dispatch of the buildings' HVAC system.

A last optimization problem was solved for the optimal dispatch of the total power of the dynamic equivalent aggregate battery to CPEV/As and then to each PEV as formulated and described in detail in [25]. The optimization problem that optimally dispatches the active power of a cluster of PEVs to each PEV it supervises in a way that the sum of PEV flexibilities is maximized and all the technical and operational constraints of the PEVs are satisfied is described next. The augmented objective function of the examined optimization problem is given in Equation (43).

$$\max_{P(i,t)} \sum_i flex_{PEV}(i, t+\Delta t) \tag{43}$$

The optimization process for the optimal operation scheduling of the examined ADN is subjected to the following constraints.

- *Building Constraints*

$$P_{HVAC,total,min}(t) \le P_{HVAC,total}(t) \le P_{HVAC,total,max} \tag{44}$$

$$P_{EL,total,min}(t) \le P_{EL,total}(t) \le P_{EL,total,max} \tag{45}$$

$$P_{HVAC,CB,min}(t) \le P_{HVAC,CB}(t) \le P_{HVAC,CB,max} \tag{46}$$

$$P_{EL,CB,min}(t) \le P_{EL,CB}(t) \le P_{EL,CB,max} \tag{47}$$

$$P_{HVAC,b,min}(t) \le P_{HVAC,b}(t) \le P_{HVAC,b,max} \tag{48}$$

$$P_{EL,b,min}(t) \le P_{EL,b}(t) \le P_{EL,b,max} \tag{49}$$

$$\sum_b P_{HVAC,b}(t) = P_{HVAC,CB}(t) \tag{50}$$

$$\sum_{CB} P_{HVAC,CB}(t) = P_{HVAC,total}(t) \tag{51}$$

$$P_{HVAC,B} = \frac{Q_{HVAC,b}}{COP}, P_{HVAC,CB} = \frac{Q_{HVAC,CB}}{COP}, P_{HVAC,total} = \frac{Q_{HVAC,total}}{COP} \tag{52}$$

$$T_{min,b} \leq T_{in,b}(t) \leq T_{max,b} \tag{53}$$

where COP is the HVAC system performance coefficient and $T_{max,b}$, $T_{min,b}$ are the internal temperature boundaries of the *b*th building (°C).

- *Plug-In Electric Vehicles Constraints*

$$SoC_{PB}(T_0) = SoC_{PB}\left(T_f\right) \tag{54}$$

$$SoC_{PB,low}(t) \leq SoC_{PB}(t) \leq SoC_{PB,high}(t) \quad \forall t \in \left[T_0 \; T_f\right] \tag{55}$$

$$P_{PB,min}(t) \leq P_{PB,total}(t) \leq P_{PB,max}(t) \quad \forall t \in \left[T_0 \; T_f\right] \tag{56}$$

$$SoC_{CEV,low}(t) \leq SoC_{CEV}(t) \leq SoC_{CEV,high}(t) \quad \forall t \in \left[T_0 \; T_f\right] \tag{57}$$

$$P_{CEV,min}(t) \leq P_{CEV}(t) \leq P_{CEV,max}(t) \quad \forall t \in \left[T_0 \; T_f\right] \tag{58}$$

$$SoC_{low}(i,t) \leq SoC(i,t) \leq SoC_{high}(i,t) \quad \forall t \in \left[T_0 \; T_f\right] \tag{59}$$

$$P_{min}(i,t) \leq P(i,t) \leq P_{max}(i,t) \quad \forall t \in \left[T_0 \; T_f\right] \tag{60}$$

$$\sum_{i} P(i,t) = P_{CEV}(t) \tag{61}$$

$$\sum_{CEV} P_{CEV}(t) = P_{PB,total}(t) \tag{62}$$

- *Power Balance Constraints*

$$\begin{aligned} P_{HVAC,total}(t) + P_{EL,total}(t) + P_{DNload}(t) = \\ P_{PB,total}(t) + P_{PV}(t) + P_{WT}(t) + P_{grid}(t) \quad \forall t \in \left[T_0 \; T_f\right] \end{aligned} \tag{63}$$

$$P_{grid,min}(t) \leq P_{grid}(t) \leq P_{grid,max}(t) \quad \forall t \in \left[T_0 \; T_f\right] \tag{64}$$

where $P_{grid,max}$, $P_{grid,min}$ are the power exchange boundaries of the ADN with the main electric grid; P_{PV} is the photovoltaic power production; P_{WT} is the wind turbine power production; and P_{DNload} is the load of the ADN.

- *Active Distribution Network Constraints*

$$\boldsymbol{V}\cdot\bar{\boldsymbol{Y}}\cdot\bar{\boldsymbol{V}} = \boldsymbol{S}_{inj} = \boldsymbol{P}_{inj} + j\cdot\boldsymbol{Q}_{inj} \tag{65}$$

$$|\boldsymbol{V}(k,t)| \leq \boldsymbol{V}_{max}(k) \tag{66}$$

$$|\boldsymbol{V}(k,t)| \geq \boldsymbol{V}_{min}(k) \tag{67}$$

$$\boldsymbol{\theta}(k,t) \geq \boldsymbol{\theta}_{min}(k) \tag{68}$$

$$\boldsymbol{\theta}(k,t) \leq \boldsymbol{\theta}_{max}(k) \tag{69}$$

$$|\boldsymbol{Y}(k,l)\cdot(\boldsymbol{V}(k,t) - \boldsymbol{V}(l,t))| \leq \boldsymbol{I}_{max}(k,l) \tag{70}$$

where $\boldsymbol{V}$ and $\boldsymbol{\theta}$ are the vectors representing the amplitudes and the angles of the distribution network voltages and $\boldsymbol{Y}$ is the network admittance matrix. $|\boldsymbol{V}(k)|$ denotes the voltage amplitude of the kth node; $\boldsymbol{\theta}(k)$ is the voltage angle at the kth node; and $\boldsymbol{I}_{max}(k,l)$ is the maximum current flowing in the electric line connecting the kth and lth node. The $\boldsymbol{S}_{inj}$ vector comprises the complex apparent power injections at the nodes of the network, while the $\boldsymbol{P}_{inj}$ and $\boldsymbol{Q}_{inj}$ vectors comprise the active and reactive power injections at all network nodes, respectively.

MatPower was used to solve the optimal power flow (OPF) problem subject to the constraints defined in Equations (65)–(70).

4. Case Study

The suggested approach was applied to a case study comprising the IEEE 33-bus radial distribution network with 1000 residential building prosumers, 32 clusters of PEVs, one wind park, and three photovoltaic parks integrated to it. The developed models were verified through detailed simulation results.

Four types of residential buildings and PEVs with different technical specifications were considered in this work. Probability density functions (PDFs) were used to obtain the arrival and dwell periods of the PEVs and can be found in [16]. Figure 7a shows the total number of connected PEVs to the distribution network, while Figure 7b shows the number of connected PEVs to each node of the network. In order to take into consideration the stochastic behavior of the ambient temperature, various trajectories were generated and used. Figure 8a displays the base trajectory of the dataset selected for demonstration purposes together with its variation area. Similarly, various trajectories of the actual electricity price were randomly generated and the respective data are shown in Figure 8b.

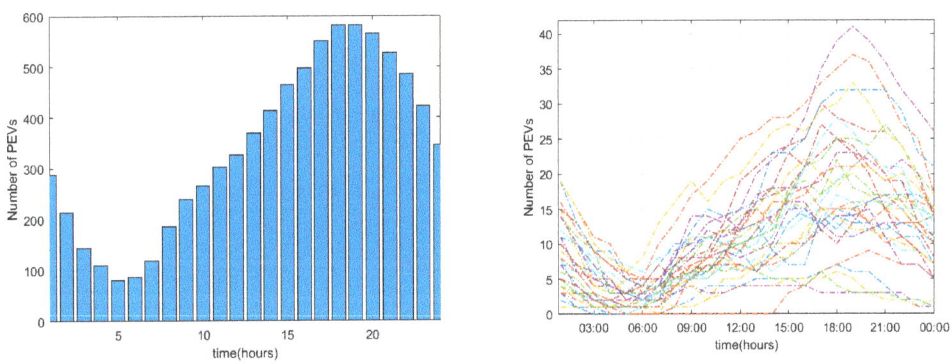

(**a**) (**b**)

Figure 7. (**a**) Total connected PEVs in the ADN. (**b**) Connected PEVs to each node of the ADN.

Various trajectories of the total ADN load consumption as well as of the total power generated by the wind park and the photovoltaics were randomly generated and the respective data are shown in Figure 9a,b, respectively.

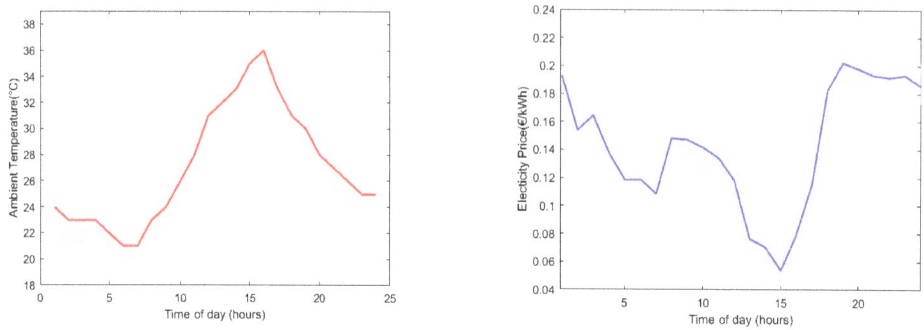

(**a**) (**b**)

Figure 8. (**a**) Outdoor temperature. (**b**) Forecasted price of electricity.

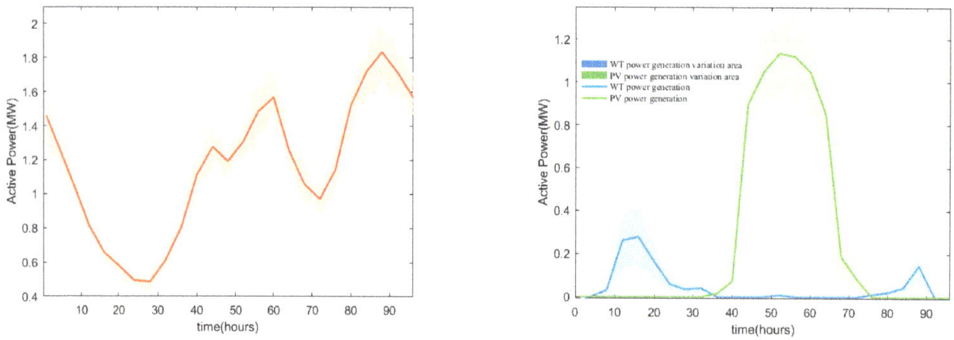

(**a**) (**b**)

Figure 9. (**a**) ADN load consumption. (**b**) WT and PV power production and their variation areas.

In addition to the proposed optimization approach, a typical ADN operation case study was considered, in which the residential building thermal loads as well as the hosted electric vehicles were not subject to any optimization. In this case, the objective of the operation of the HVAC system is to maintain the internal temperature of buildings at a predetermined set point. Electric vehicles only absorb power in order to charge their batteries and attain the desired energy level and they do not have a V2G operation ability. The examination of this case study was taken into account in order to compare its results with the respective results of the proposed strategy for the optimization of the operation of the ADN.

The total HVAC electrical power consumption and electrical power consumption of the other building loads of all residential buildings hosted in the examined active distribution network are given in Figure 10a. As can be observed, the residential buildings absorbed the largest amount of thermal power around 15:00 p.m., when the electricity price was very low, as the proposed algorithm aims to reduce the total operation cost of the examined system. Moreover, in this time period, the ambient temperature is high enough, and obviously the cooling requirements of the residences are increased. Figure 10b provides the comparison of the total buildings' HVAC power consumption between the proposed optimization examined case study and the typical scenario of the ADN's operation. It was observed that when the suggested optimization algorithm was implemented, a significant reduction in the HVAC systems' required power was achieved. Figure 11 depicts the HVAC system's power consumption of the clusters of the buildings connected to the distribution network nodes as a result of the optimal dispatch of the respective total HVAC system's power consumption. These vary due to the fact that each node comprises a different number of residential buildings with different technical specifications.

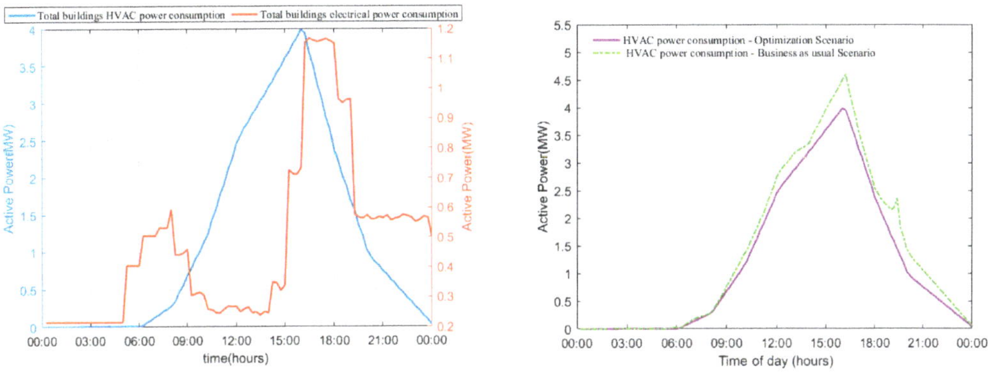

(a) (b)

Figure 10. Total buildings' HVAC and electrical power consumption. (**a**) Optimal operation. (**b**) Typical operation.

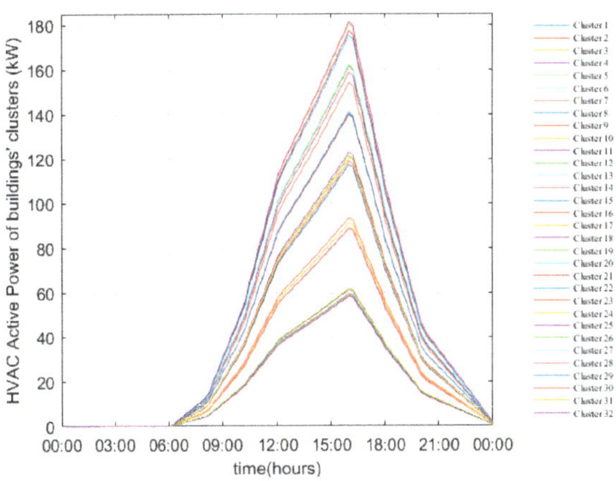

Figure 11. Building clusters' HVAC power consumption in each distribution network node.

Figure 12a,b presents the indoor temperatures of all buildings hosted by the ADN obtained by the proposed ADN optimal operation scheduling technique and the execution of the typical ADN operation, respectively. All temperatures were maintained within the considered permissible comfort range of 21–26 °C. Figure 13 depicts the power consumed by the non-flexible electrical loads of two indicative residential buildings hosted in the examined ADN. These energy consumption profiles were obtained by the building electrical agent that belongs to the category of MAS local agents.

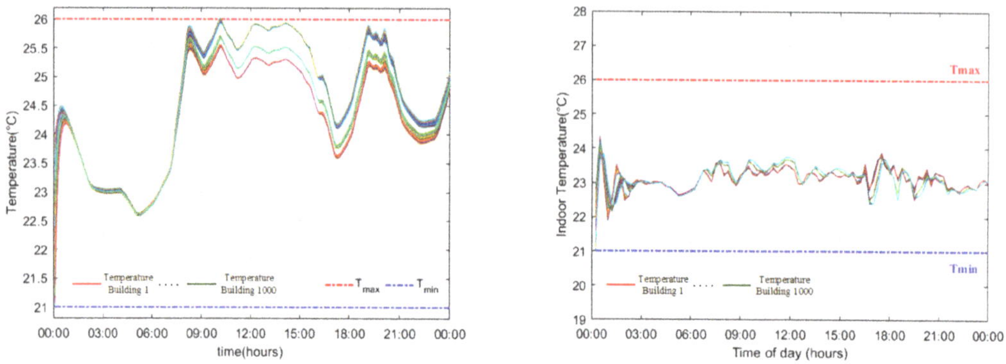

(**a**) (**b**)

Figure 12. Indoor temperatures of the buildings hosted in the distribution network. (**a**) Optimization. (**b**) Typical case.

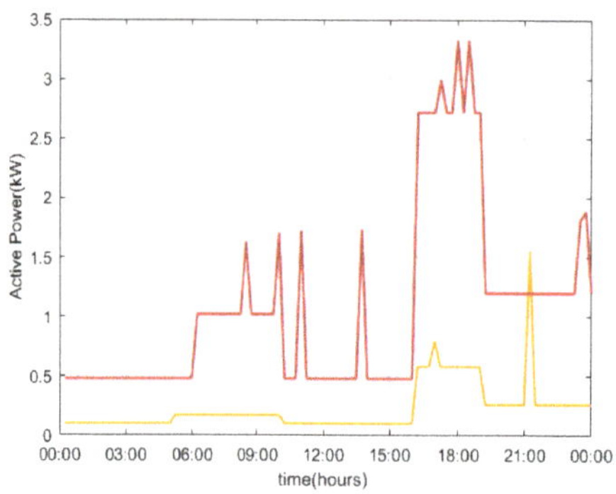

Figure 13. Electrical power consumption of two indicative buildings of the network.

Figure 14a,b depicts the total active power and the stored energy of the aggregated batteries of the PEVs together with their respective upper and lower limits. These trajectories were obtained by the application of the proposed ADN optimal operation scheduling technique and the business-as-usual ADN operation scenario. Positive power means that the equivalent aggregate battery injects power to the main grid.

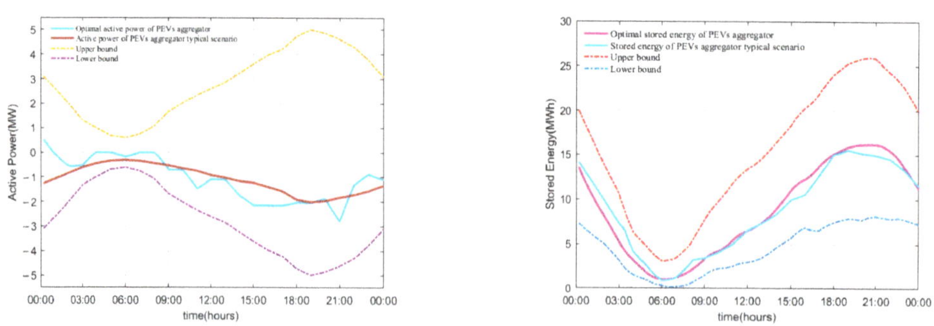

(**a**) (**b**)

Figure 14. (**a**) Total active power. (**b**) Total stored energy of the PEVs aggregate battery.

The energy stored in the equivalent aggregate battery of the PEVs as well as the respective upper and the lower limits shown in Figure 14b vary with time because of the continuously changing number of PEVs and the electricity price variations. In the optimal operation scenario, it was observed that the proposed algorithm stores energy in time periods of low electricity price (e.g., at 15:00), as the goal of the applied optimization strategy is to minimize the daily operation cost of the ADN. In the business-as-usual scenario, "dump charging" was applied to the PEVs. As shown in Figure 14a, the total power of the equivalent aggregate battery took exclusively negative values, which means that given the used generator convention, the PEVs only absorb the necessary amount of energy from the main grid in order to charge their batteries and reach the desired energy target at their disconnection time. In both of the examined scenarios, the obtained results verified that all PEVs reached their stored energy target, while all of their technical and operational constraints were satisfied.

Figures 15a and 16a depict the active power of two indicative clusters of PEVs located at nodes 21 and 31 of the examined ADN, while Figures 15b and 16b depict their stored energy, respectively. Moreover, their corresponding limits are given in the same figures. It was noticed that both the active power that clusters of PEVs share with the network and their battery energy were kept between their permitted ranges. These quantities were obtained by the implementation of the optimal dispatch of the total active power of the network's equivalent aggregate PEV battery, according to the proposed flexibilities of the PEV clusters to change their power.

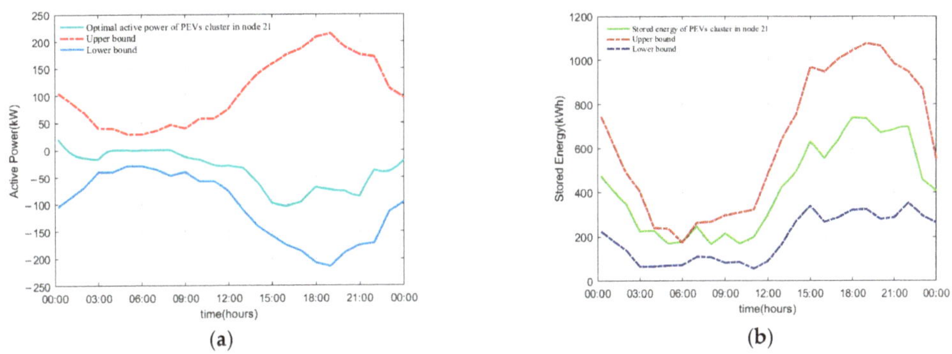

Figure 15. (**a**) Active power. (**b**) Stored energy of the cluster of PEVs in node 21 with their limits.

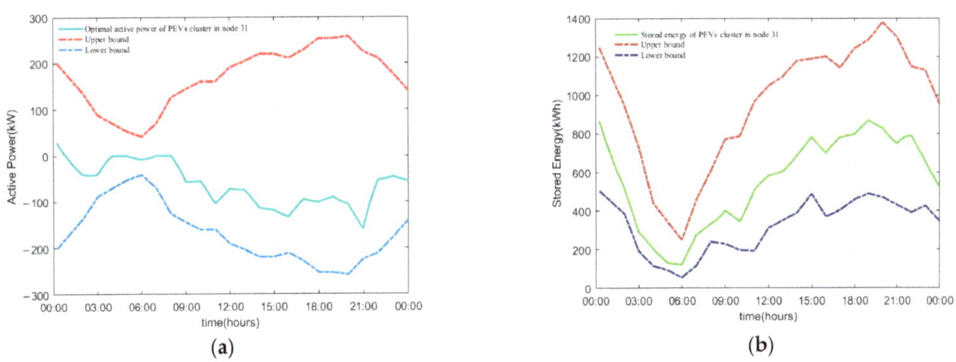

Figure 16. (**a**) Active power. (**b**) Stored energy of the cluster of PEVs in node 31 with their limits.

The total power consumptions of the buildings and PEVs together with the ADN load, the active power the distribution network exchanges with the main electric grid, and RES generation are shown in stacked form in Figure 17.

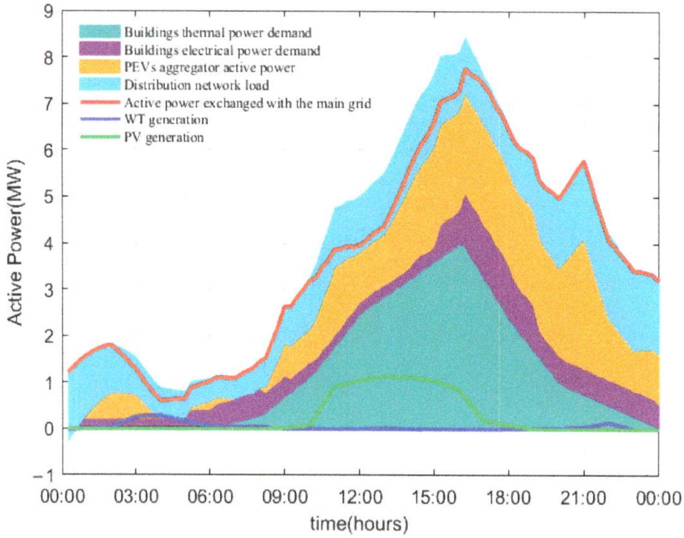

Figure 17. Buildings and PEVs' total power demand, ADN load, RES generation, and main grid power.

Figure 18a indicates that the reactive power set-points of each PEV cluster resulted from the OPF problem solution in order to keep the voltages of the distribution network busses between their permissible bounds and to meet the power flow constraints. Specifically, the reactive power set-points resulted from the OPF exploitation are forwarded back to the PEVs' cluster agents to be optimally dispatched to each PEV they supervise. The maximum and minimum voltage bounds were set to 1.1 and 0.9 (p.u.), respectively, while the nominal voltage was considered as 1 p.u. As shown in Figure 18b, all node voltages did not violate their bounds. The reactive powers of the PEV clusters and the voltages of the nodes of the examined ADN are indicated with different colors in Figures 18a and 18b, respectively.

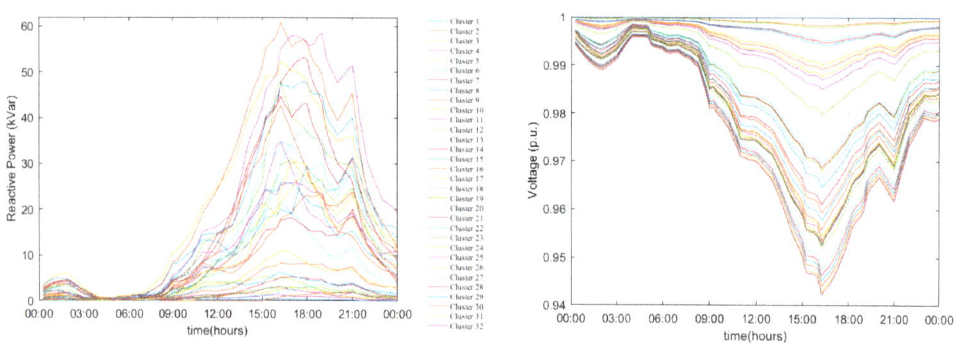

(**a**) (**b**)

Figure 18. (**a**) Reactive power of PEVs clusters. (**b**) Distribution network node voltages.

The active power the distribution network exchanges with the main electric grid both in its optimal and business as usual operation is given in Figure 19. The proposed method aims to reduce the overall operation cost of the ADN by shifting the power demand in time periods with low electricity price. It was verified in the optimal operation scenario that the main electric power grid supplies the ADN with large amounts of power when the electricity price is low, while the opposite happens when the electricity price is high. Detailed simulation results showed that significant cost reduction in the range of 17% was achieved by the application of the suggested optimization algorithm in comparison with the business as usual operation scenario of the ADN. Therefore, as shown in Figure 19, although there were time periods when the active power exchanged between the ADN and the main electric grid received higher values in the optimal operation scenario in comparison with the business-as-usual scenario, the total ADN's operation cost was significantly reduced at the end of the 24-h time period. Moreover, the obtained results show that the total active energy the ADN needs to absorb from the main electric grid to meet its 24-h energy demand is reduced in the range of 6.18%, when the suggested optimization technique is applied compared to the typical operation scenario of the ADN.

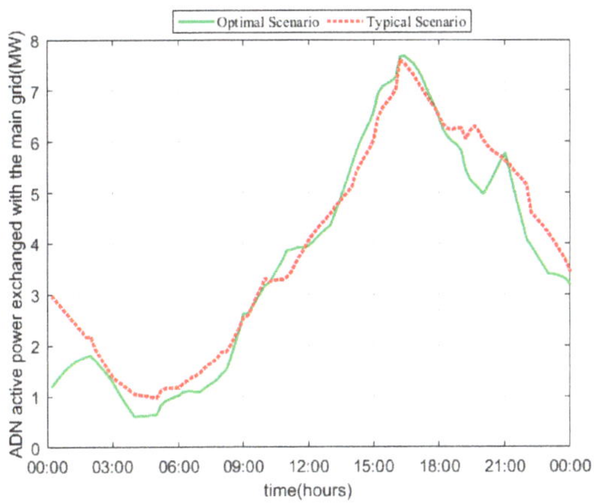

Figure 19. The total ADN's active power in both the optimal and typical operation.

The major information of the obtained cost resulting from the execution of the examined operation case studies is tabulated in Table 2. A significant cost reduction of 2478 m.u./day was achieved as a result of the implementation of the suggested optimization algorithm in comparison with a typical operation of the ADN. In addition, when the uncertainty of the ambient temperature and the electricity price were taken into consideration, and a slight increase of 0.92% in the daily operation cost of the examined system was observed.

Table 2. Examined operation case studies.

Operation Case Study	Optimal without Uncertainty	Optimal with Uncertainty	Typical
Total operation cost (m.u./day)	12,333	12,446.5	14,812

The proposed method only requires one round of communications between the agents. It was proven that the total time required for the computations and communications is very low, rendering the suggested method ideal for real-world applications in distribution networks hosting large building prosumers and PEVs. A typical average agent to agent communication latency is 0.15 s according to [31]. The required computation time to model the buildings' loads in this study was 0.1 s. Moreover, the execution of the optimal operation scheduling of the examined ADN requires on average 1517.2 s. Based on the aforementioned, the maximum expected time for the calculations and communications was $T = 4 \times 0.15\,\text{s} + 2 \times 0.1\,\text{s} + 1517.2\,\text{s} = 1518\,\text{s}$, which is absolutely satisfactory for the day-ahead operation scheduling of such complex ADN.

5. Conclusions

In this paper, an optimization method for the optimal and sustainable operation of active distribution networks hosting smart residential building prosumers, plug-in electric vehicle aggregators, and RES was proposed. The suggested power management system simultaneously satisfies a large number of operation and technical constraints of the ADN and its components. The implementation of the method was based on the exploitation of a properly designed hierarchical MAS. The goals of each type of agent and how they cooperate to accomplish them were fully outlined. Smart power dispatch techniques were also developed and applied to each cluster of buildings and electric vehicles, and additionally to each residential building and PEV hosted by the ADN to reduce the number of decision variables. Moreover, the algorithm achieved a notable reduction in the range of 17% in the daily operation cost of the examined system. The hosted PEVs and RES ensure the autonomous operation of the ADN, eliminating the need to invest in conventional polluting generators, reducing in this way greenhouse gas emissions. Consequently, the suggested approach for the optimal operation of the ADN aligns with the growing need for sustainable development. Detailed simulations of a suitably modified IEEE 33-bus test case system proved the effectiveness of the proposed strategy. As a future extension of our work, advanced control methodologies providing ancillary services by building thermal and electrical energy systems and energy storage facilities can be developed and integrated to the proposed one. Moreover, a comparison of the PSO performance application to the examined problem with other heuristic optimization methods could a future endeavor.

Author Contributions: Conceptualization, F.D.K.; Methodology, F.D.K.; Software, D.G.K.; Validation, F.D.K. and D.G.K.; Formal analysis, F.D.K. and D.G.K.; Investigation, F.D.K. and D.G.K.; Resources, F.D.K.; Data curation, F.D.K.; Writing—original draft preparation, D.G.K.; Writing—review and editing, F.D.K. and D.G.K.; Visualization, F.D.K.; Supervision, F.D.K.; Project administration, F.D.K. All authors have read and agreed to the published version of the manuscript.

Funding: This research received no external funding.

Institutional Review Board Statement: Not applicable.

Informed Consent Statement: Not applicable.

Data Availability Statement: Not applicable.

Conflicts of Interest: The authors declare no conflict of interest.

References

1. Matrawy, K.; Mahrous, A.-F.; Youssef, M. Energy management and parametric optimization of an integrated PV solar house. *Energy Convers. Manag.* **2015**, *96*, 377–383. [CrossRef]
2. Karami, N.; Moubayed, N.; Outbib, R. Energy management for a PEMFC-PV hybrid system. *Energy Convers. Manag.* **2014**, *82*, 154–168. [CrossRef]
3. Pascual, J.; Barricarte, J.; Sanchis, P.; Marroyo, L. Energy management strategy for a renewable-based residential microgrid with generation and demand forecasting. *Appl. Energy* **2015**, *158*, 12–25. [CrossRef]
4. Choudar, A.; Boukhetala, D.; Barkat, S.; Brucker, J.M. A local energy management of a hybrid PV-storage based distributed generation for microgrids. *Energy Convers. Manag.* **2015**, *90*, 21–33. [CrossRef]
5. Arcos-Aviles, D.; Pascual, J.; Marroyo, L.; Sanchis, P.; Guinjoan, F. Fuzzy logic-based energy management system design for residential grid-connected microgrids. *IEEE Trans. Smart Grid* **2018**, *9*, 530–543. [CrossRef]
6. Carli, R.; Dotoli, M. Decentralized control for residential energy management of a smart users' microgrid with renewable energy exchange. *IEEE/CAA J. Autom. Sin.* **2019**, *6*, 641–656. [CrossRef]
7. Rastegar, M.; Fotuhi-Firuzabad, M.; Zareipour, H. Home energy management incorporating operational priority of appliances. *Int. J. Electr. Power Energy Syst.* **2016**, *74*, 286–292. [CrossRef]

8. Javaid, N.; Ullah, I.; Akbar, M.; Iqbal, Z.; Khan, F.A.; Alrajeh, N.; Alabed, M.S. An Intelligent Load Management System with Renewable Energy Integration for Smart Homes. *IEEE Access* **2017**, *5*, 13587–13600. [CrossRef]
9. Shakouri, H.; Kazemi, A. Multi-objective cost-load optimization for demand side management of a residential area in smart grids. *Sustain. Cities Soc.* **2017**, *32*, 171–180. [CrossRef]
10. Alibabaei, N.; Fung, A.S.; Raahemifar, K.; Moghimi, A. Effects of intelligent strategy planning models on residential HVAC system energy demand and cost during the heating and cooling seasons. *Appl. Energy* **2017**, *185*, 29–43. [CrossRef]
11. Yoon, J.H.; Bladick, R.; Novoselac, A. Demand response for residential buildings based on dynamic price of electricity. *Energy Build.* **2014**, *80*, 531–541. [CrossRef]
12. Anvari-Moghaddam, A.; Monsef, H.; Rahimi-Kian, A. Optimal smart home energy management considering energy saving and a comfortable lifestyle. *IEEE Trans. Smart Grid* **2015**, *6*, 324–332. [CrossRef]
13. Arun, S.L.; Selvan, M.P. Intelligent Residential Energy Management System for Dynamic Demand Response in Smart Buildings. *IEEE Syst. J.* **2018**, *12*, 1329–1340. [CrossRef]
14. Estiri, H.; Zagheni, E. Age matters: Ageing and household energy demand in the United States. *Energy Res. Soc. Sci.* **2019**, *55*, 62–70. [CrossRef]
15. Csoknyai, T.; Legardeur, J.; Akle, A.A.; Horváth, M. Analysis of energy consumption profiles in residential buildings and impact assessment of a serious game on occupants' behavior. *Energy Build.* **2019**, *196*, 1–20. [CrossRef]
16. Kyriakou, D.G.; Kanellos, F.D. Optimal Operation of Microgrids Comprising Large Building Prosumers and Plug-in Vehicles Integrated into Active Distribution Networks. *Energies* **2022**, *15*, 6182. [CrossRef]
17. Turker, H.; Bacha, S. Optimal Minimization of Plug-In Electric Vehicle Charging Cost with Vehicle-to-Home and Vehicle-to-Grid Concepts. *IEEE Trans. Veh. Technol.* **2018**, *67*, 10281–10292. [CrossRef]
18. Godina, R.; Rodrigues, E.; Paterakis, N.; Erdinc, O.; Catalão, J. Innovative impact assessment of electric vehicles charging loads on distribution transformers using real data. *Energy Convers. Manag.* **2016**, *120*, 206–216. [CrossRef]
19. Qian, K.; Zhou, C.; Allan, M.; Yuan, Y. Modeling of load demand due to EV battery charging in distribution systems. *IEEE Trans. Power Syst.* **2011**, *26*, 802–810. [CrossRef]
20. Pirouzi, S.; Aghaei, J.; Niknam, T.; Khooban, M.H.; Dragicevic, T.; Farahmand, H.; Korpas, M.; Blaabjerg, F. Power Conditioning of Distribution Networks via Single-Phase Electric Vehicles Equipped. *IEEE Syst. J.* **2019**, *13*, 3433–3442. [CrossRef]
21. Xiang, Y.; Liu, J.; Liu, Y. Optimal active distribution system management considering aggregated plug-in electric vehicles. *Electr. Power Syst. Res.* **2016**, *131*, 105–115. [CrossRef]
22. Kanellos, F.D. Real-Time Control Based on Multi-Agent Systems for the Operation of Large Ports as Prosumer Microgrids. *IEEE Access* **2017**, *5*, 9439–9452. [CrossRef]
23. Sadati, S.M.B.; Moshtagh, J.; Shafie-khah, M.; Catalão, J.P.S. Smart distribution system operational scheduling considering electric vehicle parking lot and demand response programs. *Electr. Power Syst. Res.* **2018**, *160*, 404–418. [CrossRef]
24. Kanellos, F.D. Optimal Scheduling and Real-Time Operation of Distribution Networks with High Penetration of Plug-In Electric Vehicles. *IEEE Syst. J.* **2021**, *15*, 3938–3947. [CrossRef]
25. Kanellos, F.D.; Kalaitzakis, K.; Psarras, I.; Katsigiannis, Y. Efficient and robust power and energy management for large clusters of plug-in electric vehicles and distribution networks. *IET Energy Syst. Integr.* **2022**, *4*, 393–408. [CrossRef]
26. Nick, M.; Cherkaoui, R.; Paolone, M. Optimal Planning of Distributed Energy Storage Systems in Active Distribution Networks Embedding Grid Reconfiguration. *IEEE Trans. Power Syst.* **2018**, *33*, 1577–1590. [CrossRef]
27. Nick, M.; Cherkaoui, R.; Paolone, M. Optimal allocation of dispersed energy storage systems in active distribution networks for energy balance and grid support. *IEEE Trans. Power Syst.* **2014**, *29*, 2300–2310. [CrossRef]
28. Sedghi, M.; Ahmadian, A.; Aliakbar-Golkar, M. Optimal storage planning in active distribution network considering uncertainty of wind power distributed generation. *IEEE Trans. Power Syst.* **2016**, *31*, 304–316. [CrossRef]
29. Gungor, V.C.; Sahin, D.; Kocak, T.; Ergut, S.; Buccella, C.; Cecati, C.; Hancke, G.P. Smart grid technologies: Communication technologies and standards. *IEEE Trans. Ind. Inform.* **2011**, *7*, 529–539. [CrossRef]
30. Romano, P.; Paolone, M.; Chau, T.; Jeppesen, B.; Ahmed, E. A high-performance, low-cost PMU prototype for distribution networks based on FPGA. In Proceedings of the 2017 IEEE Manchester PowerTech, Manchester, UK, 18–22 June 2017.
31. Nguyen, C.P.; Flueck, A.J. Modeling of communication latency in smart grid. In Proceedings of the 2011 IEEE Power and Energy Society General Meeting, Detroit, MI, USA, 24–28 July 2011; pp. 1–6.

applied sciences

MDPI

Article

Nature-Based Solutions in Urban Areas: A European Analysis

Sara Bona [1], Armando Silva-Afonso [1,2], Ricardo Gomes [3], Raquel Matos [1] and Fernanda Rodrigues [1,*]

1 RISCO—Research Center for Risks and Sustainability in Construction, Department of Civil Engineering, University of Aveiro, 3810-193 Aveiro, Portugal
2 ANQIP—National Association for Quality in Buildings Services, Civil Engineering Department, University of Aveiro, 3810-193 Aveiro, Portugal
3 INESC Coimbra and SMAS de Leiria, Department of Civil Engineering, Polytechnic of Leiria, Campus 2 Morro do Lena—Alto do Vieiro, 2411-901 Leiria, Portugal
* Correspondence: mfrodrigues@ua.pt

Abstract: Currently, the world is facing resource scarcity as the environmental impacts of human intervention continue to intensify. To facilitate the conservation and recovery of ecosystems and to transform cities into more sustainable, intelligent, regenerative, and resilient environments, the concepts of circularity and nature-based solutions (NbS) are applied. The role of NbS within green infrastructure in urban resilience is recognised, and considerable efforts are being made by the European Commission (EC) to achieve the European sustainability goals. However, it is not fully evidenced, in an integrated way, which are the main NbS implemented in the urban environment and their effects. This article aims to identify the main and most recent NbS applied in urban environments at the European level and to analyse the integration of different measures as an innovative analysis based on real cases. For this purpose, this work presents a literature review of 69 projects implemented in 24 European cities, as well as 8 urban actions and 3 spatial scales of implementation at the district level. Therefore, there is great potential for NbS adoption in buildings and their surroundings, which are still not prioritized, given the lack of effective monitoring of the effects of NbS.

Keywords: nature-based solutions (NbS); ecosystem services; urban environment; blue–green infrastructure; green buildings; water management; urban regeneration; European Union

1. Introduction

The effects of global climate change are unquestionably among the most considerable problems facing humanity today. The current model of economic development based on fossil fuels and the indiscriminate exploitation of natural resources has brought about a situation in which negative impacts are entering a critical phase. The IPCC (Intergovernmental Panel on Climate Change) Special Report, projects that global warming will reach 1.5 °C between 2030 and 2052 if it continues to increase at the current rate [1]. Reflecting this trend, numerous regional changes in climate are projected to occur, with global warming on the planet, including increasingly extreme temperatures in cities, as well as increases in the frequency, intensity, and amount of heavy rain and droughts [1].

Cities are major contributors to climate change. Today, more than half of the world's population lives in urban settlements, and the United Nations (UN) predicts that by 2050, this urban share will reach 70% [2]. At present, cities are responsible for producing 50% of global waste [3], approximately 70% of greenhouse gas (GHG) emissions [2,4], and 75% of energy and natural resource consumption [4], generating increasing pressure on rural areas and natural ecosystems to guarantee the supply of water, energy, and food, as well as for the removal of waste [5], as the world's cities occupy only 3% of the terrestrial landscape [6].

Following this trend, the likely scenario for the coming decades is a series of global challenges related to resource depletion as a result of climate change associated with increased pollution. Due to the negative impacts that global development and consumption impose on contemporary society, new global sustainability frameworks have been adopted in recent years, such as the UN 2030 Agenda for Sustainable Development and its corresponding Sustainable Development Goals (SDGs) [2]. This and other international agreements seek to establish goals, guidelines, and action models to ensure a liveable planet and a more just and sustainable society. In this context, strategic urban planning, in association with sustainability, represents a fundamental tool for the mitigation and adaptation of cities to new and emerging challenges.

As defined by the International Union for Conservation of Nature (IUCN), NbS are actions that protect, sustainably manage, and restore natural or modified ecosystems while addressing social challenges and providing benefits for both human well-being and biodiversity [7]. This mechanism, which transfers nature to cities [5], has great potential to respond to and minimize the effects of climate change in urban spaces, creating alternatives that return natural elements to the urban environment [8] that were previously disappearing with the growth and development of cities.

In order to facilitate the conservation and recovery of ecosystems and to transform cities into more sustainable, intelligent, regenerative, and resilient environments, the concepts of circular economy (CE) and nature-based solutions (NbS) are applied. NbS and the CE are interconnected, side-by-side conceptions for the development of renaturalized and circular cities [9,10].

This article presents research on the concept and application of NbS at the urban level. Recognising the NbS theme as a key element in achieving the European sustainability goals the following research question was formulated: "*What are the recent and main NbS initiatives and practices being applied in urban environments in the European Union aiming at resilience and circularity in cities?*". To answer this question, a study was carried out through applied research with a qualitative approach around textual elements and considering the predefined content analysis methodology [11]. In a logical sequence, the fundamental phases of qualitative analysis were respected: preanalysis; exploration of the material; and treatment of results, including inference and interpretation.

In the sequence of the literature review, some knowledge gaps remain regarding the effectiveness of NbS in the urban environment, with an emphasis on short-term actions and long-term sustainability goals [12]. NbS have been promoted as a key tool in the search for innovative solutions to manage natural systems and to balance benefits for nature and society. However, the concepts and their practical applications remain imprecise and fragmented due to ambiguities linked to the integration of various scientific fields and the lack of clear standards in the NbS concept [13,14]. In recent years, several attempts have been made to define and clarify the actions inspired and driven by nature, as well as the associated benefits [12–17]. Therefore, it is possible to identify that the concept and the theme of NbS are still recent and are in an evolutionary process. Many studies point to the lack of methodology for the compilation and systematic evaluation, in a comprehensive and integrative way, of NbS knowledge [18,19]. Barriers have also been identified regarding the uncertainties concerning the performance, functionality, and implementation of NbS [20]. The ecosystem approach aligned with NbS initiatives is an example of an integrated principle and concept essential to achieving social and environmental sustainability [14]. Recognising that NbS are measures that provide multiple benefits is key to ensuring that NbS interventions deliver far-reaching benefits rather than unilateral outcomes [21]. Thus, the importance of this article consists of the integration of NbS initiatives in the urban environment, considering the performance of the provided ecosystem services. In this work, we identify and present the main and most recent NbS practices implemented in urban environments at the European scale and analyse the complementarity and integration of various measures, presenting an innovative analysis based on real-world examples. This study contributes to an improved understanding of NbS and provides a synthesis of knowledge and applicability of NbS, emphasizing the complementarity of NbS measures in the urban environment.

2. Sustainability in the Built Environment

2.1. Circular Economy

An important aspect of approaching contemporary urban systems is circularity. The transition to a CE is fundamental to face environmental challenges, as cities will have to play an essential role in this economic model. CE is a concept that directly contrasts with the traditional economic model, characterized by a predominantly linear pattern. CE can be defined as a production and consumption model that includes, whenever possible, sharing, leasing, reusing, repairing, refurbishing, and recycling of existing materials and products [22] with the aim of minimizing waste and making the most of resources [5]. In this way, the product life cycle can be extended with the expectation of fostering a closed-loop market. In a circular system, the input and waste of resources, emissions, and energy consumption are minimized by decelerating and narrowing energy and material cycles [5]. This means rethinking current paradigms, aiming at a greener and more sustainable economy.

In 2020, the EC adopted the new Circular Economy Action Plan (CEAP), which constitutes one of the main elements of the European Green Deal, Europe's agenda for sustainable growth [23]. According to CEAP, the Commission has committed itself to launch a new comprehensive Strategy for a Sustainable Built Environment to increase material efficiency and reduce the climate impacts of the built environment, particularly by promoting the principles of circularity throughout the life cycle of buildings. For this purpose, the fields of energy efficiency, resource efficiency, construction, and demolition waste management should be addressed in a sustainable strategy [23]. Implementing and addressing CE principles at the various stages of a building's life cycle is of utmost importance in the transition to a circular and dynamic built environment, as the way buildings are designed is critical to how they are used and the impact they have on their surroundings [9].

2.2. Nature-Based Solutions

NbS is a new concept that started to be used at the beginning of the 21st century and was later adopted by several global institutions [24]. NbS has evolved as an "umbrella" concept [25,26] that incorporates several existing concepts and practices [27] that can be applied to the strategic dimension, spatial planning dimension, soft engineering dimension, and performance dimension (Figure 1) [25]. The most recent EC report on NbS states that this concept embodies new ways of approaching socioecological adaptation and resilience, with equal confidence in the social, environmental, and economic domains [25].

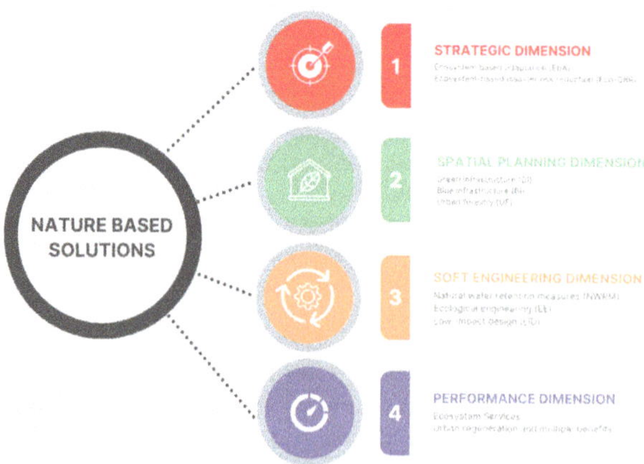

Figure 1. Overview of nature-based concepts and their relationship to existing key concepts. Adapted from [25,28].

NbS provide several beneficial ecosystem services [5,26], such as their ability to store carbon and regulate water flow [29] in order to achieve desired outcomes in urban spaces, such as disaster risk reduction, microclimate regulation, improved human health, and socially inclusive green growth [5,8,28,29].

In this sense, NbS can help in facing social and economic challenges within the sustainable development paradigm, providing benefits that meet—directly and indirectly—SDG 2, SDG 3, SDG 6, SDG 8, SDG 10, SDG 11, SDG 12, SDG 13, SDG 14, and SDG 15 [28]. Indeed, the ability to simultaneously achieve multiple SDGs is a big part of what makes NbS implementation attractive [25,30].

One of the challenges that can be addressed using NbS is the urban heat island (UHI) effect, with respect to which these solutions can be effective in mitigating air pollution and reducing air temperature in urban areas [10,31]. Systems and technologies integrated into nature-inspired construction projects (green buildings), such as green roofs and green walls—green facades, vertical gardens, or living walls—are opportunities to foster a transformation in buildings by creating effective NbS solutions [9,10,25,29], reducing thermal stress in cities, and improving air quality [28,31]. NbS also offer synergies in reducing flood and drought risks while improving water quality and quantity [29], meeting the objectives of various European regulations such as the Floods Directive and the Water Framework Directive [28,32].

In general, NbS can be grouped according to sectors and thematic areas of social importance [28,33]: (I) water management, (II) forests and forestry, (III) agriculture, (IV) urban areas, and (V) coastal areas (Table 1).

Table 1. Benefits of nature-based solutions to address climate risks in specific sectors and thematic areas.

Sectors and Areas	NbS Options	NbS Benefits	Climate Impacts
Water Management	Sustainable urban drainage systems (SUDS) Renaturalization of rivers, streams, and flood plains Restoration of ponds and lakes River buffers (vegetation strips) Water-sensitive forest management Controlled flood plains	Regulation of the water cycle Reduction in floods Improvement of the water quality Improvement of the soil quality, stability, and erosion Biodiversity Regeneration of degraded areas Recreation and aesthetic appreciation Health and quality of life	Droughts Floods
Forests and Forestry	Protection and restoration of forests Sustainable forest management Integration of trees/forests into the landscape Wetlands	Regulation of the water cycle Reduction in floods Improvement of the water quality Improvement of the soil quality, stability, and erosion Biodiversity Carbon sequestration Control of disease and pests Recreation and aesthetic appreciation Health and quality of life	Droughts Floods Fires
Agriculture	Improved soil and water management Crop-type diversification and rotation Agroforestry	Retention of water and soil retention Mitigation of heat stress Biodiversity Carbon sequestration Control of disease and pests Soil fertility	Droughts Floods Heat stress

Table 1. *Cont.*

Sectors and Areas	NbS Options	NbS Benefits	Climate Impacts
Urban Areas	Urban parks, forests, and street trees Green buildings Renaturing of abandoned areas and opportunity plots NbS for water management (bioswales, detention ponds, and rainwater harvesting) NbS for transport infrastructure (greening of streets and previous pavement and greening infrastructure) Regenerated soil Pollinator biodiversity Environmental compensation	Cooling air temperature Regulation of water runoff Improvement of the water quality Improvement of noise quality and comfort Biodiversity Carbon sequestration Regeneration of degraded areas Recreation and aesthetic appreciation Increasing the value of land and property Health and quality of life	Floods Heat stress
Coastal Areas	Rehabilitation and restoration of coastal habitats Beach regeneration, barrier islands, and beach nourishment Hybrid solutions (green dykes and vegetated levees)	Reduction in coastal flooding Stabilization of the coast Carbon sequestration Biodiversity Regeneration of degraded areas Recreation	Sea level rise Storm surges Coastal erosion

Adapted from [28,34].

The EU's ambition is to establish Europe as an inspiration and world leader in markets for NbS through research (development of a technical and scientific evidence base for NbS) and innovation (development of methods and identification of innovative best-practice approaches) [29,35]. The main objective in the field of implementation is to improve the visibility of NbS at all stages, developing demonstration sites and experiences (practical examples on a large scale) to ensure and share the relevance of these initiatives in the current market [29]. The EC has made significant efforts in promoting and disseminating NbS knowledge, experiences, and results [35] through publications, programs, and a network of NbS projects funded by Horizon 2020 (H2020 Research and Innovation Program, 2014–2020) [36]. Continuing the investments and work carried out in recent years, the EU recently launched "HORIZON Europe" (2021–2027). This is the EU's flagship research and innovation funding program (budget of EUR 95.5 billion) [37], which promises to prepare Europe to tackle climate disruption, accelerate the transition to a healthy and prosperous future, and apply solutions for resilience that will lead to the transformation of society. It also aims to accelerate soil and food health, as well as climate-neutral smart cities by 2030, among other mission areas [38]. The HORIZON Europe program emphasizes the acceptance and integration of NbS initiatives in public and private decision making [35].

2.3. Nature-Based Solutions in the Built Environment

Around the world, NbS have been developed and are being implemented to respond to current challenges and improve the quality of urbanized areas and, consequently, human health and well-being [26]. For example, in the context of the COVID-19 pandemic, NbS have been shown to provide additional health benefits, providing both physical and mental relief [39]. As a result, there has been an increase in green roof initiatives, aiding biodiversity and the expansion of urban agriculture [40]. Therefore, designing and building nature-based restorative spaces emerges as a potential response [35] in resolving public health crises [40,41].

In its annual publication, the Global Alliance for Buildings and Construction (GlobalABC) reported on the global progress of the construction and building sector towards meeting international goals with respect to climate change. According to the 2021 Global Status Report for Buildings and Construction, in 2020, the construction and operation of buildings were responsible for 37% of global energy-related carbon dioxide (CO_2) emissions, in addition to the sector's energy consumption, representing approximately 36% of global demand [42]. The document also indicates that if the effect of the COVID-19 pandemic is excluded (due to the decrease in economic activity in the period under analysis), the level of decarbonization in 2020 was only 40% of the 2050 reference path to reach the targets set out in the Paris Agreement [42]. For an alignment with the SDGs, the construction and building sector must incorporate concepts related to the CE in order to decrease the demand for building materials and reduce embodied carbon, in addition to the adoption of NbS that improve the resilience of the construction industry [43].

The use of NbS can be understood as an essential urban planning tool for cities to increase their resilience in response to climate change. In particular, in urban areas, multifunctionality makes these actions references for cities to achieve the transition from a linear to circular model, as NbS offers several benefits with respect to the main challenges highlighted in urban circularity: restoring and maintaining the water cycle (by rainwater management); water and waste treatment, recovery, and reuse; nutrient recovery and reuse; material recovery and reuse; food and biomass production; energy efficiency and recovery; and building system recovery [44].

Furthermore, the implementation of NbS in an urban environment is a central element of European strategies [25] and is directly aligned with the objectives of the 2030 Agenda [28,45] as a way to achieve sustainable cities and communities while playing an active role in the strategic implementation and fulfilment of the SDGs [28], specifically "SDG 11—Sustainable cities and communities: make cities and human settlements inclusive, safe, resilient and sustainable" and "SDG 13—Climate action: take urgent action to combat climate change and its impacts" [25,45]. In the built environment, NbS implementation can be integrated into three levels [10] characterized by:

1. Green building materials: raw natural materials taken from the biological cycle. Their processing must have minimal negative effects on the environment, with low incorporated consumption of energy, carbon, water, and chemicals. Optimal production and construction methods should allow for the safe return of nutrients to the ecosystem after the material use cycle.
2. Green building systems (systems for the greening of buildings), including green and living components integrated into structures and used for the afforestation of buildings, for example, green roofs, facade greenery, living walls, and house trees.
3. Green building sites (green urban sites): areas of land adjacent to buildings (e.g., pocket parks, urban plazas, and small community parks), which play a blue–green role in cities, emphasizing the value of open spaces with vegetation and water-sensitive urban design. These environments provide a variety of ecosystem services and reflect resilient and regenerative approaches to addressing diverse challenges such as reducing noise pollution and mitigating climate change.

Inspired and based on nature and the adoption of ecosystem principles, NbS aim to recover and regenerate natural processes and flows in the urban environment at different scales (Table 2). NbS address social challenges at a range of spatial scales—local, regional, and global [26]—which can be classified according to the level of interventions [34], representing actions from buildings or plots, including districts or neighbourhoods, to cities and beyond (other larger interventions) [46,47].

To differentiate the scale and scope of interventions, another approach used for NbS classification—for implementation in urban environments—is the proGIreg projects (funded by the EC under the Horizon 2020 programme). Specifically, NbS are classified into eight different types [48]:(1) leisure activities and clean energy on former landfills, (2) new regenerated soil, (3) community-based urban farms and gardens, (4) aquaponics, (5) green walls and roofs, (6) accessible green corridors, (7) local environmental compensation processes, and (8) pollinator biodiversity.

Table 2. Nature-based solutions and actions at different scales in urban areas.

Scale	Actions and Interventions
Micro: Building or plot	Previous pavement Urban meadow Private garden Regreening spaces between buildings Actions in community courtyards Shelter for auxiliary fauna (insects and earthworms) Green facades and vertical gardens Sustainable urban drainage systems Green roof and meadow Combined solutions—green roof with renewable energy Rooftop farming and in-height orchards
Meso: District or neighbourhood	Street furniture Previous pavement Comfortable urban places Urban microclimates (public fountains and lakes) Green wharf Spontaneous flora Street trees Community garden Allotment gardens Urban parks Permeable riverbanks and waterbodies Renaturing of abandoned areas and opportunity plots
Macro: City and beyond	Urban farming Constructed wetlands Green street networks Urban forests Ecological corridors Urban planning Renaturing of rivers and streams Controlled flood plains

Adapted from: [34,46,47].

3. Materials and Methods

An initial literature search of 69 case studies was carried out in the Oppla repository (https://oppla.eu/ (accessed on 5 September 2022)), a platform that serves as an EU information centre for knowledge and exemplary NbS cases; proGIreg (https://progireg.eu/ (accessed on 20 September 2022)) "Living Labs" projects for the implementation of innovative NbS in pioneering and strategic cities with the potential for the development of "productive Green Infrastructure for post-industrial urban regeneration"; and CLEVER Cities (https://clevercities.eu/ (accessed on 30 September 2022)) nature-based intervention projects in key city regions to promote urban regeneration. With this selection, it was possible to analyse case studies in different places in Europe, making it possible to enrich the research due to the diversity of realities. Altogether, these are projects with a particular focus on thematic coverage of the urban implementation of NbS and have received funding from the EU's Horizon 2020 innovation action programme.

Through the applied methodology, it was possible to analyse projects for 24 different European cities. The reported solutions and the knowledge acquired about NbS in different locations in the EU were analysed and acquired through the investigation of empirical knowledge, and data collection was performed through indirect and online observation and made available for public access.

The second stage of the research is to enrich the initial stage and to assist in the construction of scientific knowledge about the main strategies and current interventions with respect to urbanization in the implementation and development of more resilient and environmentally sustainable cities, consisting of the selection of keywords and adequate databases to search. The keywords used in the research were: "nature-based solutions" and the name of the cities under analysis (24 mentioned above). The sets were sequentially combined in "Scopus" and "Web of Science" by searching the "All fields" scope using the Boolean operator "AND" between sets of keywords. Then, some criteria were defined to narrow down the information, as presented below: (1) date: only articles published in the last five years were included; (2) source type: only peer-reviewed journals were considered; and (3) language: only articles written in English were included in the study. In terms of eligibility criteria, any article that provided relevant information regarding NbS in the city under study was included in the analysis, i.e., information was accepted indifferently from interested parties (for example, research institutions, local organisations, or public authorities) at all scales (building, district, or city) and during each stage of urban development (planning, design, implementation, or management).

All textual materials were gathered, compiled, and consolidated (from September to November 2022) to provide an overview of recurring NbS themes. This task was aided by the development of an Excel datasheet, which included the following elements: location and country, period or project end projections, title/name of project/program, main objectives, main impacts, project name or name of NbS actions, the spatial scale of NbS implementation, classification of types of urban actions for NbS, types of ecosystem services, classification of ecosystem services, types of benefits, and classification of multiple benefits in the implementation of NbS actions. Then, the qualitative information was synthesized and refined.

As it is a new concept, there is no unique and unambiguous identification and classification method for NbS [34]. Given the lack of standardized criteria and taking the urban context of adaptation to climate change in nature-based planning into account, it was considered appropriate to classify NbS actions implemented in an urban environment through a methodology defined by the Naturvation project (funded by the EU's Horizon 2020 research and innovation programme) [49] and the categories shown in Table 1. Therefore, eight different types of NbS classification were established for the urban sustainability challenges of a city, district, or building, namely (1) blue infrastructure, (2) community gardens and allotments, (3) green areas for water management, (4) green buildings, (5) infrastructure with green features, (6) parks and urban forests, (7) renaturing of abandoned areas and opportunity plots, and (8) urban gardens and green spaces between buildings. For the differentiation of the spatial scale, on which the impacts of NbS were evaluated, together with the type of NbS adopted and the dimension (micro, meso, or macro) in which it is implemented, and respecting the classification shown in Table 2, the following scenarios were established: (1) district or neighbourhood; (2) regional, metropolitan, or urban; and (3) street, plot, or building. Regarding the categorization of the different urban ecosystem services and multiple benefits in the implementation of NbS actions, when not made available by the program, this factor was evaluated according to the EC manuals published by and available from the "Publications Office of the European Union" [50–52].

4. Results

As the focus of the research—current initiatives to implement NbS in urban areas in the European panorama—it was possible to analyse case studies in the following cities: Amsterdam (The Netherlands), Bari (Italy), Berlin (Germany), Bilbao (Spain), Bristol (United Kingdom), Budapest (Hungary), Dresden (Germany), Dublin (Ireland), Dortmund (Germany), Edinburgh (United Kingdom), Genk (Belgium), Hamburg (Germany), Linz (Austria), Lisbon (Portugal), Ljubljana (Slovenia), London (United Kingdom), Milan (Italy), Oradea (Romania), Poznan (Poland), Rotterdam (The Netherlands), Szeged (Hungary), Turin (Italy), Utrecht (The Netherlands), and Zagreb (Croatia). The first step was to analyse the country of origin of each study (Figure 2). The colours in Figure 2 indicate the countries and the respective number of cities analysed. The most representative country was Germany, with four cities, followed by Italy, The Netherlands, and the United Kingdom, with three cities each, and Hungary, with two cities. The remaining countries include only one city under study.

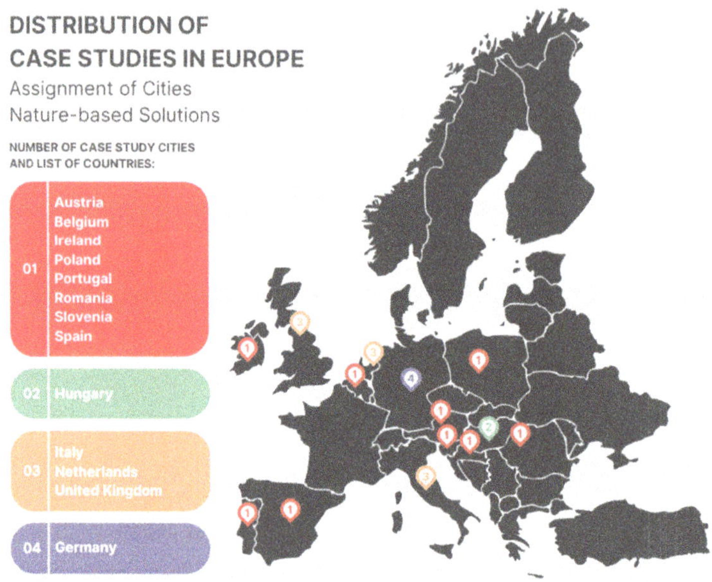

Figure 2. Spatial distribution of cities by country.

Following the preliminary analysis, Figure 3 represents the distribution of the number of NbS projects implemented by European countries. The most representative countries are Italy and the United Kingdom, followed by Germany and The Netherlands.

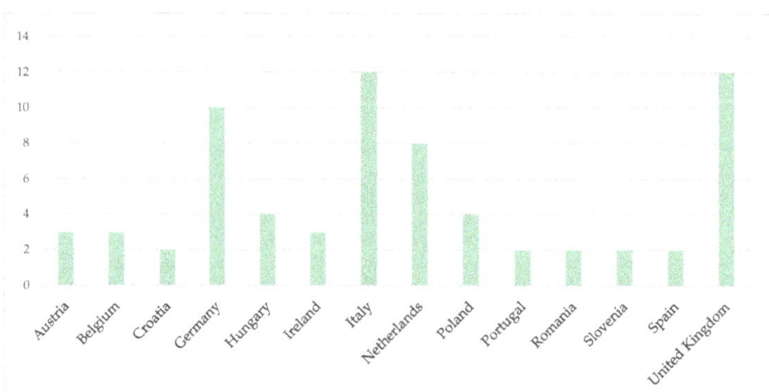

Figure 3. Number of NbS Projects by European countries.

Table 3 presents a complete list of cities with exploratory cases and experimental applications related to NbS. This list includes the identification data, main objectives, and impacts of specific and/or combined NbS initiatives and actions that support urban circular thinking and the use NbS concepts in the built environment. In the last column, the articles are identified after the selection of data and comprehensive information under the research objective and methodology implemented.

Table 3. List of cities and their projects.

Location	Name of Project/Program	Main Objectives	Main Impacts	Project Name or Name of NbS Actions	Supporting References
Amsterdam, The Netherlands	NbS for greening the city and increasing resilience	Create a creative and varied city with an integrated public transport network, high-quality urban planning, and investment in recreational green spaces, water, and renewable energy	Green spaces created/renovated; reduced flood risks; increased social awareness and political support for water diplomacy and climate change adaptation	City Parks Green Neighbourhoods Greening the city Green corridors Deltaplan	[53–56]
Bari, Italy	NbS for greening the urban space	Improve urban quality and green areas, reduce the urban heat island effect, and manage stormwater	Health benefits, mitigation of the urban heat island effect, air purification, restoration/maintenance of habitats and biodiversity, water flow regulation, and reduced impacts of heavy precipitation events	Operation Zero Degradation: Revitalizing residual urban areas as green spaces Shagree: the Green shadows programme Lama Balice Nature Park	[53]
Berlin, Germany	NbS for urban green connectivity and biodiversity	Create connectivity across the city and a 'greenbelt' as a border boundary for urban growth and protection against urban sprawl	Green adaptation to climate change, increasing urban resilience, green/blue connectivity and functionality, biodiversity, cultural wealth, water infiltration/storage, NbS social learning, and health benefits	BENE (urban greening) Green Moabit (rainwater management) Mischwaldprogramm (mixed forests) Transforming vacant urban areas into green spaces School gardens (Urban Living Lab) Twenty green walks Prinzessinnengarten (nomadic gardening)	[53,57–59]
Bilbao, Spain	NbS for dealing with extreme temperature and rainfall events	Make the city more resilient towards cold spells, heat waves, and frequent floods	Mitigation and adaptation to climate change in urban planning, social and economic cohesion, well-being, health, and a more attractive city	Zorrotzaurre project (new floodproof district) Bilbao Greenbelt Expansion (expansion and connection of green areas)	[53]
Bristol, United Kingdom	NbS for a sustainable future	Create a socially inclusive environment with facilities for sport and recreation and a natural environment with urban regeneration and renewal through green networks to minimize and mitigate climate change	Protect, provide, enhance, and expand green infrastructure; promote healthy lifestyles and social inclusion; and increase connectivity with nature	Green infrastructure Green areas for flood management	[53]
Budapest, Hungary	NbS for climate resilience and pollution control	Protection of and increase in the number of green areas, ensuring ecological connectivity and the development of 'smart' and environmentally conscious cities	Urban biodiversity, better water management (water retention), better air quality, social cohesion, public awareness of nature, health, and recreational benefits	Pocket parks Urban gardens Renewing city parks Preserving forests on the outskirts and existing green areas	[53]

Table 3. *Cont.*

Location	Name of Project/Program	Main Objectives	Main Impacts	Project Name or Name of NbS Actions	Supporting References
Dresden, Germany	NbS for sustainable urban transition	A compact city to accommodate more development and include a network of functional green spaces	Bottom-up initiatives for community gardens for diverse environmental, social, and economic benefits; a better quality of urban life; urban renewal; and increased city attractiveness	Living lab: transforming former allotment gardens into community gardens Urban landscape plan: ecological network	[53]
Dublin, Ireland	NbS for a more sustainable city by 2030	Create a sustainable and resilient city based on economy, environment, and equity	Providing an attractive place to live, work, and visit; air and water regulation; climate adaptation; and pollution reduction	Sustainable urban drainage Green infrastructure Green roofs and green walls	[53,60,61]
Dortmund, Germany	The Living Lab: From the Duesenberg to the Huckarde district	Strengthen high-quality green infrastructure and industries with a focus on urban agriculture and improved quality	Reconstruction of derelict land previously used by industry for residential areas and green leisure areas, increasing natural potential with a greenbelt	Renaturing landfill sites Urban farming and gardening on post-industrial sites Connecting Huckarde with River Emscher and Duesenberg Aquaponics as soil-less agriculture Pollinator biodiversity and citizen science	[62]
Edinburgh, United Kingdom	NbS enhancing health, wealth, and sustainability	Create a low-carbon, resource-efficient city, providing a resilient local economy and vibrant, thriving communities	New business and knowledge opportunities, public awareness, green infrastructure connectivity, biodiversity, cultural wealth, ownership, water management, and climate adaptation	Pollinator Pledge Granton Community Gardeners Duddingston Field Group	[53]
Genk, Belgium	NbS bridging green and industrial heritage	Use Genk's natural and human capital for sustainable value creation	Social cohesion, economic opportunities, green jobs, and transition to sustainability	Urban farming—Modeltuin Genk Noord Green Corridor—Stiemerbeek Valley Bee Plan—Bijenplan and Heempark	[53,63]
Hamburg, Germany	Hamburg's CLEVER Action Lab: Neugraben-Fischbek	Promote sustainable and socially integrative urban renewal through NbS and demonstrate the capacity and potential of these solutions in terms of technical, social, and economic innovation	Ecologically regenerated and healthier urban environments with greater social cohesion, more economic opportunities, and greater environmental quality	Green corridor Green roofs and green facades School playgrounds/yards	[64–68]
Linz, Austria	NbS as a motor for urban growth	Enhancing and protecting urban green areas as a way to increase the attractiveness of the city and position it as an important location for regional and international business	Recreation; biodiversity; microclimate; traffic noise control; stormwater management; and a high-quality urban landscape for real estate value, air quality, and pollution reduction	Landschaftspark Bindermichl-Spallerhof SolarCity Urban greening strategy	[53]

Table 3. *Cont.*

Location	Name of Project/Program	Main Objectives	Main Impacts	Project Name or Name of NbS Actions	Supporting References
Lisbon, Portugal	NbS enhancing resilience through urban regeneration	Increase resilience through urban regeneration and increase the importance of preserving natural, forestry, agricultural, and cultural heritage	Better ecological flows and landscape functions; green infrastructure to adapt to climate change, improve drainage, control air and water pollution, and promote leisure, contributing to healthy lifestyles and an attractive city	Green corridors and street trees Urban agriculture	[53,57,69–73]
Ljubljana, Slovenia	NbS for urban regeneration and well-being	Protect and enhance the city's natural environment	Mitigation of the urban heat island effect, accessibility for pedestrians and cyclists, gardens for recreation and leisure, and restoration of the river with an ecological corridor	Green areas (including agricultural allotments) Urban ecological zone Restoration of the river Ljubljanica (Ljubljanica Connects)	[53,56,74–77]
London, United Kingdom	NbS for a leading sustainable city	Increase green space, improve air quality, reduce the UHI effect (heat island), and prevent flash floods through climate change adaptation and mitigation	Opportunity to develop and market green skills; NbS can mitigate heat stress, flood risk, and air quality issues	Green roofs (London/Barking Riverside) Natural water retention measures—River Quaggy Brownfield restoration, Barking Riverside Olympic Park—biosolar roofs Beetle Bump, University of East London campus	[53,64,73,78,79]
	London's CLEVER Action Lab: Thamesmead	Provide opportunities to connect with natural landscapes, economic opportunities, social cohesion, and citizen well-being and advance NbS research and urban regeneration	Quality green spaces working together with education and community networks; connection with nature; social, economic, and environmental impact	Connecting people and places "Healthy Streets"—Parkview, South Thamesmead Greening Unusual and Underused Spaces—South Thamesmead Activating Southmere Lake—Southmere, South Thamesmead	
Milan, Italy	NbS for urban regeneration	Green infrastructure is the best way to achieve environmental goals; promote social development and improve social well-being	Social cohesion; promotion of multipurpose green infrastructure; offering inhabitants agricultural, forestry, cultural, and recreational activities	Parco Agricolo Sud (periurban agriculture and nature conservation) Urban gardening Bosco Verticale (vertical forest) Gorla Maggiore Water Park	[53,56,64,73,80–88]
	CLEVER Cities Milan	Experimenting with innovative green and NbS infrastructure for city regeneration, climate change mitigation, and building a better future for citizens	Diffusion of green roofs and facades; support for experimental projects; development of public green areas with innovative and shared methods in terms of planning, management, maintenance, and monitoring; and the experimental integration of green spaces into railway infrastructure	Green roofs and walls: 'Rinverdiamo Milano'—Let's re-green Milan Development of public green areas: A new park for Giambellino 129 A new green hub for Tibaldi station	

Table 3. *Cont.*

Location	Name of Project/Program	Main Objectives	Main Impacts	Project Name or Name of NbS Actions	Supporting References
Oradea, Romania	Improving the quality of life with NbS	Improve the quality of life of citizens, prioritizing increased leisure opportunities	Active utilization of new areas with improved recreation, increased connectivity, and more biodiversity; improved air quality and climate conditions; new employment and business opportunities; and positive cooperation in public–private partnerships	Green area rehabilitation and development of green infrastructure Creation of outdoor leisure areas Lake creation	[53]
Poznan, Poland	NbS for a friendly, mobile city	Transforming into a green city and improving the quality of life of all inhabitants in such a way that they have a stake in cocreating the city	Green wedge system with high cooling capacity, the transformation of parking lots into green areas, riverside areas with seasonal beaches for leisure, high tourist potential, and new housing projects close to green areas	Maintaining the green wedge system Planting 18,000 trees on the roadside and using transitional green elements; Transforming car parks into green areas Community gardens Creating seasonal beaches	[53,58,63,70,89–91]
Rotterdam, The Netherlands	NbS for building a waterproof city	Make the city 100% weatherproof by 2025, with a focus on adaptive measures including rainwater capture and delayed drainage	Blue–green corridors to facilitate natural hydrological processes, increased biodiversity, and improved quality of life	Water storage capacity Deltaplan Tidal park programme—Esch, Mallegat	[53,92–94]
Szeged, Hungary	NbS for urban regeneration and adaptation to climate change	Improve the quality of green areas, restore natural habitats and ecological corridors for social and recreational purposes, and mitigate the impacts of climate change	More recreational environments and public awareness, more stable biodiversity and ecosystems, and better air and soil quality	Green area rehabilitation Green infrastructure development Urban gardening	[53,81,85]
Turin, Italy	The Living Lab: Mirafiori Sud	Strengthen green infrastructure in areas previously dominated by industry; improve urban quality, social, and economic issues through NbS implementation; involve citizens in activities; and promote NbS-based business models	Introduction of more green space, contribution to education in schools, the inclusion of disadvantaged social groups, support for new entrepreneurship and new green jobs, and NbS regulation	New soil and plant species for urban forestry (Parco Sangone) Greenway and cycling corridor (Sangone river) Urban farming and gardening Pollinator-friendly green spaces Aquaponics tests Green roofs and walls	[62,95–97]
Utrecht, The Netherlands	NbS for urban resilience and citizens' well-being	Promoting healthy urban living, an integrated and systemic urban life that combines the cleanest local approach, recreation, and noise reduction	A water system for diverse ecosystem services, street trees, and street vegetation contributes to quality of life through the realization of a comprehensive green structure based on historical–cultural, spatial, environmental, and ecological values	Central station—a 'Smart Sustainable District' Leidsche Rijn—sustainable urban drainage systems City trees and greenery	[53,98–100]

Table 3. *Cont.*

Location	Name of Project/Program	Main Objectives	Main Impacts	Project Name or Name of NbS Actions	Supporting References
Zagreb, Croatia	The Living Lab: Sesvete	Strengthening green infrastructure in areas previously dominated by industry, social inclusion, supporting new entrepreneurship and new green jobs, and introducing NbS as a catalyst in urban regeneration	New public spaces and the introduction of NbS to sustainable urban planning, strengthening urban resilience, wellness programs (recreation and sports areas), community activities, and bioclimatic building principles	Urban gardens and green areas (educational area) Aquaponic testing New connecting cycle path (urban gardens and neighbourhood) Green roofs and/or green walls in historic buildings (former factories)	[62]

Figure 4 presents the unfolding of NbS actions classified as interventions in an urban environment. A total of 69 projects (singular or combined) were identified, corresponding to 185 actions implemented in the 24 cities under analysis. The large number of actions compared to the number of samples (cities and projects) is because, in most cases, nature-based interventions fall into more than one domain. This holistic and integrative nature of NbS has been highlighted in the literature, emphasizing its multifunctionality as the main advantage over traditional (grey) infrastructure-based solutions [101]. For example, in Budapest, in the creation of pocket parks as an initiative to increase the number of green areas in a district of the city, they are multifunctional; that is, they provide spaces for small-scale food production and the possibility of recreation and community sharing through gardening, assisting in water retention and the region's microclimate [53]. In this case, the project is characterized by all relevant types of NbS, namely "infrastructure with green features", "parks and urban forests", "community gardens and allotments", and "green areas for water management".

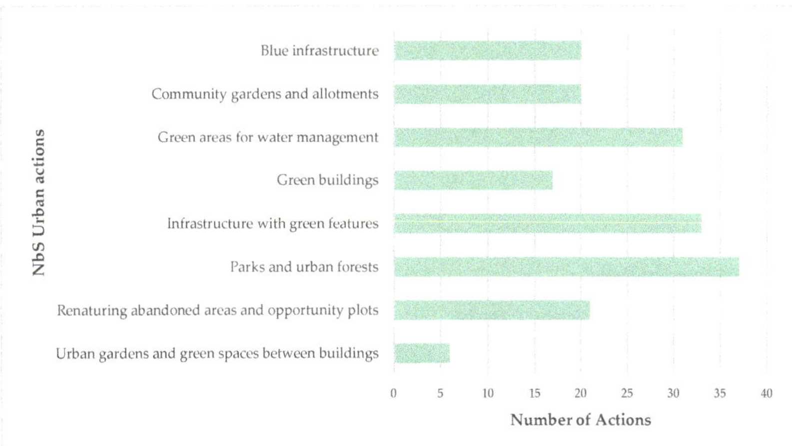

Figure 4. Distribution of NbS urban actions.

The types of NbS most represented in the built environment are "infrastructure with green features" (18%), "parks and urban forests" (20%), and "green areas for water management" (17%). On the other hand, the least represented is "urban gardens and green spaces between buildings" (3%).

Figure 5 represents the spatial distribution of NbS implementation scales in urban space. Analysis shows that the NbS implemented on the "district or neighbourhood" scale correspond to 43% of the total sample, followed by "regional, metropolitan, or urban" and "street, plot, or building", representing 35% and 22%, respectively.

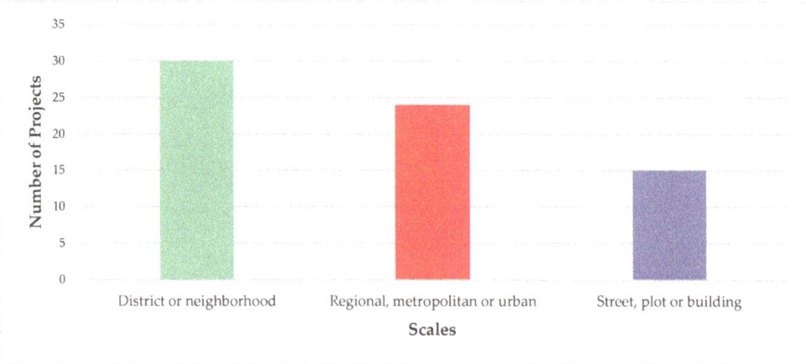

Figure 5. Spatial scale NbS of implementation.

From the classification of different urban ecosystem services and multiple benefits in the implementation of NbS actions, according to which 69 projects were analysed, it was possible to compile the data and represent them through Figures 6–9.

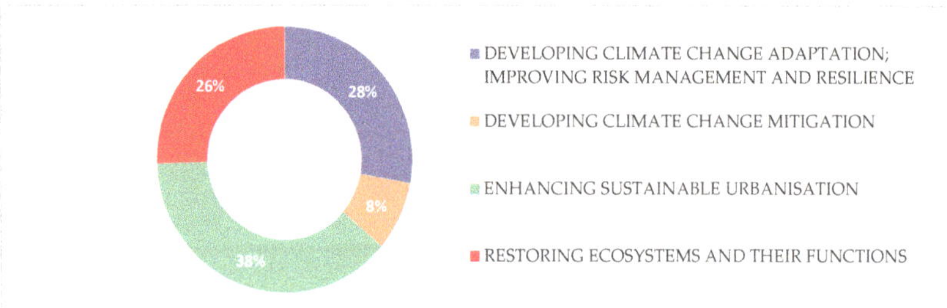

Figure 6. Distribution of the multiple benefits of NbS.

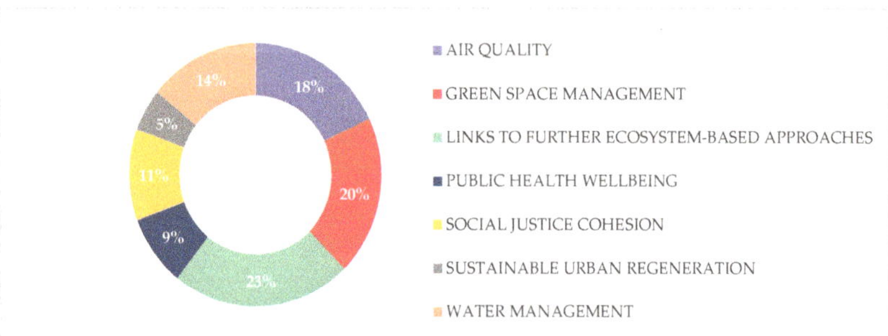

Figure 7. Distribution of NbS ecosystem services.

Figure 8. Multiple benefits NbS.

Figure 9. NbS ecosystem services.

The most prevalent NbS challenges addressed in the survey concern issues of increasing sustainable urbanization (38%), followed, in equivalence, by developing climate change adaptation and improving risk management and resilience (28%) and restoring ecosystems and their functions (26%). Challenges related to the development of climate change mitigation were less common among the analysed case studies, corresponding to 8% of NbS projects. In particular, the types of benefits most highlighted were "increasing infiltration", "reduce run-off", "reducing the temperature at meso or micro-scale", "carbon sequestration and storage", "changing image of the urban environment", and "increase biodiversity".

The last type of characterization (ecosystem services provided) includes five different categories, where, according to the results of the analyses, the most frequently provided services among the analysed NbS are related the interaction of more ecosystem-based approaches (23%), for example, natural water retention measures (NWRM) and ecosystem-based disaster risk reduction (Eco-DRR); green space management (20%); and air quality (18%). Fewer ecosystem services related to sustainable urban regeneration were provided, corresponding to 5% of NbS projects. In particular, most case studies offer services equivalent to "water flow regulation and runoff mitigation", "green infrastructure", "ecosystem-based adaptation", and "habitat and gene pool regulation", designating the crucial role of these practices for the environment.

5. Discussion

The EU has been actively engaged with the research community to better address NbS knowledge and technology gaps through its framework programs and its research and innovation strategy, namely Horizon 2020 and HORIZON Europe. Reflecting on its leading role in the spread of NbS [18,35], the EC, through its funded programs, is focused on the development and availability of a knowledge base on NbS. Indeed, many European cities and regions have undertaken NbS-inspired initiatives to address a range of societal challenges through the delivery of essential ecosystem services. This is evident in fourteen countries represented by the total sample of case studies analysed with different scenarios and applications of NbS in the urban environment.

The articles from eligible scientific journals, in their entirety, did not show similarities with the present study, confirming that the NbS theme is still evolving and that there is a lack of methods for systematic, comprehensive, and integrative compilation and assessment of NbS knowledge [18,19]. It is noted that much of the ongoing efforts are focused on addressing NbS solutions in unique ways for the evapotranspiration performance of vertical vegetation systems [57], for the emotional reaction to urban vegetation [82], for CO_2 reduction in a hermetic museum environment [83], or for its potential to attract young generations [95]. These approaches are important to answer several critical gaps in NbS knowledge; however, barriers have been identified regarding uncertainties in terms of the performance, functionality, and implementation of NbS [20]. Therefore, an integrated approach to the provided ecosystem services is fundamental, assigning the crucial role of NbS practices to the environment and mitigation of climate change impacts, as well as the performance of NbS related to the identified challenges [19].

During the second stage of the research (consultation of the "Scopus" and "Web of Science" databases), a considerable number of cities were not addressed and cited in published articles on NbS under penalty of the exclusion and eligibility criteria previously defined in the Methodology section. Thus, it was not possible to find articles that contained relevant information for the analysis of NbS actions in the following cities: Bari, Bilbao, Bristol, Budapest, Dortmund, Dresden, Edinburgh, Linz, Oradea, and Zagreb. The rest of the documents were important for classifying the different NbS actions, their benefits, and ecosystem services associated with the urban environment.

Urban environments are associated with a significant number of actions that focus on changing green infrastructure towards multifunctionality and improved quality, whereas actions to support citizens in its use are lacking [89]. Nature-based interventions require a collaborative approach to their planning and implementation [93]. The urban environment constitutes opportunities for the participation of citizens and society actors in the formation and planning of green spaces [54], with a significant importance of being at the centre of policy formulation [71]. Therefore, the implementation of NbS in urban contexts requires the cooperation of different public and private actors to manage these processes, directly contributing to learning among the participants, as evidenced in Hamburg within the scope of the CLEVER Cities project [68].

In the city of Milan, after the adoption of NbS, a study pointed out the importance of NbS interventions in citizens' perceptions of their well-being, general health, and a strong sense of belonging to the neighbourhood [80]. These benefits correspond to the increase in sustainable urbanization, as evidenced in the results presented for the multiple NbS benefits.

Studies have identified that knowledge about indicators is needed to monitor and evaluate the implementation of NbS [60,65], with the aim of transformation into a strategic green–blue link in cities, although this remains a considerable challenge to overcome [63]. Different approaches to urban planning, particularly the use and guarantee of effective NbS, shape the design of cities that are more resilient to the effects of climate change in terms of increasing the density of green areas and ensuring permeability, influencing cooling capacity [88], and the impact of floods during extreme weather events [97]. Therefore, it is of great importance to integrate ecological components into a real dynamic green infrastructure [86] throughout the design process when implementing NbS [92,102].

NbS represent an effective tool to improve ecosystem services, value environmental and sociocultural issues [84], and offer integrated benefits with environmental and water management practices [71]. Actions such as "blue infrastructure" and "green areas for water management" have been implemented in the urban environment. The highlighted benefits include "improve connectivity and functionality of green and blue infrastructures", "reduce run-off", and "increasing infiltration". With respect to the provided ecosystem services, the highlights are "water flow regulation and runoff mitigation", "flood control", and "natural water retention measures".

Many studies emphasize the potential and effectiveness of NbS adoption in urban water management [66,74–77,96,98]. The activities proposed for Hamburg, among others, focus on the management of rainwater, which will be used in irrigation systems for plantation areas, increasing the efficiency of rainwater reuse [66]. Similarly, the Leidsche Rijn water system in Utrecht is a sustainable, nature-based, closed-loop surface water system providing clean, clear surface water and supporting biodiversity and climate adaptation. The system includes NbS components in urban infrastructure regimes such as bioswales, ecological water banks, a network of canals, buffer lakes, dams with water gates, water pump stations, and permeable paving [98]. Ljubljana stands out in terms of land use planning and urban water management in the city, with retention areas that are attractive for leisure activities and have a positive impact on the microclimate [76]. Efficient waste management in Ljubljana is an example of the promotion of circularity through the processing of biological waste and mixed waste. Ljubljana Regional Center for Waste Management (RCERO) facilities are capable of producing green electricity from renewable sources of biogas composed of biological waste and electricity and heat [74,75].

Another potential for the implementation of NbS in an urban environment is the possibility of transforming abandoned land into strategic areas promoting green infrastructure and urban regeneration [79]. Many projects, such as post-industrial districts that host living laboratories [62,87], enable the development, testing, and implementation of NbS. This potential is readily identified in the Results. "Renaturing abandoned areas and opportunity plots" is prominent among interventions in urban environments and has a great influence on ecosystem services of "ecosystem-based adaptation" and the benefits of "greater ecological connectivity across urban regenerated sites" and "changing image of the urban environment".

Large green spaces with NbS have the potential to increase the volume of business, increasing revenues, in addition to creating a pleasant feeling of usability of the space for the population [81]. Vegetation contributes to real estate appreciation in the built environment, for example, in housing projects, reflecting economic benefits for the community [90]. It is worth noting that NbS actions in the social body, such as gardens in subdivisions or community gardens, are important and multifunctional areas in cities [91] that promote social cohesion and equity in the neighbourhood [99].

Although it is not the focus of this research, it is worth noting that a major advantage of implementing NbS is their high cost–benefit ratio over traditional solutions [25,28,29], as they are solutions that represent a flexible approach to sustainable inclusive growth at an affordable cost [8]. Thus, NbS represent an important tool for the development of a regenerative, shared, and circular economy in cities.

A study of the identification of NbS shares in the city of Amsterdam addressed NbS classification according to the level of applied technology (this reinforces the existence of other NbS classification methods). An example of high-tech NbS classification is the intervention of a tree-lined central square (IJburg Island), with great efforts due to the enormous amount of vegetation and the creation of a water retention system. On the other hand, an urban park (Roofpark Orly Square) classified as low-tech NbS transformed a grey space into a green space, with the intention of retaining rainwater and capillary irrigation [55]. Another example of low-tech NbS is urban gardens characterized by low-impact solutions, particularly in Lisbon, where the cost of gardening is lower than in other European regions [70].

However, with respect to differences in classification, many articles approached the initiatives in an urban environment with the same logic adopted in the present work. Research shows positive results of microscale interventions, for example, green roofs generate benefits in food production, energy savings (10-30%), and rainwater storage for cultivation and local cooling [56]. Among NbS solutions, facade vegetation and green roofs on buildings play an important role in temperature regulation, with improvements in thermal load, thermal comfort, and thermal storage of the building [67], as well as in reducing the harmful effects of heat waves on human health [85]. Therefore, buildings play an important part in these processes, not just in terms of carbon reduction but also in terms of adaptability and resilience in urban areas [103]. Furthermore, when deployed on a large scale, green roofs show great potential to develop a robust flood control network [60,103] and reduce runoff [67], in addition to potential to increase biodiversity [100]. Tree planting along streets or in urban parks has the greatest impact on heat mitigation and "greenness" (benefits in terms of restoration and mental health related to the amount of natural and seminatural areas that people experience in their surroundings, either by seeing them or directly accessing them) [100]. Small gardens also demonstrate cooling potential, as evidenced in the heavily urbanized region of Lisbon [69].

With great potential for the urban environment, as mentioned above, the actions of "green buildings" and "urban gardens and green spaces between buildings", characterized by the same scale of implementation, represent actions with less frequency in the sample of the analysed case studies. This may mean that large European investments in NbS initiatives in the urban environment occur in projects with larger scales of implementation.

On a mesoscale, studies reported projected in the city of Amsterdam, with an alternative of ditches and filter strips (water decelerating green strip), which allows for local cooling and reduced runoff and storage capacity [56]. In Ljubljana, public orchards and nectar gardens with the planting of fruit trees (mesoscale) have increased livelihoods in the city by providing additional green areas, creating recreational areas, and generating a local cooling effect during hot summer periods [56]. This scale of implementation corresponds to most actions in an urban environment, representing approximately 45% of the total cases analysed. The actions of "parks and urban forests" and "community gardens and allotments" are highly correlated with the mesoscale.

With the adoption of macroscale solutions, for example the Gorla Maggiore Water Park, there was the possibility of peak flow reduction of 86% (downstream flooding) [56]. Another example is the city of Lisbon (European Green Capital 2020), which has a network of green corridors that are part of the urban green infrastructure and actively contribute to ecological connectivity [72]. Another highlight is in Rotterdam, which is an example of a multicultural and resilient city with major green initiatives and movements (urban agriculture combined with sustainable water management systems) [94].

The application of NbS has proven to be a valuable measure to improve climate resilience and citizens' quality of life, as well as environmental justice and social coherence [67]. Urban spaces have great potential for green infrastructures [76,78], although this potential has not yet been fully discovered or used [58].

In many vulnerable regions, harnessing the power of nature is a promising and cost-effective strategy to strengthen climate resilience while promoting shared social, economic, and sustainable prosperity. Ongoing efforts are noted in Europe in scaling up and integrating NbS for mitigation of and adaptation to climate challenges, with the development of resilience in cities aiming at sustainability. Similarly, some countries in North America and East Asia have steadily advanced the use of NbS in urban planning. References to the use of NbS to combat climate change and reduce environmental degradation can be found political agendas in the United States [104], as well as in Chinese government statements supporting such solutions aimed at combating the causes of climate change [105].

To realize their full potential, NbS should be developed with reference to the expertise of all relevant stakeholders so that these solutions contribute to achieving all dimensions of sustainability in urban space. It is worth highlighting the importance of adding several NbS issues to scientific and policy agendas to address climate change adaptation and mitigation measures. With respect to the development of the present study, it is important to remark that the partners in the 69 analysed projects, beyond researchers from universities and companies, include local authorities. All projects implemented in municipalities demonstrate a real concern and interest in changing and improving the urban environment through NbS. Thus, local governments are increasing the interest and implementation of these types of innovative solutions inspired and driven by nature to face current and future challenges due to the consequences of climate change, making the urban environment more resilient.

6. Conclusions

Given the complexity of urban development and climate change, the pressure on natural resources is expected to continue. As a consequence, public agents in particular have to develop and implement long-term solutions for the sustainable and resilient development of cities based on new technologies and the establishment of development policies. Additionally, the ways of building and applying the concept of NbS, which encompasses solutions that provide a series of ecosystem services and multiple challenges and benefits, whether built permanently or environmentally, are expected to increase. Furthermore, the EU intends to continue its investment in the theorizing and operationalization of NbS, which is expected to increase with the HORIZON Europe program.

From the analysis of the results, it is possible to identify the implementation of urban actions for the development of urban parks and forests, the promotion of green areas for water management, and interventions of existing infrastructure with resources. The results also refer to the need to increase incentives for interventions in nature at a local scale, namely initiatives that promote the greening of buildings. This is essential to improve the local climate (mitigation of the heat island effect), to retain and reduce rainwater runoff, to increase biodiversity, and to improve human well-being through, for example, energy-parity solutions, such as the construction of a thermal cover to contribute to the thermal comfort of buildings through improved thermal insulation and increased potential performing in terms of energy.

The main goal of this article is to identify and present the main and most recent NbS practices applied in urban environments in the European panorama to support the analysis of the complementarity and integration of different measures, as an innovative analysis based on real cases. This study represents a systematic, comprehensive, and integrative compilation and assessment of NbS knowledge, highlighting the complementarity of NbS measures in the urban environment and the need for further research on this topic. NbS research on urban water and sewage networks is expected to continue, with the aim of reducing their inefficacity and improving both sustainability and performance on the path of urban greening.

Author Contributions: Conceptualization, F.R. and S.B.; methodology, S.B.; investigation, S.B.; writing—original draft preparation, S.B.; writing—review and editing, A.S.-A., F.R., R.G., R.M. and S.B.; visualization, F.R.; supervision, F.R. All authors have read and agreed to the published version of the manuscript.

Funding: This work was supported by the Foundation for Science and Technology (FCT)—Aveiro Research Centre for Risks and Sustainability in Construction (RISCO), Universidade de Aveiro, Portugal [FCT/UIDB/ECI/04450/2020]. This research work was partially funded by the Foundation for Science and Technology (FCT) under Doctoral Grant 2021 MPP2030-FCT, MIT Portugal Program (PRT/BD/152847/2021) awarded to the first author (S.B.). This research work was partially funded by the Portuguese Government through the FCT and European Social Fund under the PhD grant SFRH/BD/147532/2019, awarded to the fourth author (R.M.).

Institutional Review Board Statement: Not applicable.

Informed Consent Statement: Not applicable.

Data Availability Statement: Not applicable.

Conflicts of Interest: The authors declare no conflict of interest.

References

1. IPCC. Global Warming of 1.5 °C. In *An IPCC Special Report on the Impacts of Global Warming of 1.5 °C above Pre-Industrial Levels and Related Global Greenhouse Gas Emission Pathways, in the Context of Strengthening the Global Response to the Threat of Climate Change, Sustainable Development, and Efforts to Eradicate Poverty*; Cambridge University Press: Cambridge, UK; New York, NY, USA, 2018. [CrossRef]
2. United Nations. The Sustainable Development Goals Report 2022. 2022. Available online: https://unstats.un.org/sdgs/report/2022/ (accessed on 12 October 2022).
3. Williams, J. Circular Cities: Challenges to Implementing Looping Actions. *Sustainability* **2019**, *11*, 423. [CrossRef]
4. UNEP—UN Environment Programme. Global Forum on Cities Highlights Need for Sustainable Development. 2022. Available online: https://www.unep.org/news-and-stories/story/global-forum-cities-highlights-need-sustainable-development (accessed on 12 October 2022).
5. Langergraber, G.; Pucher, B.; Simperler, L.; Kisser, J.; Katsou, E.; Buehler, D.; Mateo, M.C.G.; Atanasova, N. Implementing nature-based solutions for creating a resourceful circular city. *Blue-Green Syst.* **2020**, *2*, 173–185. [CrossRef]
6. United Nations. Goal 11: Make Cities Inclusive, Safe, Resilient and Sustainable. Available online: https://www.un.org/sustainabledevelopment/cities/ (accessed on 13 October 2022).
7. IUCN. *Global Standard for Nature-Based Solutions: A User-Friendly Framework for the Verification, Design and Scaling Up of NbS*, 1st ed.; IUCN: Gland, Switzerland, 2020. [CrossRef]
8. UN-Habitat—United Nations Human Settlements Programme. World Cities Report 2022: Envisaging the Future of Cities. 2022. Available online: https://unhabitat.org/wcr/ (accessed on 13 October 2022).
9. Pineda-Martos, R.; Calheiros, C.S.C. Nature-Based Solutions in Cities—Contribution of the Portuguese National Association of Green Roofs to Urban Circularity. *Circ. Econ. Sustain.* **2021**, *1*, 1019–1035. [CrossRef]
10. Pearlmutter, D.; Theochari, D.; Nehls, T.; Pinho, P.; Piro, P.; Korolova, A.; Papaefthimiou, S.; Mateo, M.C.G.; Calheiros, C.; Zluwa, I.; et al. Enhancing the circular economy with nature-based solutions in the built urban environment: Green building materials, systems and sites. *Blue-Green Syst.* **2020**, *2*, 46–72. [CrossRef]
11. Bardin, L. *Análise de Conteúdo*, 70th ed.; Edicao Revista Actualizada; Almedian: Lisbon, Portugal, 2009.
12. Kabisch, N.; Frantzeskaki, N.; Pauleit, S.; Naumann, S.; Davis, M.; Artmann, M.; Haase, D.; Knapp, S.; Korn, H.; Stadler, J.; et al. Nature-based solutions to climate change mitigation and adaptation in urban areas: Perspectives on indicators, knowledge gaps, barriers, and opportunities for action. *Ecol. Soc.* **2016**, *21*, 39. [CrossRef]
13. Sowińska-Świerkosz, B.; García, J. What are Nature-based solutions (NBS)? Setting core ideas for concept clarification. *Nat. -Based Solut.* **2022**, *2*, 100009. [CrossRef]
14. Nesshöver, C.; Assmuth, T.; Irvine, K.N.; Rusch, G.M.; Waylen, K.A.; Delbaere, B.; Haase, D.; Jones-Walters, L.; Keune, H.; Kovacs, E.; et al. The science, policy and practice of nature-based solutions: An interdisciplinary perspective. *Sci. Total Environ.* **2017**, *579*, 1215–1227. [CrossRef]
15. Albert, C.; Schröter, B.; Haase, D.; Brillinger, M.; Henze, J.; Herrmann, S.; Gottwald, S.; Guerrero, P.; Nicolas, C.; Matzdorf, B. Addressing societal challenges through nature-based solutions: How can landscape planning and governance research contribute? *Landsc. Urban Plan.* **2019**, *182*, 12–21. [CrossRef]
16. Krauze, K.; Wagner, I. From classical water-ecosystem theories to nature-based solutions—Contextualizing nature-based solutions for sustainable city. *Sci. Total Environ.* **2019**, *655*, 697–706. [CrossRef]
17. Sarabi, S.E.; Han, Q.; Romme, A.; de Vries, B.; Wendling, L. Key Enablers of and Barriers to the Uptake and Implementation of Nature-Based Solutions in Urban Settings: A Review. *Resources* **2019**, *8*, 121. [CrossRef]

18. Calliari, E.; Castellari, S.; Davis, M.; Linnerooth-Bayer, J.; Martin, J.; Mysiak, J.; Pastor, T.; Ramieri, E.; Scolobig, A.; Sterk, M.; et al. Building climate resilience through nature-based solutions in Europe: A review of enabling knowledge, finance and governance frameworks. *Clim. Risk Manag.* **2022**, *37*, 100450. [CrossRef]
19. Grace, M.; Balzan, M.; Collier, M.; Geneletti, D.; Tomaskinova, J.; Abela, R.; Borg, D.; Buhagiar, G.; Camilleri, L.; Cardona, M.; et al. Priority knowledge needs for implementing nature-based solutions in the Mediterranean islands. *Environ. Sci. Policy* **2021**, *116*, 56–68. [CrossRef]
20. Sarabi, S.; Han, Q.; Romme, A.; de Vries, B.; Valkenburg, R.; den Ouden, E. Uptake and implementation of Nature-Based Solutions: An analysis of barriers using Interpretive Structural Modeling. *J. Environ. Manag.* **2020**, *270*, 110749. [CrossRef] [PubMed]
21. Science for Environment Policy. *Future Brief: The Solution Is in Nature*; Issue 24; EU Publications Office: Luxemburg, 2021.
22. European Parliament. Circular Economy: Definition, Importance and Benefits | News | European Parliament. 2022. Available online: https://www.europarl.europa.eu/news/en/headlines/economy/20151201STO05603/circular-economy-definition-importance-and-benefits (accessed on 12 October 2022).
23. European Commission. COM 2020—Communication from the Commission to the European Parliament, the Council, the European Economic and Social Committee and the Committee of the Regions—A New Circular Economy Action Plan—For a cleaner and More Competitive Europe. 2020. Available online: https://eur-lex.europa.eu/legal-content/EN/TXT/?qid=1583933814386&uri=COM:2020:98:FIN (accessed on 12 October 2022).
24. Giorgos, S.; Stavros, S.; Nektarios, C.; Horizon, T.N. 2020 Research and Innovation Programme under Grant Agreement No. 730338. 2019. Available online: https://www.researchgate.net/publication/339983272_ThinkNature_Nature-Based_Solutions_Handbook. (accessed on 2 November 2022).
25. European Commission. *Evaluating the Impact of Nature-Based Solutions: A Handbook for Practitioners*; Publications Office of the European Union: Luxemburg, 2021. [CrossRef]
26. Dick, J.; Miller, J.D.; Carruthers-Jones, J.; Dobel, A.J.; Carver, S.; Garbutt, A.; Hester, A.; Hails, R.; Magreehan, V.; Quinn, M. How are nature based solutions contributing to priority societal challenges surrounding human well-being in the United Kingdom: A systematic map protocol. *Environ. Evid.* **2019**, *8*, 37. [CrossRef]
27. Escobedo, F.J.; Giannico, V.; Jim, C.; Sanesi, G.; Lafortezza, R. Urban forests, ecosystem services, green infrastructure and nature-based solutions: Nexus or evolving metaphors? *Urban For. Urban Green* **2019**, *37*, 3–12. [CrossRef]
28. European Environment Agency. *Nature-Based Solutions in Europe: Policy, Knowledge and Practice for Climate Change Adaptation and Disaster Risk Reduction*; EU Publication Office: Luxembourg, 2021. [CrossRef]
29. European Commission; Directorate-General for Research, and Innovation. *Towards an EU Research and Innovation Policy Agenda for Nature-Based Solutions & Re-Naturing Cities—Final report of the Horizon 2020*, full version; EU Publications Office: Luxembourg, 2015. [CrossRef]
30. Chausson, A.; Turner, B.; Seddon, D.; Chabaneix, N.; Girardin, C.A.J.; Kapos, V.; Key, I.; Roe, D.; Smith, A.; Woroniecki, S.; et al. Mapping the effectiveness of nature-based solutions for climate change adaptation. *Glob. Chang. Biol.* **2020**, *26*, 6134–6155. [CrossRef] [PubMed]
31. Vieira, J.; Matos, P.; Mexia, T.; Silva, P.; Lopes, N.; Freitas, C.; Correia, O.; Santos-Reis, M.; Branquinho, C.; Pinho, P. Green spaces are not all the same for the provision of air purification and climate regulation services: The case of urban parks. *Environ. Res.* **2018**, *160*, 306–313. [CrossRef]
32. Brillinger, M.; Henze, J.; Albert, C.; Schwarze, R. Integrating nature-based solutions in flood risk management plans: A matter of individual beliefs? *Sci. Total Environ.* **2021**, *795*, 148896. [CrossRef]
33. McVittie, A.; Cole, L.; Wreford, A.; Sgobbi, A.; Yordi, B. Ecosystem-based solutions for disaster risk reduction: Lessons from European applications of ecosystem-based adaptation measures. *Int. J. Disaster Risk Reduct.* **2018**, *32*, 42–54. [CrossRef]
34. Ihobe Environmental Management Agency; Ministry of the Environment Territorial Planning and Housing—Basque Government. *Nature-Based Solutions for Local Climate Adaptation in the Basque Country*; Methodological Guide for Their Identification and Mapping, Donostia/San Sebastián Case Study; Ihobe: Bilbao, Spain, 2017.
35. Davies, C.; Chen, W.; Sanesi, G.; Lafortezza, R. The European Union roadmap for implementing nature-based solutions: A review. *Environ. Sci. Policy* **2021**, *121*, 49–67. [CrossRef]
36. European Commission; Directorate—General for Research and Innovation. *Nature-Based Solutions: State of the Art in EU-Funded Projects*; EU Publications Office: Luxembourg, 2020. [CrossRef]
37. European Commission. Horizon Europe. Available online: https://research-and-innovation.ec.europa.eu/funding/funding-opportunities/funding-programmes-and-open-calls/horizon-europe_en (accessed on 3 November 2022).
38. European Commission; Directorate-General for Research and Innovation. *A New Horizon for Europe: Impact Assessment of the 9th EU Framework Programme for Research and Innovation*; EU Publications Office: Luxembourg, 2018. [CrossRef]
39. Ashley, R.M.; Horton, B.; Walker, L.; Digman, C.; Shaffer, P.; van Herk, S. The benefits of Nature Based Systems in a changing and uncertain world. *Proc. Inst. Civ. Eng. Eng. Sustain.* 2022; *ahead of print*. [CrossRef]
40. Gillis, K. Nature-based restorative environments are needed now more than ever. *Cities Health* **2020**, *5*, S237–S240. [CrossRef]
41. Dubey, P.K. Biodiversity: The Nature-Based Solution for Pandemics and Human Well-being. *Clim. Change Environ. Sustain.* **2020**, *8*, 93. [CrossRef]
42. United Nations Environment Programme. *2021 Global Status Report for Buildings and Construction: Towards a Zero-Emission, Efficient and Resilient Buildings and Construction Sector*; UNEP: Nairobi, Kenya, 2021.

43. United Nations Environment Programme. *2020 Global Status Report for Buildings and Construction: Towards a Zero-Emission, Efficient and Resilient Buildings and Construction Sector*; UNEP: Nairobi, Kenya, 2020.
44. Atanasova, N.; Castellar, J.A.; Pineda-Martos, R.; Nika, C.E.; Katsou, E.; Istenič, D.; Pucher, B.; Andreucci, M.B.; Langergraber, G. Nature-Based Solutions and Circularity in Cities. *Circ. Econ. Sustain.* **2021**, *1*, 319–332. [CrossRef]
45. Anderson, V.; Gough, W.A. Enabling Nature-Based Solutions to Build Back Better—An Environmental Regulatory Impact Analysis of Green Infrastructure in Ontario, Canada. *Buildings* **2022**, *12*, 61. [CrossRef] [PubMed]
46. Herzog, C.P.; Rozado, C.A. *The EU-Brazil Sector Dialogue on Nature-Based Solutions: Contribution to a Brazilian Roadmap on Nature-Based Solutions for Resilient Cities*; EU Publication Office: Luxembourg, 2019. [CrossRef]
47. Nature4Cities. Nature Based Solutions. Available online: https://www.nature4cities.eu/nature-based-solutions (accessed on 19 October 2022).
48. proGIreg. Nature-Based Solutions. Available online: https://progireg.eu/ (accessed on 20 October 2022).
49. NAVIGATOR. Urban Nature Navigator Discovering the Potential of Nature-Based Solutions. Available online: https://naturvation.eu/sites/default/files/result/files/briefing_note_unn_purpose_and_potential (accessed on 25 October 2022).
50. European Commission; Directorate—General for Research and Innovation. *Mapping and Assessment of Ecosystems and Their Services: An EU Wide Ecosystem Assessment in Support of the EU Biodiversity Strategy*; Supplement (indicator fact sheets); EU Publications Office: Luxembourg, 2020. [CrossRef]
51. CICES—European Environment Agency (EEA). Towards a Common Classification of Ecosystem Services. Available online: https://cices.eu/resources/ (accessed on 5 November 2022).
52. Raymond, C.; Berry, P.M.; Breil, M.; Nita, M.R. *An Impact Evaluation Framework to Support Planning and Evaluation of Nature-Based Solutions Projects*; Centre of Ecology and Hydrology: Bangor, UK, 2017. [CrossRef]
53. Case Studies | Oppla. Available online: https://oppla.eu/case-study-finder (accessed on 5 October 2022).
54. Mabon, L.; Barkved, L.; de Bruin, K.; Shih, W.-Y. Whose knowledge counts in nature-based solutions? Understanding epistemic justice for nature-based solutions through a multi-city comparison across Europe and Asia. *Environ. Sci. Policy* **2022**, *136*, 652–664. [CrossRef]
55. Snep, R.P.H.; Voeten, J.; Mol, G.; van Hattum, T. Nature Based Solutions for Urban Resilience: A Distinction Between No-Tech, Low-Tech and High-Tech Solutions. *Front. Environ. Sci.* **2020**, *8*, 259. [CrossRef]
56. Carvalho, P.N.; Finger, D.C.; Masi, F.; Cipolletta, G.; Oral, H.V.; Tóth, A.; Regelsberger, M.; Exposito, A. Nature-based solutions addressing the water-energy-food nexus: Review of theoretical concepts and urban case studies. *J. Clean. Prod.* **2022**, *338*, 130652. [CrossRef]
57. Pearlmutter, D.; Pucher, B.; Calheiros, C.S.C.; Hoffmann, K.A.; Aicher, A.; Pinho, P.; Stracqualursi, A.; Korolova, A.; Pobric, A.; Galvão, A.; et al. Closing water cycles in the built environment through nature-based solutions: The contribution of vertical greening systems and green roofs. *Water* **2021**, *13*, 2165. [CrossRef]
58. Zwierzchowska, I.; Haase, D.; Dushkova, D. Discovering the environmental potential of multi-family residential areas for nature-based solutions. A Central European cities perspective. *Landsc. Urban Plan.* **2021**, *206*, 103975. [CrossRef]
59. Simperler, L.; Ertl, T.; Matzinger, A. Spatial Compatibility of Implementing Nature-Based Solutions for Reducing Urban Heat Islands and Stormwater Pollution. *Sustainability* **2020**, *12*, 5967. [CrossRef]
60. Basu, A.S.; Pilla, F.; Sannigrahi, S.; Gengembre, R.; Guilland, A.; Basu, B. Theoretical Framework to Assess Green Roof Performance in Mitigating Urban Flooding as a Potential Nature-Based Solution. *Sustainability* **2021**, *13*, 13231. [CrossRef]
61. Collier, M.J.; Bourke, M. The case for mainstreaming nature-based solutions into integrated catchment management in Ireland. *Proc. R. Ir. Acad.* **2020**, *120*, 107–113. [CrossRef]
62. Case Studies | proGIreg. Available online: https://progireg.eu/ (accessed on 5 October 2022).
63. Frantzeskaki, N.; Vandergert, P.; Connop, S.; Schipper, K.; Zwierzchowska, I.; Collier, M.; Lodder, M. Examining the policy needs for implementing nature-based solutions in cities: Findings from city-wide transdisciplinary experiences in Glasgow (UK), Genk (Belgium) and Poznań (Poland). *Land Use Policy* **2020**, *96*, 104688. [CrossRef]
64. Case Studies | CleverCities. Available online: https://clevercities.eu/ (accessed on 5 October 2022).
65. Rödl, A.; Arlati, A. A general procedure to identify indicators for evaluation and monitoring of nature-based solution projects. *Ambio* **2022**, *51*, 2278–2293. [CrossRef]
66. Tosun, J.; Leopold, L. Aligning Climate Governance with Urban Water Management: Insights from Transnational City Networks. *Water* **2019**, *11*, 701. [CrossRef]
67. Scharf, B.; Kogler, M.; Kraus, F.; Perez, I.; Garcia, L.G. NBS Impact Evaluation with GREENPASS Methodology Shown by the Case Study 'Fischbeker Höfe' in Hamburg/Germany. *Sustainability* **2021**, *13*, 9167. [CrossRef]
68. Arlati, A.; Rödl, A.; Kanjaria-Christian, S.; Knieling, J. Stakeholder Participation in the Planning and Design of Nature-Based Solutions. Insights from CLEVER Cities Project in Hamburg. *Sustainability* **2021**, *13*, 2572. [CrossRef]
69. Kumar, P.; Debele, S.E.; Sahani, J.; Rawat, N.; Marti-Cardona, B.; Alfieri, S.M.; Basu, B.; Basu, A.S.; Bowyer, P.; Charizopoulos, N.; et al. An overview of monitoring methods for assessing the performance of nature-based solutions against natural hazards. *Earth Sci. Rev.* **2021**, *217*, 103603. [CrossRef]
70. Poniży, L.; Latkowska, M.J.; Breuste, J.; Hursthouse, A.; Joimel, S.; Külvik, M.; Leitão, T.E.; Mizgajski, A.; Voigt, A.; Kacprzak, E.; et al. The Rich Diversity of Urban Allotment Gardens in Europe: Contemporary Trends in the Context of Historical, Socio-Economic and Legal Conditions. *Sustainability* **2021**, *13*, 11076. [CrossRef]

71. Cui, M.; Ferreira, F.; Fung, T.; Matos, J.S. Tale of two cities: How nature-based solutions help create adaptive and resilient urban water management practices in singapore and lisbon. *Sustainability* **2021**, *13*, 10427. [CrossRef]
72. Canet-Martí, A.; Pineda-Martos, R.; Junge, R.; Bohn, K.; Paço, T.A.; Delgado, C.; Alenčikienė, G.; Skar, S.L.G.; Baganz, G.F.M. Nature-based solutions for agriculture in circular cities: Challenges, gaps, and opportunities. *Water* **2021**, *13*, 2565. [CrossRef]
73. Mahmoud, I.H.; Morello, E.; Ludlow, D.; Salvia, G. Co-creation Pathways to Inform Shared Governance of Urban Living Labs in Practice: Lessons from Three European Projects. *Front. Sustain. Cities* **2021**, *3*, 80. [CrossRef]
74. Kisser, J.; Wirth, M.; De Gusseme, B.; Van Eekert, M.; Zeeman, G.; Schoenborn, A.; Vinnerås, B.; Finger, D.C.; Repinc, S.K.; Bulc, T.G.; et al. A review of nature-based solutions for resource recovery in cities. *Blue-Green Syst.* **2020**, *2*, 138–172. [CrossRef]
75. van Hullebusch, E.D.; Bani, A.; Carvalho, M.; Cetecioglu, Z.; De Gusseme, B.; Di Lonardo, S.; Djolic, M.; van Eekert, M.; Griessler Bulc, T.; Haznedaroglu, B.Z.; et al. Nature-based units as building blocks for resource recovery systems in cities. *Water* **2021**, *13*, 3153. [CrossRef]
76. Radinja, M.; Atanasova, N.; Lamovšek, A.Z. The water-management aspect of blue-green infrastructure in cities. *Urbani Izziv* **2021**, *32*, 98–110. [CrossRef]
77. Pagano, A.; Pluchinotta, I.; Pengal, P.; Cokan, B.; Giordano, R. Engaging stakeholders in the assessment of NBS effectiveness in flood risk reduction: A participatory System Dynamics Model for benefits and co-benefits evaluation. *Sci. Total Environ.* **2019**, *690*, 543–555. [CrossRef]
78. O'Keeffe, J.; Pluchinotta, I.; De Stercke, S.; Hinson, C.; Puchol-Salort, P.; Mijic, A.; Zimmermann, N.; Collins, A.M. Evaluating natural capital performance of urban development through system dynamics: A case study from London. *Sci. Total Environ.* **2022**, *824*, 153673. [CrossRef]
79. Song, Y.; Kirkwood, N.; Maksimović, Č.; Zheng, X.; O'Connor, D.; Jin, Y.; Hou, D. Nature based solutions for contaminated land remediation and brownfield redevelopment in cities: A review. *Sci. Total Environ.* **2019**, *663*, 568–579. [CrossRef]
80. Mahmoud, I.H.; Morello, E.; Vona, C.; Benciolini, M.; Sejdullahu, I.; Trentin, M.; Pascual, K.H. Setting the social monitoring framework for nature-based solutions impact: Methodological approach and pre-greening measurements in the case study from clever cities milan. *Sustainability* **2021**, *13*, 9672. [CrossRef]
81. Koppelaar, R.; Marvuglia, A.; Havinga, L.; Brajković, J.; Rugani, B. Is Agent-Based Simulation a Valid Tool for Studying the Impact of Nature-Based Solutions on Local Economy? A Case Study of Four European Cities. *Sustainability* **2021**, *13*, 7466. [CrossRef]
82. Piga, B.E.A.; Stancato, G.; Rainisio, N.; Boffi, M. How Do Nature-Based Solutions' Color Tones Influence People's Emotional Reaction? An Assessment via Virtual and Augmented Reality in a Participatory Process. *Sustainability* **2021**, *13*, 13388. [CrossRef]
83. Salvatori, E.; Gentile, C.; Altieri, A.; Aramini, F.; Manes, F. Nature-Based Solution for Reducing CO_2 Levels in Museum Environments: A Phytoremediation Study for the Leonardo da Vinci's 'Last Supper'. *Sustainability* **2020**, *12*, 565. [CrossRef]
84. di Biagi, P.; Basso, S.; Marin, A.; Tzortzi, J.; Guaita, L.; Kouzoupi, A. Sustainable Strategies for Urban and Landscape Regeneration Related to Agri-Cultural Heritage in the Urban-Periphery of South Milan. *Sustainability* **2022**, *14*, 6581. [CrossRef]
85. Marvuglia, A.; Koppelaar, R.; Rugani, B. The effect of green roofs on the reduction of mortality due to heatwaves: Results from the application of a spatial microsimulation model to four European cities. *Ecol. Modell.* **2020**, *438*, 109351. [CrossRef]
86. Ronchi, S.; Arcidiacono, A.; Pogliani, L. Integrating green infrastructure into spatial planning regulations to improve the performance of urban ecosystems. Insights from an Italian case study. *Sustain. Cities Soc.* **2020**, *53*, 101907. [CrossRef]
87. Conti, M.E.; Battaglia, M.; Calabrese, M.; Simone, C. Fostering Sustainable Cities through Resilience Thinking: The Role of Nature-Based Solutions (NBSs): Lessons Learned from Two Italian Case Studies. *Sustainability* **2021**, *13*, 12875. [CrossRef]
88. Ronchi, S.; Salata, S.; Arcidiacono, A. Which urban design parameters provide climate-proof cities? An application of the Urban Cooling InVEST Model in the city of Milan comparing historical planning morphologies. *Sustain. Cities Soc.* **2020**, *63*, 102459. [CrossRef]
89. Zwierzchowska, I.; Fagiewicz, K.; Poniży, L.; Lupa, P.; Mizgajski, A. Introducing nature-based solutions into urban policy—Facts and gaps. Case study of Poznań. *Land Use Policy* **2019**, *85*, 161–175. [CrossRef]
90. Gałecka-Drozda, A.; Wilkaniec, A.; Szczepańska, M.; Świerk, D. Potential nature-based solutions and greenwashing to generate green spaces: Developers' claims versus reality in new housing offers. *Urban Urban Green.* **2021**, *65*, 127345. [CrossRef]
91. Dymek, D.; Wilkaniec, A.; Bednorz, L.; Szczepańska, M. Significance of allotment gardens in urban green space systems and their classification for spatial planning purposes: A case study of poznań, poland. *Sustainability* **2021**, *13*, 1044. [CrossRef]
92. van Cauwenbergh, N.; Dourojeanni, P.; van der Zaag, P.; Brugnach, M.; Dartee, K.; Giordano, R.; Lopez-Gunn, E. Beyond TRL—Understanding institutional readiness for implementation of nature-based solutions. *Environ. Sci. Policy* **2022**, *127*, 293–302. [CrossRef]
93. Frantzeskaki, N. Seven lessons for planning nature-based solutions in cities. *Environ. Sci. Policy* **2019**, *93*, 101–111. [CrossRef]
94. Säumel, I.; Reddy, S.; Wachtel, T. Edible City Solutions—One Step Further to Foster Social Resilience through Enhanced Socio-Cultural Ecosystem Services in Cities. *Sustainability* **2019**, *11*, 972. [CrossRef]
95. Giachino, C.; Pattanaro, G.; Bertoldi, B.; Bollani, L.; Bonadonna, A. Nature-based solutions and their potential to attract the young generations. *Land Use Policy* **2021**, *101*, 105176. [CrossRef]
96. Qi, Y.; Chan, F.K.S.; Thorne, C.; O'Donnell, E.; Quagliolo, C.; Comino, E.; Pezzoli, A.; Li, L.; Griffiths, J.; Sang, Y.; et al. Addressing Challenges of Urban Water Management in Chinese Sponge Cities via Nature-Based Solutions. *Water* **2020**, *12*, 2788. [CrossRef]
97. Salata, S.; Ronchi, S.; Giaimo, C.; Arcidiacono, A.; Pantaloni, G.G. Performance-Based Planning to Reduce Flooding Vulnerability Insights from the Case of Turin (North-West Italy). *Sustainability* **2021**, *13*, 5697. [CrossRef]

98. Dorst, H.; van der Jagt, A.; Runhaar, H.; Raven, R. Structural conditions for the wider uptake of urban nature-based solutions—A conceptual framework. *Cities* **2021**, *116*, 103283. [CrossRef]
99. Dignum, M.; Dorst, H.; van Schie, M.; Dassen, T.; Raven, R. Nurturing nature: Exploring socio-spatial conditions for urban experimentation. *Environ. Innov. Soc. Transit.* **2020**, *34*, 7–25. [CrossRef]
100. Cortinovis, C.; Olsson, P.; Boke-Olén, N.; Hedlund, K. Scaling up nature-based solutions for climate-change adaptation: Potential and benefits in three European cities. *Urban Urban Green.* **2022**, *67*, 127450. [CrossRef]
101. Raymond, C.M.; Frantzeskaki, N.; Kabisch, N.; Berry, P.; Breil, M.; Nita, M.R.; Geneletti, D.; Calfapietra, C. A framework for assessing and implementing the co-benefits of nature-based solutions in urban areas. *Environ. Sci. Policy* **2017**, *77*, 15–24. [CrossRef]
102. Boros, J.; Mahmoud, I. Urban Design and the Role of Placemaking in Mainstreaming Nature-Based Solutions. Learning From the Biblioteca Degli Alberi Case Study in Milan. *Front. Sustain. Cities* **2021**, *3*, 38. [CrossRef]
103. Pimentel-Rodrigues, C.; Silva-Afonso, A. Contributions of Water-Related Building Installations to Urban Strategies for Mitigation and Adaptation to Face Climate Change. *Appl. Sci.* **2019**, *9*, 3575. [CrossRef]
104. Executive Order on Tackling the Climate Crisis at Home and Abroad | the White House. Available online: https://www.whitehouse.gov/briefing-room/presidential-actions/2021/01/27/executive-order-on-tackling-the-climate-crisis-at-home-and-abroad/ (accessed on 6 December 2022).
105. China Pushes Technical Solutions in Race to Meet Climate Goals | Reuters. Available online: https://www.reuters.com/article/us-china-climatechange/china-pushes-technical-solutions-in-race-to-meet-climate-goals-idUSKBN29I1EA (accessed on 6 December 2022).

applied sciences

MDPI

Review

Ventilation Strategies to Ensure Thermal Comfort for Users in School Buildings: A Critical Review

María Teresa Aguilar-Carrasco [1], Remedios María López-Lovillo [1], Rafael Suárez [2,*] and Ángel Luis León-Rodríguez [2]

[1] Departamento de Construcciones Arquitectónicas I, Escuela Técnica Superior de Arquitectura, Universidad de Sevilla, Av. de Reina Mercedes 2, 41012 Seville, Spain; macarrasco@us.es (M.T.A.-C.); rllovillo@us.es (R.M.L.-L.)

[2] Instituto Universitario de Arquitectura y Ciencias de la Construcción, Escuela Técnica Superior de Arquitectura, Universidad de Sevilla, Av. De Reina Mercedes 2, 41012 Seville, Spain; leonr@us.es

* Correspondence: rsuarez@us.es

Abstract: People spend most of their time indoors, where air quality is crucial to health. In this context, this study conducts a critical review of ventilation strategies in schools to ensure air quality, as well as to guarantee students' thermal comfort. Based on a bibliographical review, strategies from previous studies are identified and evaluated in order to determine their advantages. After a detailed search, a total of 19 articles were selected, which provides a thorough analysis of the ventilation strategies in school buildings considering thermal comfort. The identified strategies were categorized into natural, mechanical, and hybrid types. The results reveal a prevalence of natural ventilation, which accounts for over 50% of strategies in all climates. Mechanical ventilation is less common and is applicable to around 30% of cases. Hybrid strategies, combining natural and mechanical ventilation, are the least used and only appear in oceanic climates, with a usage rate of 20%. Most studies highlight the lack of air conditioning in many schools, making adequate thermal stress management through ventilation crucial. The results analyzed clearly show a lack of studies with optimal results whose ventilation strategies can be replicated in other similar educational buildings, ensuring thermal comfort and air quality.

Keywords: ventilation; thermal comfort; school; strategy; indoor air quality

1. Introduction

In contemporary lifestyles, people spend most of their time inside buildings, where factors such as indoor air quality and thermal comfort become essential for ensuring their health, a fundamental part of life. According to the World Health Organization (WHO), air pollution is ranked among the greatest environmental health risks [1], with prolonged exposure leading to serious health problems in humans. Numerous studies have quantified the risk of suffering different diseases affecting the respiratory [2], cardiovascular [3], immune [4], and neurological systems [5]. This also increases the risk of individuals developing chronic diseases [6–8]. Ensuring optimal indoor air quality is therefore key to mitigating these health risks and promoting overall well-being. On the other hand, the lack of thermal comfort resulting from overheating and undercooling inside buildings also poses a serious risk to health [9], causing severe illnesses [10], impairing cognitive performance [11], and leading to a decline in dexterity and manual execution capacities [12].

These effects can be even more severe in children, depending on their stage of development, negatively impacting cognitive development [13], as well as causing the diseases mentioned earlier. Children spend a considerable part of the day at school, where, in addition to the above, a high concentration of CO_2 in the environment can reduce students' concentration levels and cognitive performance [14].

Therefore, adequate ventilation is essential in order to ensure indoor air quality and to guarantee the health and proper development of children [15,16]. This has been further highlighted by the COVID-19 pandemic, emphasizing the need to ventilate spaces in order to prevent the transmission of airborne diseases [17,18].

However, the high level of interest generated in this field following the COVID-19 pandemic has sidelined an issue that had been gaining traction prior to COVID-19, that of thermal stress, which has now been relegated to a secondary position. If current rates of anthropogenic activity persist, it is highly likely that climate change will lead to an increase in the planet's average temperature of 1.5 °C above current levels between 2030 and 2052 and 4.8 °C between 2081 and 2100 [19], further exacerbating the issue of overheating and its impact on indoor environments [20]. This, in turn, negatively affects thermal comfort and indoor air quality [21], especially in warmer southern European climates [22], increasing thermal stress [23]. Educational buildings are especially vulnerable to overheating due to significant internal occupancy loads on the one hand, and on the other hand, to the fact that many fail to meet thermal conditioning requirements, as they were built prior to current regulations. In Europe, these matters are regulated by Directive 2010/31/EU (amended by Directive (EU) 2018/844) [24]. Furthermore, in most schools, ventilation is carried out naturally, helping to reduce levels of CO_2 and other pollutants while also promoting energy consumption reduction [25]. This natural ventilation misuse is arguably the main cause of discomfort and thermal stress [26], taking the form of cold in winter and heat in summer, resulting in a rise of up to 0.8 °C on average during the summer. Therefore, not only does air pollution impact negatively on children's health, but thermal stress also contributes to adverse effects such as concentration loss, fatigue, tiredness, anxiety, and stress [27].

In many cases, this increased need for ventilation has led to studies on how to solve comfort and thermal stress issues being cast aside. Furthermore, in some cases these issues have worsened due to the priority given to window opening with no protocol or control, so that ultimately these remain open throughout the day, to the detriment of thermal comfort [28,29].

Therefore, it is essential to identify ventilation strategies, whether natural or mechanical, to ensure both air quality and user thermal comfort. In this study, **indoor air quality** is assessed through **carbon dioxide (CO_2) concentration**, which is commonly used as a representative indicator of ventilation effectiveness and overall air quality in occupied spaces [30,31]. Elevated CO_2 levels are associated with insufficient ventilation, and controlling this parameter indirectly helps maintain adequate levels of other indoor pollutants. Furthermore, since proper ventilation is a key strategy to reduce airborne virus transmission, monitoring and limiting CO_2 concentrations also contributes to minimizing this risk [32,33]. As for thermal comfort, it is assessed based on temperature and relative humidity, following the criteria established in the ISO 7730 standard [34].

Reviews jointly addressing these topics—ventilation strategies in schools and thermal comfort—have been sought. However, despite the extent of this problem, few reviews have been found which present a set of ventilation strategies and their results regarding thermal comfort. In the Scopus and WoS databases, only seven reviews have been found discussing both topics combined, compared to 40 reviews found addressing the topics separately; 29 of these focus on ventilation strategies, albeit only guaranteeing CO_2 levels and air quality rather than thermal comfort, while 11 emphasized thermal comfort analysis without establishing or quantifying ventilation strategies, only analyzing the causes. As we have seen previously, this lack of reviews addressing both topics jointly may be due to the current context following the COVID-19 pandemic, where studies have primarily focused on air quality to safeguard human health by preventing disease transmission and exposure to pollutants. In addition to showing us the lack of reviews to conduct an umbrella review study, this initial search and analysis of studies highlights the need for a review of articles addressing these topics to search for ventilation strategies and analyze their results, strengths, limitations, and weaknesses in order to create protocols for use and evaluation in schools.

2. Materials and Methods

The purpose of this study is to present a critical analysis of ventilation strategies implemented in school environments and aimed at ensuring students' thermal comfort. To do so, strategies proposed in previous studies were identified and analyzed to establish their individual associated advantages and benefits. This evaluation was conducted through a review of the literature, implementing detailed search criteria as specified in Section 2.1. These criteria include the selection of relevant keywords and a search for articles by year of publication in two specific databases. Section 2.2 outlines the selection criteria employed: an initial filtering process based on the article title, followed by an analysis of the abstract, and a final detailed examination of the article content.

2.1. Search Criteria

The search criteria began with the selection of keywords related to the main objective of the research. The keywords chosen initially were "ventilation" AND "school" AND "thermal comfort" AND "strategy". To complement the information, similar keywords were proposed to broaden the search range:

- "ventilation" OR "indoor air" OR "air renovation" OR "airflow"
- "school" OR "university" OR "educational building" OR "classroom"
- "thermal comfort" OR "occupant comfort" OR "environmental comfort" OR "indoor comfort" OR "environmental quality"
- "strategy" OR "technique" OR "methodology" OR "model" OR "mode" OR "design"

Due to the focus on natural ventilation strategies during the COVID-19 pandemic, often neglecting occupants' comfort, the search was divided into two periods, before and after COVID-19, to reduce potential bias. The intermediate year was fixed at 2021. The pre-COVID-19 period ends in 2021 and goes back ten years (2011–2021). The post-COVID-19 period is defined as 2022 to the present. Given that this research focuses on schools, the search criteria were refined to include only educational buildings. Classrooms exhibit distinct characteristics, including higher occupancy densities, fixed and intensive schedules, a more vulnerable population, and predominantly natural ventilation systems with limited or no mechanical ventilation or air conditioning. These factors differentiate educational buildings from other typologies and necessitate tailored approaches.

The search was conducted in two different databases, WOS and SCOPUS, which are main databases used in other reviews [35,36]. These databases provide advanced search functionalities that enable the automated generation of combinations between keywords and their synonyms, allowing for a comprehensive and consistent search process. In both cases, the search was performed for the proposed keywords in the title, abstract, and keywords, limiting the date range for each period. These search criteria are shown in Figure 1.

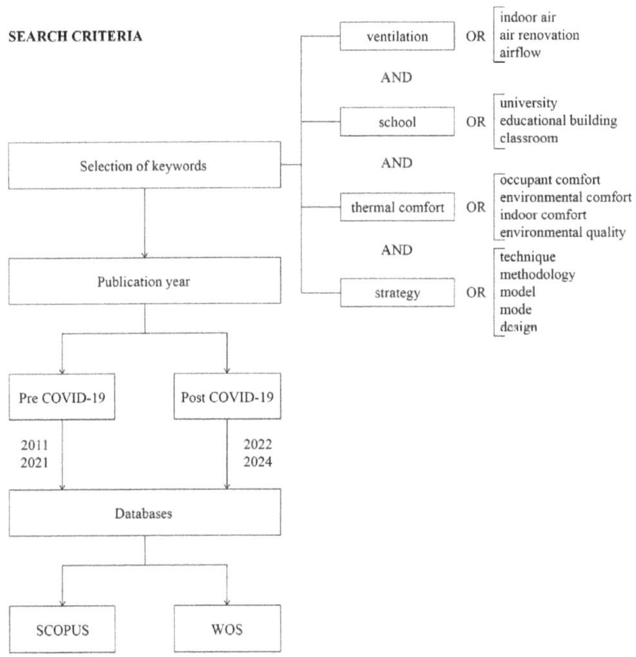

Figure 1. Methodology: search criteria.

2.2. Selection Criteria

Once the above criteria were established, the search for articles was conducted using the keywords specified. In total, 277 articles were obtained: 186 from the pre-COVID-19 period and 91 from the post-COVID-19 period. Of the 186 pre-COVID-19 articles, 50 were found in the SCOPUS database and 136 in WOS. For the post-COVID-19 articles, 32 were found in SCOPUS and 59 in WOS.

Subsequently, an analysis was carried out in order to eliminate duplicate articles from these two databases. After excluding duplicates, 163 articles from the pre-COVID-19 period and 74 from the post-COVID-19 period were selected, so that a total of 237 articles were selected according to the initial search criteria.

To support the selection process, a co-occurrence map was generated using VOSviewer (Figure 2), which revealed five main thematic clusters: thermal comfort (green), natural ventilation and architectural design (red), indoor air quality (blue), educational context (yellow), and school building retrofitting (purple). VOSviewer is developed by Leiden University, located in Leiden, The Netherlands. This mapping allowed for a clearer understanding of the thematic structure within the dataset and was used to guide the identification of the most relevant articles—those situated at the intersection of the main clusters and addressing the four core concepts: ventilation, school, thermal comfort, and strategy. From here, a filtering process began, initially selecting articles based on their titles. The titles of all extracted articles were reviewed, selecting those whose titles include aspects relating to the objectives set out in this review. This selection yielded 56 articles from the pre-COVID-19 block and 24 from the post-COVID-19 block, for a total of 80 articles.

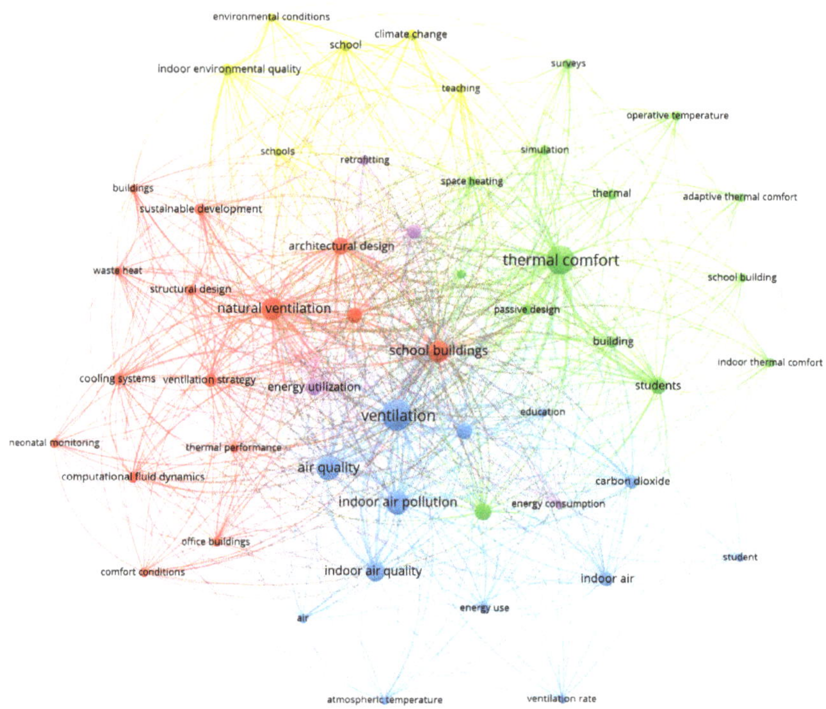

Figure 2. Keyword co-occurrence map generated by VOSviewer with five main thematic clusters: thermal comfort (green), natural ventilation and architectural design (red), indoor air quality (blue), educational context (yellow), and school building retrofitting (purple).

From these 80 articles, a new selection was carried out based on the abstracts. All the abstracts of the chosen articles were reviewed, and those which refer to ventilation strategies in schools and include thermal comfort were selected. According to the exclusion criteria followed in this part, articles focusing on other topics and failing to address users' thermal comfort or discuss ventilation strategies were excluded. Articles meeting the exclusion criteria were classified by topic, and several recurring themes were identified. In Table 1, the themes identified and the number of articles associated with each are listed. As can be seen, one of the most recurring themes is the study or modification of the building envelope to improve thermal comfort conditions (46% of articles), since one of the keywords used, "strategy", also refers to forms of retrofitting. After applying the aforementioned selection and exclusion criteria, 20 pre-COVID-19 and 9 post-COVID-19 articles were selected, for a total of 29 articles.

Table 1. Main topics of excluded articles.

Causes of discomfort	6
Changes or study of the building envelope/building retrofitting	22
Energy use	8
New simulation models	2
Air quality analysis	9
User perception	1

Finally, a selection based on the content of the article was conducted. The 29 articles selected by abstract were analyzed, identifying articles that meet the following requirements:

- propose and analyze at least one natural and/or mechanical ventilation strategy.
- are conducted in an educational building.
- consider thermal comfort through monitoring and/or surveys.

After performing these analyses, 27 articles were selected: 17 pre-COVID-19 and 10 post-COVID-19.

The number of articles extracted in each selection phase is listed in Table 2 and Figure 3.

Table 2. Number of articles according to filtering phase.

		Search	First Selection by Title	Second Selection by Abstract	Final Selection
SCOPUS	PRE	50	31	14	13
	POST	19	12	6	7
WOS	PRE	113	25	6	4
	POST	55	12	3	3
	Total	237	80	29	27

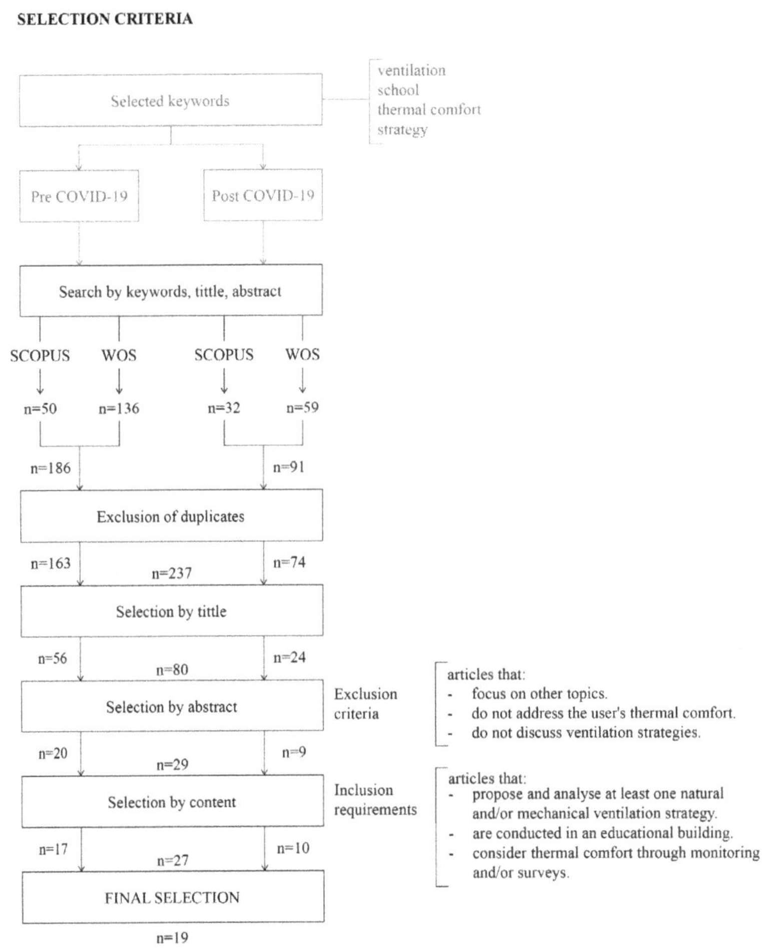

Figure 3. Methodology: selection criteria.

After reviewing the 27 articles selected, an additional 8 were excluded, as they did not provide the necessary information to meet the objectives of this study. Consequently, the analysis focuses on the results from the remaining 19 articles, which examined ventilation strategies in school buildings, while also considering thermal comfort.

3. Results

Several studies focus on different ventilation strategies to ensure thermal comfort in school buildings. Table 3 summarizes these studies, adding insight on the type of ventilation studied and the key aspects of the main selected articles (year of execution, location and climate, type of building, and season of the year when tests and measurements are carried out).

Table 3. Summary of the articles whose ventilation strategies were analyzed in depth.

Ref	Year	Location (Köppen Climatic Classification)	Type of Building	Season of the Year	Ventilation Strategy
[37]	2018	Goiânia, Brazil (Aw)	University	Summer	• Natural ventilation with constant window opening • Mechanical ventilation with variable flow according to sensors
[38]	2014	La Rochelle, France (Cfb)	University	Summer Winter	• Natural ventilation with random window opening • Natural ventilation with window opening based on sensors • Hybrid ventilation
[39]	2019	Nicosia, Cyprus (Bsh)	Secondary school	Winter	• Natural ventilation with scheduled window opening
[40]	2020	Nicosia, Cyprus (Bsh)	Secondary school	Summer Winter	• Natural ventilation with scheduled window opening
[41]	2019	Bucharest, Romania (Df)	Secondary school	Summer Winter	• Natural ventilation with scheduled window opening
[42]	2014	United Kingdom (Cfb)	Secondary school	Summer	• Natural ventilation using windcatcher
[43]	2017	Lisbon, Portugal (Csa)	Secondary school	Annually	• Natural ventilation with random window opening • Mechanical ventilation with variable flow according to sensors
[44]	2016	Drammen, Norway (Dfb)	Educational building	Winter Spring	• Mechanical ventilation with variable flow according to sensors
[45]	2019	Seville, Spain (Csa)	Secondary school	Summer Winter Spring	• Mechanical ventilation with constant flow • Mechanical ventilation with variable flow according to sensors

Table 3. *Cont.*

Ref	Year	Location (Köppen Climatic Classification)	Type of Building	Season of the Year	Ventilation Strategy
[46]	2023	Basque Country, Spain (Cfb)	University	Summer Winter Spring	• Natural ventilation with constant window opening • Natural ventilation with window opening based on sensors • Mechanical ventilation with variable flow according to sensors • Hybrid ventilation
[47]	2022	Extremadura, Spain (Csa)	University	Winter	• Natural ventilation with constant window opening • Natural ventilation with window opening based on sensors
[48]	2024	Graz, Austria (Cfb)	University	Winter Spring	• Natural ventilation with constant window opening • Natural ventilation with scheduled window opening • Mechanical ventilation with constant flow
[49]	2011	Ancona, Italy (Cfa)	Technical School	Winter	• Natural ventilation with random window opening • Natural ventilation with window opening based on sensors • Natural ventilation using windcatcher • Mechanical ventilation with constant flow
[50]	2019	Cassino, Italy (Csa)	Primary School	Winter	• Natural ventilation with scheduled window opening • Mechanical ventilation with variable flow according to sensors
[51]	2020	Jordan (Bsk)	University	Summer Winter	• Natural ventilation using stack ventilation
[52]	2023	Southern Sweden (Cfb)	Primary School	Summer	• Mechanical ventilation with constant flow • Hybrid ventilation
[53]	2023	United Kingdom (Cfb)	Educational building	Winter + Spring	• Natural ventilation with random window opening • Mechanical ventilation with constant flow • Hybrid ventilation
[54]	2023	Spain (Csa)	Primary/secondary schools	Summer Winter Spring	• Natural ventilation with random window opening
[55]	2021	Sydney (Cfa)	Secondary school	Winter Spring	• Mechanical ventilation with variable flow according to sensors

Below are the articles published by year in Figure 4 and by country in Figure 5. As can be seen, the publication trend does not increase as a result of COVID-19, which is consistent with the initial assessment made, stating that, after COVID-19, most of the articles focused on air quality, leaving aside thermal comfort. In terms of the countries where these studies have been carried out, most (84%) have been conducted in Europe, and within this group there is a notable presence of the countries of Spain and Italy.

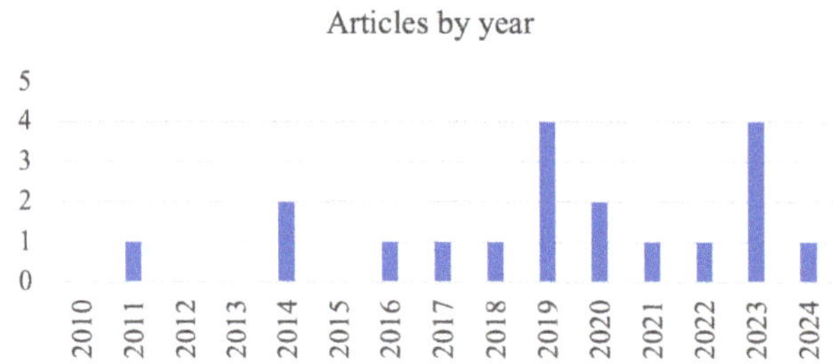

Figure 4. Distribution of the articles analyzed according to year of publication.

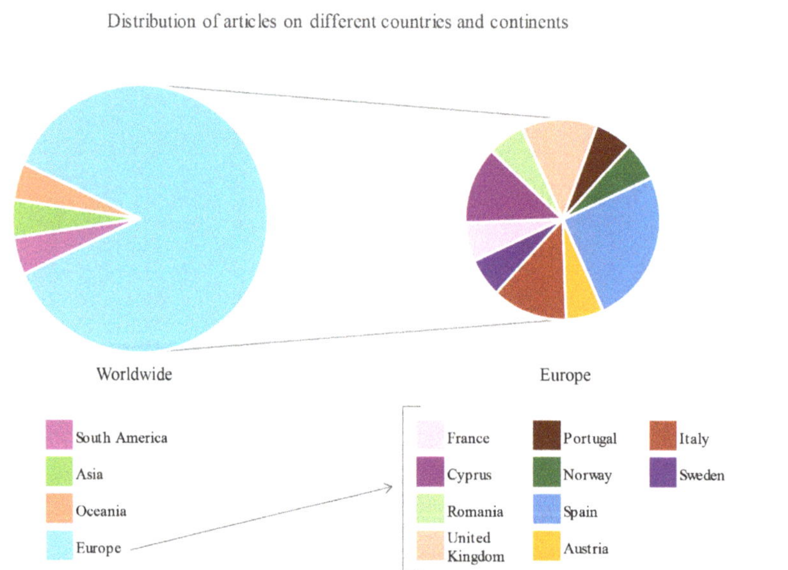

Figure 5. Classification of articles according to the country in which the study was conducted.

Based on the articles selected and analyzed, a series of ventilation strategies is presented in Table 4, covering natural, mechanical, and hybrid ventilation. It should be noted that natural ventilation is defined as ventilation in which both intake and extraction are carried out naturally, while mechanical ventilation is ventilation in which both intake and extraction are carried out mechanically, or in which intake is natural and extraction is mechanical. Finally, hybrid ventilation is ventilation in which natural and mechanical ventilation are in simultaneous operation. This means that mechanical ventilation has its own intake and mechanical extraction, with a controlled airflow and number of renewals, while, in parallel, the windows are opened, either manually by the users or mechanically. This ventilation is also referred to in some of the studies as mixed or mixed-mode.

Table 4. Classification of different ventilation strategies.

NV	1. User-dependent natural ventilation: operation relies entirely on occupant behavior; users decide when to manually open or close windows, without a specific pattern. V: No need for automation systems or technical maintenance. D: High uncertainty in performance; can be ineffective during extreme weather or when users misjudge the need for ventilation.
	1.a. With permanently open windows: windows remain open continuously, regardless of outdoor conditions or occupancy levels.
	1.b. With irregular window opening patterns: windows are opened sporadically, based on the user's perception (e.g., feeling warm or detecting odors).
	2. User-dependent natural ventilation according to established window opening patterns: users follow predefined window opening schedules or rules, based on time of day. V: Greater predictability and potentially improved performance if users are educated. D: Still relies on consistent user behavior; may not respond in real time to changing conditions.
	3. User-independent natural ventilation with parameter control through sensors: ventilation is automatically triggered based on indoor environmental parameters. V: Reliable performance; responsive to real-time indoor air quality; reduces reliance on occupant behavior. D: Requires sensor calibration and maintenance; may have higher initial installation costs.
	3.a. CO_2 sensors: ventilation activates when indoor CO_2 levels exceed a threshold, indicating occupancy and air quality deterioration.
	3.b. Temperature sensors: ventilation responds to indoor temperature levels to maintain thermal comfort.
	4. User-independent natural ventilation based on wind and thermal gradient: utilizes natural forces such as wind pressure and stack effect to drive airflow through building openings or passive devices. V: Energy-free operation; can be highly effective if well-designed. D: Highly climate-dependent; performance varies with external conditions; complex to design for year-round effectiveness.
	4.a. Windcatcher: architectural elements (typically towers) capture wind and channel it into indoor spaces, facilitating cross or downward airflow.
	4.b. Stack ventilation: warm indoor air rises and exits through high-level openings, drawing in cooler outside air through lower openings.

Table 4. *Cont.*

MV		5. Mechanical ventilation with constant airflow: a mechanical system provides a fixed rate of air exchange regardless of occupancy or indoor conditions. V: Predictable air exchange; simple to design and operate. D: Energy-intensive; may lead to over- or under-ventilation depending on actual needs; no adaptability.
		6. Mechanical ventilation with variable airflow based on CO_2 or temperature sensors: ventilation rate is modulated in response to sensor data (e.g., CO_2 concentration or temperature), providing air only as needed. V: Optimizes energy use; improves indoor air quality dynamically; aligns with demand-controlled principles. D: More complex systems; requires sensor maintenance; potential delays in system response.
HV		7. Hybrid ventilation: hybrid ventilation refers to systems where natural and mechanical ventilation operate concurrently. In this setup, mechanical ventilation functions with its own dedicated air intake and exhaust, providing a controlled airflow and defined air change rate, while at the same time, windows may be opened—either manually by occupants or automatically by a control system. V: Adaptable to a wide range of climates and building uses. D: Higher system complexity.

NV: natural ventilation, MV: mechanical ventilation, HV: hybrid ventilation, V: advantages, D: disadvantages. The shading tone in each row identifies the type of strategy. The tones are grouped into natural ventilation (warm tones), mechanical ventilation (blues), and hybrid ventilation (purples). The tones correspond to those used in Figure 6.

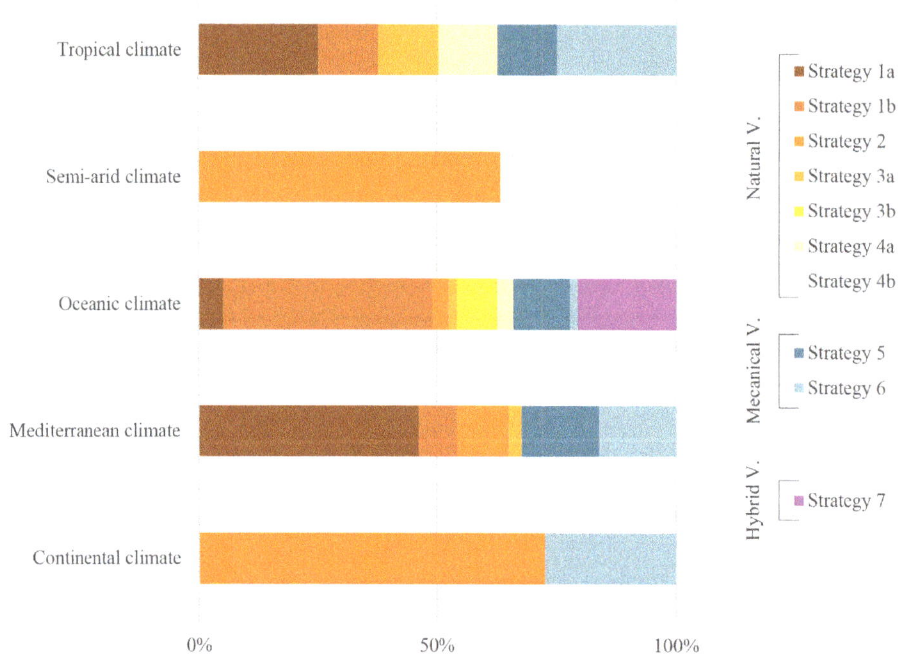

Figure 6. Relationship between the different ventilation strategies and the climates in which they are used.

Additionally, as shown in Table 5, the various existing climates have been grouped into five categories: tropical, semi-arid, oceanic, Mediterranean, and continental.

Figure 6 presents a list of the types of strategies used according to the climate, referencing the five groups previously shown in Table 5. As shown in Figure 6, natural ventilation is one of the ventilation strategies most widely used in schools, accounting for more than 50% of the total strategies used in all cases, irrespective of climate. In some cases, such as the semi-arid climate, this figure can even reach 100%. An analysis of natural ventilation strategies shows a clear majority of user-dependent strategies, with either no specific window opening patterns or scheduled timings. In all cases, these totals exceed 75%, reaching up to 95.5% or even 100% in Mediterranean and continental climates. In contrast, mechanical ventilation strategies barely reach 30% in most cases. In some climates, such as the oceanic climate, they only account for 15%, and in the semi-arid climate, they are not employed at all. Finally, hybrid ventilation strategies are the least used, and are studied solely in the oceanic climate, where they account for 20%.

Table 5. Grouping of the climates of the articles into five groups according to Köppen classification.

Climate Grouping	Köppen Climatic Classification	Number of Analyzed Articles
Tropical	Aw, Cfa	3
Semi-arid	Bsh, Bsk	3
Oceanic	Cfb	6
Mediterranean	Csa	5
Continental	Df, Dfb	2

The different strategies are analyzed independently below in order to thoroughly examine the various protocols and results. For this purpose, the analysis is divided into three groups—natural, mechanical, and hybrid ventilation—as defined previously.

3.1. Natural Ventilation Strategies

The studies with natural ventilation strategies are shown in Table 6 and Figure 7. In these, different groups have been created within each strategy in order to analyze the results by climate at a later stage, irrespective of the presence or absence of cross ventilation and of air conditioning. Additionally, the seasons during which the individual studies are conducted (Summer (S), Winter (W), Spring (SP), or Annually (A)) are specified.

Table 6. References with natural ventilation strategies.

		Air Conditioning		1		2	3		4	
Climate	**Ventilation**	**Heating**	**Cooling**	**a**	**b**		**a**	**b**	**a**	**b**
Semi-arid	Cross	Yes	No	-	-	[39] (W)	-	-	-	-
		No	No	-	-	[40] (S)	-	-	-	[51] (S, W)
	Single-sided	Yes	No	-	-	[39] (W)	-	-	-	-
		No	No	-	-	[40] (S, W)	-	-	-	-
Tropical	Cross	Yes	No	-	-	-	-	-	-	-
		No	No	-	-	-	-	-	[49] (W)	-
	Single-sided	No	Yes	[37] (S)	-	-	-	-	-	-
		No	No	[37] (S)	[49] (W)	-	[49] (W)	-	-	-
Oceanic	Cross	Yes	No	-	[53] (W + SP)	-	-	-	-	-
		No	No	-	-	-	-	-	[42] (S)	-
	Single-sided	Yes	No	[48] (W)	[53] (W + SP)	[48] (W)	-	[38] (W)	-	-
		No	No	[46] (SP) [48] (SP)	[38] (S, W)	[48] (SP)	[46] (SP)	[46] (S, W, SP) [38] (S)	-	-

Table 6. *Cont.*

		Air Conditioning		1		2	3		4	
Climate	**Ventilation**	**Heating**	**Cooling**	**a**	**b**		**a**	**b**	**a**	**b**
Mediterranean	Cross	Yes	No	-	-	-	-	-	-	-
		No	No	[47] (W)	[43] (Y)	-	[47] (W)	-	-	-
	Single-sided	Yes	No	-	-	-	-	-	-	-
		No	No	-	[54] (S, W, SP)	[50] (W)	-	-	-	-
Continental	Cross	Yes	No	-	-	[41] (W)	-	-	-	-
		No	No	-	-	[41] (S)	-	-	-	-
	Single-sided	Yes	No	-	-	-	-	-	-	-
		No	No	-	-	-	-	-	-	-

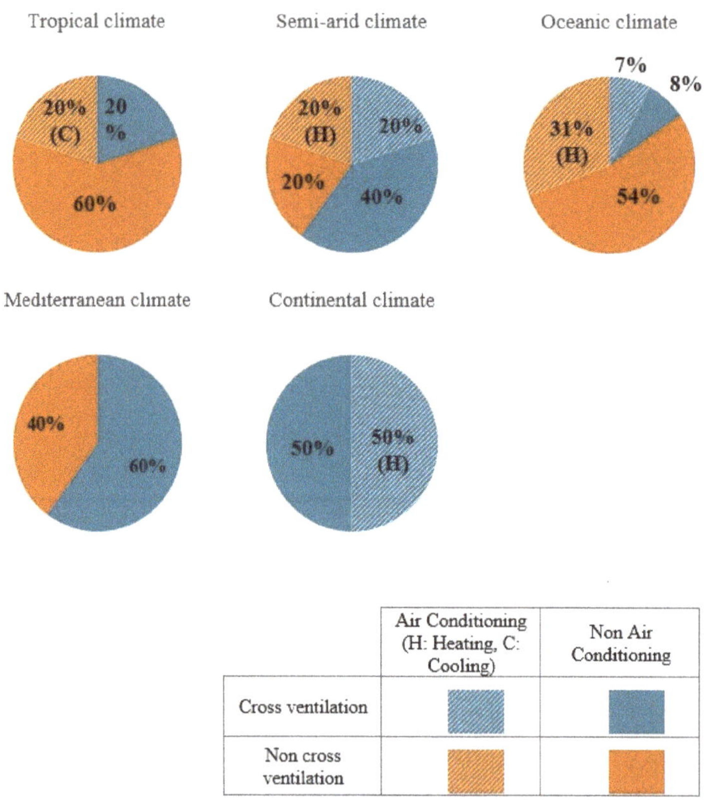

Figure 7. Detailed analysis of the implementation percentage of each natural ventilation sub-strategy.

In the analyzed articles, the study of cross ventilation predominates in semi-arid, Mediterranean, and continental climates, especially in continental climates where no cases of single-sided ventilation have been analyzed. In contrast, single-sided natural ventilation strategies are predominant in tropical and oceanic climates. Natural ventilation is the most commonly used method in the majority of climates, and in some cases it performs well, as is the case with the Mediterranean climate during periods without extreme temperatures, which is characterized by milder and more favorable temperatures. Nevertheless, in most instances, it must be complemented by active systems. It is observed that, regardless of the climate, most schools and educational buildings implementing natural ventilation strategies do not have air conditioning, whether heating or cooling, making it essential to ensure the control of thermal stress.

3.2. Mechanical Ventilation Strategies

The studies examining mechanical ventilation strategies are shown in Table 7 and Figure 8. As before, different groups have been created within each strategy to analyze the results by climate at a later stage, depending on whether the mechanical ventilation has a heat recovery unit or thermal treatment (with or without a battery) and whether there is air conditioning. The seasons during which each study is conducted (Summer (S), Winter (W), Spring (SP), or Annually (A)) are also indicated.

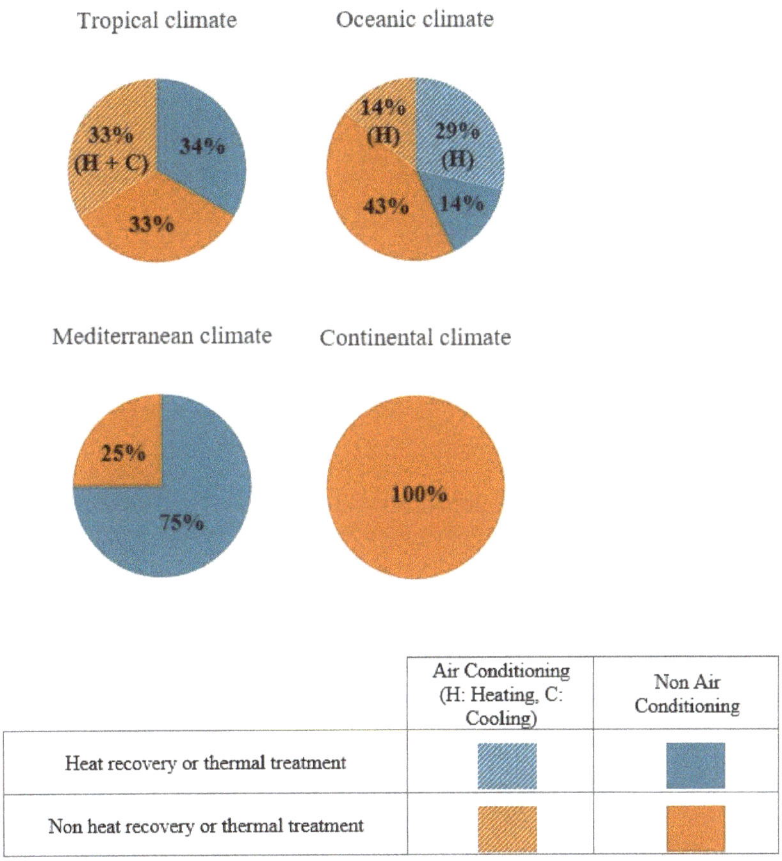

Figure 8. Detailed analysis of the implementation percentage of each mechanical ventilation sub-strategy.

As seen in Figure 8, the use of heat recovery units or batteries in mechanical ventilation systems is only predominant in the Mediterranean climate, reaching 75%. This figure is reduced to 50% in the tropical climate and to 43% in the oceanic climate. No mechanical ventilation strategies with heat recovery or batteries have been studied in the continental climate. As in the case of natural ventilation, in the selected study sample, most schools and educational buildings have no air conditioning, whether heating or cooling.

Table 7. References with mechanical ventilation strategies.

	Ventilation		Air Conditioning		Mechanical Ventilation Strategies	
Climate	**Heat Recovery**	**Thermal Treatment**	**Heating**	**Cooling**	**5**	**6**
Tropical	No	No	Yes	Yes		[37] (S)
			No	No		[55] (W, SP)
	Yes	No	Yes	No		
			No	No	[49] (W)	
Oceanic	No	No	Yes	No	[48] (W)	
			No	No	[48] (SP) [52] (S)	[46] (SP)
	Yes	Yes	Yes	No	[48] (W) [53] (W + SP)	
			No	No	[48] (SP)	
Mediterranean	No	No	Yes	No		
			No	No		[43] (A)
	Yes	No	Yes	No		
			No	No	[45] (S, W, SP)	[45] (S, W, SP)
		Yes	No	No		[50] (W)
Continental	No	No	Yes	No		
			No	No		[44] (W, SP)
	Yes	Yes	Yes	No		
			No	No		

3.3. Hybrid Ventilation Strategies

As noted at the beginning of the results section, the oceanic climate is the only climate where studies on hybrid ventilation strategies are found. As shown in Table 8, there are different combinations: natural cross ventilation with constant flow mechanical ventilation and single-sided natural ventilation with constant and variable flow mechanical ventilation. Therefore, considering the limited number of studies on hybrid strategies and the variety of combinations, no recurring pattern is observed.

Table 8. References with hybrid ventilation strategies.

Climate	Natural Ventilation	Mechanical Ventilation	AC	Hybrid Ventilation Strategies
Oceanic	Cross	Variable	Yes	[38] (S, W)
	Single-side	Constant	Yes	[53] (W + SP)
			No	[46] (S, W, SP) [52] (S)

3.4. Protocols Used

The different protocols used for each type of ventilation strategy are detailed in Tables 9–11.

Table 9. Natural ventilation protocols.

Ventilation Strategy		Ref.	Number of Different Protocols	Detailed Protocols											Occupancy	Monitored (M)/Simulated (S)	Monitored Days	Temp	Humidity	CO_2
				Season	Cross Ventilation		Air Conditioning		Different Window Opening Times Are Tested	Night Ventilation		Different Classes Studied	Different Buildings Studied	Orientation						
					Yes	No	Yes	No		Yes	No									
1. User-dependent	a	[37]	2	Summer		·	·	·			·			Southeast	25	M	3	·	·	
		[46]	1	Spring		·		·			·			Northwest	36	S	1	·	·	·
		[47]	17	Winter	·			·	·		·			North	5 to 40	M	18	·	·	·
		[48]	2	Winter		·	·				·			Southwest	20	S	1	·		·
				Spring		·		·			·			Southwest	20	S	1			·
	b	[38]	2	Summer		·		·			·			South	30	M	20	·	·	·
				Winter		·		·			·			South	30	M	20	·	·	·
		[43]	2	Annually	·			·			·			NorthSouth	25	M	550	·		·
		[49]	1	Winter		·		·			·			N/D	21	M	21	·		
		[53]	24	Winter + Spring	·	·	·				·	·	·	N/D	27 to 33	M	81	·	·	·
		[54]	1	Summer		·		·			·	·	·	N/D	11 to 28	M	150	·	·	·
			1	Winter		·		·			·	·	·	N/D	11 to 28	M	150	·	·	·
			1	Spring		·		·			·	·	·	N/D	11 to 28	M	150	·	·	·
2. User-dependent according to established window opening patterns		[39]	7	Winter	·	·	·		·	·	·	·		North + South	23 to 25	M	6/96	·	·	·
		[40]	7	Summer	·	·		·	·	·	·	·		North + South	23 to 25	M	4	·	·	
				Winter		·		·	·		·	·		North + South	23 to 25	M	7	·	·	
		[41]	8	Summer	·			·	·		·			N/D	13 to 26	S	N/D	·	·	·
				Winter	·		·		·		·			N/D	13 to 26	S	N/D	·	·	·
		[48]	2	Winter		·	·		·		·			Southwest	20	S	1	·		·
				Spring		·		·	·		·			Southwest	20	S	1			·
		[50]	2	Winter		·		·	·		·			East	25 to 27	M	60	·		·
3. User-independent with parameter control through sensors	CO_2	[46]	1	Spring		·		·			·			Northwest	36	S	1	·	·	·
		[47]	1	Winter	·			·			·			North	33	M	18	·	·	·
		[49]	1	Winter		·		·			·			N/D	21	M	21	·		·
	Temp.	[38]	2	Summer		·		·		·				South	30	M	21	·	·	·
				Winter		·	·				·			South	30	M	20	·	·	·
		[46]	3	Summer		·		·			·			Northwest	36	S	1	·	·	·
				Winter		·		·			·			Northwest	36	S	1	·	·	·
				Spring		·		·			·			Northwest	36	S	1	·	·	·
4. User-independent using windcatcher		[42]	1	Summer	·			·			·			N/D	32	S	N/D			
		[49]	1	Winter	·			·			·			N/D	21	M	21	·		
		[51]	8	Summer	·			·			·			North	48	S	31	·	·	·
				Winter	·			·			·			North	48	S	28	·	·	·

Table 10. Mechanical ventilation protocols.

Ventilation Strategy	Ref.	Number of Different Protocols	Detailed protocols											Orientation	Occupancy	Monitored (M)/Simulated (S)	Monitored Days	Temp	Humidity	CO_2
			Season	Air Conditioning		Airflow	Heat Recovery		Night Ventilation		Different Classes Studied	Different Buildings Studied								
				Yes	No		Yes	No	Yes	No										
5. Mechanical ventilation with constant airflow	[45]	6	Summer		·	6.5 ACH/6.1 ACH	·			·				Southwest	21	S	15			
			Winter		·	6.5 ACH/6.1 ACH	·			·				Southwest	21	S	15			
			Spring		·	6.5 ACH/6.1 ACH	·			·				Southwest	21	S	15			
	[48]	4	Winter	·		2.9–4.45 L/s (m²)	·	·		·				Southwest	20	S	1	·		·
			Spring		·	2.9–4.45 L/s (m²)	·	·		·				Southwest	20	S	1			·
	[49]	1	Winter		·	134 m³/h				·				N/D	21	M	21	·		
	[52]	1	Summer		·	N/D				·				N/D	N/D	M	31	·		
	[53]	2	Winter + Spring	·		N/D				·	·			N/D	30	M	81	·	·	·
6. Mechanical ventilation with variable airflow	[37]	1	Summer	·				·		·				Southeast	25	M	3	·	·	
	[43]	2	Year		·			·		·	·			North/South	25	M	550	·		·
	[44]	3	Winter		·			·		·				N/D	N/D	M	1	·		·
			Spring		·			·		·				N/D	N/D	M	1	·		·
	[45]	3	Summer		·			·		·				Southwest	21	S	15			
			Winter		·			·		·				Southwest	21	S	15			
			Spring		·			·		·				Southwest	21	S	15			
	[46]	1	Spring		·			·		·				Northeast	36	S	1	·		·
	[50]	1	Winter		·		·			·				East	25–27	M	60	·		·
	[55]	1	Winter + Spring		·						·			Northeast	7–25	M	60	·	·	·

Table 11. Hybrid ventilation protocols.

Ventilation Strategy	Ref.	Number of Different Protocols	Detailed protocols												Orientation	Occupancy	Monitored (M)/Simulated (S)	Monitored Days	Temp	Humidity	CO_2
				Natural Cross Ventilation		Air Conditioning		Mechanical Ventilation Airflow (m^3/h)	Heat Recovery		Night Ventilation		Different Classes Studied	Different Buildings Studied							
			Season	Yes	No	Yes	No		Yes	No	Yes	No									
7. Hybrid ventilation	[38]	4	Summer		·		·	250	·		·	·	·	·	South	30	M	20	·	·	·
			Winter		·	·	·	250	·			·	·	·	South	30	M	32	·	·	·
	[46]	3	Summer	·			·	250		·		·			Northwest	36	S	1	·	·	·
			Winter	·			·	250		·		·			Northwest	36	S	1	·	·	·
			Spring	·			·	250		·		·			Northwest	36	S	1	·	·	·
	[52]	1	Summer		·		·	N/D		·		·			N/D	N/D	M	31	·		
	[53]	4	Winter + Spring		·	·		N/D		·		·	·	·	N/D	31	M	81	·	·	·

4. Discussion

This article investigates different ventilation strategies in educational centers and their ability to simultaneously guarantee air quality and hygrothermal comfort conditions. The 19 articles selected were used as the starting point for an analysis of the conditions monitored in different seasons and climates for each of the ventilation strategies considered. The protocols used in each study have been analyzed and a detailed study carried out for the measurements and conditions in place. The results obtained show the main parameters conditioning thermal comfort and air quality–temperature, humidity, and CO_2.

Due to the volume of data obtained from the available literature, establishing an equivalent comparison in the different climatic zones is a difficult task. This is not only because of the lack of measurement data, but also because the differences in protocols (spot measurements versus continuous measurements, measurements only during occupied time versus measurements throughout the day) hamper comparison of the different results. Furthermore, hygrothermal sensitivity levels vary in the different climatic zones. However, a trend can be identified between the different climates and ventilation strategies. Therefore, this study does not aim to establish an optimal protocol and ventilation conditions to ensure thermal comfort and air quality but to create a guide which can be used for the purposes of reference and comparison in any future studies.

In addition to the results presented, this review reveals key trends and patterns that warrant further reflection. Natural ventilation is by far the most widely implemented strategy across all climates, likely due to its low cost and ease of implementation. However, the high dependence on user behavior—especially in strategies without predefined window opening schedules—makes it an unreliable solution for ensuring consistent indoor environmental quality. Mechanical ventilation, on the other hand, though less frequently studied, offers greater control but often lacks thermal treatment or heat recovery elements, which limits its effectiveness in colder climates. Hybrid ventilation strategies, although least represented in the literature, have shown promising results in achieving both acceptable CO_2 levels and thermal comfort, particularly when sensor-based control is involved. These findings suggest that automation and mixed-mode systems could be a direction for future development, especially in regions where outdoor conditions vary widely or where user behavior cannot be easily regulated.

The results of the measured variables—temperature, humidity, and carbon dioxide—are shown in Tables 10 and 11, detailing average, maximum, and minimum values. Only winter and summer results are analyzed, as these are the most extreme seasons in terms of thermal comfort. As the tables show, not all studies measured all variables. In Table 10, summer results, temperature values of between 23 and 26 °C, humidity between 45% and 60%, and CO_2 levels below 800 ppm are highlighted in green following the comfort parameters of the UNE EN ISO 7730:2006 [56] standard and air quality parameters according to UNE EN 16798-1 [57]. In Table 11, winter results, temperature values of between 20 and 24 °C, humidity between 40% and 50%, and CO_2 levels below 800 ppm are highlighted in green.

As observed in Table 12, there are few strategies that achieve optimal temperature and CO_2 values to ensure thermal comfort and air quality. It should be stressed that the lack of data collection in many studies complicates the analysis of the different strategies. There are three cases worth highlighting, as they achieve an average temperature of between 23 and 25 °C and average CO_2 values below 800 ppm. These three strategies were implemented in the same study by Dhalluin et al. [38]. Suitable values were achieved with user-controlled natural ventilation without any opening pattern, with natural ventilation with window opening based on temperature sensor values, and with hybrid ventilation. It should be noted that no air conditioning systems were used to achieve optimal temperature values in any of these cases. However, due to the significant lack of data, we cannot assert that the remaining strategies are unsuitable. It is therefore necessary to conduct further studies with better control of the various variables which quantify thermal comfort and air quality.

Furthermore, Table 13 shows how the situation in winter is similar to that in summer. Firstly, the lack of data encountered again in the bibliography complicates comparison of results and strategies. In this case, it is observed that almost no scenario achieves both adequate thermal comfort and air quality simultaneously. This is only found in the study by Cablé et al. [44] through a mechanical ventilation strategy with variable flow based on sensor measurements. In some scenarios where CO_2 values below 800 ppm are achieved, the temperature is compromised, with average values below 20 °C, even reaching extremely low values of 11 °C in some cases. Conversely, in scenarios where suitable average temperature values are achieved, this is at the expense of CO_2 values, which can reach up to 1400 ppm in some cases. Therefore, as in the previous case, it is concluded that further studies are necessary to increase the number of strategies studied and the measurements taken.

A major limitation found in the reviewed literature is the absence of standardized monitoring and reporting protocols. Many studies fail to specify measurement durations, occupancy levels, or whether the data correspond to peak or average conditions. This lack of uniformity significantly restricts the potential for comparative or meta-analytical approaches. Moreover, very few studies report on all three key variables (temperature, humidity, and CO_2), making it difficult to assess strategies holistically. The observed performance gaps, where either thermal comfort or air quality is sacrificed for the other, highlight the complexity of designing effective passive or semi-passive ventilation solutions.

Finally, this review does not aim to define a universal strategy, but rather to provide a comparative and critical framework for future research. It underlines the need for more robust experimental designs, broader climate coverage, and better integration of comfort and health criteria. More studies are needed, particularly in underrepresented climate zones and on underexplored strategies like hybrid systems. Incorporating real-time data collection, user behavior modelling, and post-pandemic considerations such as increased air renewal rates will also be essential in shaping future guidelines and ensuring healthy, comfortable learning environments.

Table 12. Results obtained from the measurements according to the different strategies in summer. The values shaded in green are those that meet the requirements established by regulations regarding CO_2, temperature, and humidity.

Ventilation Strategy	Climate	Ref.	Cooling	Occupancy	Monitored (M)/Simulated (S)	Monitored Days	T Average (°C)	T Max (°C)	T Min (°C)	H Average (%)	H Max (%)	H Min (%)	CO_2 Average (ppm)	CO_2 Max (ppm)	CO_2 Min (ppm)
1a	Tropical	[37]	Yes	25	M	3		29.4	25.25		74	49			
1b	Oceanic	[38]	No	30	M	20	24.4	27.3	20.9	47	73	26	800	1800	300
	Mediterranean	[54]	No	11 to 28	M	150	28.18	36.44	22.29	50.2	71.7	26.1	593	4015	341
2	Semi-arid	[40]	No	23 to 25	M	4	31.6	33.7	28.5	32	50	19			
	Continental	[41]	No	13 to 26	S	N/D		33.9	31.1		72	57		1300	875
3b	Oceanic	[38]	No	30	M	20	23.5	28.4	19.7	50	76.5	30	700	1420	380
	Oceanic	[46]	No	36	S	1	26.3	29.1	23.9	70	88	58	1079	2299	400
4	Oceanic	[42]	No	32	S	N/D									
	Semi-arid	[51]	No	48	S	31		33.75			64.5	23.75		459	
5	Mediterranean	[45]	No	21	S	15									
	Oceanic	[52]	No	N/D	M	31	23								
6	Tropical	[37]	Yes	25	M	3		29	25.3		71	40			
	Mediterranean	[45]	No	21	S	15									
7	Oceanic	[38]	No	30	M	20	24.95	24.35	24	43.5	60.75	28.75	500	965	220
	Oceanic	[46]	No	36	S	1	26	28.6	23.8	41	45	38	688	1167	400
	Oceanic	[52]	No	N/D	M	31	22.2								

Table 13. Results obtained from the measurements according to the different strategies in winter. The values shaded in green are those that meet the requirements established by regulations regarding CO_2, temperature, and humidity.

Ventilation Strategy	Climate	Ref.	Heating	Occupancy	Monitored (M)/Simulated (S)	Monitored Days	T Average (°C)	T Max (°C)	T Min (°C)	H Average (%)	H Max (%)	H Min (%)	CO_2 Average (ppm)	CO_2 Max (ppm)	CO_2 Min (ppm)
1a	Mediterranean	[47]	No	5 to 40	M	18	16.1	17.0	15.4	56.5	59.5	54.6	539.5	607.8	479.4
	Oceanic	[48]	Yes	20	S	1		17	15					770	
1b	Oceanic	[38]	No	30	M	20	20.5	23.9	15.5	53	78	26.5	1800	3000	400
	Tropical	[49]	No	21	M	21	23.16	24.82	21.96						
	Mediterranean	[54]	No	11 to 28	M	150	21.24	33.41	12.52	47	69.2	23.7	1194	4950	348
2	Semi-arid	[39]	Yes	23 to 25	M	6	20.6	22.3	18.2	58	71	45	1219	2632	419
	Semi-arid	[40]	No	23 to 25	M	7	19.3	21.2	16.7	59	68	50			
	Continental	[41]	Yes	13 to 26	S	N/D		21.9	17.1		36	18		1200	788
	Oceanic	[48]	Yes	20	S	1		21	10					1450	
	Mediterranean	[50]	No	25 to 27	M	60	23.3	24.2	21.0				1408	1943	645
3a	Mediterranean	[47]	No	33	M	18	11.09	13.1	8.2	44.22	48.6	39.9	606.99	808	494
	Tropical	[49]	No	21	M	21	23.41	23.91	22.19				920.95		
3b	Oceanic	[38]	Yes	30	M	20	20.4	24.8	11	55	76	24.5	1100	2600	400
	Oceanic	[46]	No	36	S	1	20	23.2	16.4	35	49	27	1091	2228	400
4	Tropical	[49]	No	21	M	21	21.61	23.71	21.62						
	Semi-arid	[51]	No	48	S	28		27.5			64	25.25		459	
5	Mediterranean	[45]	No	21	S	15									
	Oceanic	[48]	Yes	20	S	1	20							950	
	Tropical	[49]	No	21	M	21	23.74	24.91	22.04						
6	Continental	[44]	No	N/D	M	1	23.3						637		
	Mediterranean	[45]	No	21	S	15									
	Mediterranean	[50]	No	25 to 27	M	60	21.3	22.3	20.4				1002	1072	541
7	Oceanic	[38]	Yes	30	M	32	20.3	26.45	13.2	39.25	58	21.25	775	1850	400
	Oceanic	[46]	No	36	S	1	17.9	20.7	15.1	64	72	58	729	1158	400

5. Conclusions

This study has presented a critical review which analyzed the ventilation strategies most commonly used in schools to both guarantee indoor air quality and ensure the thermal comfort of students. After an initial selection of 277 articles and a final selection of 19, a total of nine strategies were identified in five different climates according to Köppen's classification, grouped by natural, mechanical, or hybrid ventilation. This highlighted the low number of studies which focus on ensuring both thermal comfort and air quality in schools. It is also important to highlight the geographical concentration of the analyzed studies, with most of them being located in Europe. Most of the 80 studies analyzed initially address only one of these aspects, neglecting the other, particularly thermal comfort. Most studies focus on ensuring air quality even when this means sacrificing thermal comfort. This trend has been particularly pronounced since the COVID-19 pandemic, when the focus of studies shifted to air quality in order to prevent airborne disease transmission, irrespective of the thermal comfort of the occupants. Therefore, a clear lack of studies has been identified in this area.

Within natural ventilation, the most employed strategies in studies considering air quality and thermal comfort are: user-dependent natural ventilation (with permanently open windows or with irregular window opening patterns); user-dependent natural ventilation according to established window opening patterns; user-independent natural ventilation with parameter control through sensors (CO_2 or temperature sensors); and user-independent natural ventilation using a windcatcher and stack ventilation. Mechanical ventilation strategies can be summarized as ventilation with constant airflow or ventilation with variable airflow, with the latter based on CO_2 or temperature measurements. There are no additional categories of mixed ventilation strategies.

Natural ventilation strategies are the most used in all climates: 73% in the continental climate, 68% in the Mediterranean climate, 66% in the oceanic climate, 100% in the semi-arid climate, and 63% in the tropical climate. However, only in 17% of the cases analyzed in the selected works are optimal average temperature and air quality conditions achieved with these strategies in summer. None are achieved in winter, highlighting a clear need for the implementation of hybrid or mixed systems in school spaces.

Regarding the analysis of results obtained with the different strategies, it must first be noted that the lack of data prevents us from determining whether some strategies are better than others for the simultaneous promotion of thermal comfort and air quality. After a comparison of the strategies that have sufficient data on temperature and CO_2, it is concluded that there are almost no studies where both comfort conditions and CO_2 values are ensured simultaneously; in most cases, one of the two variables is compromised. Only two studies, one in summer and one in winter, have been found to show suitable values for both parameters. It is also worth noting a greater wealth of data in winter, which may seem surprising, as some schools are heated by radiators but rarely have cooling. Therefore, summer could be more critical due to the high temperatures in some of the countries studied, including Spain, Italy, and Portugal. However, in the rest of European countries, winter is the most critical season, as they have milder summers. These issues warrant further investigation in future studies, as they represent a relevant area of interest with direct implications for indoor air quality (IAQ) and thermal comfort (TC).

Therefore, it is essential to conduct new studies testing different ventilation strategies in educational buildings to simultaneously ensure thermal comfort and air quality. This review serves as a guide to the protocols used in the various studies analyzed.

Author Contributions: Conceptualization, R.S. and Á.L.L.-R.; methodology, M.T.A.-C. and R.M.L.-L.; formal analysis, M.T.A.-C. and R.M.L.-L.; investigation, M.T.A.-C. and R.M.L.-L.; resources, M.T.A.-C. and R.M.L.-L.; writing—original draft preparation, M.T.A.-C. and R.M.L.-L.; writing—review and editing, M.T.A.-C. and R.M.L.-L.; visualization, M.T.A.-C. and R.M.L.-L.; supervision, R.S. and Á.L.L.-R.; project administration, R.S. and Á.L.L.-R.; funding acquisition, R.S. and Á.L.L.-R. All authors have read and agreed to the published version of the manuscript.

Funding: The authors wish to acknowledge the financial support provided by Grant (PID2020-117722RB-I00) "Retrofit ventilation strategies for healthy and comfortable schools within a nearly zero-energy building horizon" funded by MICIU/AEI/10.13039/501100011033).

Institutional Review Board Statement: Not applicable.

Informed Consent Statement: Not applicable.

Data Availability Statement: No new data were created.

Conflicts of Interest: The authors declare no conflicts of interest.

References

1. World Health Organization. Ambient (Outdoor) Air Pollution. Available online: https://www.who.int/health-topics/air-pollution#tab=tab_1 (accessed on 23 June 2024).
2. Krismanuel, H.; Hairunisa, N. The Effects of Air Pollution on Respiratory Problems: A Systematic Review. *Poltekita J. Ilmu Kesehat.* **2024**, *18*, 1–15. [CrossRef]
3. Brook, R.D.; Franklin, B.; Cascio, W.; Hong, Y.L.; Howard, G.; Lipsett, M.; Luepker, R.; Mittleman, M.; Samet, J.; Smith, S.C.; et al. Air pollution and cardiovascular disease—A statement for healthcare professionals from the expert panel on population and prevention science of the American Heart Association. *Circulation* **2004**, *109*, 2655–2671. [CrossRef] [PubMed]
4. Suzuki, T.; Hidaka, T.; Kumagai, Y.; Yamamoto, M. Environmental pollutants and the immune response. *Nat. Immunol.* **2020**, *21*, 1486–1495. [CrossRef] [PubMed]
5. Calderón-Garcidueñas, L.; Azzarelli, B.; Acuna, H.; Garcia, R.; Gambling, T.M.; Osnaya, N.; Monroy, S.; Tizapantzi, M.R.; Carson, J.L.; Villarreal-Calderon, A.; et al. Air pollution and brain damage. *Toxicol Pathol.* **2002**, *30*, 373–389. [CrossRef] [PubMed]
6. Raju, S.; Siddharthan, T.; McCormack, M.C. Indoor Air Pollution and Respiratory Health. *Clin. Chest Med.* **2020**, *41*, 825–843. [CrossRef]
7. Park, J.; Kim, H.-J.; Lee, C.H.; Lee, H.W. Impact of long-term exposure to ambient air pollution on the incidence of chronic obstructive pulmonary disease: A systematic review and meta-analysis. *Environ. Res.* **2021**, *194*, 110703. [CrossRef]
8. Tahery, N.; Zarea, K.; Cheraghi, M.; Hatamzadeh, N.; Farhadi, M.; Dobaradarn, S.; Mohammadi, M.J. Chronic Obstructive Pulmonary Disease (COPD) and Air Pollution: A Review. *Jundishapur J. Chronic Dis. Care* **2021**, *10*, 110273. [CrossRef]
9. Shahzad, S.; Torres, S.G.; Rijal, H.B.; Nicol, F. Lack of Thermal Comfort Is a Matter of Life and Death: A Systematic Review for Older People. *Buildings* **2025**, *15*, 1141. [CrossRef]
10. Bunker, A.; Wildenhain, J.; Vandenbergh, A.; Henschke, N.; Rocklöv, J.; Hajat, S.; Sauerborn, R. Effects of Air Temperature on Climate-Sensitive Mortality and Morbidity Outcomes in the Elderly; a Systematic Review and Meta-analysis of Epidemiological Evidence. *EBioMedicine* **2016**, *6*, 258–268. [CrossRef]
11. Liu, W.; Wu, Y.; Shi, L.; Li, J.; Wang, Z.; Tian, X. The relationship between thermal comfort, performance, and electroencephalogram during cognitive tests at normal indoor temperatures in summer. *Build. Environ.* **2024**, *257*, 111567. [CrossRef]
12. Singh, M.K.; Ooka, R.; Rijal, H.B.; Kumar, S.; Kumar, A.; Mahapatra, S. Progress in thermal comfort studies in classrooms over last 50 years and way forward. *Energy Build.* **2019**, *188–189*, 149–174. [CrossRef]
13. Midouhas, E.; Kokosi, T.; Flouri, E. Outdoor and indoor air quality and cognitive ability in young children. *Environ. Res.* **2018**, *161*, 321–328. [CrossRef]
14. Bakó-Biró, Z.; Clements-Croome, D.; Kochhar, N.; Awbi, H.; Williams, M. Ventilation rates in schools and pupils' performance. *Build. Environ.* **2012**, *48*, 215–223. [CrossRef]
15. Mendell, M.J.; Eliseeva, E.A.; Davies, M.M.; Spears, M.; Lobscheid, A.; Fisk, W.J.; Apte, M.G. Association of classroom ventilation with reduced illness absence: A prospective study in California elementary schools. *Indoor Air* **2013**, *23*, 515–528. [CrossRef] [PubMed]
16. Wargocki, P.; Wyon, D.P. Providing better thermal and air quality conditions in school classrooms would be cost-effective. *Build. Environ.* **2013**, *59*, 581–589. [CrossRef]
17. Stabile, L.; Pacitto, A.; Mikszewski, A.; Morawska, L.; Buonanno, G. Ventilation procedures to minimize the airborne transmission of viruses in classrooms. *Build. Environ.* **2021**, *202*, 108042. [CrossRef] [PubMed]

18. Ferrari, S.; Blázquez, T.; Cardelli, R.; Puglisi, G.; Suárez, R.; Mazzarella, L. Ventilation strategies to reduce airborne transmission of viruses in classrooms: A systematic review of scientific literature. *Build. Environ.* **2022**, *222*, 109366. [CrossRef]
19. IPCC—Intergovernmental Panel on Climate Change. Available online: http://www.ipcc.ch/ (accessed on 11 January 2025).
20. Santamouris, M.; Kolokotsa, D. On the impact of urban overheating and extreme climatic conditions on housing, energy, comfort and environmental quality of vulnerable population in Europe. *Energy Build.* **2015**, *98*, 125–133. [CrossRef]
21. Stazi, F.; Tomassoni, E.; Di Perna, C. Super-insulated wooden envelopes in Mediterranean climate: Summer overheating, thermal comfort optimization, environmental impact on an Italian case study. *Energy Build.* **2017**, *138*, 716–732. [CrossRef]
22. Parry, M.L.; Canziani, O.F.; Palutikof, J.P.; Van Der Linden, P.J.; Hanson, C.E. *IPCC, 2007: Climate Change 2007: Impacts, Adaptation and Vulnerability. Contribution of Working Group II to the Fourth Assessment Report of the Intergovernmental Panel on Climate Change*; Cambridge University Press: Cambridge, UK, 2007.
23. Lomas, K.J.; Porritt, S.M. Overheating in buildings: Lessons from research. *Build. Res. Inf.* **2017**, *45*, 1–18. [CrossRef]
24. European Parliament and Council of the European Union. Directive (EU) 2018/844 of the European Parliament and of the Council of 30 May 2018 amending Directive 2010/31/EU on the energy performance of buildings and Directive 2012/27/EU on energy efficiency (Text with EEA relevance). *Off. J. Eur. Union* **2018**, *L156*, 75–91.
25. Homod, R.Z.; Sahari, K.S.M.; Almurib, H.A. Energy saving by integrated control of natural ventilation and HVAC systems using model guide for comparison. *Renew. Energy* **2014**, *71*, 639–650. [CrossRef]
26. Brambilla, A.; Bonvin, J.; Flourentzou, F.; Jusselme, T. On the Influence of Thermal Mass and Natural Ventilation on Overheating Risk in Offices. *Buildings* **2018**, *8*, 47. [CrossRef]
27. Hampo, C.C.; Schinasi, L.H.; Hoque, S. Surviving indoor heat stress in United States: A comprehensive review exploring the impact of overheating on the thermal comfort, health, and social economic factors of occupants. *Heliyon* **2024**, *10*, e25801. [CrossRef] [PubMed]
28. Stazi, F.; Naspi, F.; D'Orazio, M. Modelling window status in school classrooms. Results from a case study in Italy. *Build. Environ.* **2017**, *111*, 24–32. [CrossRef]
29. Guo, F.; Chen, Z.; Wang, J. Effect of Dynamic Window Opening Behaviors on Indoor Thermal Environment and Energy Consumption in Residential Buildings of Different Chinese Thermal Climate Zones. *Buildings* **2025**, *15*, 28. [CrossRef]
30. Chatzidiakou, L.; Mumovic, D.; Summerfield, A. Is CO_2 a good proxy for indoor air quality in classrooms? Part 1: The interrelationships between thermal conditions, CO_2 levels, ventilation rates and selected indoor pollutants. *Build. Serv. Eng. Res. Technol.* **2015**, *36*, 129–161. [CrossRef]
31. Hui, P.; Wong, L.; Mui, K. Using Carbon Dioxide Concentration to Assess Indoor Air Quality in Offices. *Indoor Built Environ.* **2008**, *17*, 213–219. [CrossRef]
32. Di Gilio, A.; Palmisani, J.; Pulimeno, M.; Cerino, F.; Cacace, M.; Miani, A.; de Gennaro, G. CO_2 concentration monitoring inside educational buildings as a strategic tool to reduce the risk of SARS-CoV-2 airborne transmission. *Environ. Res.* **2021**, *202*, 111560. [CrossRef]
33. Park, S.; Song, D. CO_2 concentration as an indicator of indoor ventilation performance to control airborne transmission of SARS-CoV-2. *J. Infect. Public Health* **2023**, *16*, 1037–1044. [CrossRef]
34. *ISO 7730:2005*; Ergonomics of the Thermal Environment—Analytical Determination and Interpretation of Thermal Comfort Using Calculation of the PMV and PPD Indices and Local Thermal Comfort Criteria. AENOR: Madrid, Spain, 2005.
35. Jia, L.-R.; Han, J.; Chen, X.; Li, Q.-Y.; Lee, C.-C.; Fung, Y.-H. Interaction between Thermal Comfort, Indoor Air Quality and Comprehensive Review. *Builings* **2021**, *11*, 591.
36. Bienvenido-Huertas, D.; de la Hoz-Torres, M.L.; Aguilar, A.J.; Tejedor, B.; Sánchez-García, D. Holistic overview of natural ventilation and mixed mode in built environment of warm climate zones and hot seasons. *Build. Environ.* **2023**, *245*, 110942. [CrossRef]
37. de Abreu-Harbich, L.V.; Chaves, V.L.; Brandstetter, M.C.G. Evaluation of strategies that improve the thermal comfort and energy saving of a classroom of an institutional building in a tropical climate. *Build. Environ.* **2018**, *135*, 257–268. [CrossRef]
38. Dhalluin, A.; Limam, K. Comparison of natural and hybrid ventilation strategies used in classrooms in terms of indoor environmental quality, comfort and energy savings. *Indoor Built Environ.* **2014**, *23*, 527–542. [CrossRef]
39. Heracleous, C.; Michael, A. Experimental assessment of the impact of natural ventilation on indoor air quality and thermal comfort conditions of educational buildings in the Eastern Mediterranean region during the heating period. *J. Build. Eng.* **2019**, *26*, 100917. [CrossRef]
40. Heracleous, C.; Michael, A. Thermal comfort models and perception of users in free-running school buildings of East-Mediterranean region. *Energy Build.* **2020**, *215*, 109912. [CrossRef]
41. Chitaru, G.-M.; Istrate, A.; Catalina, T. Numerical analysis of the impact of natural ventilation on the Indoor Air Quality and Thermal Comfort in a classroom. *E3S Web Conf.* **2019**, *111*, 01023. [CrossRef]
42. Morshed, T. An investigation of naturally ventilated teaching spaces with windcathers in secondary school where site is constrained by noise and air pollution. *Build. Simul.* **2014**, *7*, 547–561. [CrossRef]

43. Duarte, R.; Gomes, M.d.G.; Rodrigues, A.M. Classroom ventilation with manual opening of windows: Findings from a two-year-long experimental study of a Portuguese secondary school. *Build. Environ.* **2017**, *124*, 118–129. [CrossRef]
44. Cablé, A.; Hammer, H.L.; Mysen, M. Comparison of Two Ventilation Control Strategies in the First Passive House Standard Norwegian School. *Int. J. Vent.* **2016**, *14*, 371–382. [CrossRef]
45. Calama-González, C.M.; León-Rodríguez, Á.L.; Suárez, R. Indoor Air Quality Assessment: Comparison of Ventilation Scenarios for Retrofitting Classrooms in a Hot Climate. *Energies* **2019**, *12*, 4607. [CrossRef]
46. Rodríguez-Vidal, I.; Otaegi, J.; Armengod, U. Evaluating the Feasibility of Hybrid Ventilation in Educational Spaces: A Simulation Study in the Basque Country Climate. *Sci. J. Riga Tech. Univ. Environ. Clim. Technol.* **2023**, *27*, 532–544. [CrossRef]
47. Miranda, M.; Romero, P.; Valero-Amaro, V.; Arranz, J.; Montero, I. Ventilation conditions and their influence on thermal comfort in examination classrooms in times of COVID-19. A case study in a Spanish area with Mediterranean climate. *Int. J. Hyg. Environ. Health* **2022**, *240*, 113910. [CrossRef] [PubMed]
48. Pollozhani, F.; McLeod, R.S.; Schwarzbauer, C.; Hopfe, C.J. Assessing school ventilation strategies from the perspective of health, environment, and energy. *Appl. Energy* **2024**, *353*, 121961. [CrossRef]
49. Di Perna, C.; Mengaroni, E.; Fuselli, L.; Stazi, A. Ventilation Strategies in School Buildings for Optimization of Air Quality, Energy Consumption and Environmental Comfort in Mediterranean Climates. *Int. J. Vent.* **2011**, *10*, 61–78. [CrossRef]
50. Stabile, L.; Buonanno, G.; Frattolillo, A.; Dell'isola, M. The effect of the ventilation retrofit in a school on CO_2, airborne particles, and energy consumptions. *Build. Environ.* **2019**, *156*, 1–11. [CrossRef]
51. Ma'Bdeh, S.N.; Al-Zghoul, A.; Alradaideh, T.; Bataineh, A.; Ahmad, S. Simulation study for natural ventilation retrofitting techniques in educational classrooms—A case study. *Heliyon* **2020**, *6*, e05171. [CrossRef]
52. Ogbuagu, T.-C.C.; Linden, E.; MacCutcheon, D.; Nilsson, E.; Persson, T.; Kabanshi, A. On the Performance of Diffuse Ceiling Ventilation in Classrooms: A Pre-Occupancy Study at a School in Southern Sweden. *Sustainability* **2023**, *15*, 2546. [CrossRef]
53. Hama, S.; Kumar, P.; Tiwari, A.; Wang, Y.; Linden, P.F. The underpinning factors affecting the classroom air quality, thermal comfort and ventilation in 30 classrooms of primary schools in London. *Environ. Res.* **2023**, *236*, 116863. [CrossRef]
54. Miao, S.; Gangolells, M.; Tejedor, B. A Comprehensive Assessment of Indoor Air Quality and Thermal Comfort in Educational Buildings in the Mediterranean Climate. *Indoor Air* **2023**, *2023*, 6649829. [CrossRef]
55. Haddad, S.; Synnefa, A.; Marcos, M.Á.P.; Paolini, R.; Delrue, S.; Prasad, D.; Santamouris, M. On the potential of demand-controlled ventilation system to enhance indoor air quality and thermal condition in Australian school classrooms. *Energy Build.* **2021**, *238*, 110838. [CrossRef]
56. *UNE-EN ISO 7730:2006*; [Recurso Electrónico]: Ergonomía del Ambiente Térmico. Determinación Analítica e Interpretación del Bienestar Térmico Mediante el Cálculo de los Índices PMV y PPD y los Criterios de Bienestar Térmico Local (ISO 7730:2005). Asociación Española de Normalización y Certificación. AENOR: Madrid, Spain, 2006.
57. *UNE-EN 16798-1:2020*; Energy Performance of Buildings. Ventilation for Buildings. Indoor Environmental Input Parameters for Design and Assessment of Energy Performance of Buildings Addressing Indoor Air Quality, Thermal Environment, Lighting and Acoustics—Module M1-6, Parte 1. AENOR: Madrid, Spain, 2020.

applied sciences

MDPI

Review

Review of Intelligent Methods and Their Potential Use towards a Smart Grid Negotiation Framework

Dimitrios K. Panagiotou and Anastasios I. Dounis *

Department of Biomedical Engineering, University of West Attica, Egaleo Park Campus, 12243 Athens, Greece; d.panagiotou@uniwa.gr

* Correspondence: aidounis@uniwa.gr

Abstract: The integration of Distributed Energy Resources utilizing Renewable Energy Sources, Energy Storage Systems, and Information and Communication Technologies is transforming traditional energy systems into adaptable, flexible, and sustainable systems, with the Smart Grid concept playing a pivotal role. This paper surveys intelligent techniques and methods applied in various markets and applications, particularly focusing on their potential adaptation for negotiation processes in Smart Grid contexts. The negotiation mechanisms, crucial for prosumers who engage in real-time transactions, are analyzed with a focus on fuzzy logic tools, specifically q-Rung Orthopair Fuzzy Sets. These tools are evaluated for their capability to handle negotiation tasks and Multi-Criteria Decision-Making problems. The paper proposes a negotiation schema for healthcare buildings, especially hospitals, given their significant environmental impact, providing insight for future research.

Keywords: Smart Grid; negotiation; Fuzzy Sets; q-Rung Orthopair Fuzzy Sets; Multi-Criteria Decision-Making; Metaheuristic Algorithms; energy transactions; smart hospitals

1. Introduction

1.1. General Context

Electricity is a fundamental element of modern civilization, significantly improving human working and living conditions [1]. Historically, the electricity industry, encompassing generation, transmission, distribution and retailing, was integrated vertically operating as a monopoly, often under government ownership. In the United States and some other countries, local electricity providers were privately owned but regulated. Throughout most of the 20th century this model prevailed until the 1980s, when market reforms were initiated to introduce competition [2], particularly in the generation and retail sectors, aiming to improve efficiency and adaptability to new generation technologies [3]. The structure of a traditional electric grid is presented in Figure 1.

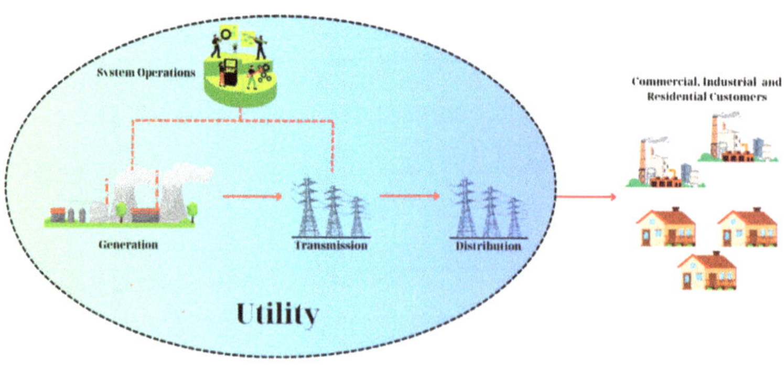

Figure 1. Traditional Electric Grid.

Recent decades have witnessed radical transformations in the electric power industry driven by environmental imperatives and technological advancements. The need for the urgent reduction of CO_2 emissions has accelerated the restructuring of the electricity sector [4], leading to the integration of Distributed Energy Resources (DER), Renewable Energy Sources (RES), Energy Storage Systems (ESS) and Information and Communication Technologies (ICT) into Smart Grids (SG). This shift moved the industry away from the traditional vertical models towards more flexible and sustainable energy systems, resulting to an incremental transition of the traditional consumers to prosumers, who actively participate in both the consumption and production of electricity [5,6]. The Smart Grid concept is presented in Figure 2.

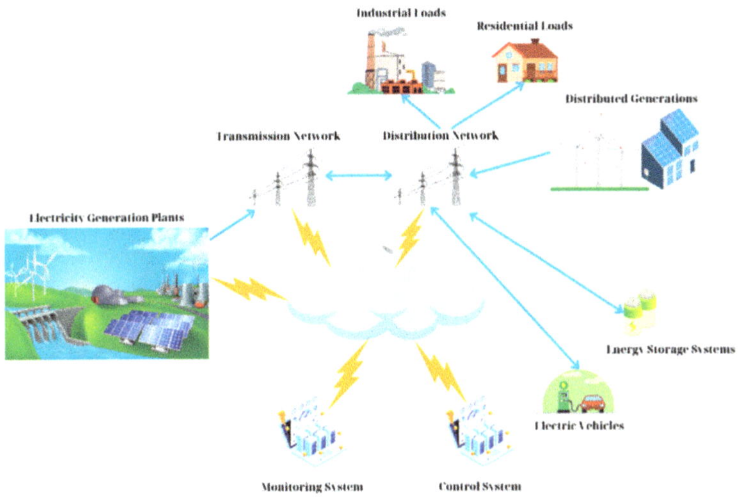

Figure 2. Smart grid concept [7].

In this new energy landscape, Demand Side Management (DSM) techniques [8], particularly Demand Response (DR) and Real-Time Pricing (RTP—also known as Dynamic Pricing, DP), play a critical role. The aforementioned techniques enhance the operational efficiency, market dynamics, and reliability of power systems by encouraging customer interaction and responsiveness. DR incentivizes consumers to shift their electricity usage to periods of surplus supply, typically through price signals or monetary incentives.

When integrated with SG and ESS, DR helps manage the impact of renewable energy variability and increases the stability of the grid [9]. RTP impacts the consumers' behavior through variable price signals that reflect real-time market conditions, contributing further to the grid's efficient operation [10].

DR and RTP techniques have operational and economic benefits. Operational benefits include load shifting as mentioned before, which optimizes the system's operation, peak load reduction, which can alleviate stress on the grid and reduce the need for expensive peaking power plants, and the integration of RES, by aligning electricity consumption with renewable generation patterns, maximizing the utilization of RES. Economic benefits include cost savings for consumers and utilities when the consumers adjust their electricity usage according to price fluctuations, revenue generation by offering incentives to the consumers to earn by curtailing their electricity usage during peak hours or shifting it to times with lower prices, and efficient resource allocation by optimizing the electricity consumption patterns based on price signals, benefitting both consumers and utilities [11].

Furthermore, the integration of DER and ESS within the SG framework improves the grid's flexibility and sustainability. As for flexibility, DER allows to generate power closer to the point of consumption, reducing the need for extensive transmission infrastructure and minimizing transmission losses. They also can be deployed strategically to meet peak demand periods, reducing the need for peaking power plants. ESS, on the other hand, can store excess energy generated in low-demand periods and release it during high-demand periods, balancing dynamically the supply and demand. They can also provide rapid response to deviations of frequency, helping to maintain the grid's stability and maintain voltage levels within the desired range, which is important for the grid's stability and efficiency. In sustainability terms, the RES integration reduces the reliance on fossil fuels, lowering greenhouse gas emissions. ESS mitigates the intermittency of RES by storing the excess energy and providing it when RES are not generating. The result is the reduction of carbon footprint and lower emissions of pollutants and greenhouse gases.

The evolution of electricity markets, especially in liberalized retail environments, fosters competition between suppliers, leading to innovative tariff structures. Consumers benefit from a wider selection of tariffs tailored to their consumption patterns and risk preferences, thereby facilitating the adoption of RTP strategies. This competitive landscape transforms the electricity market from a static, vertical structure with fixed-rate tariffs to a dynamic and resilient market where consumers are potential producers and energy transactions occur in real time, exploiting the advantages given by ICT, like smart meters and systems of energy management [12].

Prosumers in an SG contribute actively to its stability and performance. They produce and store electricity for self-consumption and engage in energy transactions when production exceeds or falls short of their needs. This requires efficient negotiation processes to determine the terms of energy transactions. This context's negotiation involves complex Decision Making (DM) where multiple parties seek to resolve conflicting interests. This process includes pre-negotiation, intention (offers and counter-offers), and post-optimization phases to achieve mutually satisfying agreements.

Artificial Intelligence (AI) technologies offer significant potential for optimizing these negotiation processes. AI techniques such as Machine Learning, Metaheuristic Algorithms (MA), and Fuzzy Logic (FL) systems can improve DM efficiency and adaptability in environments characterized as dynamic and uncertain. Especially, integrating FL systems like q-Rung Orthopair Fuzzy Sets (q-ROFS) with MA can enhance the handling of uncertainty and provide robust solutions in negotiation scenarios.

1.2. Contribution of the Review

The transition to a green energy model necessitates the integration of RES and Electric Vehicles (EVs) into the system. However, the inherent variability and unpredictability of these resources make the flexibility of the system essential for maintaining stability and security while ensuring efficient energy supply to all users. To address these challenges, the present review focuses on intelligent methods to manage variability and enhance system flexibility through demand-shifting policies.

Lower energy costs can be achieved when consumers act as prosumers forming energy communities of microgrids. These prosumers can participate collaboratively in energy exchange or trade energy directly among themselves, reducing electricity costs through local trading at mutually beneficial prices [5]. There are three primary scenarios where negotiation occurs in the SG context:

1. Energy Market Transactions: The case where prosumers sell surplus energy to the grid.
2. Peer-to-Peer (P2P) and Local Transactions: In this case, the energy transactions occur directly between prosumers.
3. Retail Market Tariff Selection: Consumers choose tariffs that best suit their needs among various options offered by retailers, as part of RTP strategy.

The first two scenarios involve direct negotiation of energy amounts and prices, while the third scenario, though not purely a negotiation, involves continuous DM as consumers evaluate and select the most suitable tariffs [13].

This review highlights the significant potential of AI techniques in optimizing these negotiation processes. By reviewing the existing literature on negotiation cases and DM scenarios using AI, including applications of q-ROFS combined with MA, the present study provides insights into effective strategies for managing complex DM processes in SG. This integration of AI enhances the robustness and reliability of negotiation processes, helping to cope with the inherent variability and uncertainty of SG where RES plays a significant role.

In summary, this review contributes to the understanding of intelligent negotiation methods in SG, emphasizing the integration of advanced AI techniques to improve DM and transaction outcomes in the evolving energy market landscape.

2. Methodology

In this survey, a total of 24 studies were examined, published from 1994 to 2023. The selected studies were chosen because of their potential applicability to the Energy Transaction context in SG. Seven of them focus on e-commerce and cloud services applications where negotiations take place, seven focus on energy transactions and allocation of sources in electricity networks and buildings, three refer to general applications without specific examples, six present examples covering a wide range of applications from negotiation in politics to car market negotiation, and one reviews the application of PSO variants in DSM applications. The majority of them introduce FS while some refer to MA, such as GA, PSO, and Bat Algorithm (BA). A summary is presented in Table 1.

Table 1. List of reviewed publications.

Year	Title/Author(s)	Commodity/ Application	Method
1994	"Fuzzy Constraint Satisfaction" (Ruttkay) [14]	Addressing problems where not all given constraints need to be satisfied, by introducing soft constraints which are adequate.	Introducing soft constraints and Fuzzy constraints to Constraint Satisfaction Problem (CSP) for which various tuples adjust the defined constraints to a different level. The heuristic search solving approaches to CSP can be performed by branch-and-bound search.
2001	"A genetic agent-based negotiation system" (Choi, Liu and Chan) [15]	e-commerce	GA-agent Automated Negotiation System (ANS). GA negotiation agents adapt to the user's preferences by dynamically modifying their mutation rate.
2002	"Using similarity criteria to make issue trade-offs in automated negotiations" (Faratin, Sierra and Jennings) [16]	General-Car dealer negotiation as an example	Linear algorithm which operates by leveraging Fuzzy similarity to estimate the preference framework of the negotiating party, subsequently employing a hill-climbing technique to navigate the spectrum of possible trade-offs and locate the most probable acceptable option.
2005	"Buyer-Supplier Negotiation by Fuzzy Logic Based Agents" (Cheng, Chan and Lin) [17]	e-marketplaces	The agent's negotiation tactics presented as Fuzzy rules also acquired through learning processes, enabling agents to infer from past offers and approximate the opponent's preference structure.
2006	"Intelligent agents for e-marketplace: Negotiation with issue trade-offs by Fuzzy inference systems" (Cheng, Chan and Lin) [18]	e-marketplaces	The agents employ tactics, explicitly formulating trade-offs as Fuzzy Inference System (FIS), which are then utilized to generate new offers in each round of negotiation.
2012	"Negotiation Agent with Adaptive Attitude Bidding Strategy for Facilitating Energy Exchanges between Smart Building and Utility Grid" (Zhu Wang and Lingfeng Wang) [19]	Smart and green buildings energy exchange	The negotiation agent, employing an Adaptive Attitude Bidding Strategy (AABS), explores different attitudes when determining bids. The negotiation model has been specifically crafted with the premise of bounded rationality rather than perfect rationality in mind.
2013	"A single-issue negotiation model for agents bargaining in dynamic electronic markets" (Ren and Zhang) [20]	e-commerce	The desire-based negotiation model facilitates adaptive negotiation behaviors in e-markets by taking into account both the e-market's conditions and the agent's preferences.

Table 1. *Cont.*

Year	Title/Author(s)	Commodity/ Application	Method
2013	"Adaptive Negotiation Agent for Facilitating Bi-Directional Energy Trading Between Smart Building and Utility Grid" (Zhu Wang and Lingfeng Wang) [21]	Smart Building and Utility Grid Bidirectional Energy Trading	AABS-optimized with PSO (PSO-AABS)
2014	"A new Fuzzy negotiation protocol for grid resource allocation" (Adabi et al.) [22]	Market-based grid resource allocation problem	Enhanced Market and Behavior Driven Negotiation Agent (EMBDNA) with a Fuzzy negotiation protocol aiming to improve Rubinstein's sequential alternating offer protocol and develop two novel Fuzzy Grid Market Pressure Determination Systems.
2014	"Sellers in e-marketplaces: A Fuzzy Logic based decision support system" (Kolomvatsos, Anagnostopoulos and Hadjiefthymiades) [23]	e-marketplaces	A decision-making model which utilizing a FL-based negotiation protocol, centering on the seller's behavior and implementing an FL-based decision-making approach without prior knowledge of the opponent's characteristics.
2014	"Development and evaluation of the many-to-many supplier negotiation strategy" (Chun-Ching Lee) [24]	Supply chain. Many-to-many supplier negotiation process under different order and competitive conditions.	The interactive bidding strategy discussed in this paper is a variation inspired by the prior research conducted by Lee and Ou-Yang [25] where a bidding strategy is proposed in a Supplier Selection Auction Market (SSAM). To fulfill its objectives of supplier selection under fluctuating order conditions, a party engaging in negotiation must attentively consider fundamental concepts such as negotiator's power, concession dynamics, target and reservation prices, time constraints, and the principle of equitable concessions. Additionally, this study integrates the competitive dimension into the evolution of bidding strategies.
2013	"Supporting Ill-Structured Negotiation Problems" (Roszkowska, Brzostowski and Wachowicz) [26]	Media Reference Model for electronic negotiation context, with four major phases of interaction, known as the Montreal Taxonomy.	Modified Fuzzy Technique for Order Preferences by Similarity to Ideal Solutions (FTOPSIS) and Analytic Hierarchy Process (AHP). AHP derives the negotiation issue weights instead of the classic approach of direct assignment of values to the issues. FTOPSIS is employed in constructing the final scoring system, enabling the assessment of all potential negotiation packages.
2015	"An adaptive Fuzzy logic system for automated negotiations" (Kolomvatsos, Trivizakis and Hadjiefthymiades) [27]	e-marketplaces	An Adaptive Fuzzy Logic System (AFLS) guides suitable actions throughout each negotiation round. Additional rules are integrated into the knowledge base through a model where the Simplex method assumes a central role. An FL controller dynamically adjusts the membership functions.
2015	"Application of Fuzzy TOPSIS to scoring the negotiation offers in ill-structured negotiation problems" (Roszkowska and Wachowicz) [28]	General buyer–seller example.	FTOPSIS facilitates the assessment of negotiation offers, creating a hierarchy of compromise solutions, formulating counteroffers, and assessing and contrasting potential concessions. Moreover, it aids in enhancing negotiation compromises by seeking Pareto optimal and dominant solutions.
2017	"A multi-demand negotiation model based on Fuzzy rules elicited via psychological experiments" (Zhan et al.) [29]	Political example. Negotiation over policies between two parties.	The extent to which a negotiating agent adjusts its preference structure based on risk, patience, and regret is determined through Fuzzy rules derived from psychological experiments.
2018	"An Atanassov intuitionistic Fuzzy constraint-based method for offer evaluation and trade-off making in automated negotiation" (Zhan, Luo and Jiang) [30]	Bussiness	An approach based on Atanassov intuitionistic Fuzzy constraints for evaluating offers throughout a negotiation.

Table 1. *Cont.*

Year	Title/Author(s)	Commodity/ Application	Method
2019	"Automated negotiation for ensuring composite service requirements in cloud computing" (Shojaiemehr, Rahmani and Qader) [31]	Cloud services	Mixed strategy ANS for Cloud Computing Service Negotiation (CCSN).
2019	"Real-time opponent learning in automated negotiation using recursive Bayesian filtering" (Eshragh, Shahbazi and Far) [32]	Two case studies: One about energy system planning decisions and the other concerning real estate purchasing negotiations	A new estimation technique is introduced for learning and modeling opponents' preferences in automated negotiations. In this technique, opponents' preference profiles are represented using Fuzzy functions.
2020	"Automated Negotiation for Peer-to-Peer Electricity Trading in Local Energy Markets" (Etukudor et al.) [33]	Electricity in local energy markets	An automated negotiation framework tailored for one-to-one P2P local electricity trading, employing the renowned Rubinstein alternating offers protocol.
2020	"Transactive Demand Response Operation at the Grid Edge using the IEEE 2030.5 Standard" (Fattahi et al.) [34]	A DR plan involving transactive negotiations within a residential network at the transformer level.	A Home Energy Management System (HEMS) agents' implementation using a Fuzzy Logic Controller (FLC) with real-world data from smart meters collected from ten anonymized residential customers.
2021	"Peer-to-peer energy trades based on multi-objective optimization" (Soriano et al.) [35]	P2P energy trading between prosumers connected to the main grid	A negotiation algorithm is proposed for interconnected DER to autonomously manage electrical exchange between producers and consumers. The algorithm combines a multi-objective BA, a Pareto front, and an Fuzzy Decision Making (FDM) approach.
2022	"The Intuitionistic Fuzzy Framework for Evaluation and Rank Ordering the Negotiation Offers" (Roszkowska) [36]	General Negotiation Model	An Intuitionistic Fuzzy Ideal Reference Point (IFIRP) multicriteria method is employed to evaluate and rank multi-issue negotiation offers, accommodating natural language, subjective considerations, and the uncertainty or imprecision in the evaluation of negotiation offers.
2023	"Algorithm for Energy Resource Selection Using Priority Degree-Based Aggregation Operators with Generalized Orthopair Fuzzy Information and Aczel-Alsina Aggregation Operators" (Akram et al.) [37]	Energy resource selection problem entailing the evaluation of various energy sources from a finite list to identify the most optimal option.	A Multi-Criteria Group Decision Making (MCGDM) technique introducing Aggregation Operators (AO) utilizing q-Rung Orthopair Fuzzy (q-ROF) information and Aczel-Alsina Triangular Norm (AATN) and Aczel-Alsina Triangular Co-Norm (AATCN).
2023	"A review of PSO-based algorithm application in Demand Side Management: Challenges & Opportunities" (Anishah et al.) [38]	DSM in electricity distribution	An examination of the utilization of PSO-based algorithms in DSM domains under certain constraints.

3. Discussion

Comprehensive Review of the Selected Studies

In [14], the author introduces soft constraints that adequately address problems where not all of the constraints need to be satisfied. More specifically, constraints or sets of constraints are compared on the basis of certain criteria, such as, for example, weights, which indicate their importance which is tied to given crisp importance. These "soft" constraints generalize conceptually the idea of crisp constraints. Since crisp constraints are characterized by their according set of tuples for which the constraints apply, it is entailed that Fuzzy sets characterize the according Fuzzy constraints, where different tuples satisfy on a different degree a given constraint. A generalization of classical solving heuristics which are used in classical CSP can be used to control the solution construction process of the Fuzzy CSP solving. Branch and bound search can replace the abovementioned heuristic search.

This study provides novelties such as a good and general framework of mathematics defining Fuzzy constraints, Fuzzy Constrained Satisfaction Problems (FCSP), and the degree of satisfaction of solutions. It also shows that the classical solution method can also be applied to FCSP cases. An example of a robot facing the problem of choosing matching pieces of clothing when trying to get dressed can be illustrated as a FCSP. The level of compatibility among potential combinations is represented by the satisfaction level of corresponding binary Fuzzy constraints. This collective satisfaction adheres to the principle of "production combination".

The suggested heuristics can be implemented with any definition of solution degree that meets the criteria of independent computation and monotonicity. Additionally, further enhancement of the branch and bound search is possible by integrating heuristics for prioritizing constraint verification.

In [15], a GA-agent ANS with application on e-commerce is proposed, in order to minimize the required user input. The negotiation model studied involves a bilateral, multi-issue negotiation. Multi-issue negotiation involves reaching an agreement encompassing multiple terms. The system acquires knowledge of the opponent's preferences through observing counteroffers and adjusts to the environment by dynamically modifying its mutation rate. The proposed structure proactively anticipates user needs, initiating a purchase process. The agent proposes a set of default parameters and seeks confirmation from the owner, considering the user's profile and schedule. Negotiation can then adopt a suitable strategy based on the deadline and user preferences.

The novelty of the suggested structure is that, unlike other GA-based negotiation agents, the proposed agent attempts to learn the opponent's preference using dynamic mutation rate modification following a rule wherein the mutation rate is directly linked to the logarithm of the population size but inversely related to the goodness value. Therefore, as the goodness value increases, a lower mutation rate should be employed. The proposed structure can be used as an online ANS.

In [16], a linear algorithm is presented to tackle the complexities of multi-issue negotiation, including the challenging task of balancing decision variables. This algorithm empowers agents to navigate trade-offs involving both discrete and continuous decision variables, particularly in scenarios involving information uncertainty and resource constraints for multi-dimensional commodities. Employing Fuzzy similarity, the algorithm approximates the opponent's structure of preferences in the negotiation and employs a hill-climbing technique to investigate the possible range of trade-offs, aiming to identify the most probable acceptable solution. Despite the algorithm's applicability being limited to linear problem domains, the abstract structure of the underlying similarity model serves as a component of the overall negotiation algorithm, making it adaptable for use by any negotiating agent. The algorithm has undergone not only theoretical analysis to assess its complexity but also empirical evaluation to gauge its operational performance. Theoretical analysis has revealed that the algorithm's average complexity is linearly correlated with the number of negotiation decision variables under consideration.

The empirical findings, using a car dealing negotiation as an illustration, showcased the algorithm's efficacy in generating trade-offs across various negotiation scenarios. Notably, it highlighted the correlation between the algorithm's exploration of potential outcomes and its capacity to yield agreements with enhanced joint gains. This amplified search for higher joint outcomes in each iteration of the algorithm, whether in a singular or multiple environment, also led to increased communication costs due to the necessity for more proposals before reaching an agreement. The crafted computational model using heuristics to represent trade-off strategies holds the potential to augment the social welfare of the system.

The opponent's preference structure is also used in [17], as the historical data from which the system learns. This study outlines the negotiation tactics of agents within e-marketplaces, presented as Fuzzy rules, wherein agents learn from past proposals aiming at estimating the opponent's preference framework This approach aims to improve subsequent negotiation rounds by enabling the formulation of more promising offers. Specifically, FIS are employed to conduct heuristic searches, facilitating the learning of opponent issue weights. Furthermore, agents' trade-off strategies are explicitly articulated as FIS, allowing for the generation of new offers in each negotiation round based on issue importance levels and the opponent's previous offer.

The experimental results indicate that the performance of the proposed approach meets satisfactory criteria regarding negotiation duration, the attained joint utility, and the Pareto efficiency of the negotiated agreements. Nevertheless, the authors propose the necessity for a more refined learning scheme to improve the accuracy of the agent's predictions regarding the opponent's preference structure.

In [18], the agents were categorized into two groups: matching agents and negotiation agents. Matching agents pair buyers and suppliers by identifying the M most analogous proposals to each participant in an e-marketplace, sorting the number of the opponents who a negotiation agent has to negotiate with, to a few who are considered the most promising. This method operates on the premise that dyads with a high matching degree are more likely to efficiently reach agreements in subsequent negotiations. This reduces the occurrence of fruitless negotiations, thereby enhancing the success rate of contracts. When matched agents inform a participant about a set of potential opponents and their offers, the participant activates its negotiation agent to engage with each of them. The agent can either decline an opponent's offer or present a counteroffer. The strategies employed by agents during trade-offs are structured as FIS, enabling the generation of new offers in each negotiation round.

The experimental findings illustrate the effectiveness of the proposed ANS regarding the number of exchanged offers, the achieved joint utility, and the Pareto efficiency of negotiated contracts. However, this study's automated negotiations did not account for the quantities demanded by buyers or provided by suppliers. In practice, sellers supply products to multiple buyers, while buyers may distribute their demand across various sellers. The quantity aspect complicates the negotiation decision-making process as it interacts with factors like price. Although the proposed approach enables agents to negotiate with multiple opponents simultaneously, each negotiation is treated independently. It could be more advantageous if agents could leverage information from one negotiation into another. The utility function employed in this approach was relatively simplistic and rigid, influencing the outcomes of automated negotiations. To prevent unfavorable contracts, a minimum acceptable value for each issue is established. In a real Business-to-Business environment, a more sophisticated utility function should be adopted to enhance the utility of the proposed approach.

A study on a topic relevant to SG and energy market concepts is presented in [19]. An AABS is devised and assessed concerning the energy exchanges between an integrated smart building and a utility grid system. The building adopts RES and ESS technologies and acts as a seller when the renewable energy has a surplus after supplying all demands. The agent bridges the smart building and the utility power grid. A Fuzzy controller is employed to estimate the necessary power under dynamic circumstances. The proposed model is not based on AI but on mathematical models expressed by equations. The strategy is adopted by a negotiation agent which explores various attitudes in bid determination. An important factor of the negotiation model is the "eagerness" of the involved sides to reach an agreement, behaving as human traders. For example, if the eagerness factor has a large value, the agent aims to accomplish the transaction within the set deadline at any cost, provided it aligns with the reservation price. Conversely, a low eagerness value suggests that the agent is willing to risk not making the sale by the deadline, opting instead to wait for a higher selling price.

Comfort level and the battery's State Of Charge (SOC) are dynamic factors affecting the outcome of the negotiation. The agents adapt their behavior according to real-time situations and should be efficient computationally, relying on the hypothesis of bounded instead of perfect rationality.

The proposed method offers customers more logical trade-off solutions by taking into account various dynamic factors across the entire system and emulating human behavior. The efficiency of the model proposed in making sound negotiation decisions is evident in its ability to lower customers' energy costs or enhance their economic profit. An enhancement of [19] is discussed in [21], where an AABS has been formulated and assessed for potential deployment within a cohesive system combining smart building technology and utility grid integration; the negotiation agent utilizes an adaptive strategy to explore various approaches in determining bids. However, a significant drawback of the AABS-based negotiation agent is its heavy reliance on assumptions regarding the negotiation environment, thus complicating generalization. A PSO-AABS negotiation agent is developed to eliminate the aforementioned limitation. The proposed agent approaches the tendency of the traders to maximize their own payoffs while ensuring the achievement of the agreement.

The models demonstrate efficacy in generating rational negotiation choices while simultaneously diminishing customers' energy expenses or enhancing their economic gains.

Intelligent technologies such as GA and NN can be utilized in the advancement of the negotiation agent.

By comparing options, the most appropriate intelligent technology for negotiation can be identified. The AABS-based negotiation model simulates traders' eagerness and behavior, employing a mathematical function for the representation and computation of eagerness. Dynamic factors like comfort level and battery SOC significantly influence the negotiation's outcome.

In [20], a desire-based negotiation model is proposed based on mathematical models. Again, the agents' behavior is influenced by their desire to complete the negotiation, similar to the eagerness concept in [19]. The model facilitates agents in exhibiting adaptive negotiation behaviors within e-markets, considering both the e-market's context and the agent's preferences. Additionally, a negotiation protocol is introduced to outline the negotiation process within e-markets.

The experimental results convincingly showed that the proposed model adeptly adapts to changes in the e-market's conditions, thereby modifying agents' negotiation behaviors accordingly.

Furthermore, drawing from the experimental results, the concept of a "trading surface" is introduced, highlighting the enhanced applicability of the proposed negotiation model compared to traditional models in e-markets. Two significant aspects of the study stand out: Firstly, the investigation into expanding the current model from single-issue negotiation to multi-issue negotiation, integrating both numerical and categorical factors. Secondly, the investigation into leveraging Fuzzy and NN approaches to bolster agents' adaptive negotiation behaviors.

A resource management of shared computing resources protocol is presented in [22], where negotiation between Grid Resource Owner (GRO) and Grid Resource Consumer (GRC) takes place to successfully reconcile the differences between GRO and GRC.

The proposed protocol emphasizes enhancing the alternating offers protocol by introducing two new Fuzzy decision controllers. One controller models the criteria of GRO, while the other models those of GRC.

Again, the concept of eagerness is underlying the concept of Grid Market Pressure (GMP). The study's distinguished features are the presentation of the GMP value determination approach, in which sets of Fuzzy rules are employed to furnish negotiation agents with a more precise GMP value, specifically in terms of the degree of relaxation. Moreover, the mentioned work introduces the EAlternating offer protocol, an improvement upon Rubinstein's sequential alternating offer protocol proposed in [39,40]. This protocol is designed to handle multiple trading opportunities and market competition, ensuring control over invalid behavior and relaxation of bargaining criteria for negotiation agents based on the GMP value. Finally, this study proposes a new design of negotiation agents, namely EMBDNA augmenting Market and Behavior Driven Negotiation Agents (MBDNA) proposed in an older work of the authors [41], which were not adopting the proposed Fuzzy negotiation model and were not relaxing their bargaining term in intense GMP conditions with the proposed negotiation protocol.

The experiments conducted for the evaluation of the performance of GRC under various market types, densities, grid loads and deadlines, demonstrated that GRC EMBDNA took shorter negotiation time on average while having higher success rates, and achieved higher expected utilities than MBDNA and GRC Enhanced Market Driven Agent (EMDA), a negotiation agent design which adapts its bargaining terms when faced with significant GMP, considering two relaxation factors. Furthermore, GRO EMBDNA typically exhibited shorter average negotiation times, higher success rates, and superior resource utilization levels compared to both MBDNA and EMDA counterparts.

A decision-making model is proposed in [23], which embraces an FL-based negotiation protocol tailored for electronic marketplaces. In a bilateral single-issue scenario, a buyer and a seller engage in exchanging offers with the aim of reaching an agreement on the price of a product. The emphasis is on the seller's behavior, and an FL-based decision-making scheme is adopted. Entities have the discretion to determine when offers should be accepted within a predefined framework. FL demonstrates capabilities which utilized without any prior knowledge of the opponent's characteristics. "Zero knowledge" entails that players are unaware of crucial parameters defining the opponent's behavior, such as the deadline, utility function, and negotiation strategy. The focus of the study is on the seller's decision-making process. The seller leverages the FL system to ascertain the deadline for each interaction and determine the appropriate course of action in every negotiation round.

The evaluation of the efficiency of the proposed system is performed in terms of the overall number of agreements across multiple negotiations, the duration needed to conclude each negotiation, and the seller's utility, the experiments indicate that the proposed system delivers notable efficiency for the seller.

A many-to-many supplier selection negotiation strategy is presented in [24] with respect to supply chain application. In this form of negotiation, each participant competes with others within the same tier of the supply chain to secure a cooperative partner. Much of the negotiation environment information remains undisclosed, except for public data like bid prices from opponents in each round and the current lineup of participating suppliers and demanders. The interactive bidding strategy design is a variation inspired by earlier research conducted by Lee and Ou-Yang [25]. Furthermore, alongside established negotiation concepts such as negotiator's power, concession, target price, reservation price, time pressure, and the principle of equal concessions, this paper extends its analysis to include the competitive factor in the formulation of bidding strategies. Specifically, the research leverages the quantities of suppliers and demanders to assess the competitive dynamics within the negotiation process, and integrates these quantitative measures into the design of the interactive bidding strategy.

The examination of experimental findings elucidates the distinct impact of individual decision variables on each parameter of negotiation performance, offering guidance to demanders in the pursuit of their supplier selection objectives across diverse negotiation scenarios. Due to the constrained information available during the negotiation process, this study highlights the competitive conditions by considering the current numbers of participants on both sides. These guidelines prove effective in providing decision support to demanders facing intricate and practical many-to-many supplier negotiation situations. The authors found it judicious to utilize Fuzzy Set Theory to introduce Fuzziness into the intricate and ambiguous negotiation conditions prevalent in many-to-many supplier scenarios. This approach facilitated the development of a Fuzzy ES tailored for enhanced practical applicability.

The AHP and FTOPSIS combination for the analysis of the negotiator's preferences, as well as the application of this approach to managing ill-structured negotiation challenges is the major contribution of the study in [26], which proposed a negotiation decision-making model designed to support negotiation problems which are ill-structured, based on the AHP and FTOPSIS techniques. The AHP technique is employed to scrutinize the structure of the negotiation conundrum and ascertain the significance weights of the negotiation issues while FTOPSIS allows the analysis of the negotiation problem representing the different kinds of data as Triangular Fuzzy Numbers (TFN) that can be aggregated further, giving the scores of any feasible package. This can be accomplished due to the mathematical characterization of offers, which entails the conversion of ambiguous expressions into Fuzzy numbers. These Fuzzy numbers facilitate straightforward arithmetic operations on offer evaluations. Furthermore, the evaluation of any package at any time during the intention negotiation phase by the negotiator in terms of overall performance is an extremely useful tool for the support of the negotiation. The proposed scoring system formation is resistant to new packages leading to a stable scoring system. FTOPSIS is computationally efficient and straightforward, while the AHP method elicits the importance of the issues rather than assigning them directly as usual. The iterative exchange of offers and counter-offers allows to reach an agreement which may be improved further during the post-agreement phase using the Nash and Kalai-Smorodinsky concepts on bargaining solutions. FTOPSIS in ill-structured negotiation problems is also analyzed in [28], as an applicable procedure to support certain tasks in negotiation, considering subjective and imprecise judgments of the negotiators. Again, FTOPSIS is used to evaluate the negotiation offers, to build the ranks of the compromise solutions, to construct the counteroffers, and to evaluate and compare the size of potential concessions. Also, it can support the improvisation of the achieved compromise by the negotiators by searching for the Pareto optimal and dominant solutions. The ability of FTOPSIS to deal with different types of values such as crisp, interval, Fuzzy, or linguistic. The FTOPSIS algorithm allows the transformation of data to a common domain of TFN and uses Fuzzy arithmetic to perform all required calculations. Again, the negotiation space for FTOPSIS calculations should be defined differently than the traditional approach, in order to avoid instabilities of the negotiation problems resulting in rank reversal. The initial negotiation space is broadened using TOPSIS-based maximum and minimum solutions, which include utopian packages that are unlikely to occur during the negotiation process, chosen on purpose by the negotiator. Nevertheless, this approach has drawbacks since the ideal and non-ideal solutions represented by max and min do not eliminate the instability problem entirely; extreme offers might still lie beyond the bounds of the feasible negotiation domain. Three approaches are suggested to deal with this problem, with the two using the TOPSIS-specific notion of distance measuring while the third one introduces the notion of deviation, shifting the viewpoint from which offers are assessed. These methodologies are not interchangeable nor universally applicable to any negotiation scenario. Their suitability hinges on the negotiator's philosophy regarding offer evaluation and the robustness of their reference points, which sets the authors' future research goals as the building of a negotiation support software and undertaking a series of experiments to determine the contextual circumstances under which each of the proposed approaches is chosen by negotiators and the underlying reasons behind these choices. In [36], a method named IFIRP multi-criteria method which belongs to the class of algorithms based on reference points such as TOPSIS is proposed; negotiation offers are assessed and ordered using an intuitionistic Fuzzy framework. Under the assumption that package selection operates within a bipolar framework, taking into account both their advantages and disadvantages across specific issues. Options' evaluation can be represented by membership and non-membership grades, respectively. Moreover, an intuitionistic Fuzzy method is proposed to apply to the Multi-Criteria Decision Making (MCDM) problem, based on Hellwig's concept of reference point to alternatives' ranking. This method is also applied to evaluate negotiation offers. The IFIRP method offers the advantage of accommodating natural language, subjective considerations, and the handling of imprecision or uncertainties in the evaluation of negotiation offers. It presents a potential alternative multi-criteria approach for supporting negotiators, assisting them in mitigating biases and heuristics during the evaluation of negotiation offers. The proposed method—according to the author—needs to be empirically verified about the usefulness of intuitionistic framework for negotiation support. DM should also testify about the behavioral decision-making aspect of the proposed approach. Important issues are also the development of techniques for the assessment of satisfaction and dissatisfaction levels for the negotiation options, and finally, the recognition of the reference point's impact or the distance measure for the obtained ranking's stability.

An AFLS that participates in electronic marketplaces is presented in [27]. The AFLS is responsible for determining suitable actions in each round of negotiation. It relies on five input parameters and allows for the incorporation of new Fuzzy rules as needed, as well as the adjustment of membership functions. The simplex method plays a key role in the addition of new rules to the knowledge base, while an FL controller is responsible for the adaptation of the membership functions. Since this method lacks a reconstruction phase, the buyer can collect knowledge during successive negotiations, even if they refer to different items, in the same manner as humans do in real scenarios. The proposed model does not require complex modeling of the behavior of the buyer. It is not necessary to define a path to equilibrium or to perform complex calculations covering all the aspects of a negotiation. It generates, in addition, actions that fully adapt to the negotiation information. Furthermore, it does not require a training process.

The negotiation model necessitated an effective decision-making mechanism capable of managing uncertainty. FL stands out as a promising solution for this task due to its ability to handle uncertainty. FL systems utilize a Fuzzy rule base that addresses various aspects of an entity's behavior, with this rule base typically defined with input from experts. Nevertheless, it remains challenging for experts to anticipate and cover all potential cases that an entity may face during automated negotiations.

In [29], a multi-demand negotiation model based on Fuzzy rules elicited via psychological experiments with an example in politics, where two parties negotiate over policies is presented. The model engages in multi-demand negotiations within discrete domains, effectively capturing human psychological traits such as risk aversion, patience, and regret. This is achieved through the utilization of Fuzzy rules to determine the extent to which a negotiating agent should adjust their preference structure, as elicited from psychological experiments.

The study elucidates how human psychological traits pertaining to risk, patience, and regret influence alterations in preference structures throughout the negotiation process, and also identifies the conditions conducive to reaching an agreement in bilateral negotiations. Fuzzy rules were deployed to represent degrees of "risk", "patience", and "regret" factors, and a standard Mamdani method of FL inference was utilized to dynamically calculate the extent to which an agent's preferences should be adjusted based on factors such as risk, patience, and regret during the negotiation process. Empirical analysis was conducted to determine the influence of attitudes towards risk on negotiation outcomes and to demonstrate the superior performance of the FL-based model in terms of negotiation success rate, efficiency, and quality compared to a well-known model. Furthermore, the model's efficacy in resolving a negotiation problem within the political domain is illustrated.

In [30], Atanassov intuitionistic Fuzzy constraints-based approach for offer evaluation during the course of a negotiation is proposed. According to the study, an offer is evaluated through three steps: First, prioritized Atanassov intuitionistic Fuzzy constraints are used to characterize the agent's goals. Next, the overall satisfaction and dissatisfaction degrees of an offer according to the beforementioned constraints reflect the hesitation degrees of the users when estimating the satisfaction degrees of all relevant constraints. Lastly, the agents' acceptability of an offer is calculated with aggregation of both overall satisfaction and dissatisfaction as well. Calculating the acceptability of the offer leads to the determination of the agents' decision to accept an offer or deny and make a counteroffer instead. Atanassov intuitionistic Fuzzy constraints can represent human-like behavior during a negotiation process, like the goals they have during the process, the human cognitive limitations, and the knowledge that leads to the hesitation to accept an offer. The contribution of this tradeoff method lies in its consideration of not only the similarity to the previous offer when crafting a counteroffer but also the satisfaction and dissatisfaction levels pertaining to its most prioritized goal. Moreover, in this method, the weights are set to embody the disparities in opinions on the importance of these three factors between different users.

Furthermore, it is necessary to develop diverse forms of satisfaction and dissatisfaction functions to represent the individual characteristics of users and to analyze how these variances impact the results of a negotiation. In this study, the functions representing satisfaction and dissatisfaction degrees concerning various negotiation objectives are presumed to be defined by users of negotiating agents or chosen from a predetermined function set within a negotiation system. Establishing the functional expression of Fuzzy constraints, or acquiring Fuzzy constraints, poses a critical challenge that must be addressed when implementing the proposed model into practical negotiation systems.

Two examples from the business sector where buyer agents in an accommodation renting scenario illustrated how the model works, showing that this method can also be applied in both cases of independent and interdependent multi-issues. As a result, integrating various concession strategies and the proposed trade-off approach into a single negotiation system may be beneficial. When integrated with the suggested trade-off decision-making method, a negotiating agent may identify a suitable compromise before conceding utility, thereby enhancing the likelihood of reaching an agreement. Consequently, in scenarios with numerous alternative trade-off proposals, it becomes necessary to devise an algorithm capable of swiftly identifying the most suitable trade-off within a multi-issue negotiation environment.

A mixed-strategy ANS for a CCSN is recommended as a strategy in [31]. In this study, a new negotiation strategy was designed and simulated aiming to maximize utility for both trading parties while the speed in reaching an agreement is increased. Moreover, a process for the aggregation of the results of the negotiations on simple task requirements to ensure end-to-end composite service requirements is proposed. Proposals are generated based on factors such as remaining time, market competition, and opponent's behavior which are environment-related factors, and preferences and utilities which are negotiator-related factors. These factors also can be divided into two categories considering their effect on the proposal generation, namely concession and tradeoff factors. A simple FIS was devised to gauge the appropriate level of concession, considering the preferences and utilities of negotiating parties, thereby striking a balance between preferences and achieved utilities to ascertain the extent of concession. Sugeno-type FIS is preferable because avoids time-consuming defuzzification in FIS which has Fuzzy consequences in its rules. The opponent's behavior is estimated using a tradeoff algorithm based on Fuzzy similarity, where a proposal that has more similarity with the opponent's proposal has a better chance of being accepted by the opponent. As a mixed strategy, the proposed framework maximizes the utilities of the negotiating parties in the shortest possible time. Simulation results show that negotiator-related factors, market competition, and opponent's behavior affect the performance's enhancement.

Opponent learning in real-time automated negotiation is the concept of the methodology presented in [32]. In the context of multi-participant and multi-issue negotiation, an estimation technique based on Fuzzy functions models the opponents' preference profiles, since Fuzzy functions are very close to the human's way of evaluating alternatives. During the negotiation, the agents learn the parameters of these models in real time recursively, using a Bayesian approach based on Unscented Particle Filtering (UPF). In UPF, the posterior distribution of the state is recursively estimated by a set of particles (samples) that evolve over time, conforming to the system's transition and observation models. In each round, the proposer agent offers a proposal to stakeholders. Stakeholders respond to the proposer by sending feedback which includes a Fuzzy score assigned to the proposal and some arguments about the proposal. Using the feedback, the preference model is recursively estimated with the UPF approach. The estimated preference model is used by the proposer to find the next proposal that is closer to opponents' preferences while at the same time meeting his criteria as well, enabling a Belief Propagation Proposal Preparation (BPPP) module. Therefore, the proposals in each round have a higher probability of being accepted by the stakeholders. This technique facilitates the process of negotiation and accelerates reaching an agreement. The proposed methodology was compared to a frequency-based approach using two different case studies: The planning of an energy system in Alberta and in real estate sales in King County, where the UPF methodology accelerated the process to 85% when realistic knowledge of the users was available and up to 95% when no prior knowledge about the users was available.

The uncertainty of the estimation is considered by the proposed approach which makes it capable of having a logical pattern in offering proposals. For example, when stakeholders show strong concerns about an attribute, the BPPP module opts for a new proposal significantly divergent from the preceding one. Moreover, the effectiveness of the proposed estimation technique relies on the initial approximations of the opponents' preferences. Enhanced initial estimations result in faster convergence of the UPF towards the actual preference profile of the user.

The authors' future work aims to provide more accurate initial approximations using natural language processing techniques. Also, another future goal is to use the auxiliary data to generate pseudo-observations for the UPF module, improving the estimations even more.

A one-to-one automated negotiation framework applied in local energy markets is proposed in [33] for P2P local trading of electricity. This framework utilizes an autonomous agent model to encompass the preferences of both sellers and buyers regarding the price and quantities of electricity to be traded across various periods within a day. A bilateral negotiation framework is established, drawing from Rubinstein's alternating offers protocol. This framework aggregates the electricity quantity and price for different time periods into daily packages, which are then negotiated between the agents representing the buyer and seller.

Buyer and seller agent models are defined by utility functions. The study examines three negotiation strategies: Zero Intelligence, Boulware, and linear heuristic. The heuristic strategies select an offer from a narrowed set of viable packages. The proposed agents, employing these strategies and the negotiation protocol, were deployed in a case study involving a prosumer equipped with a solar home system and a battery, as well as various consumers with diverse energy and price preferences. Negotiations occur sequentially between the seller and individual buyers. Simulations within this framework, coupled with agent modeling, enable prosumers to enhance their revenues while facilitating affordable electricity access for the community.

Linear heuristic strategy proved to provide similar results with Boulware but in a shorter time.

The penetration of DER and deferrable loads such as EV introduces new challenges in the operation and control of modern power systems. In [34], an event-driven autonomous distributed architecture with real-time communication between a Transactive Agent (TA) communicating with multiple HEMS agents in real time where the residential customers possess generation and storage assets as well as diverse loads is introduced. For this reason, a Transactive Demand Response (TDR) algorithm is deployed. HEMS agents negotiate with the TA to determine their best response to the DR request, using a Fuzzy inference engine to satisfy the users' comfort levels and preferences in responding to rewards offered by the TA. The DR scheme is meticulously designed to adhere fully to the IEEE2030.5 interoperability standard, enabling customers to choose their DR compliance level and offering incentives to optimize their engagement. Real-world smart meter data from Hydro Ottawa's service area was utilized to model customer behavior. Employing k-means clustering, customers were categorized, and this information was leveraged by the TA to initiate the transactive negotiation process with each HEMS agent. The TDR algorithm integrates a flexible reward function to accommodate diversity in both power consumption and TDR compliance, facilitating a more precise quantification of customer behavior.

TA engaged in negotiations with customers' HEMS agents utilizing event-based optimization techniques and a reward mechanism. The agents' response to DR requests led to a reduction in consumption, aligned with compliance levels linked to their comfort preferences, as articulated through Fuzzy rule bases. The findings indicate the effectiveness of the proposed scheme, suggesting that full capability is a viable approach for vendors and solution providers to realize the SG concept.

A negotiation algorithm is proposed in [35], where users and end-users are modeled as prosumers. A multi-agent scheme enables the modeling of each user as a generator, consumer, or both and manages energy trading within the local market. The multi-agents scheme consists of a Multi-Objective Optimization algorithm, a Pareto front, and an FDM algorithm. The Multi-Objective Optimization algorithm evaluates several scenarios of buying or selling energy to or from the main grid and other agents. Each electrical load profile and electrical producer profile is treated as a decision variable and assessed within an objective function, aimed at maximizing earnings from electrical energy sales while minimizing high-cost electrical energy purchases. The evaluation outcomes are represented in the Pareto front to determine the optimal participation strategy in a free market, facilitated by the FDM algorithm. Here, Fuzzy sets are incorporated through partial values of elements defined via membership functions. A Pareto front solution is conceptualized as a continuous, monotonically decreasing function, akin to a membership function, aiming to identify the optimal solution in the optimization process. Each membership function is characterized on a scale ranging from 0 to 1, denoting compatibility and incompatibility, respectively. Hence, the solutions identified through the Pareto front graph can be characterized as a membership function. The FDM determines the decisions regarding buying or selling electrical energy.

The proposed algorithm, employing the Multi-Objective Bat Algorithm (MOBA), was compared with a Multi-Objective Particle Swarm Optimization (MOPSO) algorithm across six instances of buying and selling electrical energy.

The findings indicate that, in three out of six cases, the proposed algorithm achieves the lowest costs for purchasing electrical energy and maximizes agent profits from electrical energy sales in the market. In one case, both algorithms act similarly.

The suggested methodology holds the potential to facilitate autonomous energy transactions and foster the development of an intelligent energy market. Subsequent endeavors should focus on analyzing and substantiating the optimality and convergence of the proposed algorithm.

In [37], q-ROF set information is used on the prioritization of AO which is based on the AATN and AATCN. q-ROF sets cover wide information, therefore they are more flexible and suitable. The study proposes two new AOs based on q-ROF, AATN, and AATCN. q-Rung Orthopair Fuzzy Aczel-Alsina Averaging (q-ROFPAAA) and q-Rung Orthopair Fuzzy Aczel-Alsina Geometric (q-ROFPAAG) operators are proposed to involve the information's priority weights. An MCGDM algorithm was developed based on the proposed operators since the prioritization of attributes and experts is of crucial importance in decision-making problems. MCGDM algorithm was applied to a real-life problem concerning the selection of energy resources and the proposed concepts compared with various q-ROF-based AO. The rival AO solved the example but failed to utilize prioritization, showcasing the contribution of the proposed approach. Additionally, AO relying on Pythagorean and Intuitionistic Fuzzy Sets, failed to solve the given problem highlighting the diverse nature of q-ROF sets and the proposed operators.

In conclusion, q-ROFPAA operators exhibit superior performance compared to existing AOs due to their consideration of prioritization relationships during information aggregation. Additionally, they have the capability to encompass a broader spectrum of information, unlike other operators that neglect certain data. Moreover, the proposed operators represent a generalization of the previous AO.

Lastly, in [38], a review of the application of PSO-based algorithms in DSM fields discusses the challenges of previous work. The variants of PSO are used to address the high complexity and uncertainties of the DSM modeling within the acceptable timeframe while considering user comfort. New prospects are discerned, paving the way for the advancement of PSO methods in forthcoming research initiatives.

In a systematic classification according to the application of PSO variants in DR programs, the Binary Particle Swarm Optimization (BPSO), MOPSO, and Enhanced Leader Particle Swarm Optimization (ELPSO) variants of PSO algorithm are proposed as suitable in RTP strategies. Specifically, BPSO qualifies in P2P trading between prosumers and consumers. Although the study is limited in DSM, the study establishes the significance of MA as a prominent component of AI in addressing optimization challenges. These methodologies demonstrate a level of precision suitable for addressing a wide array of optimization problems, yielding near-optimal solutions with minimal computational burden.

Their efficiency surpasses that of classical approaches due to their capacity to traverse the search space, approximating a globally optimal solution. MA are applicable to problems featuring numerous decision variables and can be seamlessly adapted to scenarios with multiple constraints, including real-world engineering design problems.

4. Conclusions and Remarks

The present work attempts to focus on an important aspect of the SG concept and provide a critical review of selected works which, in the authors' opinion, may give scholars and researchers the spark to adapt the presented methodologies and concepts to their work, or to enhance the reviewed methods with new features and approaches.

Some interesting points were:

- The concept of imitating eagerness and other human-like attributes like hesitation, the willingness to risk, etc. with Fuzzy sets.
- The dominance of FL in tools able to learn tactics, and make trade-offs.
- The Intuitionistic Fuzzy sets stability and applicability in ill-structured problems like negotiations.
- The Bayesian UPF approach which offers flexibility to the negotiation offers in [32].
- K-means clustering customer's classification and Fuzzy inference engine to satisfy customer's preferences in response to rewards offered by the TA in an SG context as described in [34].
- The general description of a system applied in energy trades and the use of novel metaheuristic algorithms such as BA in [35].
- The q-ROFPAA OA dominance in comparison to other q-ROF-based OA in [37].
- The potential of integrating FL, Bayesian approaches, UPF, and NLP techniques into negotiation frameworks, where agents can enhance the robustness and reliability of negotiation methods in coping with the variability and uncertainty of renewable energy sources.

In most cases, AI approaches were implemented, while in others, AI was partially employed or had the potential to integrate with the presented methodology. Some may apply to one or two of the cases referred to in Section 2, while some may, with the proper arrangements, apply in all three cases. In Table 2, a summary is given.

Table 2. A summary of the presented works and their potential use.

Reference	Method	Comments and Suggestions
[14]	Fuzzy sets to characterize Fuzzy constraints instead of crisp. Branch and Bound search to control the solution construction process of the Fuzzy CSP solving.	A metaheuristic approach may replace the Branch and Bound search.
[15]	GA is employed to train dynamically the agent to adapt on the opponent's preference.	Other metaheuristic approaches may be used instead of GA.
[16]	Fuzzy similarity approximator to approximate the opponent's structure of negotiation. Hill climbing technique to explore the potential trade-off space for the most possible acceptable solution.	Hill climbing technique may be replaced by metaheuristic approaches.

Table 2. *Cont.*

Reference	Method	Comments and Suggestions
[17]	FIS are used to learn the opponent's tactics. FIS are also used to formulate trade-offs.	Since the authors suggest that a more delicate learning scheme is required to enhance agent's prediction accuracy, an Adaptive Neuro-Fuzzy Inference System (ANFIS) or a metaheuristic algorithm to tune the FIS may apply. ANFIS are well known for their predictive accuracy along with Neural Networks (NN).
[18]	FIS formulate the tactics of the agents when making trade-offs and used to infer new offers at each negotiation round.	The authors suggest a more flexible utility function in order to be this approach more useful in real Business-to-Business environment.
[19]	Fuzzy controller calculates the required power under dynamic circumstances. The model is based on mathematical functions and not AI except the controller.	The model may be replaced by an AI approach.
[21]	Enhancement of the model in [19], integrating PSO to overcome the difficulty the previous approach had in generalization.	The authors suggest GA and NN technologies to employ, and the most suitable of them to be found by comparison.
[20]	The proposed approach didn't include any AI implementation, relying on mathematical models.	The authors consider to employ Fuzzy and NN approaches to enhance the agent's adaptive negotiation behaviors.
[22]	Two Fuzzy decision controllers were designed. Fuzzy rules used to provide negotiation agents with accurate GMP values.	According to authors, the study faces two challenges; First, to evolve the negotiation agents' structures by learning new relaxed criteria Fuzzy rules as they participate in more negotiations and secondly, to adopt the negotiation protocol to support coalition formations.
[23]	Fuzzy rules-based negotiation protocol adopting FL decision making scheme. FL determines the deadline of each interaction and the appropriate action at every round of negotiation. FL capabilities are exploited to learn the opponent's behavior in zero knowledge environment.	MA such as GA may be applied to adapt to opponent's behavior.
[24]	No AI implemented	The authors propose the utilization of Fuzzy Set Theory to introduce fuzzification to the intricate and ambiguous many-to-many supplier negotiation conditions. This approach aims to develop a Fuzzy ES conducive to more pragmatic applications.
[26,28,36]	FTOPSIS and AHP. A method which belongs to the class of algorithms based on reference points such as TOPSIS is also suggested in [36], based on IFIRP. FTOPSIS represent the different kinds of data, and to evaluate the negotiation offers, giving scores to any feasible package. It also constructs the counteroffers and evaluate and compare the size of potential concessions. Also, it can search for Pareto optimal and dominant solutions. IFIRP is the framework where negotiation offers are assessed and ordered applying on a MCDM problem.	FTOPSIS has the ability to deal with different types of data, such as crisp, interval, Fuzzy or linguistic. It is computationally efficient and straightforward. Although the system is scoring system is stable, extreme offers may cause instabilities. Therefore, FTOPSIS calculations should be defined in appropriate manner, different than the traditional approach. IFIRP method allows the use of natural language, considers of subjectivity the lack of precision or uncertainties in the evaluation of negotiation offers. Further experimental investigation is warranted to discern the negotiator's approach to offer evaluation and the robustness of their reference points. Moreover, IFIRP should be empirically verified and DM should also testify about the behavioral decision-making aspect of the proposed approach. The assessment of satisfaction and dissatisfaction levels and the recognition of reference points' impact on the distance measure for the obtained ranking's stability are also issues that may be studied thoroughly. The FTOPSIS and IFIRP approaches could be useful in ill-structured problems, such as a negotiation process.
[27]	AFLS where new rules added by Simplex method.	Metaheuristics and NN may be employed to generate new rules for the AFLS and to define the rule base.
[29]	Fuzzy rules obtained by psychological experiments. A Mamdani FIS employed to calculate agents' preference dynamically based on risk, patience and regret Fuzzy factors	Since the domain where the model applied was in politics, it is intriguing to adapt this model to a negotiation in a SG context.
[30]	Atanassov intuitionistic Fuzzy constraints method for offer evaluation and the delineation of agents' objectives through prioritized Atanassov intuitionistic Fuzzy constraints.	The problem of the function expression of Fuzzy constraints, which is a critical issue, may be solved with the implementation of the proper heuristic or metaheuristic methods.

Table 2. *Cont.*

Reference	Method	Comments and Suggestions
[31]	A Sugeno-type FIS was devised to ascertain the appropriate level of concession, weighing preferences and utilities of negotiating parties in trade-off decisions. Opponent's behavior is estimated using a tradeoff algorithm based on Fuzzy similarity.	As per the authors, certain parameters need to be appropriately initialized prior to commencing negotiations to optimize performance. This initialization may be performed by MA.
[32]	Learning opponents' preference profiles in a context of multi-participants and multi-issue negotiation, where the agents learn the models' parameters in real time recursively, based on a Bayesian approach known as UPF.	The system demonstrates flexibility, since it has the capability to have logical pattern in proposals' offering. Since the proposed technique in depending on the initial estimation of the opponent's preferences, supplying precise initial approximations utilizing techniques from natural language processing should be studied. Auxiliary data can be also used to generate pseudo-observations for the UPF module, improving further the estimations.
[33]	Buyer and seller agent models are defined by utility functions. Heuristic strategies are followed to choose an offer.	Authors future research targets include the enhancement of the method in many-to-many negotiations scenarios. Agent models may be implemented by ANFIS. Strategies may adopt MA.
[34]	Real time TDR algorithm using Fuzzy inference engine to satisfy the users' comfort levels and preferences responding to rewards offered by the TA. The customers were classified using k-means clustering,	The proposed scheme is a realistic approach for vendors and solution providers to achieve the SG vision.
[35]	A MOBA, a Pareto front and a FDM algorithm for managing the energy trading between prosumers.	The proposed methodology was tried against a similar MOPSO algorithm and promises to facilitate autonomous energy transactions, thus establishing an intelligent energy market. Subsequent research endeavors should analyze and substantiate the algorithm's optimality and convergence.
[37]	q-Rung Orthopair Fuzzy Aczel Alsina Triangular Norms and Conorms sets based Aggregator Operators implemented in a MCGDM algorithm.	The proposed approach outperforms other q-ROF based AO as well as AO relying on Pythagorean and Intuitionistic Fuzzy sets. They can be applied to a more extensive range of information. The proposed AO generalize the previous AO.
[38]	Application of BPSO in P2P trading. MA are suggested as a prevalent part of AI for solving optimization problems.	

Conclusively, the present study conducted a cross-disciplinary survey and the selected studies covered a wide range of cases where negotiation takes place. The majority of them employ FL theory as a tool to adapt to the opponent's behavior or to reach a decision-making inference. A small number of them employ or suggest metaheuristics such as PSO, GA, or BA mostly as a tool to refine Fuzzy rules allowing the agent to dynamically adapt to the opponent's strategy. A few numbers of the surveyed studies propose models where AI tools like ANFIS, NN, and metaheuristic algorithms can apply to future work. Despite the temptation to follow an already tested methodology applied in SG applications, authors recommend using these methodologies as benchmarks against novel implementations that may occur following and adapting one or more methodologies that haven't been applied yet in SG. Consulting the summary in Table 2, a scholar may step into a deeper search and propose new, more efficient models for SG and EM applications, or for markets with similar structures.

In Section 3, the detailed Discussion of the surveyed studies reveals the limitations of these methods which are:

1. The uncertainty handling, which is a primary challenge during negotiations. FL stands as a promising solution for handling uncertainty. However, it remains challenging to anticipate and cover all potential cases that may arise during a negotiation. More robust FL systems should be developed.
2. The incorporation of human psychological traits. Negotiation models often overlook the incorporation of human psychological traits such as risk aversion, patience, and regret. The authors propose the utilization of Fuzzy rules to determine the extent to which negotiating agents should adjust their preference structures based on these

psychological traits. This involves eliciting psychological experiments to inform the Fuzzy rules accurately. Future research could delve deeper into understanding how these psychological traits influence negotiation outcomes and develop more sophisticated Fuzzy rule bases to capture human behavior effectively.

3. The evaluation of negotiation offers. Evaluating offers during negotiations is a complex task, especially considering the diverse preferences and constraints of negotiating parties. The authors propose an Atanassov intuitionistic Fuzzy constraints-based approach for offer evaluation, considering satisfaction and dissatisfaction levels. This method aims to represent human-like behavior during the negotiation process and allows for the consideration of various negotiation objectives. Future research could focus on developing diverse forms of satisfaction and dissatisfaction functions to better represent individual characteristics and preferences, thereby enhancing the accuracy of offer evaluations.
4. The adaptation and learning in real time. Negotiation agents often struggle to adapt to the changing preferences and behaviors of their counterparts in real time. The authors propose a methodology based on Bayesian approaches and Unscented Particle Filtering (UPF) for opponent learning in real-time automated negotiation. This approach enables negotiation agents to dynamically estimate their counterparts' preference profiles and adjust their negotiation strategies accordingly. Future research could explore further enhancements to this methodology, such as leveraging natural language processing techniques for more accurate initial approximations of opponents' preferences.
5. The optimization in Energy Markets. Negotiation frameworks applied in energy markets face challenges such as maximizing utility for both trading parties while ensuring speed in reaching agreements. The authors propose various optimization algorithms, such as Multi-Objective Optimization and Particle Swarm Optimization (PSO) variants, to address these challenges. Future research could focus on analyzing the optimality and convergence of these algorithms further, as well as exploring new prospects for their application in energy markets.

Overall, the authors advocate for the integration of advanced AI techniques, such as FL, Bayesian approaches, and optimization algorithms, into negotiation frameworks to overcome existing limitations and improve negotiation outcomes in various domains.

Generally, by integrating flexible DR, RTP mechanisms, bi-level optimization approaches, and intelligent negotiation methods, prosumers, and the grid can optimize energy transaction negotiations, maximize economic benefits, and enhance the efficiency and reliability of the energy system.

5. Future Directions

5.1. Multi-Criteria Decision Making (MCDM)

MCDM is a vital aspect of decision theory, offering ranking outcomes for alternatives by combining diverse criteria [42]. Uncertainty affects decision-making and appears in a number of different forms. As pointed out by Zadeh, uncertainty is an attribute of information [43]. In real-world MCDM problems, decision-making information often involves imprecision and ambiguity. Therefore, the criteria are not consistently expressed as real numbers and they are better suited for representation through Fuzzy sets [44], Intuitionistic Fuzzy sets [45], Pythagorean Fuzzy sets [46], and q-ROFS [47].

Hospitals and healthcare buildings in general consist of an interesting case of study where energy consumption management and decision-making processes can reduce energy consumption. In [48], Smart Energy software achieved a reduction of 3.46% in energy consumption of a medium-sized University hospital in Brazil which equals to 100,000 kWh over a year.

Hospitals' energy consumption has a significant environmental impact, since they are responsible for the 10.3% of the total consumed energy of the commercial buildings sector while they only cover 4.8% of the total area of the commercial buildings [49].

Since the decision-making issue in electricity purchasing has multiple aspects, the authors' future work may focus on the decision-making process between different suppliers in competitive market. For example, let R_1, R_2, R_3 be three competitor retailers and C_1, C_2, C_3 and C_4 the user defined criteria concerning costs, source of energy, CO_2 emissions and value-added services such as reliability of the supply respectively. Using q-ROFN of the form $\alpha_{ij} = (\mu_{ij}, v_{ij})$, with μ representing membership and v non-membership degrees of the α, the preference matrix may have the form of the Table 3:

Table 3. q-ROFN preference matrix.

Retailers	C_1	C_2	C_3	C_4
R_1	$\alpha_{11} = (\mu_{11}, v_{11})$	$\alpha_{12} = (\mu_{12}, v_{12})$	$\alpha_{13} = (\mu_{13}, v_{13})$	$\alpha_{14} = (\mu_{14}, v_{14})$
R_2	$\alpha_{21} = (\mu_{21}, v_{21})$	$\alpha_{22} = (\mu_{22}, v_{22})$	$\alpha_{23} = (\mu_{23}, v_{23})$	$\alpha_{24} = (\mu_{24}, v_{24})$
R_3	$\alpha_{31} = (\mu_{31}, v_{31})$	$\alpha_{32} = (\mu_{32}, v_{32})$	$\alpha_{33} = (\mu_{33}, v_{33})$	$\alpha_{34} = (\mu_{34}, v_{34})$

The preference information can then be aggregated using AO of q-Rung Orthopair Fuzzy Numbers (q-ROFN) and score values for each retailer can be calculated as presented in Table 4:

Table 4. Score values of the aggregated preferences.

Retailers	Aggregation Values	Scores	Ranking
R_1	$Agg_1(\alpha_{1j})$	$S(Agg_1)$	Supposing that the selected supplier should be the retailer R_2. $R_2 > R_1 > R_3$
R_2	$Agg_2(\alpha_{2j})$	$S(Agg_2)$	
R_3	$Agg_3(\alpha_{3j})$	$S(Agg_3)$	

The authors' future work will implement, among other intelligent techniques, a decision support system based on q-ROFNs for a healthcare facility. Healthcare facilities (Hospitals, district health centers, etc.) are a special category of commercial buildings due to their social importance and their environmental impact.

5.2. Fuzzy Negotiation

Considering the aforementioned notes, the authors will furtherly enquire Fuzzy negotiation methodologies based on similarity operators and Fuzzy sets described in [17,18]. Assuming that seller and buyer act uniformly, the generalized methodology described below is common for both unless stated otherwise:

1. Firstly, an initial offer is made from the first negotiator which can be a buyer or a seller. The other participant can accept the offer ending the process or submit a counteroffer. The counteroffer creation is described as follows: The opponent negotiator evaluates the offer employing a utility function which allocates values to the available options (Equation (1)).

$$U(x) = \frac{\sum_{i=1}^{m} w_i \cdot u_i(x_i)}{\sum_{i=1}^{m} w_i}, \tag{1}$$

In Equation (1), $U(x)$ is the overall utility value for the offer $x = \begin{bmatrix} x_1 \\ \vdots \\ x_m \end{bmatrix}$ and u_i is the utility value assigned to issue i. Each utility value u_i is assigned to its corresponding weight, w_i.

Some options—the qualitative ones—may be predefined, and can be assigned to certain values in the [0,1] interval-valued q-ROFS, while the quantitative options may have more vague definitions and can be expressed by fuzzy sets and q-ROFS. For example, the option "price" which is common option for both seller and buyer has a diverse significance for them. From the seller's point of view, the price is expected to lead to a benefit and from the buyer's perspective the cost is the issue which the price option incorporates. Naturally, the option "benefit" for the seller should be the larger while for the buyer the option "cost" should be the smaller. Moreover, the participants are setting the higher and the lower bounds they are willing to accept. Thus, these bounds are forming the constraints of the problem. The weights of the options as they are set by every participant along with the respective values form the submitted offer. The negotiating agent of the participant is now set. In Figure 3, the utility function of the price issue is shown. P_l and P_h stand for the lower and the higher bounds of the price the buyer is expecting to pay while the B_l and B_h the lower and the higher benefit bounds the seller is expecting to achieve.

2. As the utility function is now formulated, the participant can guess the opponent's preferences employing a metaheuristic algorithm such as GA, PSO, or BA, assigning values to the utility functions' weights. A counteroffer is now created based on Fuzzy inference. The weights are updated at every round of the negotiation, according to the opponent's weights' deviation from the estimated weights. The offer for each issue is compared to the participant's expectation by means of similarity measure. In case of multiple opponents, a score is assigned to each of their offers, according to Section 5.1, and the selected offer is the base of the counteroffer. The issues' values are also updated by Δx_i using an inference system where the weights assigned to them by the participants (w_i^s and w_i^b for the seller and the buyer respectively) are the inputs of the system and the output is Δx_i, according to Equation (2):

$$\Delta x_i = \frac{\sum_{j=1}^{n} z_j \cdot C_j}{\sum_{j=1}^{n} z_j}, \tag{2}$$

where n is the number of rules of the FIS, z_j the firing strength of rule j as given in Equation (3):

$$z_j = \mu_{significance}(w_i^s) \otimes \mu_{significance}(w_i^b), \tag{3}$$

where the operator $\otimes$ stands for the Fuzzy "and" and C_j is the consequence of each rule.

Every issue has its significance, $\mu_{significance}$, not necessarily the same, for every participant, and is expressed in linguistic terms as insignificant, neutral and significant. Therefore, the degree of significance could be represented as membership functions of Fuzzy sets, like the one in Figure 4:

3. Depending on the type of the participant (utility grid or smart building), the negotiation strategy can be adapted by an eagerness factor representing the urge of the seller or the buyer to sell or buy [19,21]. Eagerness can be affected by multiple factors, as the time remaining, the history record of past transactions, the flexibility to curtail loads, etc. For short-time attitude, eagerness in a certain time, t for both buyer and seller can be expressed as in Equation (4):

$$EA_s = \begin{cases} C_s \; if \; Conc_{avg} < 0.01 \\ 1 \qquad Otherwise \end{cases}, \tag{4}$$

where:

$$Conc_{avg} = \frac{\sum_{t=2}^{k} (p_{opp}(t) - p_{opp}(t-1))}{k-1}$$

is the average concession rate, k is the length of time window, p_{opp} the opponent's price offer and C_s is a small positive value.

This feature simulates the human behavior during a negotiation and has an impact on their offers. An implementation proposal of the abovementioned model considers it as an issue on the negotiation offer where its significance may dynamically vary from one round to another, differentiating the overall behavior of the participant in each round.

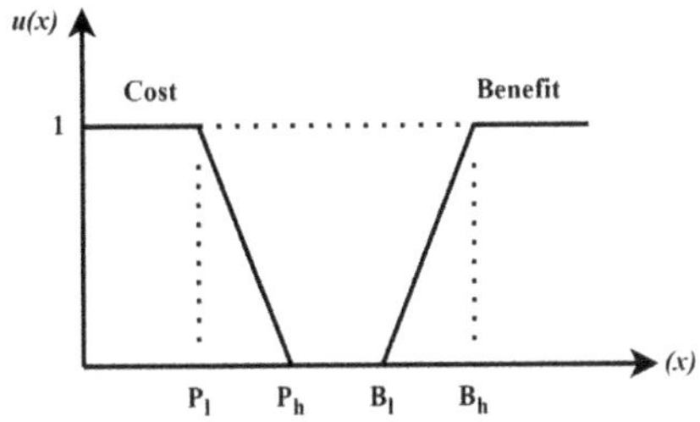

Figure 3. Utility function of the issue "price".

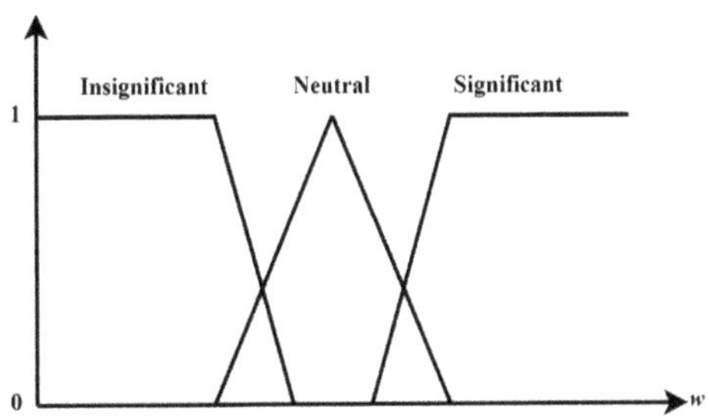

Figure 4. Membership functions of the significance levels for an issue [28,29].

Furthermore, q-ROFS can be integrated with other advanced intelligent methods to optimize negotiation outcomes in complex MCDM scenarios following the abovementioned strategies:

- Enhancement of uncertainty representation. Future research could focus on q-ROFS refining to better capture and represent uncertainty in negotiation scenarios.
- Improvement of DM precision. Since q-ROFS offer improved DM precision compared to conventional FL systems, future research could focus to the fine-tuning of the membership functions and similarity measures.
- Dynamic adaptation to changing conditions. With the ability of q-ROFS to adapt dynamically to changing conditions in real time, which is essential in negotiation scenarios, future research could focus on developing q-ROFS models that can dynamically adjust their parameters and decision-making strategies based on evolving negotiation conditions.
- Improvement of robustness and reliability. The combination of q-ROFS with other advanced intelligent methods results in decision-making systems that are more robust against errors and data inconsistencies. Future research could investigate techniques for integrating q-ROFS with machine learning algorithms to improve robustness and reliability in negotiation scenarios.
- Exploration of a Wider Range of Potential Solutions. q-ROFS enhance exploration capabilities in the decision-making process by considering a wider range of potential solutions. Future research could focus on developing optimization algorithms that leverage the exploration capabilities of q-ROFS to identify better solutions in complex negotiation scenarios.

Conclusively, integrating q-ROFS with MA offers significant advancements in DM capabilities for SG negotiation over conventional FL approaches. The key elements of this dominance are their enhanced representation of uncertainty, their improved DM precision, and the ability to search large solution spaces to find optimal or near-optimal solutions to complex, multi-objective problems representing negotiation scenarios. MA also offers dynamic adaptation to changing conditions in real time, which is essential in the SG context. Compared to conventional FL systems, q-ROFS provide higher discrimination power, differentiating closely related scenarios, and leading to more refined and accurate negotiation outcomes. Moreover, though conventional FL systems struggle with the level of ambiguity which is present in SG data, q-ROFS manage both uncertainty and vagueness offering a more robust framework for DM under ambiguous conditions. Furthermore, the combination of q-ROFS with MA result to a DM system more robust against errors and data inconsistencies. Finally, MA enhance exploration capabilities of DM process, considering a wider range of potential solutions, leading to better optimized solutions than the traditional FL systems.

Author Contributions: Conceptualization, A.I.D. and D.K.P.; methodology, A.I.D. and D.K.P.; formal analysis, A.I.D. and D.K.P.; investigation, A.I.D. and D.K.P.; resources, D.K.P.; writing—original draft preparation, D.K.P.; writing—review and editing, A.I.D. and D.K.P.; visualization, D.K.P.; supervision, A.I.D.; project administration, A.I.D. All authors have read and agreed to the published version of the manuscript.

Funding: This research received no external funding.

Conflicts of Interest: The authors declare no conflicts of interest.

References

1. Roux, E. Electricity, the decisive factor in modern civilization. *Impact Sci. Soc.* **1950**, *1*, 115–116.
2. Georgilakis, P. *Economic and Reliable Operation of Modern Electric Power Systems*; Undergraduate Textbook; Kallipos Open Academic Publications: Athens, Greece, 2023.
3. Biggar, D.R.; Hesamzadeh, M.R. *The Economics of Electricity Markets*; John Wiley & Sons Ltd.: Hoboken, NJ, USA, 2014.
4. Fu, H.; Shi, Y.; Zeng, Y. Estimating Smart Grid's Carbon Emission Reduction Potential in China's Manufacturing Industry Based on Decomposition Analysis. *Front. Energy Res.* **2021**, *9*, 681244. [CrossRef]
5. Gržanić, M.; Capuder, T.; Zhang, N.; Huang, W. Prosumers as active market participants: A systematic review of evolution of opportunities, models and challenges. *Renew. Sustain. Energy Rev.* **2021**, *154*, 111859. [CrossRef]
6. Kotilainen, K.; Saari, U.A. Policy Influence on Consumers' Evolution into Prosumers—Empirical Findings from an Exploratory Survey in Europe. *Sustainability* **2018**, *10*, 186. [CrossRef]
7. Falvo, M.C.; Martirano, L.; Sbordone, D.; Bocci, E. Technologies for Smart Grids: A brief review. In Proceedings of the 12th International Conference on Environment and Electrical Engineering, Wroclaw, Poland, 5–8 May 2013.
8. Strbac, G. Demand side management: Benefits and challenges. *Energy Policy* **2008**, *36*, 4419–4442. [CrossRef]
9. Bertoli, E.; Rozite, V.; Reidenbach, B.; Vautrin, A.; Oh, S.; Suire, M. Demand Response-IEA. IEA-International Energy Agency. 11 July 2023. Available online: https://www.iea.org/energy-system/energy-efficiency-and-demand/demand-response (accessed on 6 February 2024).
10. Panda, S.; Mohanty, S.; Rout, P.K.; Sahu, B.K.; Parida, S.M.; Samanta, I.S.; Bajaj, M.; Piecha, M.; Blazek, V.; Prokop, L. A comprehensive review on demand side management and market design for renewable energy support and integration. *Energy Rep.* **2023**, *10*, 2228–2250. [CrossRef]
11. Li, Y.; Han, M.; Yang, Z.; Li, G. Coordinating Flexible Demand Response and Renewable Uncertainties for Scheduling of Community Integrated Energy Systems with an Electric Vehicle Charging Station: A Bi-level Approach. *IEEE Trans. Sustain. Energy* **2021**, *12*, 2321–2331. [CrossRef]
12. Siano, P. Demand response and smart grids-A survey. *Renew. Sustain. Energy Rev.* **2013**, *19*, 461–478. [CrossRef]
13. Reddy, P.P.; Veloso, M.M. Negotiated Learning for Smart Grid Agents: Entity Selection Based on Dynamic Partially Observable Features. In Proceedings of the Twenty-Seventh AAAI Conference on Artificial Intelligence 1313, Bellevue, WA, USA, 14–18 July 2013.
14. Ruttkay, Z. Fuzzy Constraint Satisfaction. In Proceedings of the 1994 IEEE 3rd International Fuzzy Systems Conference, Orlando, FL, USA, 26–29 June 1994.
15. Choi, S.P.; Liu, J.; Chan, S.-P. A genetic agent-based negotiation system. *Comput. Netw.* **2001**, *37*, 195–204. [CrossRef]
16. Faratin, P.; Sierra, C.; Jennings, N. Using similarity criteria to make issue trade-offs in automated negotiations. *Artif. Intell.* **2002**, *142*, 205–237. [CrossRef]
17. Cheng, C.-B.; Chan, C.-C.H.; Lin, C.-C. Buyer-Supplier Negotiation by Fuzzy Logic Based Agents. In Proceedings of the Third International Conference on Information Technology and Applications (ICITA'05), Sydney, Australia, 4–7 July 2005.
18. Cheng, C.-B.; Chan, C.-C.H.; Lin, K.-C. Intelligent agents for e-marketplace: Negotiation with issue trade-offs by fuzzy inference systems. *Decis. Support Syst.* **2006**, *42*, 626–638. [CrossRef]
19. Wang, Z.; Wang, L. Negotiation agent with adaptive attitude bidding strategy for facilitating energy exchanges between smart building and utility grid. In Proceedings of the PES T&D 2012, Orlando, FL, USA, 7–10 May 2012.

20. Ren, M.Z.F. A single issue negotiation model for agents bargaining in dynamic electronic markets. *Decis. Support Syst.* **2014**, *60*, 55–67. [CrossRef]
21. Wang, Z.; Wang, L. Adaptive Negotiation Agent for Facilitating Bi-Directional Energy Trading Between Smart Building and Utility Grid. *IEEE Trans. Smart Grid* **2013**, *4*, 702–710. [CrossRef]
22. Adabi, S.; Movaghar, A.; Rahmani, A.M.; Beigy, H.; Dastmalchy-Tabrizi, H. A new fuzzy negotiation protocol for grid resource allocation. *J. Netw. Comput. Appl.* **2014**, *37*, 89–126. [CrossRef]
23. Kolomvatsos, K.; Anagnostopoulos, C.; Hadjiefthymiades, S. Sellers in e-marketplaces: A Fuzzy Logic based decision support system. *Inf. Sci.* **2014**, *278*, 267–284. [CrossRef]
24. Lee, C.-C. Development and evaluation of the many-to-many supplier negotiation strategy. *Comput. Ind. Eng.* **2014**, *70*, 90–97. [CrossRef]
25. Lee, C.C.; Ou-Yang, C. Development and evaluation of the interactive bidding strategies for a demander and its suppliers in supplier selection auction market. *Int. J. Prod. Res.* **2008**, *46*, 4827–4848. [CrossRef]
26. Roszkowska, E.; Brzostowski, J.; Wachowicz, T. Supporting Ill-Structured Negotiation Problems. In *Human-Centric Decision-Making Models for Social Sciences*; Guo, P., Pedrycz, W., Eds.; Springer: Berlin/Heidelberg, Germany, 2013; pp. 339–367.
27. Kolomvatsos, K.; Trivizakis, D.; Hadjiefthymiades, S. An adaptive fuzzy logic system for automated negotiations. *FUZZY Sets Syst.* **2015**, *269*, 135–152. [CrossRef]
28. Roszkowska, E.; Wachowicz, T. Application of fuzzy TOPSIS to scoring the negotiation offers in ill-structured negotiation problems. *Eur. J. Oper. Res.* **2014**, *242*, 920–932. [CrossRef]
29. Zhan, J.; Luo, X.; Feng, C.; He, M. A multi-demand negotiation model based on fuzzy rules elicited via psychological experiments. *Appl. Soft Comput.* **2018**, *67*, 840–864. [CrossRef]
30. Zhan, J.; Luo, X.; Jiang, Y. An Atanassov intuitionistic fuzzy constraint based method for offer evaluation and trade-off making in automated negotiation. *Knowl.-Based Syst.* **2018**, *139*, 170–188. [CrossRef]
31. Shojaiemehr, B.; Rahmani, A.M.; Qader, N.N. Automated negotiation for ensuring composite service requirements in cloud computing. *J. Syst. Archit.* **2019**, *99*, 101632. [CrossRef]
32. Eshragh, F.; Shahbazi, M.; Far, B. Real-time opponent learning in automated negotiation using recursive Bayesian filtering. *Expert Syst. Appl.* **2019**, *128*, 28–53. [CrossRef]
33. Etukudor, C.; Couraud, B.; Robu, V.; Früh, W.-G.; Flynn, D.; Okereke, C. Automated Negotiation for Peer-to-Peer Electricity Trading in Local Energy Markets. *Energies* **2020**, *13*, 920. [CrossRef]
34. Fattahi, J.; Samadi, M.; Erol-Kantarci, M.; Schriemer, H. Transactive Demand Response Operation at the Grid Edge using the IEEE 2030.5 Standard. *Engineering* **2020**, *6*, 801–811. [CrossRef]
35. Soriano, L.A.; Avila, M.; Ponce, P.; Rubio, J.D.J.; Molina, A. Peer-to-peer energy trades based on multi-objective optimization. *Int. J. Electr. Power Energy Syst.* **2021**, *131*, 107017. [CrossRef]
36. Roszkowska, E. The Intuitionistic Fuzzy Framework for Evaluation and Rank Ordering the Negotiation Offers. In Proceedings of the INFUS 2021 Conference, Istanbul, Turkey, 24–26 August 2021.
37. Akram, M.; Ullah, K.; Ćirović, G.; Pamucar, D. Algorithm for Energy Resource Selection Using Priority Degree-Based Aggregation Operators with Generalized Orthopair Fuzzy Information and Aczel–Alsina Aggregation Operators. *Energies* **2023**, *16*, 2816. [CrossRef]
38. Zaini, F.A.B.; Sulaima, M.F.; Razak, I.A.W.A.; Zulkafli, N.I.; Mokhlis, H. A review of PSO-based algorithm application in Demand Side Management: Challenges & Opportunities. *IEEE Access* **2023**, *11*, 53373–53400.
39. Rubinstein, A. Perfect Equilibrium in a Bargaining Model. *Econometrica* **1982**, *50*, 97–109. [CrossRef]
40. Sim, K.; Ng, K. Relaxed-criteria negotiation for grid resource allocation. *Int. Trans. Syst. Sci. Appl.* **2007**, *3*, 105–117. [CrossRef]
41. Adabi, S.; Movaghar, A.; Rahmani, A.; Beigy, H. Market-based grid resource allocation using new negotiation model. *J. Netw. Comput. Appl.* **2013**, *36*, 543–565. [CrossRef]
42. Liu, P.W.P. Some q-rung orthopair fuzzy aggregation operators and their applications to multiple-attribute decision making. *Int. J. Intell. Syst.* **2018**, *33*, 259–280. [CrossRef]
43. Zadeh, L.A. Toward a generalized theory of uncertainty (GTU)—-An outline. *Inf. Sci.* **2005**, *172*, 1–40. [CrossRef]
44. Zadeh, L. Fuzzy sets. *Inf. Control* **1965**, *8*, 338–353. [CrossRef]
45. Atanassov, K.T. Intuitionistic fuzzy sets. *Fuzzy Sets Syst.* **1986**, *20*, 87–96. [CrossRef]
46. Yager, R.R.; Abbasov, A.M. Pythagorean Membership Grades, Complex Numbers, and Decision Making. *Int. J. Intell. Syst.* **2013**, *28*, 436–452. [CrossRef]
47. Yager, R.R. Generalized Orthopair Fuzzy Sets. *IEEE Trans. Fuzzy Syst.* **2017**, *25*, 1222–1230. [CrossRef]
48. Oliveira, K.B.D.; Santos, E.F.D.; Neto, A.F.; Santos, V.H.D.M.; Oliveira, O.J.D. Guidelines for efficient and sustainable energy management in hospital buildings. *J. Clean. Prod.* **2021**, *329*, 12964.
49. Bawaneh, K.; Nezami, F.; Rasheduzzaman, M.; Deken, B. Energy Consumption Analysis and Characterization of Healthcare Facilities in the United States. *Energies* **2019**, *12*, 3775. [CrossRef]

applied sciences

MDPI

Review

Advanced Digital Tools for Data-Informed and Performance-Driven Design: A Review of Building Energy Consumption Forecasting Models Based on Machine Learning

Andrea Giuseppe Di Stefano *, Matteo Ruta and Gabriele Masera

Department of Architecture, Built Environment and Construction Engineering, Politecnico di Milano, Via Ponzio 31, 20133 Milano, Italy; matteo.ruta@polimi.it (M.R.); gabriele.masera@polimi.it (G.M.)
* Correspondence: andreagiuseppe.distefano@polimi.it

Abstract: Cities and buildings represent the core of human life, the nexus of economic activity, culture, and growth. Although cities cover less than 10% of the global land area, they are notorious for their substantial energy consumption and consequential carbon dioxide (CO_2) emissions. These emissions significantly contribute to reducing the carbon budget available to mitigate the adverse impacts of climate change. In this context, the designers' role is crucial to the technical and social response to climate change, and providing a new generation of tools and instruments is paramount to guide their decisions towards sustainable buildings and cities. In this regard, data-informed digital tools are a viable solution. These tools efficiently utilise available resources to estimate the energy consumption in buildings, thereby facilitating the formulation of effective urban policies and design optimisation. Furthermore, these data-driven digital tools enhance the application of algorithms across the building industry, empowering designers to make informed decisions, particularly in the early stages of design. This paper presents a comprehensive literature review on artificial intelligence-based tools that support performance-driven design. An exhaustive keyword-driven exploration across diverse bibliographic databases yielded a consolidated dataset used for automated analysis for discerning the prevalent themes, correlations, and structural nuances within the body of literature. The primary findings indicate an increasing emphasis on master plans and neighbourhood-scale simulations. However, it is observed that there is a lack of a streamlined framework integrating these data-driven tools into the design process.

Keywords: machine learning-based tools; building energy forecasting; data-informed design; performance-driven design; operational carbon reduction

1. Introduction

Since the dawn of the industrial revolution, urban environments have emerged as significant influencers of energy consumption patterns. Today, as urbanisation continues its relentless march, over half of the global population resides in cities. These urban hubs, while being vibrant centres of culture and commerce, are also responsible for a staggering 75% of global carbon emissions [1]. This urban-centric carbon footprint underscores cities' dual role: they are both major contributors to the climate challenge and, simultaneously, pivotal players in its solution. A testament to this is the commitment of over 100 cities worldwide to achieve net-zero carbon emissions by 2050 [2]. Within this urban landscape, buildings stand out, playing an indispensable role in the carbon equation. Yet, while the importance of buildings in this narrative is clear, the roadmap to achieving substantial carbon emission reductions remains shrouded in complexity.

The 21st century presents a defining challenge for the built environment: constructing energy-efficient urban spaces while retrofitting existing ones to meet modern standards. Addressing this challenge demands a granular understanding of how buildings consume energy and the strategies required to minimise this consumption. Building designers, policymakers, administrators, and even tenants, must come together, pooling their expertise to devise and implement impactful, cost-efficient measures [3].

Building performance simulations (BPSs) have long been the bedrock of decision-making in building design. As shown in Figure 1, BPS plays a critical role from the initial planning stage through to occupancy, with its influence peaking during the design phase. As legislative frameworks evolve—pushing for greener, more efficient buildings—BPS has adapted, integrating richer, more reliable data streams into the design process. The digital age has ushered in an era of unprecedented data availability, paving the way for the integration of machine learning and other data-driven tools into BPS. These advancements promise not only enhanced simulation and modelling methods but also their seamless integration right from the earliest design stages. In this scenario, the recent revision of the Energy Performance of Buildings Directive (EPBD) [4] echoes this sentiment, emphasising the integration of data-informed and performance-driven tools into BPSs.

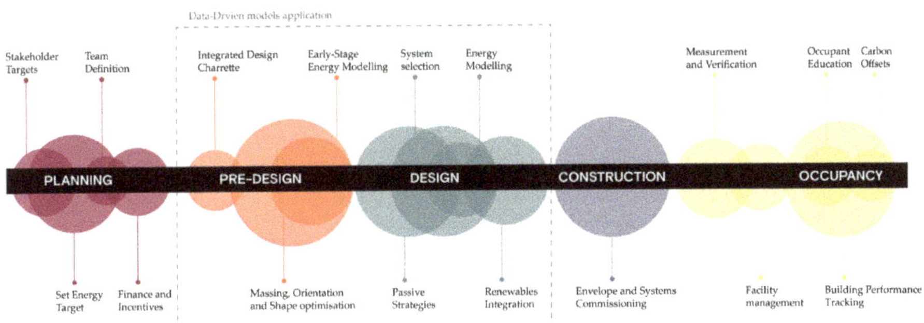

Figure 1. Phases of the building design lifecycle—each phase is depicted with a distinct colour to denote the corresponding phase, while the circle dimensions indicate their level of influence on the project's energy efficiency and sustainability outcomes.

Yet, the path is not without its hurdles. The BPS realm offers a vast array of modelling methods, creating a maze that often leaves stakeholders, from researchers to policymakers, perplexed [5]. On the other hand, the data, the lifeblood of these models, are often fragmented, unstructured, or locked behind gates of inaccessibility.

The early design stage of a building is a period of immense potential. It is during these phases that critical design parameters are set, influencing over 40% of a building's energy-saving potential [6]. To harness this potential, there is a pressing need to optimise these parameters from the earliest stages.

While the allure of data-driven models, with their promise of rapid and reasonably accurate energy consumption estimates, is undeniable, traditional BPS methods cannot be overtaken with ease. Even as data-driven BPSs evolve, challenges persist, from computational intensity to potential overestimations of the energy savings [7], from underlying biases to unclear data processing. However, recent research [8,9] shines a spotlight on the potential of data-driven methods, with techniques like Artificial Neural Networks (ANNs) [10,11], Support Vector Machine (SVM) [12,13], and Decision Tree-based models (DT) [14] gaining traction in building energy modelling [15].

While several reviews have analysed specific tools [8,16], simulation techniques and algorithms, and related indices and evaluation tools [7,17,18], this study pivots to explore the current state of the art concerning the integration of data-driven tools in the architecture, engineering, and construction (AEC) sector. The main contribution of this study lies in offering a comprehensive review of the Artificial Intelligence (AI)-based tools supporting performance-driven design and identifying an increasing focus on master plans and neighbourhood-scale simulations. Most significantly, it uncovers the absence of a streamlined framework for embedding these data-driven tools within the early stages of the design process.

Through this work, we aim to clear the mist surrounding the potential of data-driven BPS, mapping out a direction for future research and practical application, with a vision to empower stakeholders to make informed, impactful decisions for the sustainable transformation of our urban landscapes.

2. Research Methodology

Informed by the understanding that urban environments significantly influence energy consumption patterns, and the pivotal role buildings play in carbon emissions, our methodology sought to delve deep into the integration of digital tools with BPS. The pressing need to optimise design parameters from the earliest stages underpinned our analytical approach. This imperative led us to adopt a data-driven perspective, considering both traditional BPS and emerging data-driven methodologies.

To structure our exploration, we adopted a systematic literature review approach, as presented in Figure 2. Our methodology comprised of keyword identification, which was steered by the overarching aim of the paper. This process ensured a comprehensive review, encompassing advanced digital tools for data-informed design, performance-driven design concepts, and early stage energy and operational carbon assessment strategies. Recognising the importance of interdisciplinary research, multiple bibliography databases were utilised to achieve a holistic view. Further, the extraction process was not limited to a simplistic keyword search. Given the complexity and vastness of the topic, bibliometrics and data processing techniques were used to discern patterns, correlations, and thematic structures in the available literature. This approach allowed us to identify both dominant themes and gaps in the research landscape, which served as a foundation for our subsequent analyses and discussions.

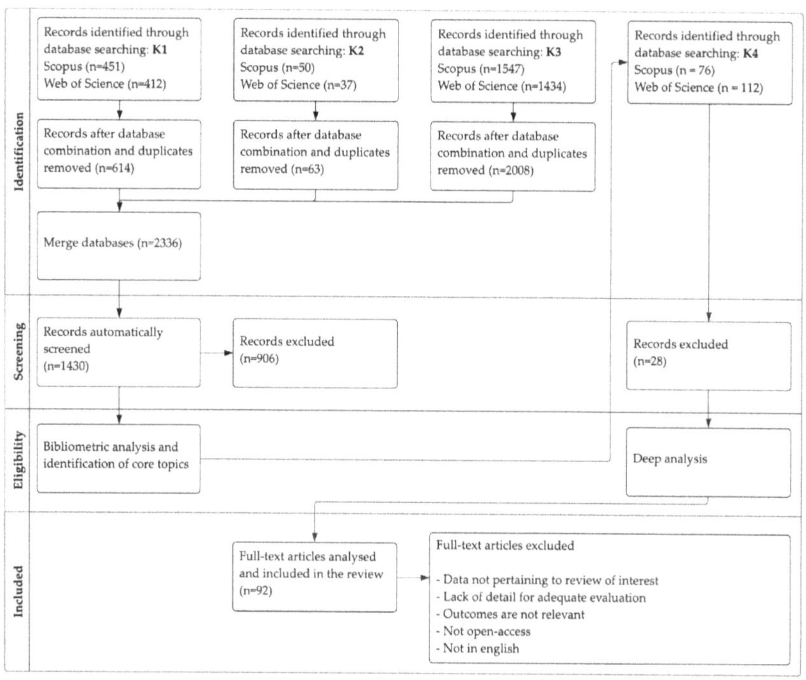

Figure 2. Literature review methodology process.

It is pertinent to note that while data-driven techniques and their application in building performance simulations are undoubtedly significant, this review will not focus on a systemic analysis of the tools and instruments, including their indexes and measurements.

Instead, the emphasis will be on the broader integration of data-driven tools into BPS, especially in the early design stages of buildings. The methodology's robustness lies in its multi-faceted approach, combining a traditional literature review with advanced data analysis techniques. It provides a balanced perspective, ensuring both breadth and depth in the exploration of the topic.

The methodology adopted for this research presents a systematic review process, ensuring a structured approach to the literature analysis. Initially, three distinct keyword searches were conducted, respectively, K1, K2, and K3. The records identified from the database searches were subsequently collated, resulting in a combined total of 2336 records. To maintain the quality and uniqueness of the data, duplicate records were removed.

Following the de-duplication process, the merged database underwent an automated analysis using specialised tools, such as VosViewer [19], to identify notable correlations, patterns, and significant themes within the literature. The insights derived from this automated analysis informed the final keyword search, designated as K4. This culminated in the identification of articles from Scopus (n = 76) and Web of Science (n = 112).

The final phase involved a scrupulous screening of the K4 search results based on their titles and abstracts to ensure their alignment with the research objectives. Out of the screened articles, 92 were identified as being of the utmost relevance, forming the foundational dataset for the review. These selected articles were incorporated into the research bibliography, providing the basis for subsequent discussions and findings.

2.1. Application of the Methodology

This section outlines how the described methodology was employed, from keyword identification to the analysis of the results. We commenced with a precise definition of the primary research question, which, in turn, naturally steered our selection of keywords by pinpointing the essential concepts relevant to the query. As described in the introductory paragraph, the aim of this paper is to analyse the current scenario in the predictive energy- and carbon-related analyses and machine learning tools for data-informed building design.

A preliminary review of the existing literature provided a baseline understanding of the dialogue within the field, identifying the terms frequently used by authors on the subject. This initial scan was instrumental in building a foundation of commonly used language and jargon in the domain of energy-efficient building design. These selected keywords were categorised into three sub-topics:

1. Advanced digital tools for data-informed design: including terms like Artificial Neural Networks, machine learning, genetic algorithm, sensitivity analysis, multi-objective optimisation, and Metamodel.
2. Performance-driven design: encompassing performance, energy consumption prediction, and building performance simulation, among others.
3. Early stage energy and operational carbon assessment: featuring early stage life-cycle assessment and operational carbon.

Bibliography databases, such as Web of Science and Scopus, along with the European Commission research project platform CORDIS, were utilised. Searches were conducted considering the intersection of different keywords to evaluate both the quantity and quality of the results. The filters applied to refine the search included open access results, specific subject areas (engineering, energy, environmental science, etc.), all document types, time periods, and affiliations.

The results from the database searches were analysed to understand the significance and temporal impact of each sub-topic. Tools like VosViewer and Bibliometrix [20] facilitated a deep semantic analysis. Preliminary findings indicate a rising interest in the confluence of these sub-topics. While performance-driven design for energy optimisation is a prevalent theme, words such as "framework", "integration", and "requirements" integrated with carbon assessment elements remains sparse, indicating a potential gap in the literature.

2.2. Bibliometrics and Data Processing

The starting phase of our research was marked by a detailed and sensitive analysis, an approach designed to identify primary sources and ensure a comprehensive, cross-disciplinary exploration across the research landscape. In this brainstorming phase, we cast a wide net for potential keywords and phrases, considering a broad range of related terms, synonyms, and variations in terminologies that might impact the search results' breadth and depth.

We formulated search strings by synthesising and Boolean combining the previously delineated keywords. Examples of these combinations include:

1. K1: (("Artificial Neural Network" OR "ANN" OR "Neural Network") AND ("Building Energy consumption" OR "Building Energy performance"));
2. K2: (("Metamodel" OR "Surrogate model") AND "Building" AND ("Energy performance" OR "Operational carbon"));
3. K3: (("Artificial Neural Network" OR "Artificial Intelligence") AND "Building" AND "Energy" AND "Performance").

Using these strings, a pilot search was conducted to test the effectiveness of the keywords chosen. Based on the relevance and quality of the articles retrieved, we fine-tuned our search terms, ensuring that the literature we collated was of the utmost relevance to our research aims.

Following our preliminary review, we conducted a correlation analysis to assess the relative prominence of each keyword within the domain of our study. In the network visualisation in Figure 3, a nuanced constellation of the thematic clusters emerges, each representing a topic within the scientific literature. At the centre of the visualisation, substantial nodes in red highlight the pivotal concepts of 'model', 'performance', and 'design'. This central clustering indicates an intensive focus on theoretical issues and performance metrics, which are foundational to the field's discourse. These nodes are not only larger, suggesting their higher frequency or importance, but are also densely interconnected, signifying a robust dialogue within the literature concerning the development and evaluation of systemic models and performance-based design methodologies. Adjacent to this core, we observe nodes in shades of green, representing the applied facets of the domain, including specific models and analytical techniques. Another discernible cluster, in yellow, punctuates the visualisation with the themes associated with energy consumption and systems' efficiency. The variance in node size and the dispersion within this group suggests a range of subtopics of variable research intensity focused on building energy-related aspects. The more dispersed light-green clusters highlight discussions on multiobjective optimisation, sensitivity analysis, and meta-modelling. This subdomain, combined with the orange one more focused on algorithms and networks for consumption predictions, signifies a rich exploration of abstract modelling processes and transformations, and their overarching significance in the broader research context.

Informed by these correlations and aiming for a more focused review, we refined our search strings. Notably, the term "Artificial intelligence" was set aside, given its ubiquity, which risked overshadowing more specific topics. The refined, primary search string was:

4. K4: (("Metamodel" OR "Surrogate model") AND "Building" AND ("Energy performance" OR "Operational carbon")).

It is paramount to emphasise that this refined, targeted analysis is designed to complement, not supplant, the broader exploration based on a diverse array of keyword combinations.

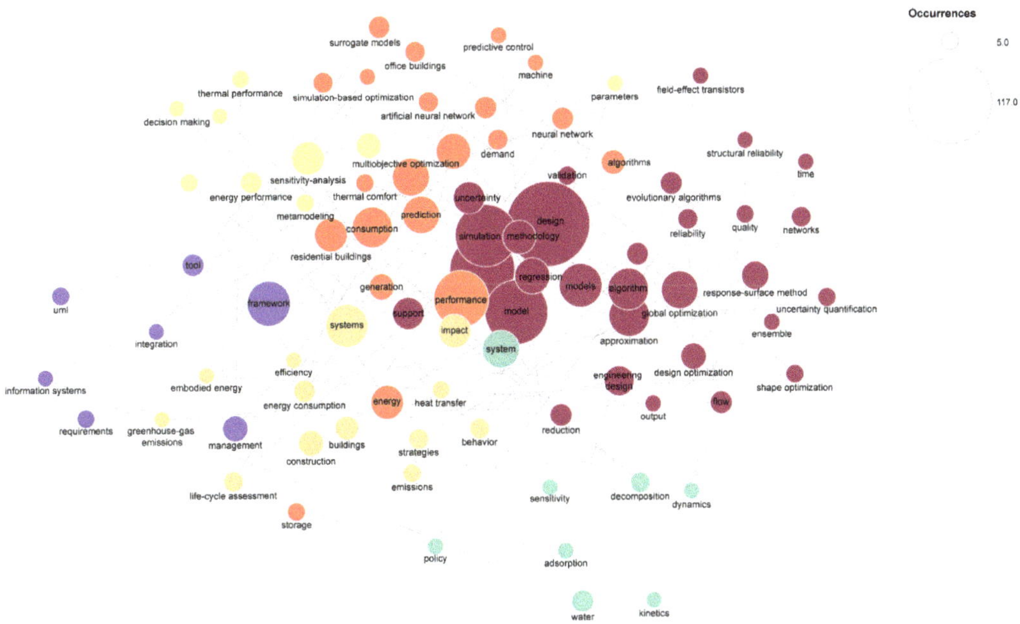

Figure 3. Correlation analysis—interconnected keyword network for AI-driven design.

3. Results

The results obtained from the analysis of the papers retrieved from the string K4 reveal that the annual production of scientific papers regarding metamodelling and surrogate modelling techniques for the energy performance and operational carbon assessment in buildings has an average annual growth rate of 8%.

Two notable observations from Figure 4 are the exponential growth around 2010 and the peak in publications on the topic in 2019. Several factors could have contributed to this surge.

The early 2010s saw a rise in environmental advocacy groups and movements, bringing attention to the importance of sustainable building practices to combat climate change. Moreover, after the 2010s—called, by Chaillou [21], the "Deep Learning Revolution" period—several other events could have influenced this growth. For instance, the Deep Learning Breakthroughs. The early 2010s marked significant advancements in deep learning. The introduction of architectures like AlexNet in 2012 revolutionised the field of computer vision. The triumph of this convolutional neural network in the ImageNet Large Scale Visual Recognition Challenge served as a catalyst, accelerating the embrace of deep learning methodologies across diverse fields, building performance simulations included. Furthermore, in 2014, Ian Goodfellow, alongside his team, unveiled Generative Adversarial Networks (GANs). GANs opened up new possibilities for generating synthetic data. In the context of building simulations, this could mean generating synthetic building designs or performance metrics to augment datasets.

On the other hand, 2019 marked another significant step in the field thanks to several factors such as:

1. Technological Advancements: By 2019, there were significant breakthroughs in computational capabilities, particularly in the realm of AI and machine learning. The development and dissemination of tools like TensorFlow and PyTorch, which made deep learning more accessible, played a pivotal role. Additionally, the proliferation of cloud computing platforms, such as AWS, provided researchers with affordable,

scalable computational resources, enabling more intricate simulations and models related to building energy performance.

2. Regulatory Changes: Around 2019, the European Union introduced the 'Clean Energy for All Europeans' package—which includes the EPBD.
3. Funding and Grants: In 2019, international bodies like the United Nations and the European Union emphasised the sustainable development goals, leading to increased funding opportunities for research on sustainable infrastructure and green building practices. Such financial injections often catalyse academic research, leading to a surge in publications.

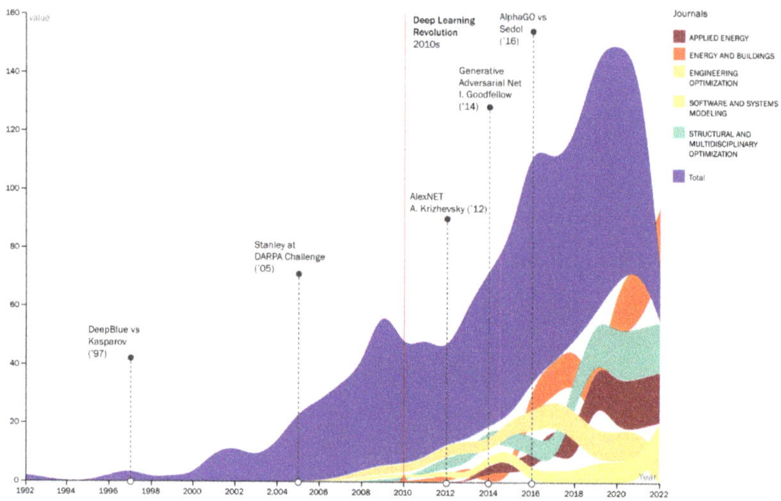

Figure 4. Growth rate and annual publications analysis by journal, with indication of the key milestones in AI development influencing scientific production.

After noting the growing interest in the researched topics, we proceeded with a more in-depth semantic analysis of the word occurrences across the titles, keywords, and abstracts of the papers obtained from the string K4, returning an initial identification of the most widely addressed topics. The treemap in Figure 5 shows the frequency of terms in the surveyed literature, helping in identifying areas of concentrated research and emerging trends.

For clarity and structured interpretation, we categorised these keywords into six thematic clusters, as depicted in the list below and visually represented in Figure 6.

- Performance—incorporating terms such as simulation, global optimisation, and parameters.
- Optimisation—covering concepts like framework, algorithm, regression, and reliability.
- Carbon—highlighting terms like impact, construction, life-cycle assessment, and emissions.
- Consumption—encompassing terms like consumption, prediction, neural network, and surrogate model.
- System—including descriptors like system, water, and storage.
- Sensitivity analysis—capturing concepts like calibration, tool, metamodelling techniques, and decision-making.

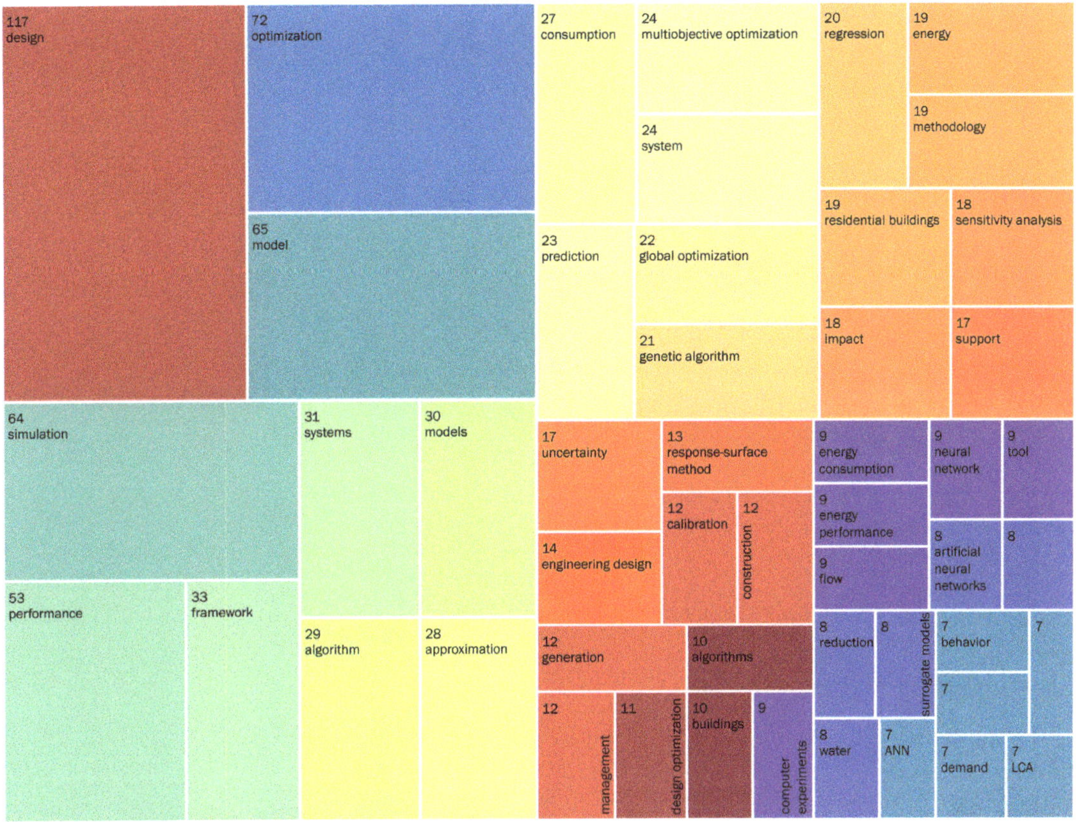

Figure 5. Keywords co-occurrence analysis of selected papers.

To evaluate the significance and coherence of these clusters, a Density–Centrality diagram was utilised (Figure 7). In this analysis, Centrality represents the degree of interaction a cluster has with other parts of the network [22]. It essentially measures the strength of the links from one research theme or community to other research themes or communities, and is an indicator of the significance of a theme or community in the development of an entire field [23]. As a cluster obtains stronger links in a network, the more central its position becomes [24]. On the other hand, Density is the measurement of a cluster's development [23]. It can be understood as the strength of all internal ties (edges) linking together the nodes that make up a theme or community [25]. Density provides a good representation of a cluster's ability to maintain itself and grow over time [26]. As a cluster increases in density, the more coherent it becomes and the more likely it is to contain inseparable nodes [22].

The analysis spotlighted "Consumption", "Carbon", and "Performance" as core themes. Within the "Consumption" cluster, emphasis is placed on neural networks and predictive models. The "Carbon" cluster seeks to merge digital tools with carbon and emission reduction analyses, while the "Performance" cluster underscores foundational elements such as global optimisation and simulation.

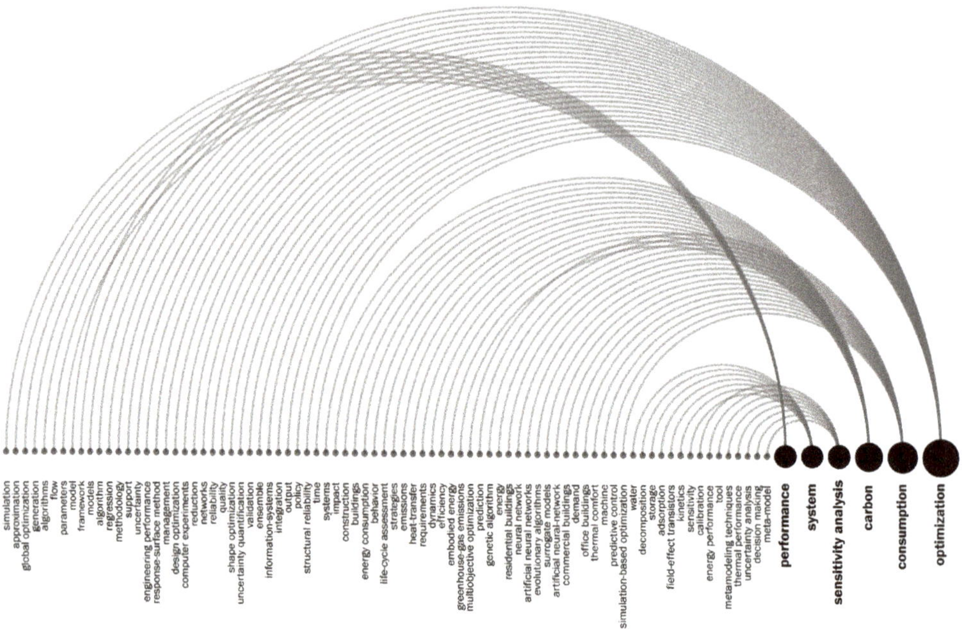

Figure 6. Keywords clustering.

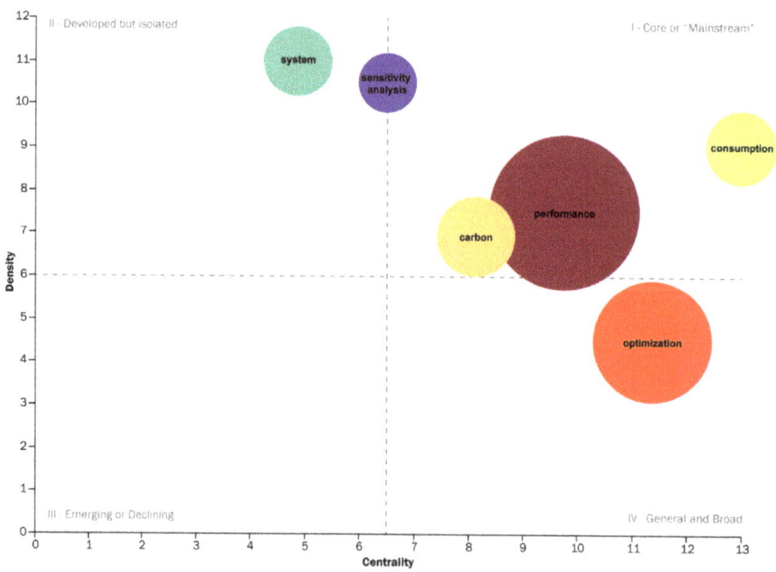

Figure 7. Density–Centrality diagram with keyword clusters.

4. Discussion

4.1. Background

The projected increase in global built floor area, projected to reach an overwhelming 235 billion m^2 by 2050 [27], stands as one of the main challenges stemming from population growth and ever-growing living standards [28]. Such a projection carries profound implications, spanning environmental, economic, and social dimensions. Given that buildings account for roughly 40% of the total energy consumption and 38% of the CO_2 emissions within the European Union [29], the global energy-saving potential is estimated at a staggering 53 Hexajoules annually by 2050 [30]. In this context, the role of building designers becomes paramount in harnessing this vast energy conservation potential. Moreover, as illustrated in Figure 8, sustainable building and city planning must face the pressing need for a holistic approach to energy and climate goals, integrating strategies for sustainable infrastructure that encompass fuel depletion, escalating energy prices, and climate finance, while also addressing the imperatives for healthier infrastructure and the urgent transition towards Net Zero Energy and Carbon buildings.

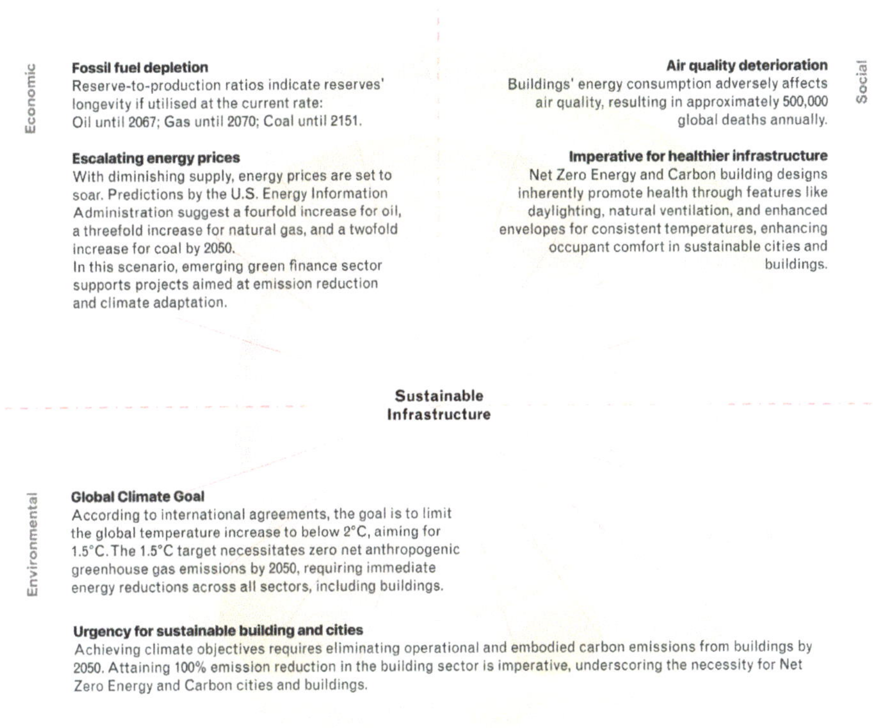

Figure 8. Sustainable infrastructure diagram.

Urban centres globally, in their pursuit of net-zero emissions, are tasked with the dual responsibility of curtailing greenhouse gas (GHG) emissions whilst also compensating for residual emissions. It is worth noting that emission reductions yield a plethora of benefits, encompassing enhanced community health, improved air and water quality, the mitigation of urban heat islands, and efficient resource stewardship.

Over the recent decades, BPSs have ascended as indispensable tools, assisting designers in navigating a multitude of design alternatives. Yet, these simulations encompass numerous variable parameters, giving rise to an expansive multi-dimensional "design space". Navigating this space necessitates significant modelling and computational endeavours, escalating costs and perpetuating inherent uncertainties [31]. Consequently, the feasibility of large-scale applications, encompassing design space exploration, uncertainty analysis, sensitivity analysis, and optimisation, remains constrained. However, the swift technological evolution, especially in the realms of data-centric and AI-driven BPS, presents unparalleled opportunities [32]. Such advancements hold promise to bolster sustainability initiatives, further carbon reduction agendas, and pave the way for more sustainable and inclusive urban development strategies.

The assimilation of AI systems emerges as a pivotal factor in addressing the multifaceted challenges of urbanisation, spanning social, economic, and ecological dimensions. In this scenario, the European Green Deal [33] underscores the urgency for comprehensive, robust policy responses to the climate crisis, with an emphasis on optimising health, quality of life, resilience, and competitiveness dividends. Recognising the pivotal role of digital innovations, including AI, the Green Deal envisions them as instrumental catalysts in realising sustainability objectives across diverse sectors [33]. Concurrently, the EU's sustainable investment framework [34] and the revised EPBD [4] are geared towards incentivising energy sobriety in the built environment. These strategic policies are anchored in the ambition to curtail the GHG emissions and energy consumption in buildings by 2030, whilst setting sights on achieving EU-wide climate neutrality by 2050.

4.2. AI in the AEC Sector

The genesis of what we understand as AI today can be traced back to the 1940s. Pioneering work by American scientists Warren McCulloch and Walter Pitts in 1943 introduced a mathematical representation of the biological neuron, offering an early conceptualisation of an "artificial network" [35]. Over the decades, AI has evolved from isolated research endeavours in the 1950s to its current interdisciplinary nature, influencing numerous fields, including architecture [21].

As Song, Ghaboussi, and Kwon [36] have indicated, in the realm of architecture, the design process is uniquely multifaceted. The design exploration phase, often derived from abstract concepts, significantly influences various performance metrics, including energy consumption, daylight utilisation, and acoustics [37]. This initial design phase is pivotal, as it sets the trajectory for the subsequent construction and the building's entire lifecycle. AI's role in this context is not to provide definitive solutions but to aid in exploring potential design avenues [38]. While some design aspects can be quantified, many remain intangible, making it challenging to establish explicit evaluation criteria [39,40].

Beyond technological advancements, the successful implementation of AI in the AEC sector requires robust interdisciplinary collaboration. It is not just about integrating AI tools but ensuring that AI experts, environmental scientists, urban planners, and architects collaboratively harness these tools to create sustainable and efficient designs [41]. This collaborative approach ensures that the technological solutions align with the real-world complexities of urban environments and architectural challenges.

4.3. Metamodels and Data-Driven Models for Energy Forecasting

There has been a notable transition within the field of architectural design, moving away from conventional problem-solving instruments towards embracing optimisation as a tool for exploration, predominantly in the early design stages [42–48]. However, the real-world application of optimisation during these early stages remains limited. This limited application can be ascribed to various factors, including the intensive time requirements, difficulties in interpretation, constraints inherent to parametric models, and the frequently indistinct character of performance objectives in the realm of architectural design.

Buildings, throughout their lifecycle, contribute to both embodied and operational emissions. While construction primarily results in embodied emissions, operational emissions persist throughout the building's life. These operational emissions are influenced by evolving factors like climate change, electricity grid decarbonisation, and user behaviour. Notably, a discrepancy often exists between a building's design performance and its actual operational performance [49–51]. This 'performance gap' arises from complex interdependencies not accounted for during optimisation, such as building construction, supply systems, environmental parameters, and occupant behaviour. Considering the influence of global warming on the future performance of buildings, planning based exclusively on historical weather data proves to be inadequate. Therefore, it is of paramount importance to integrate uncertainty into building energy evaluations and meticulously assess the robustness of building energy systems in the face of the forthcoming conditions [50,52–54].

Metamodels, also known as data-driven models, act as approximate mathematical depictions crafted to explore the intricate relationships between inputs and outputs that are displayed by more sophisticated physics-based models [55]. Essentially, while mathematical models abstract the real world, data-driven models or metamodels abstract these mathematical models further [56]. Metamodels boast the advantage of being constructible from a modest dataset obtained through the simulation of physics-based models. These models are not only user-friendly but also allow for analytical expression in their formulations. After being adequately trained with a selected set of inputs and outputs from a physics-based model, metamodels possess the capability to forecast outputs for input values that were not part of the initial training set.

Conversely, data-driven models, which are trained using data harvested from real buildings, aim to emulate real-world scenarios rather than another model. For a data-driven approach to be effective, extensive long-term measurements across various building types are essential. Nevertheless, this method may predominantly be applicable to existing building stocks, thereby restricting the examination of alternative energy-efficient constructions. In contrast, the metamodelling approach is capable of efficiently producing the necessary database, thereby facilitating the use of virtually unlimited values for design variables.

All supervised metamodelling approaches follow a comparable procedure, which includes constructing representative samples of inputs for both training and validation purposes. For each set of inputs, a physics-based model is executed to generate the respective outputs. The metamodel is then trained with the gathered input–output samples utilising a selected technique, followed by a validation process using multiple performance indicators [32,57,58]. Upon the completion of these stages, the metamodel is ready for quick deployment in future BPS analyses.

Within the sphere of BPS applications, various techniques are prominently utilised, including polynomial regression, multivariate adaptive regression splines (MARS), the Gaussian process (also known as Kriging), SVM, and ANNs. For example, Romani et al. [59] applied polynomial models to metamodel the requirements of heating and cooling energy, contributing to the optimisation of the envelope design for a low-energy building situated in Morocco. In a similar vein, Cheng and Cao [60] devised a method for forecasting the energy performance of buildings, leveraging evolutionary multivariate adaptive regression splines. Furthermore, Rackes et al. [61] employed SVM to provide design guidance and performance labelling for passive commercial structures located in regions with hot climates. Additionally, Yuan et al. [62] utilised Gaussian processes to craft a method that simultaneously calibrates and ranks parameters for building energy models.

While the promise of AI-driven metamodels in energy forecasting is evident, it is essential to recognise their inherent challenges and limitations. The accuracy of the data, the dynamic and ever-evolving nature of urban environments, and the unpredictability of human behaviour can pose significant challenges to even the most advanced models. As the field progresses, the continuous refinement and understanding of these limitations will be crucial. Looking ahead, the intersection of AI and the AEC sector promises even more revolutionary changes. Emerging technologies and methodologies on the horizon could further refine and enhance the capabilities of AI in this field. As AI models become more sophisticated and data collection becomes more nuanced, the next decade could witness a transformative shift in how cities are planned, designed, and experienced.

4.4. AI and Cities

The European Commission's recent proposal for a regulation, known as the Artificial Intelligence Act, holds significant implications for the construction and urban development sectors. This act is a testament to the growing recognition of AI's transformative potential, especially within the construction ecosystem. It seeks to strike a harmonious balance in the deployment of AI systems, particularly for professionals in the construction sector who are increasingly relying on data-driven systems within the European Union. Key areas of concern for the AEC sector encompass the establishment of harmonised rules for AI systems, the delineation of specific requirements for high-risk AI systems, and the introduction of supportive measures to foster innovation [63].

In the broader urban context, cities are emerging as focal points in AI discourse. As underscored by the United Nations Habitat in their report titled "AI and Cities" [64], cities are uniquely positioned to harness the capabilities of AI to address a myriad of socio-economic and ecological challenges. As urban centres grapple with multifaceted issues ranging from resource constraints to governance complexities and mounting environmental threats, the infusion of AI-driven innovations is becoming indispensable. However, to truly capitalise on the transformative potential of AI, a concerted effort is required from various stakeholders to create an ecosystem that promotes sustainable and inclusive development. This necessitates a judicious balance between leveraging the opportunities presented by AI and mitigating the associated risks.

While numerous countries have proactively rolled out national AI guidelines [65], local governments, urban planners, and policymakers are navigating the intricate maze of developing, implementing, and evaluating regulatory frameworks tailored to their unique urban contexts. The rapid pace of AI advancements has opened up a plethora of potential applications in urban settings. The United Nations, in its deliberations, has spotlighted several sectors ripe for AI integration, including "energy, mobility, public safety, water and waste management, healthcare, urban planning, and city governance" [64].

One sector in which the impact of AI is particularly pronounced is energy. Data-driven systems, underpinned by AI, are poised to revolutionise energy management, propelling cities towards a more sustainable, low-carbon future. These systems are adept at forecasting the energy supply from various sources, facilitating predictive maintenance, and optimising energy distribution [64]. As urban landscapes increasingly pivot towards renewable energy sources, the role of data-driven models in forecasting, planning, and optimising energy consumption and distribution becomes paramount. These models can harness a diverse array of data to generate forecasts, offering insights into emissions and consumption patterns in intricate urban matrices.

However, for several years, climate datasets tailored to building performance simulations have been accessible, primarily due to the inception of an effective data format known as the typical meteorological year (TMY) [66]. This was further bolstered by the global availability of data in this format. Beyond merely expanding the global reach of these datasets, recent research endeavours have delved into techniques used to simulate local microclimatic events within urban landscapes, such as the urban heat island effect [67]. In a research study focused on London, Mavrogianni et al. amalgamated temperature profiles obtained locally with an Urban Building Performance Simulation (U-BPS). This integration aimed to elucidate the influence exerted by the urban heat island effect on not only the energy consumption of buildings but also on the well-being of the residents within these structures [68]. Concurrently, there is a burgeoning interest in forecasting local wind trajectories [69] and in aligning current TMYs with climate change projections from the Intergovernmental Panel on Climate Change (IPCC) [70]. These are pivotal research areas with direct ramifications for U-BPSs.

In many instances, researchers have amalgamated the export/import functionalities of existing platforms, such as Geographic Information Systems (GIS) and Building Information Modelling (BIM), with bespoke scripts. This approach facilitates the creation of a thermal model, oversees the simulations, and presents the findings through tools like spreadsheets or GIS applications [68,71–74]. Some teams have further refined and automated these simulation workflows, incorporating additional urban performance indicators to make U-BPS more accessible to urban planners and designers.

The efficacy of U-BPS models in informing design choices or policy decisions hinges significantly on the accuracy of the simulation outcomes. Considering the potential discrepancies between individual BPS predictions and actual measurements, due to variables like infiltration rates and occupant behaviours, one might question the U-BPS's capacity to authentically forecast energy consumption across numerous buildings. Yet, when juxtaposing aggregated yearly measured data against simulated energy usage spanning multiple structures, the individual model deviations appear to neutralise. This results in documented error margins of merely 7% to 21% for heating loads [73,75,76] and a range of 1% to 19% for the overall Energy Use Intensity (EUI) [77–80].

The integration of AI and machine learning models into U-BPSs has garnered considerable attention in recent years. Nutkiewicz and Jain (2019) delved into the amalgamation of physics-based building simulation methodologies and machine learning techniques, with a particular focus on transfer learning. Their exploration aimed to evaluate the effects of retrofit policies on urban structures [81]. This integrated methodology, dubbed Data-driven Urban Energy Simulation (DUE-S), showcased its potential in identifying the energy implications urban environments have on buildings undergoing retrofit processes. In a similar vein, Neumann et al. (2021) conducted an examination into the viability of establishing Positive Energy Districts (PEDs) within various urban typologies located in Wien [82]. Their work underscored the necessity of implementing extensive energy efficiency measures, promoting electrification, and harnessing renewable energy sources, as pivotal steps in the transformation of existing building stocks into PEDs. Moreover, the imperative of understanding the building stock at a larger scale, with a spotlight on building geometry, was accentuated by Dai et al. (2022) [83]. Their research unveiled a novel methodology designed to automatically gauge the building dimensions derived from remote sensing data, employing unsupervised machine learning algorithms in the process. Concluding their series of studies, Hey et al. (2022) stressed the crucial role of modelling in the adoption of energy retrofits within urban residential building stocks. They introduced an innovative concept in which carbon valuations are assigned to households, serving as a determinant for identifying optimal retrofit solutions [84]. Their approach combined surrogate models, optimisation procedures, and neural networks to evaluate building performance, offering insights into the potential of such models in informing policy decisions.

These AI-driven models are equipped to provide granular insights into the energy demand within buildings [85], elucidate energy dynamics in urban microclimates [86], and categorise distinct energy demand patterns [87], thereby facilitating more informed decision-making.

4.5. Frameworks and Workflows

The integration of artificial intelligence in the AEC sector has seen the emergence of various frameworks in recent times. Geraldi and Ghisi [88] introduced a seven-step framework that seamlessly integrates surveying, data collection, and ex-post analysis. This framework trains an Artificial Neural Network to predict a building's performance with high precision.

In contrast, Mouakher et al. [89] delved into the realm of explainable deep learning models (XAI) to address the "black box" nature of traditional neural networks. Their approach aimed to provide more transparency and an understanding of how these models make decisions.

Dong et al. [90] underscored the imperative for a framework that can intelligently automate the design process with a concentration on daylight and energy performances. Within the workflow of energy-efficient design, they identified three predominant knowledge gaps. The initial gap is the ambiguous correlations existing between different building components, resulting in repetitive modelling in the course of adjusting parameters within building information models. The subsequent gap involves the necessity for the generation of optimal alternatives that exhibit the desired performances. The third and final gap is the indispensable need for a method of decision-making that is effective, serving to sift through and select the most appropriate design plan. The task of assessing a framework's efficacy in implementing an energy-efficient design is fraught with challenges. This scenario necessitates the need for a process that is automated from start to finish, spanning from building information modelling to the final stages of decision-making. Furthermore, there is a call for a method of decision-making that is not only effective but is also seamlessly integrated with an algorithm dedicated to optimisation.

In the urban context, the rise of U-BPS tools has been notable. These tools expand the application of BPS to the urban level, leveraging physics-based methods and incorporating statistical approximations to address computational, data, and statistical challenges. The surge in U-BPS frameworks [91] can be attributed to global carbon reduction commitments, increased accessibility of urban datasets [92], advancements in computational tools, and affordable hardware.

U-BPSs are designed to provide data-driven insights applicable to a variety of urban-level scenarios, including urban planning initiatives, the development of new neighbourhoods, strategies for reducing carbon at the stock level, and the integration of buildings into the electrical grid [93]. The simulation results from U-BPSs can be aggregated and displayed at various temporal and spatial scales, contingent upon the use case in question. For example, a U-BPS that is crafted for the purpose of formulating policies for retrofitting buildings city-wide may employ an archetype approach. This approach involves simulating the distributions of energy usage for different types of residential buildings. However, to obtain accurate results at the level of individual buildings, a calibration process that is conducted building by building and is based on actual data is necessary [94]. There are various approaches available for the development of U-BPSs [95]. At its most basic, a U-BPS will incorporate the fundamental geometries of buildings, files pertaining to the relevant weather, and templates for building simulations that represent various categories of buildings [86].

Tardioli et al. (2019) introduced a novel feature engineering procedure that combines results from calibrated physics-based building energy models with traditional predictors in a forecasting framework, demonstrating improved forecasting outcomes for the heating demand in urban districts [96]. Xu and Wang (2023) presented a comprehensive methodological framework for urban decarbonisation strategies, emphasising "the integration of multi-scale energy performance evaluation within the design development process" [97]. Their data-driven methodology, utilised in a Sheffield, UK, case study, assesses scenarios of energy demand and supply at the urban level. Monien et al. (2017) highlighted the challenges in building energy evaluation tools, particularly the balance between data availability and the assumptions made [98]. Their research compared two tools based on 3D models, demonstrating the scalability of their urban simulation tool, SimStadt, from single buildings to city districts. Moreover, as encapsulated by Cappelletti and Ballarini (2021), the Building Simulation 2019 conference spotlighted recent trends in the applications of building simulation. It underscored the amalgamation of various pre-existing tools aimed at evaluating comprehensive building performance while aiding both the design and control processes of buildings [99].

5. Conclusions

This paper has conducted a literature review to chart the progression of artificial intelligence within the AEC sector, focusing on its integration into building performance simulations and its integration into the broader design process.

The review has identified a marked shift in the application of AI tools. While the use of AI for BPS at the building scale is becoming increasingly established, there is a notable emerging trend in its application for larger-scale projects, such as neighbourhood-scale developments or master plans. In these contexts, the literature highlights AI's value in facilitating complex calculations and analyses that surpass the capabilities of traditional BPS methods. This trend underscored a growing recognition of AI's potential to handle the complexities and nuances associated with achieving energy efficiency and decarbonisation objectives within the AEC sector.

Despite these advances, several open questions persist. The research has identified a large number of systematic methods for incorporating data-driven BPS in the early design stages and a need for clarity on how these tools can best support the sustainability goals. Central to the conclusions of this review is the advocacy for a coherent framework that delineates how AI-driven BPS can be embedded into design workflows. Such a framework would streamline the adoption of AI, ensuring it is effectively leveraged to optimise design choices, particularly those impacting sustainability and energy efficiency. This proposed framework would serve as a cornerstone for future research, guiding the development of actionable strategies that harness AI's full potential in the design process.

Author Contributions: Conceptualisation, A.G.D.S. and M.R.; methodology, G.M.; formal analysis, G.M.; investigation, A.G.D.S.; resources, A.G.D.S.; writing—original draft preparation, A.G.D.S.; writing—review and editing, M.R. and G.M.; visualisation, A.G.D.S.; supervision, M.R. and G.M.; All authors have read and agreed to the published version of the manuscript.

Funding: This research received no external funding.

Institutional Review Board Statement: Not applicable.

Informed Consent Statement: Not applicable.

Data Availability Statement: Not applicable.

Conflicts of Interest: The authors declare no conflict of interest.

References

1. Intergovernmental Panel on Climate Change (IPCC). *Climate Change 2014: Synthesis Report. Contribution of Working Groups I, II and III to the Fifth Assessment Report of the Intergovernmental Panel on Climate Change*; IPCC: Geneva, Switzerland, 2015; p. 151.
2. International Institute for Sustainable Development. *77 Countries, 100+ Cities Commit to Net Zero Carbon Emissions by 2050 at Climate Summit*; International Institute for Sustainable Development: Winnipeg, MB, Canada, 2019.
3. Ang, Y.Q. Using Urban Building Energy Modeling to Develop Carbon Reduction Pathways for Cities. Ph.D. Thesis, Technology, Massachusetts Institute of Technology, Cambridge, MA, USA, 2022.
4. European Commission Directive of the European Parliament and of the Council on the Energy Performance of Building (Recast) 2021. Available online: https://eur-lex.europa.eu/legal-content/EN/TXT/?uri=CELEX%253A52021PC0802&qid=1641802763889 (accessed on 14 March 2022).
5. Attia, S.G.; De Herde, A. Early Design Simulation Tools for Net Zero Energy Buildings: A Comparison of Ten Tools. In Proceedings of the Building Simulation 2011, Sydney, Australia, 14–16 November 2011; pp. 94–101.
6. Attia, S.; Hamdy, M.; O'Brien, W.; Carlucci, S. Assessing Gaps and Needs for Integrating Building Performance Optimisation Tools in Net Zero Energy Buildings Design. *Energy Build.* **2013**, *60*, 110–124. [CrossRef]
7. Wei, Y.; Zhang, X.; Shi, Y.; Xia, L.; Pan, S.; Wu, J.; Han, M.; Zhao, X. A Review of Data-Driven Approaches for Prediction and Classification of Building Energy Consumption. *Renew. Sustain. Energy Rev.* **2018**, *82*, 1027–1047. [CrossRef]
8. Sun, Y.; Haghighat, F.; Fung, B.C.M. A Review of The-State-of-the-Art in Data-Driven Approaches for Building Energy Prediction. *Energy Build.* **2020**, *221*, 110022. [CrossRef]
9. Ali, U.; Shamsi, M.H.; Hoare, C.; Mangina, E.; O'Donnell, J. Review of Urban Building Energy Modeling (UBEM) Approaches, Methods and Tools Using Qualitative and Quantitative Analysis. *Energy Build.* **2021**, *246*, 111073. [CrossRef]
10. Javeed Nizami, S.; Al-Garni, A.Z. Forecasting Electric Energy Consumption Using Neural Networks. *Energy Policy* **1995**, *23*, 1097–1104. [CrossRef]
11. González, P.A.; Zamarreño, J.M. Prediction of Hourly Energy Consumption in Buildings Based on a Feedback Artificial Neural Network. *Energy Build.* **2005**, *37*, 595–601. [CrossRef]
12. Dong, B.; Cao, C.; Lee, S.E. Applying Support Vector Machines to Predict Building Energy Consumption in Tropical Region. *Energy Build.* **2005**, *37*, 545–553. [CrossRef]
13. Li, Q.; Meng, Q.; Cai, J.; Yoshino, H.; Mochida, A. Applying Support Vector Machine to Predict Hourly Cooling Load in the Building. *Appl. Energy* **2009**, *86*, 2249–2256. [CrossRef]
14. Touzani, S.; Granderson, J.; Fernandes, S. Gradient Boosting Machine for Modeling the Energy Consumption of Commercial Buildings. *Energy Build.* **2018**, *158*, 1533–1543. [CrossRef]
15. Wu, J.; Nguyen, S.; Alahakoon, D.; Silva, D.D.; Mills, N.; Rathnayaka, P.; Moraliyage, H.; Jennings, A. A Computational Study to Evaluate and Compare Machine Learning-Based Energy Baseline Models Across Multiple Building Types. 38p. Available online: https://papers.ssrn.com/sol3/papers.cfm?abstract_id=4195468 (accessed on 14 March 2022).

16. Cordeiro-Costas, M.; Villanueva, D.; Eguía-Oller, P.; Martínez-Comesaña, M.; Ramos, S. Load Forecasting with Machine Learning and Deep Learning Methods. *Appl. Sci.* **2023**, *13*, 7933. [CrossRef]
17. Runge, J.; Zmeureanu, R. A Review of Deep Learning Techniques for Forecasting Energy Use in Buildings. *Energies* **2021**, *14*, 608. [CrossRef]
18. Farzaneh, H.; Malehmirchegini, L.; Bejan, A.; Afolabi, T.; Mulumba, A.; Daka, P.P. Artificial Intelligence Evolution in Smart Buildings for Energy Efficiency. *Appl. Sci.* **2021**, *11*, 763. [CrossRef]
19. Waltman, L.; van Eck, N.J. VOSviewer—Visualizing Scientific Landscapes. Available online: https://www.vosviewer.com/ (accessed on 14 March 2022).
20. Aria, M.; Cuccurullo, C. Bibliometrix: An R-Tool for Comprehensive Science Mapping Analysis. *J. Informetr.* **2017**, *11*, 959–975. [CrossRef]
21. Chaillou, S. *Artificial Intelligence and Architecture: From Research to Practice*; Birkhauser Verlag GmbH: Basel, Switzerland; Boston, MA, USA, 2022; ISBN 978-3-0356-2400-7.
22. Liu, Y.; Goncalves, J.; Ferreira, D.; Hosio, S.; Kostakos, V. Identity Crisis of Ubicomp?: Mapping 15 Years of the Field's Development and Paradigm Change. In Proceedings of the 2014 ACM International Joint Conference on Pervasive and Ubiquitous Computing, Seattle, WA, USA, 13–17 September 2014; pp. 75–86.
23. Muñoz-Leiva, F.; Viedma-del-Jesús, M.I.; Sánchez-Fernández, J.; López-Herrera, A.G. An Application of Co-Word Analysis and Bibliometric Maps for Detecting the Most Highlighting Themes in the Consumer Behaviour Research from a Longitudinal Perspective. *Qual. Quant.* **2012**, *46*, 1077–1095. [CrossRef]
24. Nielsen, A.E.; Thomsen, C. Sustainable Development: The Role of Network Communication. *Corp. Soc. Responsib. Environ. Manag.* **2011**, *18*, 1–10. [CrossRef]
25. He, Q. Knowledge Discovery Through Co-Word Analysis. *Libr. Trends* **1999**, *48*, 133–159.
26. Callon, M.; Courtial, J.P.; Laville, F. Co-Word Analysis as a Tool for Describing the Network of Interactions between Basic and Technological Research: The Case of Polymer Chemsitry. *Scientometrics* **1991**, *22*, 155–205. [CrossRef]
27. United Nations; Department of Economic and Social Affairs; Population Division. *World Population Prospects Highlights, 2019 Revision Highlights, 2019 Revision*; United Nations: Nairobi, Kenya, 2019; ISBN 978-92-1-148316-1.
28. Sachs, J.; Schmidt-Traub, G.; Kroll, C.; Lafortune, G.; Fuller, G. *Sustainable Development Report 2019*; Bertelsmann Stiftung and Sustainable Development Solutions Network (SDSN): New York, NY, USA, 2019.
29. United Nations Environment Programme Building Sector Emissions Hit Record High, but Low-Carbon Pandemic Recovery Can Help Transform Sector—UN Report. Available online: http://www.unep.org/news-and-stories/press-release/building-sector-emissions-hit-record-high-low-carbon-pandemic (accessed on 1 August 2023).
30. United Nations Environment Programme. *2022 Global Status Report for Buildings and Construction: Towards a Zero-Emission, Efficient and Resilient Buildings and Construction Sector*; United Nations Environment Programme: Nairobi, Kenya, 2022.
31. Hong, T.; Langevin, J.; Sun, K. Building Simulation: Ten Challenges. *Build. Simul.* **2018**, *11*, 871–898. [CrossRef]
32. Østergård, T.; Jensen, R.L.; Maagaard, S.E. A Comparison of Six Metamodeling Techniques Applied to Building Performance Simulations. *Appl. Energy* **2018**, *211*, 89–103. [CrossRef]
33. European Commission. *The European Green Deal*; European Commission: Brussels, Belgium, 2019.
34. European Parliament and the Council of European Union Regulation (EU) 2020/852 of the European Parliament and of the Council of 18 June 2020 on the Establishment of a Framework to Facilitate Sustainable Investment, and Amending Regulation (EU) 2019/2088 2020. Available online: https://eur-lex.europa.eu/legal-content/EN/TXT/?uri=celex%253A32020R0852 (accessed on 14 March 2022).
35. McCulloch, W.; Pitts, W. A Logical Calculus of the Ideas Immanent in Nervous Activity. *Bull. Math. Biophys.* **1943**, *5*, 115–133. [CrossRef]
36. Song, H.; Ghaboussi, J.; Kwon, T.-H. Architectural Design of Apartment Buildings Using the Implicit Redundant Representation Genetic Algorithm. *Autom. Constr.* **2016**, *72*, 166–173. [CrossRef]
37. Agirbas, A. Façade Form-Finding with Swarm Intelligence. *Autom. Constr.* **2019**, *99*, 140–151. [CrossRef]
38. Logan, B.; Smithers, T. Creativity and Design as Exploration. In *Modeling Creativity and Knowledge-Based Creative Design*; Gero, J.S., Maher, M.L., Eds.; Lawrence Erlbaum Associates, Inc.: Hillsdale, NJ, USA, 1993; p. 31.
39. Choudhary, R.; Michalek, J. Design Optimization in Computer-Aided Architectural Design. In Proceedings of the 10th International Conference on Computer Aided Architectural Design Research in Asia, New Delhi, India, 28 April 2005; TVB School of Habitat Studies. Volume 1, p. 10.
40. Castro Pena, M.L.; Carballal, A.; Rodríguez-Fernández, N.; Santos, I.; Romero, J. Artificial Intelligence Applied to Conceptual Design. *A Review of Its Use in Architecture. Autom. Constr.* **2021**, *124*, 103550. [CrossRef]
41. Brown, N.C.; Mueller, C.T. Designing with Data: Moving beyond the Design Space Catalog. In Proceedings of the ACADIA 2017: DISCIPLINES & DISRUPTION, Cambridge, MA, USA, 2–4 November 2017; pp. 154–163.
42. Bradner, E.; Iorio, F.; Davis, M. Parameters Tell the Design Story: Ideation and Abstraction in Design Optimisation. In Proceedings of the Symposium on Simulation for Architecture & Urban Design (SimAUD '14), San Diego, CA, USA, 13–16 April 2014; pp. 1–8.
43. Evins, R. A Review of Computational Optimisation Methods Applied to Sustainable Building Design. *Renew. Sustain. Energy Rev.* **2013**, *22*, 230–245. [CrossRef]

44. Tian, Z.; Zhang, X.; Jin, X.; Zhou, X.; Si, B.; Shi, X. Towards Adoption of Building Energy Simulation and Optimization for Passive Building Design: A Survey and a Review. *Energy Build.* **2018**, *158*, 1306–1316. [CrossRef]
45. Caldas, L.G. An Evolution-Based Generative Design System: Using Adaptation to Shape Architectural Form. Ph.D. Thesis, Building Technology, Department of Architecture, Massachusetts Institute of Technology, Cambridge, MA, USA, 2001.
46. Mueller, C.T.; Ochsendorf, J.A. Combining Structural Performance and Designer Preferences in Evolutionary Design Space Exploration. *Autom. Constr.* **2015**, *52*, 70–82. [CrossRef]
47. Turrin, M.; von Buelow, P.; Stouffs, R. Design Explorations of Performance Driven Geometry in Architectural Design Using Parametric Modeling and Genetic Algorithms. *Adv. Eng. Inform.* **2011**, *25*, 656–675. [CrossRef]
48. Project Discover: An Application of Generative Design for Architectural Space Planning. In Proceedings of the 2017 Symposium on Simulation for Architecture and Urban Design (SimAUD 2017), Toronto, ON, Canada, 22–24 May 2017; Society for Modeling and Simulation International (SCS): Toronto, ON, Canada, 2017.
49. van Dronkelaar, C.; Dowson, M.; Spataru, C.; Mumovic, D. A Review of the Regulatory Energy Performance Gap and Its Underlying Causes in Non-Domestic Buildings. *Front. Mech. Eng.* **2016**, *1*, 17. [CrossRef]
50. Kotireddy, R.; Hoes, P.-J.; Hensen, J.L.M. A Methodology for Performance Robustness Assessment of Low-Energy Buildings Using Scenario Analysis. *Appl. Energy* **2018**, *212*, 428–442. [CrossRef]
51. Frei, B.; Sagerschnig, C.; Gyalistras, D. Performance Gaps in Swiss Buildings: An Analysis of Conflicting Objectives and Mitigation Strategies. *Energy Procedia* **2017**, *122*, 421–426. [CrossRef]
52. Tian, W.; Heo, Y.; de Wilde, P.; Li, Z.; Yan, D.; Park, C.S.; Feng, X.; Augenbroe, G. A Review of Uncertainty Analysis in Building Energy Assessment. *Renew. Sustain. Energy Rev.* **2018**, *93*, 285–301. [CrossRef]
53. Pourarian, S.; Kearsley, A.; Wen, J.; Pertzborn, A. Efficient and Robust Optimization for Building Energy Simulation. *Energy Build.* **2016**, *122*, 53–62. [CrossRef] [PubMed]
54. Huang, P.; Huang, G.; Sun, Y. A Robust Design of Nearly Zero Energy Building Systems Considering Performance Degradation and Maintenance. *Energy* **2018**, *163*, 905–919. [CrossRef]
55. Fang, K.; Li, R.; Sudjianto, A. *Design and Modeling for Computer Experiments*; Computer science and data analysis series; Chapman & Hall/CRC: Boca Raton, FL, USA, 2006; ISBN 978-1-58488-546-7.
56. Roman, N.D.; Bre, F.; Fachinotti, V.D.; Lamberts, R. Application and Characterization of Metamodels Based on Artificial Neural Networks for Building Performance Simulation: A Systematic Review. *Energy Build.* **2020**, *217*, 109972. [CrossRef]
57. Zhao, H.; Magoulès, F. A Review on the Prediction of Building Energy Consumption. *Renew. Sustain. Energy Rev.* **2012**, *16*, 3586–3592. [CrossRef]
58. Van Gelder, L.; Das, P.; Janssen, H.; Roels, S. Comparative Study of Metamodelling Techniques in Building Energy Simulation: Guidelines for Practitioners. *Simul. Model. Pract. Theory* **2014**, *49*, 245–257. [CrossRef]
59. Romani, Z.; Draoui, A.; Allard, F. Metamodeling the Heating and Cooling Energy Needs and Simultaneous Building Envelope Optimization for Low Energy Building Design in Morocco. *Energy Build.* **2015**, *102*, 139–148. [CrossRef]
60. Cheng, M.-Y.; Cao, M.-T. Accurately Predicting Building Energy Performance Using Evolutionary Multivariate Adaptive Regression Splines. *Appl. Soft Comput.* **2014**, *22*, 178–188. [CrossRef]
61. Rackes, A.; Melo, A.P.; Lamberts, R. Naturally Comfortable and Sustainable: Informed Design Guidance and Performance Labeling for Passive Commercial Buildings in Hot Climates. *Appl. Energy* **2016**, *174*, 256–274. [CrossRef]
62. Yuan, J.; Nian, V.; Su, B.; Meng, Q. A Simultaneous Calibration and Parameter Ranking Method for Building Energy Models. *Appl. Energy* **2017**, *206*, 657–666. [CrossRef]
63. European Commission. *Transition Pathway for Construction*; European Commission: Brussels, Belgium, 2023.
64. Kizhakkethottam, C.; Lahoud, C.; Pienaar, J.; Thung, I.; Owigar, J.; Sommer, K.; Nkuidje, L.; Pennanen-Rebeiro-Hargrave, P.; Westerberg, P.; Sietchiping, R.; et al. *AI and Cities: Risks, Applications and Governance*; UN Habitat: Nairobi, Kenya, 2023.
65. Schmitt, L. Mapping Global AI Governance: A Nascent Regime in a Fragmented Landscape. *AI Ethics* **2022**, *2*, 303–314. [CrossRef]
66. Crawley, D.B.; Hand, J.W.; Lawrie, L.K. Improving the Weather Information Available to Simulation Programs. In Proceedings of the Building Simulation 1999, Kyoto, Japan, 13–15 September 1999.
67. Bueno, B.; Norford, L.; Pigeon, G.; Britter, R. A Resistance-Capacitance Network Model for the Analysis of the Interactions between the Energy Performance of Buildings and the Urban Climate. *Build. Environ.* **2012**, *54*, 116–125. [CrossRef]
68. Mavrogianni, A.; Davies, M.; Kolokotroni, M.; Hamilton, I. A GIS-Based Bottom-up Space Heating Demand Model of the London Domestic Stock. In Proceedings of the Building Simulation 2009, Glasgow, UK, 27–30 July 2009.
69. Liu, J.; Heidarinejad, M.; Gracik, S.; Srebric, J. The Impact of Exterior Surface Convective Heat Transfer Coefficients on the Building Energy Consumption in Urban Neighborhoods with Different Plan Area Densities. *Energy Build.* **2015**, *86*, 449–463. [CrossRef]
70. Jentsch, M.F.; James, P.A.B.; Bourikas, L.; Bahaj, A.S. Transforming Existing Weather Data for Worldwide Locations to Enable Energy and Building Performance Simulation under Future Climates. *Renew. Energy* **2013**, *55*, 514–524. [CrossRef]
71. Huber, J.; Nytsch-Geusen, C. Development of Modeling and Simulation Strategies for Large-Scale Urban Districts. In Proceedings of the Building Simulation 2011, Sydney, Australia, 14–16 November 2011.
72. Mastrucci, A.; Baume, O.; Stazi, F.; Salvucci, S.; Leopold, U. A GIS-Based Approach to Estimate Energy Savings and Indoor Thermal Comfort for Urban Housing Stock Retrofitting. In Proceedings of the Fifth German-Austrian IBPSA Conference (BauSIM 2014), Aachen, Germany, 22–24 September 2014.

73. Dall'O', G.; Galante, A.; Torri, M. A Methodology for the Energy Performance Classification of Residential Building Stock on an Urban Scale. *Energy Build.* **2012**, *48*, 211–219. [CrossRef]
74. Ascione, F.; De Masi, R.F.; De Rossi, F.; Fistola, R.; Sasso, M.; Vanoli, G.P. Analysis and Diagnosis of the Energy Performance of Buildings and Districts: Methodology, Validation and Development of Urban Energy Maps. *Cities* **2013**, *35*, 270–283. [CrossRef]
75. Filogamo, L.; Peri, G.; Rizzo, G.; Giaccone, A. On the Classification of Large Residential Buildings Stocks by Sample Typologies for Energy Planning Purposes. *Appl. Energy* **2014**, *135*, 825–835. [CrossRef]
76. Orehounig, K.; Mavromatidis, G.; Evins, R.; Dorer, V.; Carmeliet, J. Predicting Energy Consumption of a Neighbourhood Using Building Performance Simulations. In Proceedings of the 2nd IBPSA-England Conference BSO 2014, London, UK, 10–11 June 2014; p. 72.
77. Heiple, S.; Sailor, D.J. Using Building Energy Simulation and Geospatial Modeling Techniques to Determine High Resolution Building Sector Energy Consumption Profiles. *Energy Build.* **2008**, *40*, 1426–1436. [CrossRef]
78. Shimoda, Y.; Fujii, T.; Morikawa, T.; Mizuno, M. Residential End-Use Energy Simulation at City Scale. *Build. Environ.* **2004**, *39*, 959–967. [CrossRef]
79. Caputo, P.; Costa, G.; Ferrari, S. A Supporting Method for Defining Energy Strategies in the Building Sector at Urban Scale. *Energy Policy* **2013**, *55*, 261–270. [CrossRef]
80. Fonseca, J.A.; Schlueter, A. Integrated Model for Characterisation of Spatiotemporal Building Energy Consumption Patterns in Neighborhoods and City Districts. *Appl. Energy* **2015**, *142*, 247–265. [CrossRef]
81. Nutkiewicz, A.; Jain, R.K. Exploring the Integration of Simulation and Deep Learning Models for Urban Building Energy Modeling and Retrofit Analysis. In Proceedings of the Building Simulation 2019, Rome, Italy, 2–4 September 2019; pp. 3209–3216.
82. Neumann, H.-M.; Hainoun, A.; Stollnberger, R.; Etminan, G.; Schaffler, V. Analysis and Evaluation of the Feasibility of Positive Energy Districts in Selected Urban Typologies in Vienna Using a Bottom-Up District Energy Modelling Approach. *Energies* **2021**, *14*, 4449. [CrossRef]
83. Dai, M.; Ward, W.O.C.; Arbabi, H.; Densley Tingley, D.; Mayfield, M. Scalable Residential Building Geometry Characterisation Using Vehicle-Mounted Camera System. *Energies* **2022**, *15*, 6090. [CrossRef]
84. Hey, J.; Siebers, P.-O.; Nathanail, P.; Ozcan, E.; Robinson, D. Surrogate Optimization of Energy Retrofits in Domestic Building Stocks Using Household Carbon Valuations. *J. Build. Perform. Simul.* **2023**, *16*, 16–37. [CrossRef]
85. Robinson, C.; Dilkina, B.; Hubbs, J.; Zhang, W.; Guhathakurta, S.; Brown, M.A.; Pendyala, R.M. Machine Learning Approaches for Estimating Commercial Building Energy Consumption. *Appl. Energy* **2017**, *208*, 889–904. [CrossRef]
86. Nutkiewicz, A.; Yang, Z.; Jain, R.K. Data-Driven Urban Energy Simulation (DUE-S): A Framework for Integrating Engineering Simulation and Machine Learning Methods in a Multi-Scale Urban Energy Modeling Workflow. *Appl. Energy* **2018**, *225*, 1176–1189. [CrossRef]
87. Zhang, W.; Robinson, C.; Guhathakurta, S.; Garikapati, V.M.; Dilkina, B.; Brown, M.A.; Pendyala, R.M. Estimating Residential Energy Consumption in Metropolitan Areas: A Microsimulation Approach. *Energy* **2018**, *155*, 162–173. [CrossRef]
88. Geraldi, M.S.; Ghisi, E. Data-Driven Framework towards Realistic Bottom-up Energy Benchmarking Using an Artificial Neural Network. *Appl. Energy* **2022**, *306*, 117960. [CrossRef]
89. Mouakher, A.; Inoubli, W.; Ounoughi, C.; Ko, A. Expect: EXplainable Prediction Model for Energy ConsumpTion. *Mathematics* **2022**, *10*, 248. [CrossRef]
90. Dong, Y.; Sun, C.; Han, Y.; Liu, Q. Intelligent Optimization: A Novel Framework to Automatize Multi-Objective Optimization of Building Daylighting and Energy Performances. *J. Build. Eng.* **2021**, *43*, 102804. [CrossRef]
91. Shivaram, R.; Yang, Z.; Jain, R.K. Context-Aware Urban Energy Analytics (CUE-A): A Framework to Model Relationships between Building Energy Use and Spatial Proximity of Urban Systems. *Sustain. Cities Soc.* **2021**, *72*, 102978. [CrossRef]
92. Wang, C.; Wei, S.; Du, S.; Zhuang, D.; Li, Y.; Shi, X.; Jin, X.; Zhou, X. A Systematic Method to Develop Three Dimensional Geometry Models of Buildings for Urban Building Energy Modeling. *Sustain. Cities Soc.* **2021**, *71*, 102998. [CrossRef]
93. Ang, Y.Q.; Berzolla, Z.M.; Reinhart, C.F. From Concept to Application: A Review of Use Cases in Urban Building Energy Modeling. *Appl. Energy* **2020**, *279*, 115738. [CrossRef]
94. Nagpal, S.; Mueller, C.; Aijazi, A.; Reinhart, C.F. A Methodology for Auto-Calibrating Urban Building Energy Models Using Surrogate Modeling Techniques. *J. Build. Perform. Simul.* **2019**, *12*, 1–16. [CrossRef]
95. Reinhart, C.F.; Cerezo Davila, C. Urban Building Energy Modeling—A Review of a Nascent Field. *Build. Environ.* **2016**, *97*, 196–202. [CrossRef]
96. Tardioli, G.; Kerrigan, R.; Oates, M.; O'Donnell, J.; Finn, D.P. A Novel Hybrid Technique For Building Demand Forecasting Based on Data-Driven and Urban Scale Simulation Approaches. In Proceedings of the Building Simulation 2019, Rome, Italy, 2–4 September 2019; pp. 3722–3729.
97. Xu, H.; Wang, T.-H. A Generative Computational Workflow to Develop Actionable Renovation Strategies for Renewable Built Environments: A Case Study of Sheffield. *Int. J. Archit. Comput.* **2023**, *21*, 516–535. [CrossRef]

98. Monien, D.; Strzalka, A.; Koukofikis, A.; Coors, V.; Eicker, U. Comparison of Building Modelling Assumptions and Methods for Urban Scale Heat Demand Forecasting. *Future Cities Environ.* **2017**, *3*, 2. [CrossRef]
99. Cappelletti, F.; Ballarini, I. Building Simulation 2019: 16th IBPSA International Conference and Exhibition. *Sci. Technol. Built Environ.* **2021**, *27*, 1017. [CrossRef]

MDPI AG
Grosspeteranlage 5
4052 Basel
Switzerland
Tel.: +41 61 683 77 34

Applied Sciences Editorial Office
E-mail: applsci@mdpi.com
www.mdpi.com/journal/applsci

www.ingramcontent.com/pod-product-compliance
Lightning Source LLC
Chambersburg PA
CBHW061602220926
51850CB00005B/400
9783725857678